Mitosis/Cytokinesis

This is a volume in
CELL BIOLOGY
A series of monographs

Editors: D. E. Buetow, I. L. Cameron, G. M. Padilla, and A. M. Zimmerman

A complete list of the books in this series appears at the end of the volume.

Mitosis / Cytokinesis

Edited by

ARTHUR M. ZIMMERMAN

Department of Zoology
Ramsay Wright Zoological Laboratories
University of Toronto
Toronto, Ontario, Canada

ARTHUR FORER

Department of Biology
York University
Downsview, Ontario, Canada

1981

ACADEMIC PRESS

A Subsidiary of Harcourt Brace Jovanovich, Publishers

New York   London   Toronto   Sydney   San Francisco

ACADEMIC PRESS, INC.
111 Fifth Avenue, New York, New York 10003

United Kingdom Edition published by
ACADEMIC PRESS, INC. (LONDON) LTD.
24/28 Oval Road, London NW1 7DX

Library of Congress Cataloging in Publication Data
Main entry under title:

Cellular dynamics.

(Cell biology)
Includes bibliographies and index.
1. Mitosis. 2. Cytokinesis. I. Zimmerman,
Arthur M., Date. II. Forer, Arthur. III. Series.
QH605.2.C46   574.87'623   81-14976
ISBN 0-12-781240-7   AACR2

PRINTED IN THE UNITED STATES OF AMERICA

81 82 83 84   9 8 7 6 5 4 3 2 1

Contents

v

3   The Movements of the Nuclei during Fertilization

GERALD SCHATTEN

4   The Architecture of and Chromosome Movements within the Premeiotic Interphase Nucleus

KATHLEEN CHURCH

5   Chromosome Movements within Prophase Nuclei

GEOFFREY K. RICKARDS

II Mitotic Mechanisms and Approaches to the Study of Mitosis

6 Light Microscopic Studies of Chromosome Movements in Living Cells

ARTHUR FORER

7 Chromosome Micromanipulation Studies

GORDON W. ELLIS AND DAVID A. BEGG

8 Mitotic Mutants

BERL R. OAKLEY

9   Mutants as an Investigative Tool in Mammalian Cells

V. LING

10   Immunofluorescence Studies of Cytoskeletal Proteins during Cell Division

J. E. AUBIN

11   Mitosis through the Electron Microscope

I. BRENT HEATH

12 Mitosis: Studies of Living Cells—A Revision of Basic Concepts

A. S. BAJER AND J. MOLÈ-BAJER

13 Studies of Mitotic Events Using Lysed Cell Models

JUDITH A. SNYDER

14 The Isolated Mitotic Apparatus: A Model System for Studying Mitotic Mechanisms

ARTHUR M. ZIMMERMAN AND ARTHUR FORER

15 Calmodulin and ATPases in the Mitotic Apparatus

BARBARA W. NAGLE AND JOAN C. EGRIE

III Mechanisms of Cytokinesis

16 Mechanisms of Cytokinesis in Animal Cells

GARY W. CONRAD AND RAYMOND RAPPAPORT

17 Mechanical Properties of Dividing Cells

YUKIO HIRAMOTO

18 Electron Microscope Studies of Cytokinesis in Metazoan Cells

JOLANTA KARASIEWICZ

19 Inhibitors and Stimulators in the Study of Cytokinesis

JESSE E. SISKEN

20 Cell Division: A Commentary

E. W. TAYLOR

List of Contributors

J. E. Aubin (211), Medical Research Council Group in Periodontal Physiology, Faculty of Dentistry, University of Toronto, Toronto, Ontario M5S 1A8, Canada

A. S. Bajer (277), Department of Biology, College of Arts and Sciences, University of Oregon, Eugene, Oregon 97403

David A. Begg (155), Department of Anatomy, and Laboratory of Human Reproduction and Reproductive Biology, Harvard Medical School, Harvard University, Boston, Massachusetts 02115

Kathleen Church (83), Department of Zoology, Arizona State University, Tempe, Arizona 85281

Gary W. Conrad (365), Division of Biology, Kansas State University, Manhattan, Kansas 66506

Joan C. Egrie (337), Department of Zoology, University of California, Berkeley, California 94720

Gordon W. Ellis (155), Department of Biology, University of Pennsylvania, Philadelphia, Pennsylvania 19104

Arthur Forer (135, 327), Department of Biology, York University, Downsview, Ontario M3J 1P3, Canada

Patricia J. Harris (29), Department of Biology, College of Arts and Sciences, University of Oregon, Eugene, Oregon 97403

I. Brent Heath (245), Department of Biology, York University, Downsview, Ontario M3J 1P3, Canada

Yukio Hiramoto (398), Biological Laboratory, Tokyo Institute of Technology, Ookayama, Meguro-ku, Tokyo 152, Japan

Jolanta Karasiewicz (419), Department of Embryology, Zoological Institute, University of Warsaw, 00-927 Warsaw, Poland

V. Ling (197), Department of Medical Biophysics, University of Toronto and The Ontario Cancer Institute, Toronto, Ontario M4X 1K9, Canada

J. Molè-Bajer (277), Department of Biology, College of Arts and Sciences, University of Oregon, Eugene, Oregon 97403

Barbara W. Nagle (337), Department of Zoology, University of California, Berkeley, California 94720

Berl R. Oakley (181), Department of Pharmacology, College of Medicine and Dentistry of New Jersey–Rutgers Medical School, Piscataway, New Jersey 08854

John R. Pringle (3), Department of Cellular and Molecular Biology, Division of Biological Sciences, The University of Michigan, Ann Arbor, Michigan 48109

Raymond Rappaport (365), Department of Biological Sciences, Union College, Schenectady, New York 12308, and The Mount Desert Island Biological Laboratory, Salsbury Cove, Maine 04672

Geoffrey K. Rickards (103), Botany Department, Victoria University of Wellington, Private Bag, Wellington, New Zealand

Gerald Schatten (59), Department of Biological Science, The Florida State University, Tallahassee, Florida 32306

Jesse E. Sisken (437), Department of Pathology, and Department of Physiology and Biophysics, College of Medicine, University of Kentucky, Lexington, Kentucky 40536

Judith A. Snyder (301), Department of Biological Science, University of Denver, Denver, Colorado 80208

E. W. Taylor (461), Department of Biology, The University of Chicago, Chicago, Illinois 60637

Arthur M. Zimmerman (327), Department of Zoology, Ramsay Wright Zoological Laboratories, University of Toronto, Toronto, Ontario M5S 1A1, Canada

Preface

Mitosis and cytokinesis are two activities of fundamental importance to eukaryotic cells. This book reflects the current knowledge of investigators whose chief concern has been to understand mitosis and cytokinesis. Even though various aspects of mitosis and cytokinesis have been covered in separate chapters or review articles, no comprehensive treatment of these subjects has appeared since the classic monograph of Franz Schrader in 1953 and the extended review of Dan Mazia in 1961. We have attempted to fill this gap by providing in one book an extended treatment of cell division, from the formation of chromosomes in the nucleus until the end of cell cleavage.

The chapters in this book cover various aspects of mitosis and cytokinesis as studied from different points of view by various authors. The chapters summarize work at different levels of organization, including phenomenological, molecular, genetic, and structural levels. In many cases we asked the contributors to restrict themselves to studies at one particular level of organization or to studies using one particular approach. The authors were asked to include an overview of the field, to develop a main theme in their area of expertise, and to describe the conclusions so that they could be understood by a broad range of biologists. They were also encouraged, if and where appropriate, to speculate somewhat on potential developments and to include in their contributions new and previously unpublished material. Thus we anticipate that this volume will provide background and perspective into research on mitosis and cytokinesis that will be of use and of interest to a broad range of scientists and advanced students interested in basic cellular events, including cell biologists, molecular biologists, developmental biologists, geneticists, biochemists, and physiologists.

The book is organized into three general sections. The chapters in Part I deal with premeiotic and premitotic events, in Part II with mitosis, and in Part III with cytokinesis. We hope that the book will give readers some appreciation of how workers in the field presently understand and approach mitosis and cytokinesis, two processes of prime importance to the eukaryotic cell.

Arthur M. Zimmerman
Arthur Forer

Premeiotic / Premitotic Events

1

The Genetic Approach to the Study of the Cell Cycle

JOHN R. PRINGLE

I. INTRODUCTION

During the past decade, a promising start has been made on the genetic analysis of the eukaryotic cell division cycle. Mutations affecting particular steps of the cell cycle have been isolated in the budding yeast *Saccharomyces cerevisiae* (Pringle and Hartwell, 1981), the fission yeast *Schizosaccharomyces pombe* (Minet *et al.*, 1979; Nurse and Thuriaux, 1980), the mycelial fungus *Aspergillus nidulans* (Trinci and Morris, 1979; Oakley, 1981), the smut fungus *Ustilago maydis* (Jeggo *et al.*, 1973), the slime mold *Physarum polycephalum* (Laffler *et al.*, 1979), the green alga *Chlamydomonas reinhardtii* (Howell *et al.*, 1977), the ciliated protozoan *Tetrahymena thermophila* (Frankel *et al.*, 1980), the fruit fly *Drosophila melanogaster* (Baker *et al.*, 1978), and various lines of mammalian cells

3

MITOSIS/CYTOKINESIS
Copyright © 1981 by Academic Press, Inc.
All rights of reproduction in any form reserved.
ISBN 0-12-781240-7

(Simchen, 1978; Liskay and Prescott, 1978; Ling, 1981). Given the limitations of space and the existence of other recent reviews (Hartwell, 1978; Simchen, 1978; Oakley, 1981; Ling, 1981), I have made no attempt to summarize here the particulars of these various studies. Instead, I have attempted to distill from the practical experiences and theoretical discussions of the past 10 years some generally applicable ideas about the nature and uses of cell cycle mutants. Most of my specific examples are taken from studies of S. cerevisiae, both because I know this organism best and because its cell cycle has been the most extensively analyzed genetically. A more systematic review of the S. cerevisiae cell cycle is presented elsewhere (Pringle and Hartwell, 1981).

II. THE NATURE OF CELL CYCLE MUTANTS

A. Continuous and Discontinuous (Stage-Specific) Processes

As has often been noted, successful completion of a cell cycle requires a cell to integrate the processes that duplicate the cellular material with the processes that partition the duplicated material into two viable daughter cells. An alternative formulation that is probably more instructive as to mechanism is that successful cellular reproduction requires a cell to integrate the discontinuous (or stage-specific) events of the cell cycle proper with the continuous processes of metabolism, maintenance, and growth. This formulation is similar to Mazia's (1978) view of the cell cycle as a "bicycle" with a "reproductive wheel" and a "growth wheel."

The stage-specific events occur once or a few times per cycle. They include aspects of the duplication of cellular materials (e. g., the duplication of microtubule-organizing centers and of chromosomal DNA) as well as aspects of the partitioning of material into daughter cells (e. g., mitosis and cytokinesis). Both types of stage-specific events occur as elements of a well-defined program (Section III,D; Pringle and Hartwell, 1981) that has a definite beginning, in the events leading up to duplication of the microtubule-organizing centers and the initiation of chromosomal DNA synthesis, and a definite end, in the completion of mitosis and cytokinesis. Cells that because of nutrient limitation or differentiation are not actively engaged in reproduction are generally arrested between the end of one transit of the program and the beginning of the next (Baserga, 1976; Prescott, 1976; Pardee et al., 1978; Tucker et al., 1979; Pringle and Hartwell, 1981).

In contrast, the continuous processes occur throughout the program of stage-specific events and also, for the most part, in cells that have not undertaken this program or that are blocked in its execution. Whether or not a cell is actively growing or reproducing, it must continuously expend energy for such purposes as protein turnover (Goldberg and St. John, 1976) and the maintenance of intracellular pH (Navon *et al.*, 1979) and ionic composition (De Luise *et al.*, 1980). When net growth does occur, it generally proceeds continuously. Small daughter cells increase steadily in volume prior to initiating the program of stage-specific events (Johnston *et al.*, 1977), and total cell mass seems to increase steadily during the cell cycle (Mitchison, 1971). Moreover, the increase of volume and mass continues unabated when cell cycle progress is blocked by any of a variety of mutations or inhibitors that affect particular stage-specific events (Johnston *et al.*, 1977). Presumably, the steady increase in total cell mass reflects a more or less continuous generation of ATP and formation of biosynthetic precursors, as well as a continuous accumulation of macromolecules. In yeast, at least, it seems clear that synthesis of the major cell wall polysaccharides, mitochondrial DNA, the major classes of RNA, total protein, and most individual proteins is continuous throughout the program of stage-specific events and also continues when this program is blocked (reviewed by Pringle and Hartwell, 1981).

Thus, the picture that emerges is of a continuous background of metabolism, maintenance, and (environmental and developmental circumstances permitting) net growth, on which can be superimposed the program of discontinuous events (including some special aspects of growth) that leads to cell division. The problems of coordination facing a cell are basically three. First, when reproduction occurs, the stage-specific events must be coordinated with each other, so that they occur in a proper order (Section III, D). Second, when growth occurs, the various continuous processes must be coordinated with each other, so that growth is "balanced" (Warner and Gorenstein, 1978; Swedes *et al.*, 1979). Third, reproduction must be coordinated with growth. For continuously proliferating cells, this means that division must occur approximately once per doubling in mass achieved by balanced growth, a goal that seems met primarily by having one (Johnston *et al.*, 1977; Hartwell and Unger, 1977; Pringle and Hartwell, 1981) or two (Nurse and Thuriaux, 1980) particular stage-specific events dependent on the achievement of a threshold amount of growth. However, as Mazia (1978) has emphasized, growth and reproduction can be geared together in many ways, as exemplified most dramatically by many animal eggs, which first grow extensively without dividing and then divide repeatedly without growing.

B. Cell Cycle Mutants

It is clear that mutations producing defects in continuous processes such as ATP generation, protein synthesis, or mitochondrial DNA replication can prevent or impair cellular reproduction. However, the discussion of Section II,A should make clear that it is both logical and heuristically valuable to restrict the term *cell cycle mutation* to mutations that lead to defects in particular stage-specific functions of the cell cycle, and this has, in fact, been the conventional definition (Hartwell, 1974, 1978; Pringle, 1978; Simchen, 1978). The immediate effect of the mutation is said to be on the *primary defect event*, which can be either synthesis or function of the cell cycle gene product (Pringle, 1978). However, the effects of the mutation can be analyzed only in terms of events that can be monitored biochemically or morphologically with presently available techniques [i.e., the so-called *landmark events* (Hartwell, 1974, 1978; Pringle, 1978)]. The first such event known to be affected by the mutation is called the *diagnostic landmark*. Unless the molecular nature of the primary defect is known, it cannot be decided whether this event is directly involved in, or merely a prerequisite for, the diagnostic landmark. Indeed, the diagnostic landmark may occur considerably later in the cycle than the primary defect event and may require revision as new landmark events are recognized. For example, discovery of the microfilament ring at the mother-bud neck in *S. cerevisiae* (Byers and Goetsch, 1976a) led to the recognition that mutants whose diagnostic landmark was originally cytokinesis (Hartwell, 1971) were also defective in the much earlier event of microfilament ring formation (Byers and Goetsch, 1976b).

In principle, a cell cycle mutation could affect a function that was either essential for division or only helpful (e. g., in increasing the fidelity of DNA replication), and could either block completely, or only produce abnormalities in, the affected function. In practice, most cell cycle mutations studied to date have produced complete (although not necessarily immediate) blockage of events that are essential for cell cycle progress. Since such mutations are lethal, conditional mutants of some type must be used (Pringle, 1975). To date, ordinary temperature-sensitive mutants have been used in most studies, but there have also been some recent successes with *S. cerevisiae* in isolating cold-sensitive cell cycle mutants (D. Moir, S. Stewart, and D. Botstein, personal communication) and suppressible nonsense cell cycle mutants (Reed, 1980b; Rai and Carter, 1981).

When conditional-lethal cell cycle mutants are placed under restrictive conditions, each cell ceases normal development at the same point in the cell cycle (i. e., the time at which the defective gene product would normally function). Thus, from an initially asynchronous population there develops a population of cells that is homogeneous with respect to the stage-specific events that have and have not been completed; the arrested cells are said to exhibit the characteristic *terminal phenotype* for the mutation (Hartwell, 1974; Pringle, 1978). If the normal cell cycle involves a sequence of easily recognized morphological stages, as in yeasts and ciliates, then the terminal phenotype includes a characteristic morphology, a fact that can be exploited in screening collections of conditional-lethal mutants for cell cycle mutants (Hartwell *et al.*, 1973; Minet *et al.*, 1979; Frankel *et al.*, 1980). It must be emphasized that both continuous processes and discontinuous processes not dependent on the primary defect event continue while this event is blocked; thus, the terminal phenotype cannot be regarded simply as a normal stage of the cell cycle, and experiments involving the return of arrested cells to permissive conditions can be very difficult to interpret (Pringle, 1978).

Two additional complications affecting these concepts and definitions should be noted. First, the primary defect event of a cell cycle mutant need not itself be a stage-specific function. For example, a mutation blocking the normally continuous synthesis of a gene product whose function is stage specific will in general be identified as a cell cycle mutation. Also, the role of G_1 events in regulating the rate of cell proliferation (Baserga, 1976; Prescott, 1976; Pardee *et al.*, 1978; Pringle and Hartwell, 1981) makes it clear that various mutations affecting continuous aspects of metabolism and growth could lead to G_1 arrest; the discovery that the *cdc19* mutation of *S. cerevisiae* (Hartwell *et al.*, 1973) is a temperature-sensitive pyruvate kinase mutant (Kawasaki, 1979) is a case in point. Second, it seems likely that some gene products are responsible for two, or a few, stage-specific functions. Such gene products are not less important to an understanding of cellular reproduction than those participating in single stage-specific functions, yet the corresponding mutants have probably been excluded from consideration by the definition and screening criterion discussed above.

III. THE USES OF CELL CYCLE MUTANTS

In considering the uses of cell cycle mutants, it is important to recognize that different conditionally defective products of the same gene will often behave differently under permissive, restrictive, or intermediate conditions (Pringle, 1975). Thus, comparison of the properties of different mutants of the same gene will often yield useful information directly or allow the choice of the mutant best suited to a particular use (Hartwell *et al.*, 1973; Hereford and Hartwell, 1974; Pringle, 1975, 1978; Newlon and Fangman, 1975; Hartwell, 1978; Sloat *et al.*, 1981).

A. Miscellaneous Uses

1. How Does Cellular Reproduction Occur?

One important class of questions about the cell cycle is concerned with *how* a cell reproduces itself once it has undertaken to do so. Cell cycle mutants contribute in a variety of ways to attempts to deal with such questions. For example, cell cycle mutants can be used to generate synchronous cultures by incubating under restrictive conditions and then releasing the cell cycle block by shifting to permissive conditions (Zakian et al., 1979). Such synchronous cultures can then be used to study the timing and biochemistry of landmark events. However, the fact that some stage-specific processes may continue while others are blocked (Section III,D; see also comments on the nature of the terminal phenotype in Section II,B) means that the synchrony induced may apply only to a subset of the stage-specific events, so that caution is necessary in interpreting the results obtained with such cultures.

Since the known landmark events represent only a small fraction of the stage-specific events comprising the cell cycle (Section III,B), a major incentive for isolating cell cycle mutants is their potential usefulness in identifying the individual molecular steps of the cell cycle. Unfortunately, progress to date in identifying the primary defect events of cell cycle mutants has been slow. Of the 50 cell cycle genes identified in S. cerevisiae (Pringle and Hartwell, 1981), the gene products affected by the mutations have been identified in only three cases [cdc9, DNA ligase (Johnston and Nasmyth, 1978); cdc19, pyruvate kinase (Kawasaki, 1979); cdc21, thymidylate synthetase (Bisson and Thorner, 1977)], all of which are, in a sense, trivial in that they represent functions already well known from biochemical studies. Fortunately, the development of procedures that make possible the molecular cloning of cell cycle genes once these have been identified by mutations (Clarke and Carbon, 1980; Nasmyth and Reed, 1980) should greatly facilitate the identification of the protein products of such genes (e. g., by the translation in vitro of mRNAs that hybridize to the cloned genes), although identification of the functions of these proteins will doubtless be more difficult. The availability of the cloned genes will also make possible analysis of the control of transcription of these genes during the cell cycle.

Whether or not the primary defect event of a cell cycle mutant is known, the mutant can be used to explore the range of functions in which a particular gene product participates. For example, the cdc9 mutants of S. cerevisiae have provided evidence that a single DNA ligase is involved in DNA replication, recombination, and the repair of damage due to ultraviolet and X-irradiation in this organism (Johnston and Nasmyth, 1978; Fabre and Roman, 1979). Perhaps the most important use of this type has been the demonstration that some, but not all, of the gene products involved in mitosis or meiosis function in both of these processes (Simchen, 1974; Baker et al., 1976, 1978; Simchen and Hirschberg, 1977; Zamb and Roth, 1977; Schild and Byers, 1978).

Finally, cell cycle mutants can be used to determine where inhibitor-treated cells are arrested, relative to the program of stage-specific events (Bücking-Throm et al., 1973; Wilkinson and Pringle, 1974; Hartwell, 1976), or to ask what cells blocked at particular positions in the cell cycle are capable of doing (Reid and Hartwell, 1977; Sloat *et al.*, 1981), although the nature of the terminal phenotype (Section II,B) suggests caution in interpreting experiments of the latter type.

2. How Is Cell Proliferation Controlled?

The other important class of questions about the cell cycle deals with the factors that determine *when* a cell will undertake to reproduce itself. Here also, cell cycle mutants are a vital tool. For example, mutants have been instrumental in clarifying both the mechanisms coordinating successive mitotic cycles (Hartwell *et al.*, 1974) and the mechanisms coordinating the mitotic cycle with the alternative developmental pathways of meiosis (Hirschberg and Simchen, 1977) and sexual conjugation (Bücking-Throm *et al.*, 1973; Wilkinson and Pringle, 1974; Reid and Hartwell, 1977) in *S. cerevisiae*. Cell cycle mutants have also helped define the mechanisms coordinating cell proliferation with the availability of essential nutrients in *S. cerevisiae* (Hartwell, 1974); particularly striking is the conclusion, from the G_1 arrest of certain methionyl-tRNA synthetase mutants, that the cell monitors the availability of sulfate by means of its effect on the rate of protein synthesis (Unger, 1977). Cell cycle mutants have also been critical to the development of current models of how cell growth is coordinated with cell division (Johnston *et al.*, 1977; Nurse and Thuriaux, 1980; Pringle and Hartwell, 1981). Finally, mutants are providing major clues as to the nature of the event or events in the G_1 phase that seem to be the major site for the overall control of cell proliferation (Hereford and Hartwell, 1974; Liskay and Prescott, 1978; Reed, 1980a); in particular, the behavior of certain mutants has focused attention on the microtubule-organizing center as the likely physical site at which the control is exerted (Byers and Goetsch, 1974; Dutcher, 1980; Byers, 1981; Pringle and Hartwell, 1981).

3. Other Uses

The cell cycle mutants already in hand can be extremely valuable in allowing the selection of additional cell cycle mutants (Reid, 1979; Reed, 1980a, 1980b; D. Moir, S. Stewart, and D. Botstein, personal communication). Also, certain kinetic analyses of cellular behavior are simplified if cells are prevented from dividing by a cell cycle mutation (Hartwell, 1973). Finally, the chromosome loss sustained by some cell cycle mutants can be useful in mapping genes to chromosomes (Kawasaki, 1979).

B. Estimation of the Number of Stage-Specific Functions

A reliable estimate of the number of discontinuous functions comprising the eukaryotic cell cycle, or (what is almost the same question) of the total number of cell cycle genes, would have important implications for research strategies. Our general ignorance of the molecular details of cell cycle processes makes such estimates unapproachable biochemically. However, since cell cycle mutants can be isolated by procedures that demand no *a priori* knowledge of these molecular details, it should be possible to obtain mutations affecting any cell cycle gene. Thus, genetic analysis should allow the desired estimates of the molecular complexity of the cell cycle. Unfortunately, studies to date in yeast and other organisms do not provide a clear picture.

Extensive studies in *Drosophila* have suggested that the total number of genes is approximately 5000–6000 (Lefevre, 1974; Lewin, 1974; Young and Judd, 1978; Ripoll and Garcia-Bellido, 1979; Hilliker *et al.*, 1980), that about 90% of these genes are essential for survival of the organism (Lefevre, 1974; Ripoll and Garcia-Bellido, 1979), but that only about 10–12% of these essential genes (some 500–600) are "cell lethal" (i. e., essential for the growth and division of individual somatic cells). The last conclusion was based on the behavior in mosaic spots (produced by induced mitotic recombination) of cells homozygous either for ethylmethane sulfonate-induced lethal point mutations (Ripoll, 1977) or for small deletions (Ripoll and Garcia-Bellido, 1979). In studies both of temperature-sensitive (Hartwell *et al.*, 1973) and of cold-sensitive (D. Moir, S. Stewart, and D. Botstein, personal communication) conditional-lethal mutants of *S. cerevisiae*, about 10% of "cell-lethal" mutants are cell cycle mutants. If a similar figure applies to *Drosophila*, then this organism contains only some 50–60 cell cycle genes. This argument suggests that the 50 cell cycle genes already characterized in *S. cerevisiae* (Pringle and Hartwell, 1981) are a majority of the total, a conclusion that at first glance seems supported by the observation that most newly isolated temperature-sensitive cell cycle mutants now prove to carry alleles of previously known genes (Hartwell *et al.*, 1973; Johnston and Game, 1978; J. Pringle, unpublished observations).

However, a variety of considerations suggests that the foregoing argument must, in some nonobvious way, be grossly misleading. The stripped-down macronuclear genome of vegetatively growing *Oxytricha* (a ciliated protozoan) contains some 17,000 genes (Lawn *et al.*, 1978); vegetatively growing amoebae of the slime mold *Dictyostelium* display some 4000–5000 polysome-associated mRNAs (Blumberg and Lodish, 1980a), while more than 3000 additional polysomal messages appear during the differentiation of stalk cells and spores (Blumberg and Lodish, 1980b); and sea urchins seem to require the expression of at least 25,000 structural genes during early development, and of many additional thousands of structural genes in differentiated cells of adult organisms (Galau *et al.*, 1976; Davidson and Britten, 1979; Lee *et al.*, 1980). This impression of complexity based on studies of nucleic acids seems supported by recent studies demonstrating surprisingly large numbers of proteins in such seemingly simple structures as the *Chlamydomonas* flagellar axoneme ($\geq$ 200 polypeptides: Piperno and Luck, 1979; Luck *et al.*, 1981), the silkmoth eggshell ($\geq$ 186 polypeptides: Regier *et al.*, 1980), and the vitelline layer of the sea urchin egg ($\geq$ 60 polypeptides: Glabe and Vacquier, 1977; B. Hough-Evans and E. H. Davidson, personal communication). It is possible that the estimate of 5000–6000 genes in *Drosophila* is several-fold too low (O'Brien, 1973; Hough-Evans *et al.*, 1980).

Given such data, it seems unlikely that the many critical processes of the cell cycle would require only some 50–60 polypeptides. Indeed, a close look at the information available on cell cycle mutants in *S. cerevisiae* supports the view that many more than 50 cell cycle genes will ultimately be identified. First, the nonrandom distribution of known cell cycle mutations among known cell cycle genes implies that these genes differ greatly in their susceptibility to conventional temperature-sensitive mutations, and thus that some (perhaps many) cell cycle genes are difficult or impossible to identify using such mutations exclusively. For example, the 148 independently isolated cell cycle mutants of Hartwell *et al.* (1973) defined 32 genes, represented by numbers of alleles ranging from 1 (10 genes) to 18 (1 gene). Moreover, this set of mutants contained only three that arrested as large, unbudded, multinucleate cells; all three were mutant in gene *CDC24*. Screening of an additional 5000 temperature-sensitive lethal clones for mutants of similar phenotype yielded 31 independently isolated mutants; 25 were defective in gene *CDC24*, while 5 defined the new gene *CDC43* and just 1 defined the new gene *CDC42* (A. Adams, R. Longnecker, and J. Pringle, unpublished results). Similarly, the original set of 148 mutants included only 1 rather leaky mutant of gene *CDC28* and 2 genetically intractable mutants (*cdc22* and *cdc32*) with apparently similar phenotypes. Screening several thousand additional temperature-sensitive lethal clones yielded only one additional *cdc28* mutant and no other mutants with similar phenotypes. However, use of a powerful selective method did allow Reed (1980a) to isolate 33 mutants with this phenotype, including 17 additional independent isolates defective in gene *CDC28*, 14 independent isolates defining the new gene *CDC36*, and 1 isolate each defining the new genes *CDC37* and *CDC39*.

Second, the studies of D. Moir, S. Stewart, and D. Botstein (personal communication) on cold-sensitive cell cycle mutants strongly support the conclusion that many genes may be missed in studies relying exclusively on conventional temperature-sensitive mutations. Among the nine cold-sensitive complementation groups identified by these workers, there is one that appears to be a gene previously identified by temperature-sensitive mutations, five that appear to be newly identified genes, and three about which no statement can be made.

Third, it seems clear that only a minority of the *S. cerevisiae* genes whose products are directly involved in DNA replication has been identified. Since DNA replication in *Escherichia coli* involves at least 14 proteins (Alberts and Sternglanz, 1977; Wickner, 1978), it hardly seems likely that the *CDC9* gene product (DNA ligase: Johnston and Nasmyth, 1978), the *CDC8* gene product (Fangman and Zakian, 1981), and possibly the *CDC2* gene product (Hartwell, 1976; Fangman and Zakian, 1981) are the only proteins involved in this process in yeast. It has proven extremely difficult to identify additional temperature-sensitive DNA replication mutants (Johnston and Game, 1978), but a recent intensive search in a different genetic background has apparently identified several additional genes (A. Thomas and L. H. Johnston, personal communication). Unless the DNA replication mutants are somehow atypical, these results suggest that only a minority of all cell cycle genes has been identified in *S. cerevisiae*.

Finally, there is good evidence that *S. cerevisiae* cells growing exponentially on rich media contain about 4000 different mRNA species (Hereford and Rosbash, 1977). If it is true that approximately 10% of the genes essential for vegetative growth in rich media are cell cycle genes (Hartwell *et al.*, 1973), then the total number of cell cycle genes in yeast may be close to 400 rather than to 50.

If it is true that only a minority of cell cycle genes has been identified in *S. cerevisiae*, then continuing efforts to identify new genes clearly remain important. Such efforts must take account of at least six factors that may be contributing to the recent difficulties in extending the list of known cell cycle genes.

1. As discussed above, it seems clear that one important factor is the difficulty of obtaining temperature-sensitive alleles of some genes. This difficulty can probably be circumvented by working with a variety of additional types of conditional mutations (Pringle, 1975). Cold-sensitive mutations and suppressible nonsense mutations (obtained in strains carrying temperature-sensitive suppressor genes) should be especially valuable in this regard.

2. Some yeast structural genes are represented by two or three nontandem copies (or near copies) per haploid genome (Sherman *et al.*, 1978; Hereford *et al.*, 1979; Woolford *et al.*, 1979; Holland and Holland, 1980). If any cell cycle genes are in this category, it may be difficult or impossible to detect recessive mutations affecting single copies of these genes. The only apparent hope for dealing with this nasty possibility is that such genes can be detected as extragenic suppressors of previously identified cell cycle mutations (Jarvik and Botstein, 1975; Morris *et al.*, 1979). The altered gene products possessing suppressor activity are expected to be epistatic to the

normal gene products coded for by any other, nonmutant copies of the gene, an expectation that is supported by the limited experimental evidence available (D. Moir, S. Stewart, and D. Botstein, personal communication).

3. It is possible that the products of most cell cycle genes are not essential for cell cycle progress but only helpful (Section II,B), a speculation that is encouraged by the observation that the majority of genes in the bacteriophage T4 are "nonessential" for growth on conventional strains of host bacteria (Wood and Revel, 1976). Mutations in such genes would not have been detected by the screening procedures used to date with S. cerevisiae but could presumably be studied using approaches modeled on those used successfully to study such mutations in Drosophila (Baker et al., 1978).

4. As noted above (Section II,B), it is likely that genes whose products are responsible for two or a few essential stage-specific functions have been overlooked in the screening done to date for cell cycle mutants of S. cerevisiae. It would seem possible to deal with this problem by using a more flexible screening criterion, in which mutants giving a mixture of two or three distinct morphological cell types under restrictive conditions would be saved for analysis.

5. It is also conceivable that there is a flaw in the prevailing assumption that any mutation directly affecting a single essential stage-specific function will yield a morphologically homogeneous terminal phenotype in S. cerevisiae. Johnston and Game (1978) considered this to be a possible explanation for the paucity of mutants isolated to date with primary defects in DNA synthesis, and consequently screened directly for mutants with reduced incorporation of precursors into DNA. The only mutations identified with pronounced effects on DNA synthesis and heterogeneous terminal phenotypes also had pronounced effects on RNA synthesis; presumably, the effects on the continuous process of RNA synthesis account for the morphologically heterogeneous arrest. Although such apparent involvement of a single gene product both in a stage-specific function and in a continuous process may not be rare, at least as regards nucleic acid synthesis (Fangman and Zakian, 1981), it is noteworthy that the more recent successes in isolating new DNA synthesis mutants (A. Thomas and L. H. Johnston, personal communication) did involve a search for mutants with morphologically homogeneous terminal phenotypes.

6. It has recently become clear that reactions catalyzed by distinct gene products in prokaryotes are often catalyzed by the separate domains of multifunctional polypeptides in eukaryotes (Kirschner and Bisswanger, 1976; Pringle, 1979). If the reactions catalyzed are sequential, and if products are nondiffusible, an entire gene coding for a multifunctional protein should behave as a single complementation group. Thus, it is possible that the number of stage-specific functions is significantly greater than the number of

complementation groups of cell cycle mutants. This possibility seems difficult to test by purely genetic methods, but answers should emerge as the products of cell cycle genes are identified and characterized in detail.

C. Temporal Mapping of Cell Cycle Events

A *temporal map* is simply a diagram that summarizes the temporal order of events without attempting to indicate the mechanisms that determine this order or even whether the order shown is fixed or flexible (Pringle, 1978). Of the several ways in which it has been proposed to use cell cycle mutants in generating temporal maps of cell cycle events, only two seem logically sound (Pringle, 1978). First, the relative temporal order of events is sometimes obtained as a corollary of the functional relationships that can be deduced from the behavior of cell cycle mutants (Section III,D). Because other methods for temporal mapping (e. g., monitoring landmark events in synchronous cultures) have limited resolving power, this approach is especially valuable in considering events that occur at about the same time. Second, it is sometimes possible to learn about the timing of gene product synthesis and of gene product function from the *execution points* that all conditional cell cycle mutants display. The execution point is the time in the cell cycle beyond which a shift from permissive to restrictive conditions can no longer prevent a mutant cell from successfully completing the cycle in question (Hartwell, 1974; Pringle, 1978). Thus, for a mutant exhibiting *first-cycle arrest*, a cell that is prior to the execution point at the time of the shift to restrictive conditions arrests, with the appropriate terminal phenotype, without having divided under the restrictive conditions; a cell that is past the execution point divides once successfully under the restrictive conditions and then arrests, with the appropriate terminal phenotype, in the subsequent cell cycle (see Hartwell, 1974, Figs. 3 and 4). A mutant that fails to exhibit first-cycle arrest has an execution point in the cycle previous to the cycle of reference (or in an even earlier cycle); a cell prior to the execution point will divide *n-1* times under restrictive conditions, and a cell past the execution point will divide *n* times, before both arrest with the appropriate terminal phenotype (see Hartwell, 1974, Fig. 3).

It is crucial to realize that the execution point is defined operationally; it is a number that can be determined experimentally by any of several straightforward techniques (reviewed by Pringle, 1978). However, such numbers are of no discernible intrinsic interest, and the desired interpretation of execution point data in terms of times of gene product function and times of gene product synthesis, although possible in principle and potentially valuable, is not straightforward and must be approached with great caution. Consider a cell cycle gene whose product is synthesized continuously but carries out a function essential for division only during a discrete, short interval of the cell cycle. A *tight conditional-labile* (or *conditional-for-function*) mutation in this gene (in which the mutant gene product loses its activity instantaneously and completely upon a shift to restrictive conditions) will display an execution point just at the end of the time of gene product function. A *leaky* conditional-labile mutant of the same gene will display an earlier execution point; how much earlier will depend on (1) how slowly or incompletely the mutant gene product loses activity under restrictive conditions and (2) the degree to which the gene product is present in cells growing under permissive conditions in excess of the minimal amount required to complete the essential function and progress toward division. [The significance of point (2) is demonstrated by the recent experiments of Byers and Sowder (1980), who formed *S. cerevisiae* cells whose nuclei contained temperature-sensitive cell cycle mutations and whose cytoplasms contained normal levels of the wild-type products of the genes in question. In all cases tested but one, such hybrid cells could complete several cell cycles at restrictive temperature, suggesting that the wild-type gene products are normally present continuously in considerable excess over what is required for completion of a cell cycle, although the presence of excess amounts of exceptionally stable mRNAs for these gene products could also conceivably explain the results obtained.] A tight *conditional-for-synthesis* mutation in the same gene (in which the mutant gene product retains its activity under restrictive conditions, but no new active gene product can be formed) will display an execution point at whatever time enough of the gene product has been accumulated to allow successful completion of the cell cycle of reference; a leaky conditional-for-synthesis mutant will display a correspondingly earlier execution point.

Thus, the execution point is clearly an allele-specific, rather than a gene-specific, parameter (see Hartwell *et al.*, 1973, Table 3; and Hartwell, 1974, Fig. 3), and the extraction of useful information from execution point data requires that we know which mutants are tight conditional-labile, which are tight conditional-for-synthesis, and which are leaky. Until such discriminations can be made by direct assays of the gene products, it is necessary to rely on indirect arguments, of which the following seem likely to be useful.

1. In the absence of direct assays of gene product activity, it seems difficult or impossible to distinguish between conditional-for-synthesis and leaky conditional-labile alleles. I have suggested previously (Pringle, 1975, 1978) that nonsense mutants isolated in a strain carrying a tight temperature-sensitive nonsense suppressor would aid in making this distinction. However, the efficiencies of suppression in such strains at permissive temperature will presumably always be much less than 100%. Thus, the time at which sufficient gene product for cell cycle completion has been accumu-

lated in such a strain growing at permissive temperature (measurable as the execution point of the strain) will presumably be quite different from the corresponding time in a wild-type strain. A direct assay of gene product activity would make possible not only the identification of tight conditional-for-synthesis alleles but also direct determinations of the times of gene product synthesis. However, execution-point determinations would still be useful in defining when *sufficient* gene product had been synthesized for the completion of the cell cycle of reference.

2. It is possible that some cell cycle gene products complete their essential functions for cell cycle *n* during cell cycle *n-1* (or even earlier). However, both our general concept of the cell cycle as a program of stage-specific events with a definite beginning (Sections II, A and III, D) and the existence for most known cell cycle genes of alleles showing first cycle arrest (Hartwell *et al.*, 1973) suggest that such genes will be exceptions. Thus, if the one or few available mutant alleles of a particular gene all fail to exhibit first cycle arrest, they should be regarded as conditional-for-synthesis and/or leaky until proven otherwise.

3. If it is known that a particular gene product is normally present in substantial excess over what is required for cell cycle completion, then mutants displaying first cycle arrest must carry conditional-labile alleles. An indication that a gene product is normally present in excess is provided by the existence for the gene in question both of alleles displaying first cycle arrest and of alleles not displaying first cycle arrest (a common situation: see Hartwell *et al.*, 1973). Stronger evidence has been provided for many genes in *S. cerevisiae* by Byers and Sowder (1980; see above).

4. An indication of which conditional-labile, first-cycle-arrest mutants are tight can come from a comparison of strains carrying various mutant alleles of a gene of interest. The allele with the latest execution point sets a limit on the time of gene product function; the gene product cannot complete its essential function earlier in the cycle than this latest execution point. The observation of a group of alleles with the same late execution point would suggest that these mutants are tight and that their execution point is indeed the end of the time of gene product function in this strain.

Unfortunately, it may often be invalid to infer the end of the time of gene product function in a wild-type strain from that determined in a cell cycle mutant strain growing under permissive conditions. This is predictable from the observation that many conditionally labile gene products have reduced activity even under permissive conditions (Pringle, 1975; Johnston and Nasmyth, 1978) and has been documented experimentally by the observation that *cdc8* and *cdc21* strains (both of which are defective in DNA synthesis at restrictive temperatures) have significantly later execution points for hydroxyurea arrest than do related strains (Hartwell, 1976, Table

3). A further complication is the possibility that genetic background differences may affect the execution points observed for a particular mutant allele. If strong conclusions are to be drawn from execution point data, this possibility must be assessed directly. A final point is that some gene products may complete functions that are essential for division but then continue to function later in the cycle in ways that are not essential for division (Sloat *et al.*, 1981).

D. Functional Sequence Mapping of Cell Cycle Events

It is clear that the stage-specific events of the cell cycle cannot occur in a random temporal order if viable daughter cells are to be produced. There are at least two possible mechanisms by which the normal temporal order of cell cycle events might be achieved (Mitchison, 1971). First, a central timer, or "clock," might trigger events at appropriate times. Second, the temporal order of events might be a consequence of their functional interrelationships; that is, the temporal order A, then B, might result from the functional dependence of event B upon the prior occurrence of event A. While possible cell cycle clocks remain hypothetical, the study of cell cycle mutants has provided solid evidence for the existence of networks of functional interrelationships that can explain both the fixities and the flexibilities of the timing of cell cycle events.

1. Functional Sequence Maps

A functional sequence map is a diagram summarizing the functional interrelatedness of cell cycle events. In the usual form of such a diagram, symbols for the events of interest are associated with arrows. If two events are *interdependent* (i. e., neither can be completed while the other is blocked), their symbols are associated with the same arrow (A,B). If event B is *dependent* upon event A (i.e., A can be completed while B is blocked, but not vice versa), their symbols are associated with arrows connected head-to-tail or by an unbroken sequence of arrows connected head-to-tail (A B or A X Y B; the latter example summarizes a *dependent series*, or *dependent sequence*, of B upon Y upon X upon A). If two events are *independent* (i.e., either can be completed while the other is blocked), their symbols are associated with arrows "in parallel" (i.e., arrows that cannot be connected by an unbroken sequence of arrows connected head-to-tail). In addition to suggesting how the normal temporal order may be achieved, a functional sequence map can provide important clues to the molecular nature of cell cycle events (Hartwell, 1976).

The construction and interpretation of functional sequence maps are subject to a variety of constraints and potential complications (Pringle, 1978). A

major constraint is that in most cases we do not know the functions of cell
cycle gene products (Section III, A). Thus, cell cycle events must be treated
as if they are in two largely nonoverlapping categories: landmark events,
which we can monitor directly but block only indirectly (by blocking primary
defect events on which they are dependent), and the primary defect events
of cell cycle mutants or of stage-specific inhibitors, which we can block
directly but monitor only indirectly (by monitoring landmark events that are
dependent on them). This dichotomy has profound consequences for the
methods that can be used to infer functional sequence maps; a principal
consequence is that these methods can provide direct evidence for a dependent sequence of primary defect events but not for a dependent sequence of
landmark events (Pringle, 1978). A further constraint is that in many cases,
we do not know whether particular mutant alleles are defective in the synthesis or in the function of the mutant gene products; thus, we do not know
whether the steps defined in the functional sequence map are gene product
function events or gene product synthesis events. Fortunately, the results of
Byers and Sowder (1980; see Section III,C) mitigate this problem, at least
for studies of S. cerevisiae, by suggesting both that most first-cycle-arrest
mutants are conditional-for-function and that most cell cycle gene products
are synthesized continuously, so that few gene product synthesis events
require inclusion in a functional sequence map.

Another problem relates to the fact that major landmark events such as
DNA synthesis and nuclear division are presumably composed of many individual steps of gene product function and (perhaps) gene product synthesis.
Thus, in an ideal functional sequence map, such landmarks would be seen to
correspond to *groups* of arrows representing these individual steps. At present, however, the composite nature of the landmark events is obscured
both by our general ignorance of the molecular details of cell cycle events
and by the nature of the methods available for functional sequence mapping.
Since these methods all ask which events can be *completed* while a particular
event is blocked, events that are chemically sequential but more or less
contemporaneous (e. g., the synthesis of DNA precursors and the polymerization of these into DNA) are likely to behave as if interdependent rather
than sequential; the apparent interdependence of the thymidylate
synthetase-catalyzed step with the *CDC8*-controlled step and the
hydroxyurea-sensitive step in S. cerevisiae (Hartwell, 1976) is probably a
case in point. Moreover, a landmark event that is in fact composed of a
dependent sequence of gene product function events will presumably behave formally as if dependent on the earlier events, and interdependent only
with the last event, in that sequence. The composite nature of major landmark events means that statements about the dependence or independence

of landmarks, although formally allowed by the methods of functional sequence mapping (Pringle, 1978) and seemingly informative (Hartwell *et al.*, 1974; Hartwell, 1978), are naive if not misleading. For example, the statement that nuclear division is dependent on DNA synthesis must really mean that one or more of the individual molecular steps comprising nuclear division are dependent on one or more of the individual molecular steps comprising DNA synthesis.

Subject to these constraints, three main methods are presently available for functional sequence mapping. The application of these methods has been treated in detail elsewhere (Pringle, 1978) and will here be summarized only briefly and illustrated with pertinent examples from studies of *S. cerevisiae*. In the *single mutant method*, strains carrying single cell cycle mutations are shifted to restrictive conditions, and the landmarks that do and do not occur are monitored and compared for the various mutants. This method can provide information about the functional relations both among landmark events and among the mutants' primary defect events. For example, the facts that *cdc4* mutants bud but do not initiate DNA synthesis, while *cdc24* mutants initiate DNA synthesis but do not bud (Hartwell *et al.*, 1974), allow the strong conclusion that bud emergence and the initiation of DNA synthesis are independent, as are the primary defect events of the *cdc4* and *cdc24* mutants. In other cases, the single mutant method provides only suggestive evidence about the relationships among landmark events and no information at all about the relationships among primary defect events. For example, the facts that *cdc8* mutants complete neither DNA synthesis nor nuclear division, while *cdc13* mutants complete DNA synthesis but not nuclear division (Hartwell *et al.*, 1974; Hartwell, 1976), suggest but do not prove that nuclear division is dependent on DNA synthesis and provide no information about the relationships between the primary defect events of these mutants.

In some cases where the single mutant method provides little or no information, the *double mutant method* can provide further information about the relationship between the two primary defect events. In this method, the terminal phenotypes of two single mutants are compared to that of the constructed double mutant. (Note that it must be possible to apply restrictive conditions for both mutations simultaneously. Thus, two temperature-sensitive mutants can be used, but not a temperature-sensitive and a cold-sensitive mutant.) For example, although the *cdc4* and *cdc7* mutations both prevent the initiation of DNA synthesis while allowing budding, their terminal phenotypes are not identical (Hartwell *et al.*, 1974). Thus, the fact that the terminal phenotype of the *cdc4 cdc7* double mutant is indistinguishable from that of the *cdc4* single mutant suggests that the *cdc4* primary defect event precedes the *cdc7* primary defect event in a dependent series

(Hereford and Hartwell, 1974). However, the double mutant method cannot rule out the possibility that the *cdc4* and *cdc7* primary defect events are independent (Hereford and Hartwell, 1974; Pringle, 1978).

In other cases, further information about the primary defect events can be provided by the *reciprocal shift method.* This method requires two reversible blocks that can be applied independently; thus, a temperature-sensitive and a cold-sensitive mutation can be used (so long as some intermediate temperature is permissive for both mutations), but not two temperature-sensitive mutations. Alternatively, a conditional mutation can be used in conjunction with a stage-specific inhibitor. The method requires two separate experiments. In the first experiment, block A alone is applied during a first incubation; when the cells are arrested, block A is removed and block B is applied, and the ability of the cells to complete some "downstream" landmark event during this second incubation is monitored. The second experiment is the reciprocal of the first: block B alone is applied during the first incubation, and block A alone is applied during the second. Although beset by a variety of possible complications (Jarvik and Botstein, 1973; Hereford and Hartwell, 1974; Pringle, 1978), the reciprocal shift method can, in principle, distinguish among the four possible functional relationships between two primary defect events. The different behavior of the *cdc28, cdc4,* and *cdc7* mutants in reciprocal shift experiments employing mating pheromone as a stage-specific inhibitor has provided the prime evidence for a sequence of steps preceding the initiation of DNA synthesis (Hereford and Hartwell, 1974).

A potential complication of the reciprocal shift method that deserves special mention is that in the form described above, what the reciprocal shift method really measures is whether the execution point of block B can be passed while block A is applied, and vice versa. If either block is leaky, this "order of executability" (a result of no discernible intrinsic interest) may not be the same as the functional relationship between the two primary defect events (see the discussion in Section III,C of the interpretation of execution points). Fortunately, this problem can generally be dealt with (1) by a systematic effort to work only with tight mutant alleles (Pringle, 1978) and (2) by imposing in each of the reciprocal experiments an intermediate incubation, of variable length, in which *both* blocks are imposed (to give any gradually inactivated gene product time to be inactivated). The latter approach must be applied with caution because of the danger of exacerbating another complication, viz, that B function *is* completed during the first incubation but results in a product or state that is transient and can be dissipated before the incubation permissive for A but restrictive for B is applied (Jarvik and Botstein, 1973; Pringle, 1978).

In addition to the three methods just discussed, other types of observations may contribute to the development of functional sequence maps. For example, if the relative temporal order of two events varies under different growth conditions, those two events cannot be linked in a dependent sequence.

2. A Functional Sequence Map of the Saccharomyces cerevisiae Cell Cycle

Fig. 1 presents a tentative functional sequence map of the *S. cerevisiae* cell cycle that has been developed using the methods described above. A similar map has been presented elsewhere, with detailed documentation (Pringle and Hartwell, 1981); here I want only to call attention to several points of special interest.

a. **Dependent Sequences of Landmark Events Are Uncertain.**   That the difficulty of obtaining direct evidence for a dependent sequence of landmark events (Section III,D,1; Pringle, 1978) is not a purely hypothetical concern has recently been emphasized by an improvement in our understanding of the functional relations of events involving the spindle pole body (the microtubule-organizing center in the nuclear envelope). The behavior of the G_1-arrest mutants *cdc28* (which fails to duplicate its spindle pole bodies), *cdc4* (whose spindle pole bodies duplicate but do not separate), and *cdc7* (whose spindle pole bodies separate, although DNA synthesis is not initiated) had suggested the functional sequence $\overrightarrow{SPBD}\ \overrightarrow{SPBS}\ \overrightarrow{iDS}$ (Byers and Goetsch, 1974; Hartwell, 1978; abbreviations as in Fig. 1). However, we now realize that ultrastructural observations (Byers and Goetsch, 1975) and DNA measurements (Hartwell, 1974) on wild-type strains are not compatible with the separation of spindle pole bodies being a prerequisite for the initiation of DNA synthesis. Moreover, *cdc31* mutants have been discovered to synthesize DNA even though their spindle pole bodies enlarge without achieving morphological duplication (Byers, 1981). Thus, the functional sequence map has been redrawn as shown in Fig. 1. Other dependence relationships among landmark events depicted in the current map may also prove to be naive, although the confidence in such relationships increases when many mutants show behavior consistent with the relationship (Pringle, 1978) or if the primary defect event for one or more mutants is known. For example, the number of mutants and inhibitors involved and the knowledge of the *CDC9* and *CDC21* gene products (Section III,A,1) make the putative dependence of spindle elongation upon DNA synthesis (Fig. 1) seem solidly established.

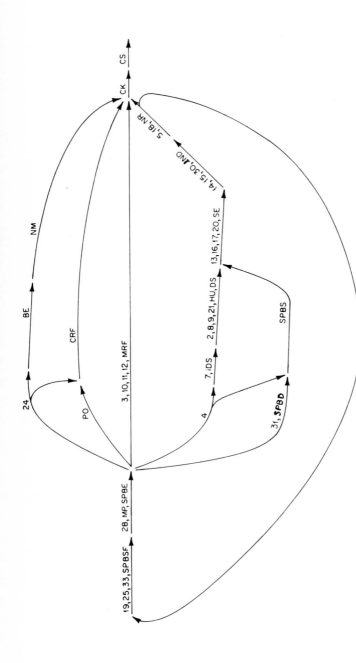

Fig. 1. A tentative functional sequence map of the *Saccharomyces cerevisiae* cell cycle. The basic form of the diagram and some of the reservations about it are described in the text; a map of somewhat different form, but embodying essentially the same substance, has been justified in detail by Pringle and Hartwell (1981). Abbreviations: numbers represent the corresponding *CDC* genes; SPBSF, formation of the spindle pole body (Byers and Goetsch, 1975); MP, mating pheromone (Bücking-Throm *et al.*, 1973; Wilkinson and Pringle, 1974); SPBE, enlargement of the spindle pole body (Byers, 1981); PO, polyoxin D (Cabib and Bowers, 1975); SPBD, morphological duplication of the spindle pole body (Byers and Goetsch, 1975); iDS, initiation of chromosomal DNA synthesis; MRF, formation of the microfilament ring (Byers and Goetsch, 1976a); CRF, formation of the chitin ring (Sloat and Pringle, 1978); BE, bud emergence; NM, nuclear migration (Hartwell *et al.*, 1974); SPBS, separation of the duplicate spindle pole bodies and formation of a complete spindle (Byers and Goetsch, 1975); HU, hydroxyurea (Hartwell, 1976); SE, spindle elongation (Byers and Goetsch, 1975); IND, late nuclear division (Hartwell *et al.*, 1974); NR, nuclear reorganization (Pringle and Hartwell, 1981); CK, cytokinesis; CS, cell separation (Hartwell *et al.*, 1974).

b. The Map Contains Parallel Pathways as Well as Dependent Sequences.
The arguments that the *CDC4* and *CDC24* steps are independent, as are
the *CDC4* and *CDC31* steps, have been outlined above (Sections III,
D,1 and 2,a). In addition, the facts that *cdc24* mutants make abundant
chitin while failing either to bud or to form a normal chitin ring (Sloat and
Pringle, 1978; Sloat *et al.*, 1981), while glucosamine auxotrophs (Ballou *et
al.*, 1977) and cells treated with the inhibitor polyoxin D (Cabib and Bowers,
1975) apparently bud despite their inability to synthesize chitin, suggest the
independence of the *CDC24* and polyoxin-sensitive steps, as shown in Fig.
1. The putative independence of the step of microfilament ring formation
(Fig. 1) is essentially speculative (Pringle and Hartwell, 1981). Since the
conclusion that two steps are independent is weakened by the possibility
that a normal control mechanism might be missing in a particular mutant
(Pringle, 1978), our confidence in the branch involving the *CDC24* step has
recently been greatly increased by the isolation of mutants of two additional
genes that are also unable to bud although chitin synthesis, spindle pole
body duplication, and DNA synthesis continue (A. Adams and J. Pringle,
unpublished results). A branched functional sequence map of cell cycle
events has also recently been reported for the prokaryote *Caulobacter crescentus* (Osley and Newton, 1980).

**c. The Functional Sequence Map Can Explain Both the Fixed and Flexi-
ble Aspects of the Temporal Map.**   The occurrence both of dependent
sequences and of parallel pathways of events (Fig. 1) allows the relative
timing of some events (e. g., DNA synthesis and nuclear division) to be
fixed, while the relative timing of other events (e. g., bud emergence and the
initiation of DNA synthesis) can vary with strain and growth conditions, as is
in fact observed (Pringle and Hartwell, 1981). Since the parallel pathways
are all dependent on a single precursor event (the *CDC28* and mating
pheromone-sensitive step) and are all prerequisite for the completion of the
cycle by cytokinesis, events in the parallel pathways are prevented from
getting grossly out of phase. Thus, there seems little justification, at present,
for postulating a central timer to control cell cycle events, at least for *S.
cerevisiae.*

IV. GENERAL CONCLUSIONS

The genetic analysis of the eukaryotic cell cycle is in its infancy. Only a few
of the gene products affected by known cell cycle mutations have been
identified (Section III,A,1), and it seems virtually certain that the known
cell cycle genes are but a minority of the total, even in the relatively well
studied *S. cerevisiae* (Section III,B). Nonetheless, the genetic studies have

already provided very important insights such as the overlap of molecular functions between mitosis and meiosis (Section III,A,1), the probable significance of the duplication of microtubule-organizing centers as a control step for cell proliferation (Section III,A,2), and the organizational complexity of the program of stage-specific events (Section III,D). Indeed, given the complexity of a process such as eukaryotic cellular reproduction, it is difficult to see any hope of substantial progress in understanding unless the ability of mutational analysis to disrupt any cellular function, to do so specifically, and to do so without the investigator's needing to know in advance what function will be disrupted is fully utilized.

REFERENCES

Alberts, B., and Sternglanz, R. (1977). Recent excitement in the DNA replication problem. *Nature (London)* **269**, 655–661.

Baker, B. S., Carpenter, A. T. C., Esposito, M. S., Esposito, R. E., and Sandler, L. (1976). The genetic control of meiosis. *Annu. Rev. Genet.* **10**, 53–134.

Baker, B. S., Carpenter, A. T. C., and Ripoll, P. (1978). The utilization during mitotic cell division of loci controlling meiotic recombination and disjunction in *Drosophila melanogaster. Genetics* **90**, 531–578.

Ballou, C. E., Maitra, S. K., Walker, J. W., and Whelan, W. L. (1977). Developmental defects associated with glucosamine auxotrophy in *Saccharomyces cerevisiae. Proc. Natl. Acad. Sci. U.S.A.* **74**, 4351–4355.

Baserga, R. (1976). "Multiplication and Division in Mammalian Cells." Dekker, New York.

Bisson, L., and Thorner, J. (1977). Thymidine 5′-monophosphate-requiring mutants of *Saccharomyces cerevisiae* are deficient in thymidylate synthetase. *J. Bacteriol.* **132**, 44–50.

Blumberg, D. D., and Lodish, H. F. (1980a). Complexity of nuclear and polysomal RNAs in growing *Dictyostelium discoideum* cells. *Dev. Biol.* **78**, 268–284.

Blumberg, D. D., and Lodish, H. F. (1980b). Changes in the messenger RNA population during differentiation of *Dictyostelium discoideum. Dev. Biol.* **78**, 285–300.

Bücking-Throm, E., Duntze, W., Hartwell, L. H., and Manney, T. R. (1973). Reversible arrest of haploid yeast cells at the initiation of DNA synthesis by a diffusible sex factor. *Exp. Cell Res.* **76**, 99–110.

Byers, B. (1981). Multiple roles of the spindle pole bodies in the life cycle of *Saccharomyces cerevisiae. In* "Alfred Benzon Symposium 16, Molecular Genetics of Yeast" (D. von Wettstein, J. Friis, M. Kielland-Brandt, and A. Stenderup, eds.), Munksgaard, Copenhagen (in press).

Byers, B., and Goetsch, L. (1974). Duplication of spindle plaques and integration of the yeast cell cycle. *Cold Spring Harbor Symp. Quant. Biol.* **38**, 123–131.

Byers, B., and Goetsch, L. (1975). Behavior of spindles and spindle plaques in the cell cycle and conjugation of *Saccharomyces cerevisiae. J. Bacteriol.* **124**, 511–523.

Byers, B., and Goetsch, L. (1976a). A highly ordered ring of membrane-associated filaments in budding yeast. *J. Cell Biol.* **69**, 717–721.

Byers, B., and Goetsch, L. (1976b). Loss of the filamentous ring in cytokinesis-defective mutants of budding yeast. *J. Cell Biol.* **70**, 35a.

Byers, B., and Sowder, L. (1980). Gene expression in the yeast cell cycle. *J. Cell Biol.* **87**, 6a.

Cabib, E., and Bowers, B. (1975). Timing and function of chitin synthesis in yeast. *J. Bacteriol.* **124**, 1586–1593.

Clarke, L., and Carbon, J. (1980). Isolation of a yeast centromere and construction of functional small circular chromosomes. *Nature (London)* **287**, 504–509.

Davidson, E. H., and Britten, R. J. (1979). Regulation of gene expression: Possible role of repetitive sequences. *Science* **204**, 1052–1059.

De Luise, M., Blackburn, G. L., and Flier, J. S. (1980). Reduced activity of the red-cell sodium-potassium pump in human obesity. *N. Engl. J. Med.* **303**, 1017–1022.

Dutcher, S. K. (1980). Genetic control of karyogamy in *Saccharomyces cerevisiae*. Ph. D. Dissertation, University of Washington, Seattle.

Fabre, F., and Roman, H. (1979). Evidence that a single DNA ligase is involved in replication and recombination in yeast. *Proc. Natl. Acad. Sci. U.S.A.* **76**, 4586–4588.

Fangman, W. L., and Zakian, V. A. (1981). Genome structure and replication. *In* "Molecular Biology of the Yeast *Saccharomyces*" (J. Strathern, E. Jones, and J. Broach, eds.), Cold Spring Harbor Lab., Cold Spring Harbor, New York (in press).

Frankel, J., Mohler, J., and Frankel, A. K. (1980). Temperature-sensitive periods of mutations affecting cell division in *Tetrahymena thermophila*. *J. Cell Sci.* **43**, 59–74.

Galau, G. A., Klein, W. H., Davis, M. M., Wold, B. J., Britten, R. J., and Davidson, E. H. (1976). Structural gene sets active in embryos and adult tissues of the sea urchin. *Cell* **7**, 487–505.

Glabe, C. G., and Vacquier, V. D. (1977). Isolation and characterization of the vitelline layer of sea urchin eggs. *J. Cell Biol.* **75**, 410–421.

Goldberg, A. L., and St. John, A. C. (1976). Intracellular protein degradation in mammalian and bacterial cells: Part 2. *Annu. Rev. Biochem.* **45**, 747–803.

Hartwell, L. H. (1971). Genetic control of the cell division cycle in yeast. IV. Genes controlling bud emergence and cytokinesis. *Exp. Cell Res.* **69**, 265–276.

Hartwell, L. H. (1973). Synchronization of haploid yeast cell cycles: A prelude to conjugation. *Exp. Cell Res.* **76**, 111–117.

Hartwell, L. H. (1974). *Saccharomyces cerevisiae* cell cycle. *Bacteriol. Rev.* **38**, 164–198.

Hartwell, L. H. (1976). Sequential function of gene products relative to DNA synthesis in the yeast cell cycle. *J. Mol. Biol.* **104**, 803–817.

Hartwell, L. H. (1978). Cell division from a genetic perspective. *J. Cell Biol.* **77**, 627–637.

Hartwell, L. H., and Unger, M. W. (1977). Unequal division in *Saccharomyces cerevisiae* and its implications for the control of cell division. *J. Cell Biol.* **75**, 422–435.

Hartwell, L. H., Mortimer, R. K., Culotti, J., and Culotti, M. (1973). Genetic control of the cell division cycle in yeast. V. Genetic analysis of *cdc* mutants. *Genetics* **74**, 267–286.

Hartwell, L. H., Culotti, J., Pringle, J. R., and Reid, B. J. (1974). Genetic control of the cell division cycle in yeast. *Science* **183**, 46–51.

Hereford, L. M., and Hartwell, L. H. (1974). Sequential gene function in the initiation of *Saccharomyces cerevisiae* DNA synthesis. *J. Mol. Biol.* **84**, 445–461.

Hereford, L. M., and Rosbash, M. (1977). Number and distribution of polyadenylated RNA sequences in yeast. *Cell* **10**, 453–462.

Hereford, L., Fahrner, K., Woolford, J., Jr., Rosbash, M., and Kaback, D. B. (1979). Isolation of yeast histone genes H2A and H2B. *Cell* **18**, 1261–1271.

Hilliker, A. J., Clark, S. H., Chovnick, A., and Gelbart, W. M. (1980). Cytogenetic analysis of the chromosomal region immediately adjacent to the rosy locus in *Drosophila melanogaster*. *Genetics* **95**, 95–110.

Hirschberg, J., and Simchen, G. (1977). Commitment to the mitotic cell cycle in yeast in relation to meiosis. *Exp. Cell Res.* **105**, 245–252.

Holland, J. P., and Holland, M. J. (1980). Structural comparison of two nontandemly repeated yeast glyceraldehyde-3-phosphate dehydrogenase genes. *J. Biol. Chem.* **255**, 2596–2605.

Hough-Evans, B. R., Jacobs-Lorena, M., Cummings, M. R., Britten, R. J., and Davidson, E. H. (1980). Complexity of RNA in eggs of *Drosophila melanogaster* and *Musca domestica*. *Genetics* **95**, 81–94.

Howell, S. H., Posakony, J. W., and Hill, K. R. 1977. The cell cycle program of polypeptide labeling in *Chlamydomonas reinhardtii*. *J. Cell Biol.* **72**, 223–241.

Jarvik, J., and Botstein, D. (1973). A genetic method for determining the order of events in a biological pathway. *Proc. Natl. Acad. Sci. U.S.A.* **70**, 2046–2050.

Jarvik, J., and Botstein, D. (1975). Conditional-lethal mutations that suppress genetic defects in morphogenesis by altering structural proteins. *Proc. Natl. Acad. Sci. U.S.A.* **72**, 2738–2742.

Jeggo, P. A., Unrau, P., Banks, G. R., and Holliday, R. (1973). A temperature-sensitive DNA polymerase mutant of *Ustilago maydis*. *Nature (London)* **242**, 14–16.

Johnston, G. C., Pringle, J. R., and Hartwell, L. H. (1977). Coordination of growth with cell division in the yeast *Saccharomyces cerevisiae*. *Exp. Cell Res.* **105**, 79–98.

Johnston, L. H., and Game, J. C. (1978). Mutants of yeast with depressed DNA synthesis. *Mol. Gen. Genet.* **161**, 205–214.

Johnston, L. H., and Nasmyth, K. A. (1978). *Saccharomyces cerevisiae* cell cycle mutant *cdc9* is defective in DNA ligase. *Nature (London)* **274**, 891–893.

Kawasaki, G. (1979). Karyotypic instability and carbon source effects in cell cycle mutants of *Saccharomyces cerevisiae*. Ph. D. Dissertation, University of Washington, Seattle.

Kirschner, K., and Bisswanger, H. (1976). Multifunctional proteins. *Annu. Rev. Biochem.* **45**, 143–166.

Laffler, T. G., Wilkins, A., Selvig, S., Warren, N., Kleinschmidt, A., and Dove, W. F. (1979). Temperature-sensitive mutants of *Physarum polycephalum:* Viability, growth, and nuclear replication. *J. Bacteriol.* **138**, 499–504.

Lawn, R. M., Heumann, J. M., Herrick, G., and Prescott, D. M. (1978). The gene-size DNA molecules in *Oxytricha*. *Cold Spring Harbor Symp. Quant. Biol.* **42**, 483–492.

Lee, A. S., Thomas, T. L., Lev, Z., Britten, R. J., and Davidson, E. H. (1980). Four sizes of transcript produced by a single sea urchin gene expressed in early embryos. *Proc. Natl. Acad. Sci. U.S.A.* **77**, 3259–3263.

Lefevre, G., Jr. (1974). The relationship between genes and polytene chromosome bands. *Annu. Rev. Genet.* **8**, 51–62.

Lewin, B. (1974). Lethal mutants and gene numbers. *Nature (London)* **251**, 373–375.

Ling, V. (1981). Mutants as an investigative tool (this volume).

Liskay, R. M., and Prescott, D. M. (1978). Genetic analysis of the G_1 period: Isolation of mutants (or variants) with a G_1 period from a Chinese hamster cell line lacking G_1. *Proc. Natl. Acad. Sci. U.S.A.* **75**, 2873–2877.

Luck, D. J. L., Huang, B., and Piperno, G. (1981). Genetic and biochemical analysis of the eukaryotic flagellum. *Symp. Soc. Exper. Biol.*, in press.

Mazia, D. (1978). Origin of twoness in cell reproduction. *In* "Cell Reproduction: In Honor of Daniel Mazia" (E. R. Dirksen, D. M. Prescott, and C. F. Fox, eds.), pp. 1–14. Academic Press, New York.

Minet, M., Nurse, P., Thuriaux, P., and Mitchison, J. M. (1979). Uncontrolled septation in a cell division cycle mutant of the fission yeast *Schizosaccharomyces pombe*. *J. Bacteriol.* **137**, 440–446.

Mitchison, J. M. (1971). "The Biology of the Cell Cycle." Cambridge Univ. Press, London and New York.

Morris, N. R., Lai, M. H., and Oakley, C. E. (1979). Identification of a gene for α-tubulin in *Aspergillus nidulans*. *Cell* **16**, 437–442.

Nasmyth, K. A., and Reed, S. I. (1980). Isolation of genes by complementation in yeast: Molecular cloning of a cell-cycle gene. *Proc. Natl. Acad. Sci. U.S.A.* **77**, 2119–2123.

Navon, G., Shulman, R. G., Yamane, T., Eccleshall, T. R., Lam, K.-B., Baronofsky, J. J., and Marmur, J. (1979). Phosphorus-31 nuclear magnetic resonance studies of wild-type and glycolytic pathway mutants of *Saccharomyces cerevisiae. Biochemistry* **18**, 4487–4499.

Newlon, C. S., and Fangman, W. L. (1975). Mitochondrial DNA synthesis in cell cycle mutants of *Saccharomyces cerevisiae. Cell* **5**, 423–428.

Nurse, P., and Thuriaux, P. (1980). Regulatory genes controlling mitosis in the fission yeast *Schizosaccharomyces pombe. Genetics* **96**, 627–637.

Oakley, B. (1981). Mutants of mitosis (this volume).

O'Brien, S. J. (1973). On estimating functional gene number in eukaryotes. *Nature (London), New Biol.* **242**, 52–54.

Osley, M. A., and Newton, A. (1980). Temporal control of the cell cycle in *Caulobacter crescentus:* Roles of DNA chain elongation and completion. *J. Mol. Biol.* **138**, 109–128.

Pardee, A. B., Dubrow, R., Hamlin, J. L., and Kletzien, R. F. (1978). Animal cell cycle. *Annu. Rev. Biochem.* **47**, 715–750.

Piperno, G., and Luck, D. J. L. (1979). An actin-like protein is a component of axonemes from *Chlamydomonas* flagella. *J. Biol. Chem.* **254**, 2187–2190.

Prescott, D. M. (1976). "Reproduction of Eukaryotic Cells." Academic Press, New York.

Pringle, J. R. (1975). Induction, selection, and experimental uses of temperature-sensitive and other conditional mutants of yeast. *Methods Cell Biol.* **12**, 233–272.

Pringle, J. R. (1978). The use of conditional lethal cell cycle mutants for temporal and functional sequence mapping of cell cycle events. *J. Cell. Physiol.* **95**, 393–405.

Pringle, J. R. (1979). Proteolytic artifacts in biochemistry. *In* "Limited Proteolysis in Microorganisms" (G. N. Cohen and H. Holzer, eds.), pp. 191–196. U. S. Govt. Printing Office, Washington, D.C.

Pringle, J. R., and Hartwell, L. H. (1981). The *Saccharomyces cerevisiae* cell cycle. *In* "Molecular Biology of the Yeast *Saccharomyces*" (J. Strathern, E. Jones, and J. Broach, eds.), Cold Spring Harbor Laboratory, Cold Spring Harbor, New York (in press).

Rai, R., and Carter, B. L. A. (1981). The isolation of nonsense mutations in cell division cycle genes of the yeast *Saccharomyces cerevisiae. Mol. Gen. Genet.* (in press).

Reed, S. I. (1980a). The selection of *S. cerevisiae* mutants defective in the start event of cell division. *Genetics* **95**, 561–577.

Reed, S. I. (1980b). The selection of amber mutations in genes required for completion of start, the controlling event of the cell division cycle of *S. cerevisiae. Genetics* **95**, 579–588.

Regier, J. C., Mazur, G. D., and Kafatos, F. C. (1980). The silkmoth chorion: Morphological and biochemical characterization of four surface regions. *Dev. Biol.* **76**, 286–304.

Reid, B. J. (1979). Regulation and integration of the cell-division cycle and the mating reaction in the yeast *Saccharomyces cerevisiae.* Ph. D. Dissertation, University of Washington, Seattle.

Reid, B. J., and Hartwell, L. H. (1977). Regulation of mating in the cell cycle of *Saccharomyces cerevisiae. J. Cell Biol.* **75**, 355–365.

Ripoll, P. (1977). Behavior of somatic cells homozygous for zygotic lethals in *Drosophila melanogaster. Genetics* **86**, 357–376.

Ripoll, P., and Garcia-Bellido, A. (1979). Viability of homozygous deficiencies in somatic cells of *Drosophila melanogaster. Genetics* **91**, 443–453.

Schild, D., and Byers, B. (1978). Meiotic effects of DNA-defective cell division cycle mutations of *Saccharomyces cerevisiae. Chromosoma* **70**, 109–130.

Sherman, F., Stewart, J. W., Helms, C., and Downie, J. A. (1978). Chromosome mapping of the *CYC7* gene determining yeast iso-2-cytochrome *c*: Structural and regulatory regions. *Proc. Natl. Acad. Sci. U.S.A.* **75**, 1437–1441.

Simchen, G. (1974). Are mitotic functions required in meiosis? *Genetics* **76**, 745–753.

Simchen, G. (1978). Cell cycle mutants. *Annu. Rev. Genet.* **12**, 161–191.

Simchen, G., and Hirschberg, J. (1977). Effects of the mitotic cell-cycle mutation *cdc4* on yeast meiosis. *Genetics* **86**, 57–72.

Sloat, B. F., and Pringle, J. R. (1978). A mutant of yeast defective in cellular morphogenesis. *Science* **200**, 1171–1173.

Sloat, B. F., Adams, A., and Pringle, J. R. (1981). Roles of the *CDC24* gene product in cellular morphogenesis during the *Saccharomyces cerevisiae* cell cycle. *J. Cell Biol.* **89**, 395–405.

Swedes, J. S., Dial, M. E., and McLaughlin, C. S. (1979). Regulation of protein synthesis during energy limitation of *Saccharomyces cerevisiae*. *J. Bacteriol.* **138**, 162–170.

Trinci, A. P. J., and Morris, N. R. (1979). Morphology and growth of a temperature-sensitive mutant of *Aspergillus nidulans* which forms aseptate mycelia at non-permissive temperatures. *J. Gen. Microbiol.* **114**, 53–59.

Tucker, R. W., Pardee, A. B., and Fujiwara, K. (1979). Centriole ciliation is related to quiescence and DNA synthesis in 3T3 cells. *Cell* **17**, 527–535.

Unger, M. W. (1977). Methionyl-transfer ribonucleic acid deficiency during G_1 arrest of *Saccharomyces cerevisiae*. *J. Bacteriol.* **130**, 11–19.

Warner, J. R., and Gorenstein, C. (1978). Yeast has a true stringent response. *Nature (London)* **275**, 338–339.

Wickner, S. H. (1978). DNA replication proteins of *Escherichia coli*. *Annu. Rev. Biochem.* **47**, 1163–1191.

Wilkinson, L. E., and Pringle, J. R. (1974). Transient G_1 arrest of *S. cerevisiae* cells of mating type α by a factor produced by cells of mating type *a*. *Exp. Cell Res.* **89**, 175–187.

Wood, W. B., and Revel, H. R. (1976). The genome of bacteriophage T4. *Bacteriol. Rev.* **40**, 847–868.

Woolford, J. L., Jr., Hereford, L. M., and Rosbash, M. (1979). Isolation of cloned DNA sequences containing ribosomal protein genes from *Saccharomyces cerevisiae*. *Cell* **18**, 1247–1259.

Young, M. W., and Judd, B. H. (1978). Nonessential sequences, genes, and the polytene chromosome bands of *Drosophila melanogaster*. *Genetics* **88**, 723–742.

Zakian, V. A., Brewer, B. J., and Fangman, W. L. (1979). Replication of each copy of the yeast 2 micron DNA plasmid occurs during the S phase. *Cell* **17**, 923–934.

Zamb, T. J., and Roth, R. (1977). Role of mitotic replication genes in chromosome duplication during meiosis. *Proc. Natl. Acad. Sci. U.S.A.* **74**, 3951–3955.

2

Calcium Regulation of Cell Cycle Events

PATRICIA J. HARRIS

I. INTRODUCTION

A. Background

The importance of calcium in a wide range of cellular processes, including mitosis, was appreciated long before the structural components of the mito-

MITOSIS/CYTOKINESIS
Copyright © 1981 by Academic Press, Inc.
All rights of reproduction in any form reserved.
ISBN 0-12-781240-7

tic apparatus began to be identified and subjected to chemical analysis. Its possible role in muscle contraction, cell adhesion, clotting reactions, and gelation of cytoplasm, parthenogenetic activation of various eggs, and numerous other effects had already been anticipated by Heilbrunn (1937, 1956) in his classic physiology texts. In the past 20 years there has been an explosion of information identifying calcium as a key regulator of cellular function. (For recent reviews, see Berridge, 1975; Duncan, 1976; Rasmussen and Goodman, 1977; Scarpa and Carafoli, 1978; Berridge *et al.*, 1979.) At the same time, we have acquired a great deal of information about the structural elements involved in cell division, and it is increasingly apparent that strict calcium regulation is necessary for the polymerization of spindle microtubules and for the interaction of actin, myosin, and associated proteins in contraction, gelation, and cytoplasmic streaming. With the discovery of the calcium regulator protein "calmodulin" (reviewed by Klee *et al.*, 1980) and its presence within the mitotic spindle (Marcum *et al.*, 1978; Andersen *et al.*, 1978), the possibilities for calcium regulation have been broadened even further. Current models of mitosis which invoke polymerization–depolymerization of microtubules, association of actin and myosin in contraction, or microtubule–microtubule interaction with the aid of dynein arms or zippers all would require the cell to regulate its internal calcium within a restricted range and against an enormous gradient. (For pertinent symposia and reviews, see Inoue and Stephens, 1975; Soifer, 1975; Goldman *et al.*, 1976; Dirksen *et al.*, 1978; Roberts and Hyams, 1979; Raff, 1979.)

B. Regulation of Intracellular Calcium

Our understanding of calcium homeostasis and the role of intracellular calcium in the regulation of cell function comes from diverse sources. Studies on nerve and muscle have provided the basis for our understanding of cell excitability, but it is now known that many, if not most, cells possess a degree of excitability based on an intracellular calcium-regulating system (Baker, 1976). Furthermore, many excitable systems show repeated oscillations; these occur normally, for example, in beating cardiac muscle and in smooth muscle peristalsis, but also under experimental conditions in mesenchymal cell cultures (Nelson and Henkart, 1979). Studies on oscillatory contractions in skinned muscle fibers (Endo *et al.*, 1970; Fabiato and Fabiato, 1978) have demonstrated that calcium itself can trigger a regenerative release of calcium from intracellular stores. An understanding of the role of the sarcoplasmic reticulum in skeletal muscle in the sequestering and triggered release of calcium has led to the discovery of analogous sequestering systems in a wide range of cells, for example, in microsomes of mouse and chick fibroblasts (Moore and Pastan, 1977a, b), in the vesicles associated with the mitotic apparatus (Silver *et al.*, 1980; Wick and Hepler, 1980), and in the smooth endoplasmic reticulum of presynaptic nerve terminals (McGraw *et al.*, 1980). The mitochondria, which are equally important storage sites of intracellular calcium, possess a large storage capacity as well as a triggered release (Carafoli and Crompton, 1978). The plasma membrane also plays an important role; its outwardly directed calcium pumps have been studied extensively in mammalian red blood cells (reviewed by Schatzmann and Bürgin, 1978). Many other factors, as well, such as cyclic nucleotides and feedback systems that may modulate the calcium fluxes, have been reviewed in some detail by Rasmussen and Goodman (1977), Durham (1974), and Berridge (1976). These control mechanisms allow living cells to use calcium as a transient signal to activate a wide range of cellular processes. Among these processes are those related to the cell division cycle and to the mechanism of mitosis itself.

This chapter is not meant to be a comprehensive review, but rather an attempt to bring together information and ideas from various approaches. Much of the evidence is indirect and conclusions are necessarily speculative, but the objective is to offer new ways of looking at old problems that seem to have become lost in a forest of details.

II. ACTIVATION OF QUIESCENT CELLS AND ENTRY INTO THE DIVISION CYCLE

Studies of wound healing, regeneration, and initiation of division in quiescent cells provide observations on a vast number of serum components, hormones, and growth factors that apparently are capable of stimulating DNA synthesis and subsequent cell division. There are many reasons to believe that these various first messengers operate by raising the cytoplasmic calcium ion concentration, first by increasing the permeability of the plasma membrane to calcium and then in most cases by a subsequent release of calcium from internal stores (Berridge, 1976). A classic example of such entry from quiescence into the division cycle is the fertilization reaction, which has been studied in detail in sea urchin eggs. Although egg activation may seem at first glance to be a highly specialized case, with egg polarization and mechanisms to orient the early cleavage divisions superimposed on the division process itself, it is more likely that eggs simply provide an exaggerated example of conditions influencing any mitotic division.

A. Calcium Release from Intracellular Stores

One of the first visible manifestations of the fertilization reaction is the wavelike breakdown of the cortical granules and the subsequent raising of the fertilization membrane. The involvement of calcium in this reaction has been known for a long time (reviewed by Tyler, 1941; Jaffe, 1980). Quantitative measurements by Mazia (1937) showed that 15% of the bound calcium of unferilized *Arbacia* eggs was released into the cytosol immediately after fertilization. Vacquier (1975) has shown that the cortical granules require calcium for exocytosis, and it was initially believed that the reaction was self-propagated by calcium release from the cortical granules themselves. However, a partial cortical reaction can result from marginal concentrations of activating agents (Sugiyama, 1956; Chambers and Hinkley, 1979), demonstrating that initiation of the cortical reaction in a localized region of the egg cortex does not necessarily lead to a propagated response.

The propagated release of calcium at fertilization is a transient one, most beautifully demonstrated by Gilkey *et al.* (1978) during the cortical reaction of fish eggs. By preloading medaka eggs with the calcium-sensitive photoprotein aequorin, they showed a pattern of luminescence traveling as a bright narrow band from the point of sperm entry to the opposite side of the egg. The calcium released at the advancing front of the wave was rapidly sequestered behind, but within the band itself the calcium concentration was calculated to be about 30 micromolar, a 300-fold increase over that prior to fertilization. When aequorin preloaded eggs were treated with medium containing the calcium ionophore A23187, calcium waves were initiated at numerous points on the egg surface, eventually coalescing and producing an apparently completely activated egg.

B. Separating the Components of the Cell Cycle

The mitotic cycle, set in motion by the fertilization reaction, is a composite of numerous other cycling activities: DNA synthesis and the chromosome cycle, periodic formation and breakdown of asters, cyclic changes in cortical stiffness, changes in the activity of Ca-ATPase, centriole duplication and separation, and presumably many other activities as yet undiscovered or unrecognized. Mitchison (1971), has discussed in some detail the possible links between these markers of the cell cycle and has separated them into two different fixed sequences or cycles: the DNA division cycle and the growth cycle. In fertilized eggs the growth cycle is negligible during the first several divisions, and the normal synchrony of the divisions facilitates the dissection of the division cycle into its component parts and identification of their causal relationships. A number of elegant experiments using various kinds of eggs have given considerable insight into what these causal relationships might be. Of particular interest were studies by Loeb (1913) on artificial parthenogenesis.

C. Independence of the Nuclear DNA Cycle of "Early" Events

Among the activating agents used by Loeb (1913) were weak bases such as ammonia. More recently, Mazia and his colleagues have reexamined the partial activation of sea urchin eggs with ammoniacal sea water, noting the biochemical and structural differences between artificially activated and normally fertilized eggs (see Epel, 1978 for review). They found that the ammonia apparently bypasses the early responses of normal fertilization; i. e., there is no discharge of cortical granules, no formation of the fertilization membrane, and no release of calcium or respiratory burst. On the other hand, the ammonia initiates the "late" events, which include development of K conductance and membrane potential, polyadenylation of RNA, increased permeation of thymidine, and initiation of DNA synthesis and the chromosome cycles. However, while the ammonia-activated cell can express the periodicity of the mitotic cycle, it never divides. It is apparently unable to produce a bipolar mitotic apparatus. Fine structure studies (Paweletz and Mazia, 1979) have shown the presence of a centered clear zone characterized by its astral structure, containing an aggregation of membrane-bounded vesicles and a relatively sparse array of microtubules, which appear to center on clumps of osmiophilic granules or foci. The microtubules appear and disappear with the same periodicity as the chromosome cycle, suggesting that there may be some degree of calcium or pH cycles as well.

D. Initiation of Rhythmic Activity Independent of the Nuclear Cycle

Although the DNA cycle of the nucleus can be turned on by late events of fertilization, bypassing the early cortical reaction, the question arises of whether cyclic events turned on by the cortical reaction and early changes are dependent on the nucleus for their maintenance. Bell (1962) was the first to observe that non-nucleate cytoplasm squeezed from a fertilized ascidian egg continued a cyclic rounding up in phase with the cleavage cycle of the egg from which it was derived. The rhythmic rounding up was obviously independent of the zygote nucleus. Kojima (1960) found cyclic changes in the thickness of the hyaline layer of non-dividing activated sea urchin eggs that corresponded to the cleaving stage of normally fertilized eggs. Even activated non-nucleate fragments showed the cyclic changes, and the cycles persisted when the activated fragments were treated with colchicine. The cyclic activity was apparently independent of aster formation and nuclear activities. Hiramoto (1962b) enucleated fertilized eggs at various times after sperm–egg interaction and showed that in all cases a cyclic activity was initiated. If the sperm centrioles were not present, only a monaster would form at the time of normal mitosis. If the egg pronucleus was removed, leaving only the sperm pronucleus and centrioles, two asters appeared, but they were unable to form a functional mitotic apparatus without some spindle-forming factor from the egg nucleus. But even when both pronuclei were removed, the rhythmic formation and breakdown of the monaster remained, accompanied by a cyclic rounding of the cell and an increase in the space between the cell surface and the hyaline layer. Hiramoto's conclusion was that the sperm plays two different roles in fertilization: induction of rhythmic activity in the egg cytoplasm and contribution of a pair of aster centers. Yoneda *et al.* (1978) confirmed Kojima's earlier work on activated non-nuclear sea urchin egg fragments, showing periodic thickening of the intrahyaloplasmic space, and in addition made quantitative measurements of cyclic tension changes whose period corresponded with the division cycle. Again it was shown that a periodic activity could be triggered without nuclear control. Similar conclusions can be drawn from Hara's studies on periodic contraction waves observed in fertilized or activated amphibian eggs (Hara, 1971; Hara *et al.*, 1977; Hara *et al.*, 1980 and by Yoneda *et al.*, (1978) on activated non-nucleate sea urchin egg fragments.

The relative contributions of the cell cortex and the endoplasm to changes in cell stiffness or rounding up at mitosis are difficult to sort out. Even more difficult to evaluate is the role that calcium plays in this process. A gradual increase in cytoplasmic calcium might increase cell stiffness by affecting actin, myosin, and their associated proteins: the interactions of these proteins are highly dependent on the ionic species and concentrations present and may be reflected as changes in streaming, gelation, or contraction (Pollard and Weihing, 1974; Condeelis and Taylor, 1977; Bryan and Kane, 1978). For example, the shuttle streaming of the plasmodium *Physarum*, which stops during the periodic synchronous nuclear division, is believed to be affected by a drop in ATP or changes in cytoplasmic free calcium (Sachsenmaier and Hansen, 1973). Similarly, cytoplasmic streaming is stopped or greatly reduced during plant cell division in tobacco (Das *et al.*, 1966) and sycamore (Roberts and Northcote, 1970) callus cultures, and probably indicates ionic changes in the cytoplasm.

There are other observations of calcium-related periodicities during sea urchin cleavage divisions, some of which are difficult to correlate with known morphological changes. Clothier and Timorian (1972) reported cyclic calcium uptake and release in that cells grown continuously in sea water containing ^{45}Ca and harvested at intervals through the cell cycle show cyclic variations in their ^{45}Ca content. Three peaks were identified, one immediately after fertilization and two others later in the cycle, but the proposed relationship of the latter two with certain mitotic stages suggests that these workers had mistaken the interphase asters for early stages of mitosis.

Cyclic changes in Ca-dependent ATPase have also been reported (Petzelt, 1972a, b). Two peaks, one at interphase and one at metaphase, could be related to the interphase asters (or cytoplasmic microtubule complex) and the mitotic apparatus, both of which require low cytoplasmic calcium for the polymerization of microtubules (Weisenberg, 1972).

III. DETERMINATION OF STRUCTURAL PATTERNS

A. Fertilization and Formation of the Monaster: The Nature of the Signal

If the signal that starts a cell into the division cycle is an increase of cytoplasmic calcium ions to a certain threshold level, either by entry through the cell membrane from outside or by release from an intracellular store, we must now ask, how is the signal carried and what receives it? If the message is a transient calcium wave, either it may act, in passing, to trigger other events which then become independent of the calcium concentration, or it may be a "permit" which allows certain activities and disallows others within a certain delimited region of the cell and for a certain period of time. Most important, the message moves away from its initiation point and thus provides a directionality. As a regenerative calcium-triggered calcium release, the wave front might be propagated without the necessity of an electrical coupling of the membranes by a combination of diffusion and triggered release, either from vesicles of the smooth endoplasmic reticulum or from mitochondria.

The initial rise in calcium ions in the cortical region is followed closely by an increase in pH, a respiratory burst, and activation of various enzymes (see the review by Epel, 1978); the calcium provides an environment favoring actin polymerization and the breakdown of any polymerized tubulin which may exist in the unfertilized egg. There may also be an activation of small, scattered microtubule-initiating centers. Studies using fluorescently labeled actin in living cells (Wang and Taylor, 1979) show that there is a transient accumulation of actin in the cortex which peaks at about 4–10 after fertilization and then tapers off. Light and electron microscopy have shown that during this period (i. e., 4–10 minutes after fertilization) there is a burst of elongation of surface microvilli and polymerization of actin from the microvillus cores (Eddy and Shapiro, 1976; Burgess and Schroeder, 1977; Begg and Rebhun, 1979). Hiramoto (1969) has shown a corresponding increase in cortical stiffness at this time.

During this period, pigment vesicles randomly distributed in the cytoplasm of unfertilized eggs of at least several species of sea urchin begin to move to the cell surface, and by about 15 minutes after fertilization the outermost region of the egg, extending to a depth of about 20–25 μm, is cleared of pigment vesicles, which are now positioned close to the plasma membrane. By this time the calcium wave has moved on, and conditions again begin to favor microtubule polymerization as calcium is again sequestered. The sperm aster begins to form, and a system of microtubules more or less radially oriented begins to grow from some as yet unidentified nucleating centers near the cell surface toward the center of the egg (Harris, 1979; Mar, 1980; Harris *et al.*, 1980), functioning at least in part to move the pronuclei toward the egg center (Mar, 1980). The monaster is thus formed by both the sperm aster and the microtubules originating in the cortex, and is the product of a permissive and directional signal or message which has reached its ultimate recipient, the fusion nucleus; DNA synthesis now begins.

This rather simplistic picture obviously has some serious flaws. We still do not know the causal relationships of all the obvious markers of the activation process, and the fusion nucleus is not necessarily the key recipient of the message but only the last of many that read the message as it goes by. The most obvious identifiable response from the unknown key recipient is the initiation of a contraction–relaxation rhythm that sets the timing for the mitotic cycle. The cyclic structural changes associated with mitosis depend on the presence of many participants and their ability to respond at their appointed times. If they miss their cue during one cycle, they must wait until the next time around.

The repeated cycles that are initiated by the original signal are not necessarily repetitions of the original signal, but are more likely a separate message, which no doubt uses the same "postal system" or means of transmission. The first message was, "Turn on" the next, "Perform your act." Evidence that these are separate messages comes from observations of fertilized *Xenopus* eggs. Hara *et al.* (1977) have shown that the postfertilization wave is initiated at the point of sperm entry, but the rhythmic surface contraction waves that follow begin at the animal pole and proceed toward the vegetal pole. In activated sea urchin egg merogones (non-nucleate fragments), Yoneda *et al.* (1978) have shown that the latent period between activation and initiation of the rhythmic activity can vary widely from one merogone to another, but once the beat starts, the cycle length is the same in all cases, on the order of 40 minutes. It is the cell's reply to the initial stimulus that sets off the mitotic oscillations, and the time for the reply may vary depending on metabolic or other factors, for example, the cellular constituents that may or may not be included in the anucleate merogones as a result of centrifugation.

B. Organization of the Interphase Asters

The cell's reply to the activating signal is probably in the form of another release of calcium from intracellular stores. What triggers this release is not known. It could be a drop in pH. Gerson (1978) has shown pH oscillations in *Physarum*, from 5.9 in interphase to 6.8 during mitosis, and has suggested that a rise in pH may stimulate DNA synthesis and division. The rise in pH, however, might come as a result of the release of calcium, which itself is triggered by low pH. But this is only speculation. The second release of calcium may depend on an initial general increase of intracellular calcium due to depolarization of the smooth endoplasmic reticulum and subsequent calcium leakage. It is known that automatic calcium oscillations can be elicited in skinned muscle preparation by the presence of caffeine, which causes calcium release from the sarcoplasmic reticulum. If the depolarization occurs at many scattered points throughout the cell, possibly brought about by individual cytoplasmic structures acting as local depolarizers—which we will call "senders" of the return message—the result would be a general calcium release and a subsequent overall contraction or rounding up of the cell, similar to the overall cortical contraction of *Xenopus* eggs brought about by the calcium ionophore A23187 (Schroeder and Strickland, 1974). However, if the senders are aggregated to such a degree that their combined depolarizing effect can raise the free calcium concentration in a localized region to the threshold level for the all-or-none release, a wave could be initiated, just as the postfertilization wave is initiated from the point of sperm entry or by a single needle prick in *Xenopus* (Hara *et al.*, 1977). In normally fertilized eggs, the senders or organizing centers are the sperm centrioles. In ammonia-activated eggs, Paweletz and Mazia (1979) have shown that small accumulations of osmiophilic material aggregate in the center or around the periphery of the spherical clear zone in the center of the egg and are probably the origin of a small number of microtubules that grow outward from this region to form a monaster.

Miki-Noumura (1977) has demonstrated the presence of a large number of organizing centers capable of forming cytasters in butyric acid–hypertonic sea water-activated eggs. Under the conditions she used, the number of cytasters formed was constant within a species but varied from one species to another. The supplementary treatment of the partially activated egg with hypertonic sea water apparently somehow supplies organizing centers, which normally are provided by the sperm. Zucker *et al.* (1978) have shown that the hypertonic sea water brings about another release of calcium, but this time from a different store than the one supplying the initial calcium burst. What process is taking place is not really known, but one might suspect that a change affecting bound water is a factor, since not only hypertonic sea water also but D_2O and ethanol are effective (Kuriyama and Borisy, 1978).

The degree of aggregation of these organizing centers apparently depends on the concentrations of the activating agents and the duration of the first and second treatments (Kuriyama and Borisy, 1978), resulting in the formation of different numbers and sizes of cytasters under any given condition. These individual, initially scattered, organizing centers thus are the equivalent of the proposed senders of the return message.

If we apply the model proposed for the organization of the sperm monaster to the formation of successive asters that appear during the cell cycle, i. e., the directed regrowth of microtubules in the wake of a calcium wave, we might expect the following events. An initial depolarization of the smooth endoplasmic reticulum at the center of the aster is brought about by the presence of the centriole or an accumulation of organizing granules. Of course, we do not know what actually causes the depolarization, but it could involve changes in pH, ATP, etc. A threshold level of calcium sets off an all-or-none release of calcium, progressing outward as a trigger wave from the monaster center. Because this release regenerates itself, i. e., removes free calcium behind it and reloads the intracellular stores, the breakdown of the sperm monaster, proceeding from the aster center outward, would be followed by the growth of a new aster or asters from the active organizing centers.

Indirect evidence that this actually occurs comes from observations by Harris *et al.* (1980) using indirect immunofluorescence microscopy to follow the distribution of microtubules in sea urchin eggs throughout the first cleavage division cycle. This sequence of aster breakdown and re-formation is shown in Fig. 1. The drawings accompanying the fluorescence pictures indicate the regions where it is presumed that the calcium level is too high to support microtubule polymerization. The monaster is fully formed by 25–30 minutes after sperm entry and extends from the cell periphery to the nucleus. The sperm centrioles are separated and appear as bright spots on either side of the nucleus. Between 30 and 40 minutes this monaster breaks down, beginning at the cell center. Even before the peripheral remnants of the monaster are completely gone, new asters begin to grow from the organizing centers to form an enormous array of microtubules of the interphase asters, reaching a peak at the streak stage at 75 to 80 minutes. Following the streak stage, there is a breakdown of the interphase asters, again beginning at the center of the cell and progressing outward. Microtubules remaining associated with the aster centers lose their rather limp appearance and become very straight, suggesting that they may be under tension, perhaps being pulled into the centrosphere region. This might explain the transient defor-

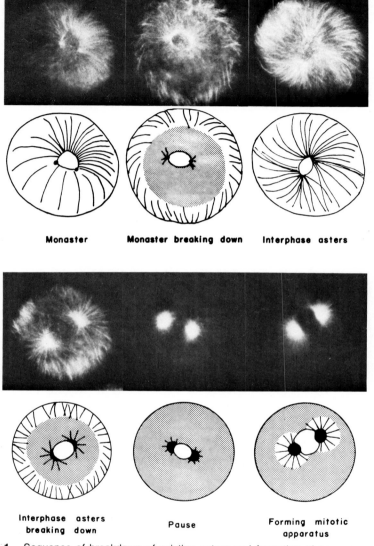

Fig. 1. Sequence of breakdown of existing asters and formation of new ones at two transition points in the sea urchin egg first division. A drawing below each immunofluorescence micrograph indicates regions where microtubules are stable or growing (unshaded) and regions where the microtubules are depolymerizing (shaded). Upper row: transition from monaster to interphase asters. .ower row: transition from interphase asters to mitotic apparatus. reproduced with permission from Harris *et al.*(1980). Rockefeller University Press.

mation of the nucleus, which becomes quite elongate at this time. The loss of microtubules is almost complete at the "pause" stage, after which the mitotic apparatus begins to form. At metaphase the asters are still relatively small but continue to grow through telophase. However, already at anaphase (Fig. 2), the breakdown of the aster microtubules is apparent, with loss of staining in the centrosphere, indicating that another round of aster breakdown has begun, to be followed by cytokinesis and regeneration of new asters for the interphase of the daughter cells. Actually, the new asters begin to form before the breakdown of the mitotic asters is complete. The two apparently merge, causing what appears to be a collapse of the old centrosphere to form the interphase asters.

The intervals between what are probably calcium pulses or waves, beginning with the fertilization reaction and continuing on through at least the second division, are between 35 and 40 minutes in *Strongylocentrotus purpuratus*. Measurements on the rate of reloading of the emptied intracellular calcium stores in aequorin-loaded sea urchin eggs (Zucker *et al.*, 1978) showed that a period of 40 minutes or more was required to elicit a response equal to the release at fertilization in *Lytechinus pictus*, which has a somewhat shorter division time than *S. purpuratus*. Certainly this period will vary from one species to another, and there are many factors which may influence the calcium pumping. The times of the pulses observed in all of the examples reported so far, while not identical, are of the same order of magnitude and suggest that the early cleavage divisions could be timed by a calcium-driven oscillator.

Although calcium oscillators can be self-contained, as demonstrated by the self-induced oscillations in skinned muscle fiber preparations, in whole cells one would expect some depolarization of the cell membrane and exchange with the surrounding medium resulting from the periodic rise in intracellular calcium. If this depolarization permitted external calcium to enter an egg, preferentially at the animal pole, a surface contraction wave similar to that seen in *Xenopus* eggs could be propagated in the cortex. Differences in membrane permeability between the animal and vegetal poles are known to exist during *Xenopus* egg maturation as a result of a calcium-controlled chloride current and may result in persistent permeability differences (Robinson, 1979). Many other eggs demonstrate cortical contraction waves oriented with respect to the animal-vegetal axis, as seen in *Xenopus* (reviewed by Schroeder, 1975). A number of eggs are known to be subjected to ionic currents during maturation (Jaffe and Nuccitelli, 1977), and it may be that an electrophoretic segregation of cell contents, together with cortical signals, determine the orientation of the spindle and ultimate polarity of the embryo.

C. Activation and Multiplication of the Centers

As noted earlier, artificial activation of an egg will start the mitotic clock, but without organizing centers to respond to the calcium message, the cell is unable to produce a functional mitotic apparatus. The introduction of a pair of centrioles by the sperm at fertilization provides the necessary organizing centers, but if the cell continues to divide, the centers must duplicate sometime during the course of the cell cycle. Mazia *et al.* (1960) used 2-mercaptoethanol as a mitotic blocking agent and showed that if sea urchin zygotes were blocked before first metaphase, and were removed from the block when the control cells were in their second division, the recovering cells would divide directly from one to four cells. Further studies to investigate the multiplicity of the mitotic centers revealed that when the cells resulting from the one-to-four division later attempted to divide, they were capable of forming only a half-spindle, and since they could not divide, they re-formed their nuclei and returned to interphase. At the next division time, apparently after they had duplicated their centers, they were able to form a bipolar figure and carried out an apparently normal division.

By applying the block at various times in the cell cycle to determine when the centers were duplicated, it was found that there was a transition point before which the cells divided normally from one to two cells and after which they divided directly to four cells. In the first division cycle, this occurred about the time of pronuclear fusion. A similar transition point at cytokinesis, just following telophase, was found for the second division. It should be noted that the term "center" was used to include the centrioles, pericentriolar material, vesicles, and any other components that constitute a functional pole. Although it was recognized that many organisms, including higher plants, do not have organized centrioles at the spindle poles, the generative mode of centriolar duplication was assumed to apply for the centers as well. That is, the initial step was the establishment of a "seed" or bud from which the new centriole grew (see Mazia, 1961).

The interpretation of the mercaptoethanol-blocking experiments, therefore, was that the centers were normally composed of two units which could split apart and separate in the mercaptoethanol block but could not initiate duplication, that is, could not form a bud. The transition points indicated the times at which the centers duplicated or became capable of duplicating, a process that, once started, could proceed during the blocking period. The process of reproduction of the centers was believed to consist of three steps: (1) initiation of a bud which was capable of growing into a full-sized center, (2) splitting of the original center from the new one, each becoming capable of functioning independently, and (3) physical separation of the centers following splitting apart.

Subsequent work by Hinegardner *et al.* (1964) showed that DNA synthesis in sea urchin zygotes, at least for the first three synthetic periods after fertilization, coincided with the transition points that indicated a duplication of the centers. The beginning of centriole replication in the late G_1 or early S period of the division cycle has also been documented in other cell types, using electron microscopy (Robbins *et al.*, 1968; Rattner and Phillips, 1973) and, more recently, indirect immunofluorescence with antitubulin antibody (Tucker *et al.*, 1979). The question of a possible causal relationship between these two events was addressed by Bucher and Mazia (1960), who showed that sea urchin eggs blocked with mercaptoethanol before the transition point for new center initiation could incorporate [^{3}H]thymidine into DNA during the blocking period, which prevented duplication of the centers. Although the validity of this work has been questioned (Tucker *et al.*, 1979), the facts that DNA synthesis in unfertilized sea urchin eggs can be turned on by NH_3 (Mazia and Ruby, 1974) and that regularly occurring chromosome cycles can be initiated (Mazia, 1974) without the presence of either centrioles or organizing centers fully support the conclusion drawn from the mercaptoethanol studies. Blocking of DNA synthesis with cytosine arabinoside in mouse L929 cells, on the other hand, does not prevent centriole duplication, although division is blocked at a later stage (Rattner and Phillips, 1973).

The apparent relationship between centriole duplication and DNA synthesis may be only coincidental and need not be causally linked. In sea urchin zygotes these events also mark the time of a rise in cytoplasmic calcium, judging from the loss of microtubule structures, and suggest that periodic calcium pulses may be driving both the centriole and the chromosome cycles. The calcium pulses occur at two transition points in the cell cycle: entry into the S period and entry into mitosis. In each case, there is breakdown of an existing astral structure and formation of a new one.

The breakdown of the interphase cytoplasmic microtubule complex in mammalian tissue culture cells just before entry into mitosis has been documented by antitubulin immunofluorescence (Weber *et al.*, 1975; Brinkley *et al.*, 1976; reviewed by Brinkley *et al.*, 1980). These studies also indicate that this is a time of centriole separation. Later, at telophase, the mitotic asters apparently give rise directly to the new cytoplasmic microtubule complex. In sea urchin zygotes, which do not have a G_1 period, a new round of centriole duplication and DNA synthesis begins at late telophase. At this time, there is a partial breakdown of existing mitotic asters before new interphase asters are formed. If there is some causal relationship among centriole duplication, initiation of DNA synthesis, and the proposed calcium pulse, one might expect to find in cultured cells some degree of breakdown of the cytoplasmic microtubule complex at the transition from the G_1 to the S phase. Fine structure observations are suggestive. Robbins *et al.* (1968) noted that at this time there was nothing in the pericentriolar region of HeLa cells to indicate any centriole activity except the splitting apart of the centriole pair from their orthogonal configuration, and only an occasional microtubule. Shortly afterward DNA synthesis began, followed by the initiation of new procentrioles and increasing numbers of microtubules.

Similarities between the stimulation of ciliated quiescent tissue culture cells to enter the division cycle (Tucker *et al.*, 1979) and the activation of sea urchin eggs at fertilization further suggest that calcium may be involved in the centriole cycle. Quiescent 3T3 cells possess a pair of unduplicated centrioles, one of which is ciliated. Serum stimulation results in three-step response: first, an immediate initial deciliation, then regeneration of the cilia over a period of about 13 hours, and finally a second deciliation that coincides with the beginning of DNA synthesis. It was suggested that changes in intracellular calcium might be associated with the deciliation. Moreover, there is considerable evidence that calcium is the intracellular second messenger which acts in the stimulus–division coupling (Berridge, 1975, 1976). In eggs, the fertilization reaction is accompanied by a calcium release, followed by formation of the monaster, and finally breakdown of the monaster at the time of DNA synthesis. A similar three-step response of quiescent cells to a mitogenic stimulus, involving two random transitions separated by a lag phase, has been proposed by Brooks *et al.* (1980) on the basis of a mathematical analysis of cell culture kinetics. As these authors point out, there are many similarities between the model and the observed centriole cycles in stimulated cultured cells (Tucker *et al.*, 1979) and sea urchin eggs (Mazia *et al.*, 1960).

D. Formation of the Mitotic Apparatus

During the cell cycle, two sets of microtubule-containing astral structures are formed: (1) the interphase cytoplasmic microtubule complex (Brinkley *et al.*, 1980) of tissue culture cells or the interphase asters of sea urchin zygotes and (2) the mitotic apparatus assembled at the time of division. Basically, the interphase structures of cultured cells and sea urchin eggs are similar. They are formed following the breakdown of a previous astral structure, either a mitotic apparatus or a sperm monaster, presumably as a result of calcium release which provides available subunits for the assembly of new asters around existing organizing centers. The microtubules of the interphase structures grow to very long lengths; in tissue culture they originate from the centriolar region and grow outward to the cell periphery (Osborn and Weber, 1978), often following the plasma membrane or doubling back to form a massive tangle that is no longer identifiable as an aster. The sea urchin

interphase asters are also massive structures, with aggregates of rather relaxed or floppy microtubule bundles making up the astral rays.

The asters of the mitotic apparatus differ from the interphase asters in
several respects. The astral rays of the mitotic apparatus are not so extensive
and, unlike the relaxed appearance of the interphase microtubules, appear
stiff and straight. The aster centers begin to accumulate a large mass of
smooth endoplasmic reticulum, especially exaggerated in sea urchin eggs,
forming a large clear area from which yolk and mitochondria are for the most
part excluded. As the asters grow, changes take place in other cytoplasmic
membranes; the annulate lamellae disappear, and finally the nuclear membrane breaks down (Harris, 1967; Longo, 1971). A large number of microtubules rapidly invades the nuclear region, and microtubules are soon
observed attached to kinetochores of the chromosomes.

This greater density of microtubules in the spindle region compared to
that of the asters deserves a comment here, as it appears that there might be
something in the zygote nucleus which catalyzes microtubule polymerization. For some reason, the male pronucleus alone is not capable of forming a
fully functional spindle, as Hiramoto (1962b) has shown with his enucleation
studies of sea urchin eggs. Among the long list of nuclear components, actin
has been shown to be present in the nuclei of various species (LeStourgeon
et al., 1975; Clark and Rosenbaum, 1979), but it is conceivable that the
nucleus may also contain microtubule-associated proteins (MAPs) (Sloboda
et al., 1976) or other substances that catalyze microtubule polymerization.
Such a catalyst could thus ensure that there are enough microtubules in the
spindle to move the chromosomes, even under adverse conditions. Stephens
(1972) has shown that sea urchin eggs developing at low temperatures have
greatly reduced or no asters, although a functional spindle is formed. The
presence of some factor in the nucleus is suggested by the results of
Heidemann and Kirschner (1975), who found that basal bodies injected into
Xenopus oocytes did not form asters, but asters were formed after the germinal vesicle broke down and nuclear contents were mixed with the cytoplasm (Heidemann and Kirschner, 1978). In ammonia-activated eggs, however, Paweletz and Mazia (1979) have shown that the condensed chromosomes that are present following nuclear membrane breakdown have no
microtubules associated with kinetochores, even though the contents of the
zygote nucleus are now available and the presence of a few microtubules
shows that some tubulin is present. It is possible in this case that the tubulin
concentration is too low for adequate polymerization (Weisenberg, 1978) and
that the large calcium release at fertilization, which does not occur with
ammonia activation, is necessary to release tubulin subunits from
polymerized sources. Supporting this idea is the fact that live sperm injected
into unfertilized but mature sea urchin eggs do not form an aster, but if the

egg is subsequently fertilized, both the injected and fertilizing sperm form asters and eventually a polyspermic mitotic apparatus (Hiramoto, 1962a). Godfrey *et al.* (1980) have shown that about 10% of the tubulin in the unfertilized egg behaves as if it were polymerized, and this could be a source of tubulin made available by the calcium burst at fertilization.

E. *In Vitro* "Possible" versus *in Vivo* "Permitted" Reactions

The volume of information concerning the structure and function of the proteins which are most likely to be involved in mitosis is growing at a staggering rate, yet there are few satisfactory explanations of how these constituents of the mitotic apparatus are able to function in a coordinated fashion. *In vitro* systems tell us a great deal about what is possible, but in a coordinated activity only some of these possibilities may be allowed.

Most of the models proposed for mitosis have assumed that the cytoplasmic medium in which the mitotic apparatus sits is homogeneous and unchanging, and that when the mitotic apparatus is isolated and presented with the proper broth, it will perform its act of chromosome movement. Evidence is now accumulating that the cytoplasm is not homogeneous and that there is not necessarily a free diffusion of ions. In cells preloaded with the calcium sensitive photoprotein, aequorin, Rose and Lowenstein (1975), Kiehart and Inoué (1976) and Kiehart (1981) have shown that injected calcium ions are restricted to limited domains by energizing sequestering. Moreover, these domains may move, as demonstrated by the calcium trigger wave of the fertilized fish egg (Gilkey *et al.*, 1978). Tubulin immunofluorescence studies of sea urchin eggs (Harris *et al.*, 1980) provide evidence that such moving domains also occur in mitotic cells, as apparent waves of some condition, possibly calcium ions, favoring microtubule breakdown move outward from the cell or aster centers. Similar observations were previously reported by Swann (1951a, b), who measured *in vivo* changes in the coefficient of birefringence of sea urchin spindle and asters, using a technique of quantitative birefringence analysis. The anaphase aster, which continues to grow in size and birefringence at its periphery, shows at the same time a decrease in birefringence which begins at the center and moves outward. The presence of such restrictive domains and their spatial and temporal characteristics allow a sorting out of actual permitted reactions from the vast number that are possible under various *in vitro* conditions.

During the early stages of formation of the mitotic asters, before the breakdown of the nuclear membrane, astral rays at the growing front of the aster are reading the signal to polymerize. At anaphase, however, as seen in Fig. 2 and diagrammed in Fig. 3, the outer region of the asters may still be

Fig. 2.   Tubulin immunofluorescence micrograph of a sea urchin egg at anaphase. Note the termination of astral rays at the boundary of the centrosphere and loss of tubulin staining within the centrosphere.

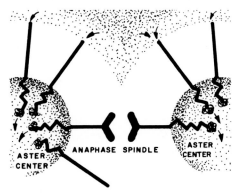

Fig. 3.   An interpretation of what might be occurring in the anaphase cell in Fig. 2. The boundaries between shaded and unshaded areas represent the wave fronts traveling outward from the aster centers. The upper or distal shaded area is a region where calcium is being sequestered but where the level is not yet low enough to permit microtubule polymerization (arrows in). The shaded aster centers represent a new wave of calcium release; here a contracting actomyosin network pulls microtubules into the centrosphere (zigzag lines), where they depolymerize (arrows out). Microtubules which are capped by attachment to a kinetochore are unable to lengthen and can only be pulled to the centrosphere boundary.

responding to the signal to polymerize, but a different message is now being read in the aster centers. Evidence from electron microscopy (Harris, 1975), as welfl as the loss of tubulin fluorescence or bifringence, suggests that this is now a region of microtubule breakdown. The rather abrupt periphery of the growing centrosphere delimits a region where the microtubules are greatly disorganized and where small fragments of microtubules can be seen associated with the dense amorphous pericentriolar material.

Although microtubules are the most conspicuous and consistent components of the mitotic apparatus, there is a body of evidence that actin is also present. Heavy meromyosin (HMM) binding to fibers in glycerinated cells has been demonstrated in sectioned or negative-stained preparations for electron microscopy (reviewed by Forer, 1978). Fluorescence studies using fluorescein-labeled HMM (Sanger and Sanger, 1976) or indirect immunofluorescence with antiactin antibodies (Cande et al., 1977) have similarly shown localization of actin at the spindle poles and the chromosome-to-pole region, but unlike the glycerinated preparations, no actin was found in the interzone between separating chromosomes. Injection of fluorescently labeled actin into living sea urchin eggs (Wang and Taylor, 1979), however, showed a diffuse presence of actin in the region of the mitotic apparatus but no specific identification with structural elements. Furthermore, Aubin et al. (1979 and this volume), in a critical study of immunofluorescence and HMM binding, have shown that previous positive reports of actin in the spindle are subject to criticism on the basis of likely preparational artifacts. However, actin is a widely occurring major protein of non-muscle cells, and myosin, though not so extensively studied, has also been found in various non-muscle actin-containing cells (Pollard and Weihing, 1974). It would thus be surprising if actin, and probably myosin as well, were not distributed throughout the cell. If these proteins take part in any contractile activity during mitosis, it would depend largely on the ionic or other conditions in localized regions of the cell.

Returning now to the idea of permitted reactions within the restrictive conditions in the living cell, actin and myosin, like tubulin, can be expected to respond to localized changes in ionic concentrations or other conditions resulting from a calcium wave. Where microtubules are breaking down in the centrosphere regions of anaphase asters, presumably as a result of increasing calcium, any actin or myosin in that region would be given the signal to contract.

One problem which might be resolved by the notion of *in vivo* permitted versus *in vitro* possible events is that of kinetochore microtubule polarity. Although *in vitro* experiments have demonstrated the ability of kinetochores to initiate microtubule assembly (Telzer et al., 1975; Gould and Borisy, 1978) and the fact that these microtubules grow with the fast-growing end away from the chromosomes (Summers and Kirschner, 1979; Bergen *et al.*, 1980), there is evidence to show that this does not necessarily happen *in vivo*. Fine structure studies of mitotic stages from such diverse sources as higher plants (Bajer and Mole-Bajer, 1972), mammalian cells (Roos, 1973a), and sea urchin eggs (Harris, 1975) show no evidence of kinetochore fibers until the nuclear region is invaded by polar microtubules following nuclear membrane breakdown. During the intranuclear mitosis of the slime mold *Physarum*, the spindle, composed almost entirely of kinetochore-to-pole tubules, clearly originates from a tubule organizer or division center, which splits and migrates to opposite poles at prophase (reviewed by Goodman, 1980). Furthermore, Roos (1973b) has shown that during prometaphase in PtK2 cells, chromosomes are occasionally oriented initially toward only one pole and only one of the kinetochores has a fiber attached, while the other one is bare. This is a situation one would not expect if the kinetochores are generating microtubules. The directional constraints of a wave or moving domain which permits microtubule assembly, progressing outward from the aster center or pole, will of necessity require the direction of microtubule growth to be from the pole to the kinetochore, not the reverse. Recent studies consistent with this idea have shown a uniform polarity of kine-tochore microtubules in each half-spindle of *Spisula* eggs (Telzer and Haimo, 1981) and in lysed PtK$_1$ cells, where it was also determined that essentially all kinetochore microtubules originate at the poles (Euteneuer and McIntosh, 1981).

Fig. 3 is a simplified diagram which attempts to tie these ideas together, showing what might be going on in the anaphase mitotic apparatus presented in Fig. 2. It suggests how regions of permitted activities, moving outward as a wave from the aster centers, can affect the shape of the mitotic apparatus as well as its functioning. The unshaded area represents the region where microtubules are stable or are permitted to polymerize. The upper or distal shaded area is a region from which calcium is being sequestered, but where the level is not yet low enough to support microtubule polymerization. As the boundary between the shaded and unshaded areas moves away from the aster centers, microtubules can lengthen by the addition of subunits to their distal ends. The shaded area at the aster centers represents a new wave of calcium release; here a contracting actomyosin network pulls microtubules into the centrosphere (zigzag lines), where they depolymerize. Microtubules which are capped by attachment to a kinetochore are unable to lengthen and can only be pulled toward the centrosphere boundary.

A more detailed discussion of the implications of such a model with re-spect to pole-directed particle movements and other phenomena of spindle and aster behavior has been presented elsewhere (Harris, 1978). It is offered here simply as a new way of looking at the mitotic process that might provide more insight into its overall control mechanisms than most current approaches have done so far.

IV. BASIS FOR THE MITOTIC OSCILLATOR

A. Division Protein or Factor: A Relaxation Oscillator

Attempts to analyze the mitotic cycle have employed a wide variety of assaults, from inhibitors of specific enzymes to general irradiation with ultraviolet light or x-rays to induce mitotic delays. Mitchison (1971) has reviewed the vast amount of confusing and often conflicting data, comparing results from heat shock, irradiation, and inhibition of protein synthesis on a wide range of cell types, from prokaryotes to mammalian tissue culture cells. He has presented a number of different possible models for division control, but one which appears to fit the data best is derived from heat shock synchronization of division in the ciliate *Tetrahymena*. In this model, one or more division proteins are synthesized through the cell cycle and are necessary for division. The division protein is labile and can be affected by inhibitors up to a certain transition point, after which the cell continues through mitosis but is blocked in the next division. The division protein is presumed to be used up during mitosis and must be synthesized anew during the next division cycle. The plot of the amount of division protein against time is a "sawtooth" wave form, with a gradual rise to a threshold level which triggers mitosis and a sudden drop as the cells enter the next cycle. Such an oscillator, in which variables change slowly over one part of the cycle and rapidly over another, is known as a "relaxation oscillator."

Rusch *et al.* (1966) and Sachsenmaier *et al.* (1972) have developed a similar model with plasmodia of the slime mold *Physarum polycephalum*. The nuclei of the multinucleate plasmodium undergo synchronous mitoses with a division period of about 12 hours. By fusing plasmodia one-half cycle out of phase at different times in the division cycle they found that the cycle was shortened in one and delayed in the other, and that all nuclei divided synchronously at an intermediate time. Furthermore, differences in the sizes of the respective plasmodia that were fused affected the intermediate point that was reached, strongly suggesting that there is some factor that builds up during the cycle and triggers mitosis at a critical level. The synchrony throughout the plasmodium rules out a nuclear origin of the mitotic clock but suggests that it relies on an accumulation of a cytoplasmic factor that is dependent on continuous RNA and protein synthesis. In the model, the trigger depends on a critical ratio of division factor to nuclear sites, and with doubling of the nuclei at mitosis, the ratio dropped back to its starting point. As with Mitchison's model, a sawtooth wave form is generated to describe the oscillation.

B. Coupled Limit Cycle Oscillators: Mitotic Synchronization or Contact Inhibition

Kauffman and Wille (1975), also using *Physarum*, have analyzed the effects of fusing plasmodia at almost all possible phases and phase differences and have investigated the nature of heat shock delays of mitosis. An analysis of their own data, as well as reexamination of those of Sachsenmaier *et al.* (1972), shows that a sawtooth wave form derived from the slow increase and sudden drop in mitogen concentration is not the only possible interpretation; the results can be described as a two-component limit cycle oscillator. On the basis of several lines of evidence, including an apparent uncoupling of mitosis from the mitotic clock by heat shock, they propose that mitosis in *Physarum* is controlled by but is not part of a continuous biochemical oscillator.

Mathematical models in which such cellular oscillators are coupled show some interesting possible results. If the initial phase difference is small, coupling leads to a nearly synchronous large-amplitude oscillation; but if the initial phase difference is large, the amplitude of one or both oscillators may be damped to such a degree that neither reaches the threshold to trigger mitosis. The authors argue that this could be an explanation for the phenomenon of contact inhibition of confluent cell cultures. Using a similar model of coupled oscillators, Burton and Canham (1973) have shown that the number of intercellular contacts or the position of a cell in a population, as well as the degree of synchrony, may play a role in contact inhibition of cell division. Parisi *et al.* (1978) have applied this model of coupled limit cycle oscillators to explain the gradient-like distribution of cell divisions in sea urchin embryos at the early blastula stage, and they propose that mitotic oscillators exist within the cells of the embryo from which periodic chemical signals are released and transmitted from cell to cell, with the micromeres acting as pacemakers. The expectations based on theory show a close correlation with what is actually observed in the developing embryo.

The role of calcium as a mitogen, as well as its role in a number of cellular membrane oscillators, makes it a likely candidate for the periodically transmitted message. However, the frequency and amplitude of the oscillations in the calcium ion concentrations are apparently dependent on other metabolic reactions. Parisi *et al.* (1979) found that sea urchin embryos grown in the presence of actinomycin D from fertilization showed a completely disrupted mitotic gradient, although cell division itself was not seriously inhibited. If the actinomycin D treatment was begun after the fourth cleavage, when micromeres are formed, the gradient appeared normally, as in the controls. Actinomycin treatment, to be effective, had to be started before the micromeres were segregated. It was argued that a rapidly labeled RNA, which is synthesized uniquely in the micromeres at the time of their segregation, may be responsible for the gradient. Inhibition of required syntheses at this time might greatly alter the frequency of the micromere oscillations. Entrainment occurs only if the pacemaker frequency is fairly close to that of the individual oscillators to be entrained. Too great a difference from the natural frequency can result in chaotic behavior. The driving oscillator for the mitotic clock thus appears to reside in some metabolic circuit, possibly ultimately the glycolytic oscillator (Chance et al., 1973; reviewed by Hess, 1979).

V. CONCLUSIONS

The importance of calcium in the regulation of cellular activities is reflected in the widespread occurrence of mechanisms to control the intracellular concentration of free calcium ions. Calcium sequestering by mitochondria and smooth endoplasmic reticulum, as well as outward-directed calcium pumps in the cell membrane, maintain low levels of intracellular calcium against a large gradient. With the aid of regulator proteins and various feedback systems, small increments in calcium ion concentration can signal large changes in cellular function.

In the cell division cycle, calcium can act directly as a mitogen or as a second messenger in the stimulus–division coupling. Evidence strongly suggests that it starts a mitotic clock, whose rhythmic calcium pulses govern either directly or indirectly the activation and duplication of mitotic centers, DNA synthesis, and formation and breakdown of the interphase asters and the mitotic apparatus.

The calcium signal may be a localized one, restricted to a limited domain by energized sequestering, or if the calcium concentration reaches a threshold level, a triggered all-or-none release can start a calcium wave, introducing not only a time element but direction as well. The directional aspect of a calcium wave may serve to coordinate activities by imposing a temporal and spatial sequence, thus restricting the many possible reactions in the formation of the mitotic structures to a few permitted ones, and these only in their proper sequence. These restrictions need to be considered in any models designed to explain the mechanisms of mitosis and chromosome movement.

REFERENCES

Andersen, B., Osborn, M., and Weber, K. (1978). Specific visualization of the distribution of the calcium regulatory protein of cyclic nucleotide phosphodiesterase (modulator protein) in tissue culture cells by immunofluorescence microscopy: Mitosis and intercellular bridge. *Cytobiologie* **17**, 354–364.

Aubin, J. E., Weber, K., and Osborn, M. (1979). Analysis of actin and microfilament associated proteins in the mitotic spindle and cleavage furrow of PtK2 cells by immunofluorescence microscopy. A critical note. *Exp. Cell Res.* **124**, 93–109.

Bajer, A. S., and Mole-Bajer, J. (1972). "Spindle Dynamics and Chromosome Movement." Academic Press, New York.

Baker, P. F. (1976). The regulation of intracellular calcium. *Symp. Soc. Exp. Biol.* **30**, 67–88.

Begg, D. A., and Rebhun, L. I. (1979). pH regulates the polymerization of actin in the sea urchin egg cortex. *J. Cell Biol.* **83**, 241–248.

Bell, L. G. E. (1962). Some mechanisms involved in cell division. *Nature (London)* **193**, 190–191.

Bergen, L. G., Kuriyama, R., and Borisy, G. G. (1980). Polarity of microtubules nucleated by centrosomes and chromosomes of Chinese hamster ovary cells *in vitro*. *J. Cell Biol.* **84**, 151–159.

Berridge, M. J. (1975). The interaction of cyclic nucleotides and calcium in the control of cellular activity. *Adv. Cyclic Nucleotide Res.* **6**, 1–98.

Berridge, M. J. (1976). Calcium, cyclic nucleotides, and cell division. *Symp. Soc. Exp. Biol.* **30**, 219–231.

Berridge, M. J., Rapp, P. E., and Treherne, J. E., eds. (1979). "Cellular Oscillators." *J. Exp. Biol.* **81**. Cambridge University Press, Cambridge.

Brinkley, B. R., Fuller, G. M., and Highfield, D. P. (1976). Tubulin antibodies as probes for microtubules in dividing and non-dividing mammalian cells. *In* "Cell Motility" (R. Goldman, T. Pollard, and J. Rosenbaum, eds.), pp. 435–456. Cold Spring Harbor Lab., Cold Spring Harbor, New York.

Brinkley, B. R., Fistel, S. H., Marcum, J. M., and Pardue, R. L. (1980). Microtubules in cultures cells; Indirect immunofluorescence staining with tubulin antibody. *Int. Rev. Cytol.* **63**, 59–95.

Brooks, R. F., Bennett, D. C., and Smith, J. A. (1980). Mammalian cell cycles need two random transitions. *Cell* **19**, 493–504.

Bryan, J., and Kane, R. E. (1978). Separation and interaction of the major components of sea urchin actin gel. *J. Mol. Biol.* **125**, 207–224.

Bucher, N. L. R., and Mazia, D. (1960). Deoxyribonucleic acid synthesis in relation to duplication of centers in dividing eggs of the sea urchin, *Strongylocentrotus purpuratus*. *J. Biophys. Biochem. Cytol.* **7**, 651–655.

Burgess, D. R., and Schroeder, T. E. (1977). Polarized bundles of actin filaments within microvilli of fertilized sea urchin eggs. *J. Cell Biol.* **74**, 1032–1037.

Burton, A. C., and Canham, P. B. (1973). The behavior of coupled biochemical oscillators as a model of contact inhibition of cellular division. *J. Theor. Biol.* **39**, 555–580.

Cande, W. Z., Lazarides, E., and McIntosh, J. R. (1977). A comparison of the distribution of actin and tubulin in the mammalian mitotic spindle as seen by indirect immunofluorescence. *J. Cell Biol.* **72**, 552–567.

Carafoli, E., and Crompton, M. (1978). The regulation of intracellular calcium by mitochondria. *Ann. N. Y. Acad. Sci.* **307**, 269–283.

Chambers, E. L., and Hinkley, R. E. (1979). Non-propagated cortical reactions induced by the divalent ionophore A 23187 in eggs of the sea urchin, *Lytechinus variegatus*. *Exp. Cell Res.* **124**, 441–446.

Chance, B., Williamson, G., Lee, I. Y., Mela, L., DeVault, D., Ghosh, A., and Pye, E. K. (1973). Synchronization phenomena in oscillations of yeast cells and isolated mitochondria. *In* "Biological and Biochemical Oscillators" (B. Chance, E. K. Pye, A. K. Ghosh, and B. Hess, eds.), pp. 285–300. Academic Press, New York.

Clark, T. G., and Rosenbaum, J. L. (1979). An actin filament matrix in hand-isolated nuclei of *X. laevis* oocytes. *Cell* **18**, 1101–1108.

Clothier, G., and Timourian, H. (1972). Calcium uptake and release by dividing sea urchin eggs. *Exp. Cell Res.* **75**, 105–110.

Condeelis, J. S., and Taylor, D. L. (1977). The contractile basis of amoeboid movement. V. The coltrol of gelation, solation, and contraction in extracts from *Dictyostelium discoideum. J. Cell Biol.* **74**, 901–927.

Das, N. K., Hildebrandt, A. C., and Riker, A. J. (1966). Cine photomicrography of low temperature effects on cytoplasmic streaming, nucleolar activity, and mitosis in single tobacco cells in microculture. *Am. J. Bot.* **53**, 253–259.

Dirksen, E. R., Prescott, D. M., and Fox, C. F., eds. (1978). "Cell Reproduction. In Honor of Daniel Mazia." Academic Press, New York.

Duncan, C. J., ed. (1976). "Calcium in Biological Systems." *Symp. Soc. Exp. Biol.* **30**. Cambridge University Press, Cambridge.

Durham, A. C. H. (1974). A unified theory of the control of actin and myosin in nonmuscle movements. *Cell* **2**, 123–135.

Eddy, E. M., and Shapiro, B. M. (1976). Changes in the topography of the sea urchin egg after fertilization. *J. Cell Biol.* **71**, 35–48.

Endo, M., Tanaka, M., and Ogawa, Y. (1970). Calcium induced release of calcium from the sarcoplasmic reticulum of skinned skeletal muscle fibres. *Nature (London)* **228**, 34–36.

Epel, D. (1978). Mechanisms of activation of sperm and egg during fertilization of sea urchin gametes. *Curr. Top. Dev. Biol.* **12**, 185–246.

Euteneuer, U., and McIntosh, J. R. (1981). Structural polarity of kinetochore microtubules in PtK$_1$ cells. *J. Cell Biol.* **89**, 338–345.

Fabiato, A., and Fabiato, F. (1978). Calcium-induced release of calcium from the sarcoplasmic reticulum of skinned cells from adult human, dog, cat, rabbit, rat, and frog hearts and from fetal and new-born rat ventricles. *Ann. N. Y. Acad. Sci.* **307**, 491–522.

Forer, A. (1978). Electron microscopy of actin. *In* Principles and Techniques of Election Microscopy, Biological Applications" (M. A. Hayat, ed.), Vol. 9, pp. 126–174. Van Nostrand-Reinhold, Princeton, New Jersey.

Gerson, D. F. (1978). Intracellular pH and the mitotic cycle in *Physarum* and mammalian cells. *In* "Cell Cycle Regulation" (J. R. Jeter, Jr., I. L. Cameron, G. M. Padilla, and A. M. Zimmerman, eds.), pp. 105–131. Academic Press, New York.

Gilkey, J. C., Jaffe, L. F., Ridgway, E. B., and Reynolds, G. T. (1978). A free calcium wave traverses the activating egg of the medaka, *Oryzias latipes. J. Cell Biol.* **76**, 448–466.

Godfrey, C. B., Lee, V. D., and Wilson, L. (1980). Quantitation of tubulin pools in fertilized sea urchin eggs. *In* "Control of Cellular Development and Division." ICN-UCLA Symp. on Mol. and Cell. Biol. 1980. (Abstr.)

Goldman, R., Pollard, T., and Rosenbaum, J., eds. (1976). "Cell Motility," Cold Spring Harbor Conferences on Cell Proliferation, Vol. 3. Cold Spring Harbor Lab. Cold Spring Harbor, New York.

Goodman, E. M. (1980). *Physarum polycephalum:* A review of a model system using a structure-function approach. *Int. Rev. Cytol.* **63**, 1–58.

Gould, R. R., and Borisy, G. G. (1978). Quantitative initiation of microtubule assembly by chromosomes from Chinese hamster ovary cells. *Exp. Cell Res.* **113**, 369–371.

Hara, K. (1971). Cinematographic observation of "surface contraction waves" (SCW) during the early cleavage of axolotl eggs. *Wilhelm Roux' Arch. Entwicklungsmech. Org.* **167**, 183–186.

Hara, K., Tydeman, P., and Hengst, R. T. M. (1977). Cinematographic observation of "post-fertilization waves" (PFW) on the zygote of *Xenopus laevis*. *Wilhelm Roux' Arch. Entwicklungsmech. Org.* **181**, 189–192.

Hara, K., Tydeman, P., and Kirschner, M. (1980). A cytoplasmic clock with the same period as the division cycle in *Xenopus* eggs. *Proc. Natl. Acad. Sci. U.S.A.* **77**, 462–466.

Harris, P. (1967). Structural changes following fertilization in the sea urchin egg. Formation and dissolution of heavy bodies. *Exp. Cell Res.* **48**, 569–581.

Harris, P. (1975). The role of membranes in the organization of the mitotic apparatus. *Exp. Cell Res.* **94**, 409–425.

Harris, P. (1978). Triggers, trigger waves, and mitosis: A new model. *In* "Cell Cycle Regulation" (J. R. Jeter, Jr., I. L. Cameron, G. M. Padilla, and A. M. Zimmerman, eds.), pp. 75–104. Academic Press, New York.

Harris, P. (1979). A spiral cortical fiber system in fertilized sea urchin eggs. *Dev. Biol.* **68**, 525–532.

Harris, P., Osborn, M., and Weber, K. (1980). Distribution of tubulin-containing structures in eggs of the sea urchin *Strongylocentrotus purpuratus* from fertilization through first cleavage. *J. Cell Biol.* **84**, 668–679.

Heidemann, S. R., and Kirschner, M. W. (1975). Aster formation in eggs of *Xenopus laevis*. Induction by isolated basal bodies. *J. Cell Biol.* **67**, 105–117.

Heidemann, S. R., and Kirschner, M. W. (1978). Induced formation of asters and cleavage furrows in oocytes of Xenopus laevis during *in vitro* maturation. *J. Exp. Zool.* **204**, 431–444.

Heilbrunn, L. V. (1937). "An Outline of General Physiology." Saunders, Philadelphia, Pennsylvania.

Heilbrunn, L. V. (1956). "The Dynamics of Living Protoplasm." Academic Press, New York.

Hess, B. (1979). The glycolytic oscillator. *J. Exp. Biol.* **81**, 7–14.

Hinegardner, R. T., Rao, B., and Feldman, D. E. (1964). The DNA synthetic period during early development of the sea urchin egg. *Exp. Cell Res.* **36**, 53–61.

Hiramoto, Y. (1962a). Microinjection of the live spermatozoa into sea urchin eggs. *Exp. Cell Res.* **27**, 416–426.

Hiramoto, Y. (1962b). An analysis of the mechanism of fertilization by means of enucleation of sea urchin eggs. *Exp. Cell Res.* **28**, 323–334.

Hiramoto, Y. (1969). Mechanical properties of the protoplasm of the sea urchin egg. II. Fertilized egg. *Exp. Cell Res.* **56**, 209–218.

Inoué, S., and Stephens, R. E., eds. (1975). "Molecules and Cell Movement." Raven, New York.

Jaffe, L. F. (1980). Calcium explosions as triggers of development. *Ann. Rev. N.Y. Acad. Sci.* **339**, 86–101.

Jaffe, L. F., and Nuccitelli, R. (1977). Electrical controls of development. *Annu. Biophys. Bioeng.* **6**, 445–476.

Kauffman, S., and Wille, J. J. (1975). The mitotic oscillator in *Physarum polycephalum*. *J. Theor. Biol.* **55**, 47–93.

Kiehart, D. P. (1981). Studies on the *in vivo* sensitivity of spindle microtubules to calcium ions and evidence for a vesicular calcium-sequestering system. *J. Cell Biol.* **88**, 604–617.

Kiehart, D. P., and Inoué, S. (1976). Local depolymerization of spindle microtubules by microinjection of calcium ions. *J. Cell Biol.* **70**, 230a.

Klee, C. B., Crouch, T. H., and Richman, P. G. (1980). Calmodulin. *Annu. Rev. Biochem.* **49**, 489–515.

Kojima, M. K. (1960). Cyclic changes of the cortex and the cytoplasm of the fertilized and activated sea urchin egg. I. Changes in the thickness of the hyaline layer. *Embryologia* **5**, 1–7.

Kuriyama, R., and Borisy, G. G. (1978). Cytasters induced within unfertilized sea urchin eggs. *J. Cell Biol.* **79**, 294. (Abstr.)

LeStourgeon, W. M., Forer, A., Wang, Y. -Z., Bertram, J. S., and Rusch, H. P. (1975). Contractile proteins: Major components of nuclear and chromosome nonhistone proteins. *Biochim. Biophys. Acta* **379**, 529–552.

Loeb, J. (1913). "Artificial Parthenogenesis and Fertilization." Univ. of Chicago Press, Chicago, Illinois.

Longo, F. J. (1971). An ultrastructural analysis of mitosis and cytokinesis in the zygote of the sea urchin *Arbacia punctulata. J. Morphol.* **138**, 207–238.

McGraw, C. F., Somlyo, A. V., and Blaustein, M. P. (1980). Localization of calcium in presynaptic nerve terminals. *J. Cell Biol.* **85**, 228–241.

Mar, H. (1980). Radial cortical fibers and pronuclear migration in fertilized and artificially activated eggs of *Lytechinus pictus. Dev. Biol.* **78**, 1–13.

Marcum, J. M., Dedman, J. R., Brinkley, B. R., and Means, A. R. (1978). Control of microtubule assembly-disassembly by calcium dependent regulator protein. *Proc. Natl. Acad. Sci. U.S.A.* **75**, 3771–3775.

Mazia, D. (1937). The release of calcium in *Arbacia* eggs on fertilization. *J. Cell. Comp. Physiol.* **10**, 291–308.

Mazia, D. (1961). Mitosis and the physiology of cell division. *In* "The Cell" Vol. III (J. Brachet and A. E. Mirsky, eds.), pp. 77–412. Academic Press, New York.

Mazia, D. (1974). Chromosome cycles turned on in unfertilized sea urchin eggs exposed to NH_4OH. *Proc. Natl. Acad. Sci. U.S.A.* **71**, 690–693.

Mazia, D., and Ruby, A. (1974). DNA synthesis turned on in unfertilized sea urchin eggs by treatment with NH_4OH. *Exp. Cell Res.* **85**, 167–172.

Mazia, D., Harris, P. J., and Bibring, T. (1960). The multiplicity of the mitotic centers and the time-course of their duplication and separation. *J. Biophys. Biochem. Cytol.* **7**, 1–20.

Miki-Noumura, T. (1977). Studies on the de novo formation of centrioles: Aster formation in the activated eggs of sea urchin. *J. Cell Sci.* **24**, 203–216.

Mitchison, J. M. (1971). "The Biology of the Cell Cycle." Cambridge Univ. Press, London and New York.

Moore, L., and Pastan, I. (1977a). Energy-dependent calcium uptake activity in cultured mouse fibroblast microsomes. *J. Biol. Chem.* **252**, 6304–6309.

Moore, L., and Pastan, I. (1977b). Regulation of intracellular calcium in chick embryo fibroblast: Calcium uptake by the microsomal fraction. *J. Cell Physiol.* **91**, 289–296.

Nelson, P. G., and Henkart, M. P. (1979). Oscillatory membrane potential changes in cells of mesenchymal origin: The role of an intracellular calcium regulating system. *J. Exp. Biol.* **81**, 49–61.

Osborn, M., and Weber, K. (1978). Cytoplasmic microtubules in tissue culture cells appear to grow from an organizing structure towards the plasma membrane. *Proc. Natl. Acad. Sci. U.S.A.* **73**, 867–871.

Parisi, E., Filosa, S., DePetrocellis, B., and Monroy, A. (1978). The pattern of cell division in the early development of the sea urchin *Paracentrotus lividus. Dev. Biol.* **65**, 38–49.

Parisi, E., Filosa, S., and Monroy, A. (1979). Actinomycin D: Disruption of the mitotic gradient in cleavage stages of the sea urchin embryo. *Dev. Biol.* **72**, 167–174.

Paweletz, N., and Mazia, D. (1979). Fine structure of the mitotic cycle of unfertilized eggs activated by ammoniacal sea water. *Eur. J. Cell Biol.* **20**, 37–44.

Petzelt, C. (1972a). Ca^{2+}-activated ATPase during the cell cycle of the sea urchin *Strongylocentrotus purpuratus. Exp. Cell Res.* **70**, 333–339.

Petzelt, C. (1972b). Further evidence that a Ca^{++}-activated ATPase is connected with the cell cycle. *Exp. Cell Res.* **74**, 156–162.

Pollard, T. D., and Weihing, R. R. (1974). Actin and myosin and cell movement. *Crit. Rev. Biochem.* **2**, 1–65.

Raff, E. C. (1979). The control of microtubule assembly *in vivo*. *Int. Rev. Cytol.* **59**, 1–96.

Rasmussen, H., and Goodman, B. P. (1977). Relationships between calcium and cyclic nucleotides in cell activation. *Physiol. Rev.* **57**, 421–509.

Rattner, J. B., and Phillips, S. G. (1973). Independence of centriole formation and DNA synthesis. *J. Cell Biol.* **57**, 359–372.

Robbins, E., Jentzsch, G., and Micali, A. (1968). The centriole cycle in synchronized HeLa cells. *J. Cell Biol.* **36**, 329–339.

Roberts, K., and Hyams, J. S., eds. (1979). "Microtubules." Academic Press, New York.

Roberts, K., and Northcote, D. H. (1970). The structure of sycamore callus cells during division in a partially synchronized suspension culture. *J. Cell Sci.* **6**, 299–321.

Robinson, K. R. (1979). Electrical currents through full-grown and maturing *Xenopus* oocytes. *Proc. Natl. Acad. Sci. U.S.A.* **76**, 837–841.

Roos, U.-P. (1973a). Light and electron microscopy of rat kangaroo cells in mitosis. I. Formation and breakdown of the mitotic apparatus. *Chromosoma* **40**, 43–82.

Roos, U.-P. (1973b) Light and electron microscopy of rat kangaroo cells in mitosis. II. Kinetochore structure and function. *Chromosoma* **41**, 195–220.

Rose, B., and Loewenstein, W. R. (1975). Calcium ion distribution in cytoplasm visualized by aequorin: Diffusion in the cytosol is restricted due to energized sequestering. *Science* **190**, 1204–1206.

Rusch, H. P., Sachsenmaier, W., Behrens, K., and Gruter, V. (1966). Synchronization of mitosis by the fusion of the plasmodium of *Physarum polycephalum*. *J. Cell Biol.* **31**, 204–209.

Sachsenmaier, W., and Hansen, K. (1973). Long- and short-period oscillations in a myomycete with synchronous nuclear divisions. *In* "Biological and Biochemical Oscillators" (B. Chance, E. K. Pye, A. K. Ghosh, and B. Hess, eds.), pp. 429–447. Academic Press, New York.

Sachsenmaier, W., Remy, U., and Plattner-Schobel, R. (1972). Initiation of synchronous mitosis in *Physarum polycephalum*. A model of the control of cell division in eukariots. *Exp. Cell Res.* **73**, 41–48.

Sanger, J. W., and Sanger, J. M. (1976). Actin localization during cell division. *In* "Cell Motility" (R. Goldman, T. Pollard, and J. Rosenbaum, eds.), pp. 1295–1316. Cold Spring Harbor Lab., Cold Spring Harbor, New York.

Scarpa, A., and Carafoli, E., eds. (1978). "Calcium Transport and Cell Function." *Ann. N.Y. Acad. Sci.* **307**. N.Y. Acad. Sci., New York.

Schatzmann, H. J., and Bürgin, H. (1978). Calcium in human red blood cells. *Ann. N.Y. Acad. Sci.* **309**, 125–147.

Schroeder, T. E. (1975). Dynamics of the contractile ring. *In* "Molecules and Cell Movement" (S. Inoué and R. E. Stephens, eds.), pp. 305–334. Raven, New York.

Schroeder, T. E., and Strickland, D. L. (1974). Ionophore A 23187, calcium, and contractility in frog eggs. *Exp. Cell Res.* **83**, 139–142.

Silver, R. B., Cole, R. D., and Cande, W. Z. (1980). Isolation of mitotic apparatus containing vesicles with calcium sequestering acitivity. *Cell* **19**, 505–516.

Sloboda, R. D., Dentler, W. L., Bloodgood, R. A., Telzer, B. R., Granett, S., and Rosenbaum, J. L. (1976). Microtubule associated proteins (MAPs) and the assembly of micrtubules *in vitro*. *In* "Cell Motility" (R. Goldman, T. Pollard, and J. Rosenbaum, eds.), pp. 1171–1212. Cold Spring Harbor Lab., Cold Spring Harbor, New York.

Soifer, D., ed. (1975). "The Biology of Cytoplasmic Microtubules." *Ann. N.Y. Acad. Sci.* **253.** N.Y. Acad. Sci., New York.

Steinhardt, R. A., and Epel, D. (1974). Activation of sea urchin eggs by a calcium ionophore. *Proc. Natl. Acad. Sci. U.S.A.* **71,** 1915–1919.

Stephens, R. E. (1972). Studies on the development of the sea urchin *Strongylocentrotus droebachiensis.* II. Regulation of mitotic spindle equilibrium by environmental temperature. *Biol. Bull.* **142,** 145–159.

Sugiyama, M. (1956). Physiological analysis of the cortical response of the sea urchin egg. *Exp. Cell Res.* **10,** 364–376.

Summers, K., and Kirschner, M. (1979). Characteristics of the polar assembly and disassembly of microtubules observed *in vitro* by darkfield light microscopy. *J. Cell Biol.* **83,** 205–217.

Swann, M. M. (1951a). Protoplasmic structure and mitosis. I. The birefringence of the metaphase spindle and asters of the living sea urchin egg. *J. Exp. Biol.* **28,** 417–433.

Swann, M. M. (1951b). Protoplasmic structure and mitosis. II. The nature and cause of birefringence changes in the sea urchin egg at anaphase. *J. Exp. Biol.* **28,** 434–444.

Telzer, B. R., and Haimo, L. T. (1981). Decoration of spindle microtubules with dynein: evidence for uniform polarity. *J. Cell Biol.* **89,** 373–378.

Telzer, B. R., Moses, M. J., and Rosenbaum, J. L. (1975). Assembly of microtubules onto kinetochores of isolated mitotic chromosomes of HeLa cells. *Proc. Natl. Acad. Sci. U.S.A.* **72,** 4023–4027.

Tucker, R. W., Pardee, A. B., and Fujiwara, K. (1979). Centriole ciliation is related to quiescence and DNA synthesis in 3T3 cells. *Cell* **17,** 527–535.

Tyler, A. (1941). Artificial parthenogenesis. *Biol. Rev. Camb. Philos. Soc.* **16,** 291–336.

Vacquier, V. D. (1975). The isolation of intact cortical granules from sea urchin eggs: Calcium ions trigger granule discharge. *Dev. Biol.* **43,** 62–74.

Wang, Y.-L., and Taylor, D. L. (1979). Distribution of fluorescently labeled actin in living sea urchin eggs during early development. *J. Cell Biol.* **82,** 672–679.

Weber, K., Bibring, T., and Osborn, M. (1975). Specific visualization of tubulin-containing structures in tissue culture cells by immunofluorescence. *Exp. Cell Res.* **95,** 111–120.

Weisenberg, R. C. (1972). Microtubules formation *in vitro* in solutions containing low calcium concentrations. *Science* **177,** 1104–1105.

Weisenberg, R. C. (1978). Assembly of sea urchin egg asters *in vitro. In* "Cell Reproduction: In Honor of Daniel Mazia" (E. R. Dirksen, D. M. Prescott, and D. F. Fox, eds.), pp. 359–366. Academic Press, New York.

Wick, S. M., and Hepler, P. K. (1980). Localization of Ca^{2+}-containing antimonate precipitates during mitosis. *J. Cell Biol.* **86,** 500–513.

Yoneda, M., Ikeda, M., and Washitani, S. (1978). Periodic change in the tension at the surface of activated non-nucleate fragments of sea urchin eggs. *Dev. Growth Differ.* **20,** 329–336.

Zucker, R. S., Steinhardt, R. A., and Winkler, M. M. (1978). Intracellular calcium release and the mechanisms of parthenogenetic activation of the sea urchin egg. *Dev. Biol.* **65,** 285–295.

3

The Movements of the Nuclei during Fertilization

GERALD SCHATTEN

I. FERTILIZATION: REQUIREMENTS, CONSEQUENCES, AND PROSPECTIVES

The fundamental significance of the fertilization process is the merging of the maternal and paternal genomes. For fertilization to be successful, the

59

MITOSIS/CYTOKINESIS
Copyright © 1981 by Academic Press, Inc.
All rights of reproduction in any form reserved.
ISBN 0-12-781240-7

sperm must first enter the egg cytoplasm, and then the sperm nucleus (male pronucleus) and the egg nucleus (female pronucleus) must move together. The contact between the pronuclei, and their subsequent fusion, signal the conclusion of fertilization by restoring the diploid state necessary for further development. In this chapter, the movements during fertilization and each associated system of motility will be traced from the first sperm–egg interactions through fusion, sperm incorporation, the pronuclear migrations, syngamy, and the establishment of the first cleavage axis. The results presented will be of sea urchins, and readers interested in other animals are directed to the reviews of Austin (1968) and Longo (1973) for various phyla and of Bedford (1970) for mammals.

The importance of the study of fertilization for understanding the processes during cell division is both in tracing the components of each gamete which ultimately participate in the first division and in the utility of fertilization as a model experimental system. Similar events occur during both fertilization and cell division, including a cycle of microtubule assembly and disassembly required for the proper functioning of the sperm aster and the mitotic apparatus, respectively. Cortical microfilaments are involved in both sperm incorporation and cytokinesis, and examples of plasma and nuclear membrane fusions occur, as do changes in the state of chromatin condensation. Unlike cell division, in which chromosome movement occurs almost simultaneously with pole separation and cytokinesis, the movements during fertilization are typically unidirectional and temporally well separated. Sperm incorporation occurs well before the centripetal migration of the male pronucleus and the later movement of the female pronucleus. The aim in this chapter is to review our understanding of the movements necessary for fertilization and to generate an appreciation for the relative simplicity of these events as models to explore cellular motility in general and cell division in particular.

II. SPERM INCORPORATION

The manner of sperm entry has been investigated by classical cytological methods (Wilson, 1925; Chambers, 1933), phase contrast microscopy (Dan, 1950), transmission electron microscopy (Longo and Anderson, 1968), scanning electron microscopy (Schatten and Mazia, 1976; Schatten and Schatten, 1980) and differential interference contrast microscopy (Hamaguchi and Hiramoto, 1980; Schatten, 1979, 1981b). Scanning electron microscopy of sperm incorporation, viewed from the surfaces of the vitelline layer, the plasma membrane, and the egg cortex, will be reviewed and the manner in which the sperm enters the egg, as observed with live material recorded by time-lapse video microsocpy, will be presented.

A. SEM Observations of the Vitelline Layer, Plasma Membrane, and Cortical Surfaces during Incorporation

Eggs were prepared in three different manners to study the entering sperm with scanning electron microscopy. Untreated eggs afforded a view of the sperm–vitelline layer interactions and the earliest stages of incorporation prior to the elevation of the fertilization coat. The plasma membrane surfaces of eggs at fertilization were investigated in denuded eggs. To observe the events occurring at the egg cortex, the egg surface was isolated immediately after insemination, permitting observations of the discharge of the successful sperm into the egg cytoplasm.

Observations at the vitelline layer during fertilization (Schatten and Mazia, 1976) demonstrate that sperm attach to the egg surface by the extended acrosomal process. Fusion occurs with the spermatozoon held in an erect position, and egg microvilli adjacent to the attached sperm elongate to form the fertilization cone (Fig. 1a). The sperm is next observed to undergo a rotation so that it lies on the egg surface. The elevation of the fertilization coat obliterates the sperm head and midpiece, as a portion of the sperm tail projects from the elevation coat.

To avoid the obstruction of the elevating fertilization coat, unfertilized eggs were denuded of their vitelline layers with dithiothreitol (DTT; Epel *et al.*, 1970). Scanning electron microscopy of the egg surface alterations following fertilization have been presented by Eddy and Shapiro (1976). Sperm incorporation of DTT-treated eggs (Schatten and Schatten, 1980) confirms the results obtained at the vitelline layer, i. e., sperm attach to the egg surface by the extended acrosomal process, and fusion occurs with the sperm held in an erect position. The activity at the egg plasma membrane inferred in Fig. 1a is clearly apparent in these naked eggs. Fig. 1b, an early stage of fusion, illustrates the elongation of adjacent microvilli during incorporation. These microvilli (Fig. 1c) extend up and around the sperm head, midpiece, and even the sperm tail to form the fertilization cone. During the later stages of sperm incorporation, the entire sperm enters the egg, leaving a small patch of presumably sperm-derived membrane in the egg plasma membrane. As will be discussed in Section IV, A, 1, DTT-treated eggs exposed to cytochalasin B are not able to extend the microvilli apparently necessary for incorporation.

To observe sperm incorporation at the cortical face of the egg surface, 30 seconds after insemination egg surfaces were isolated in 0.3 M KCl, 0.35 M

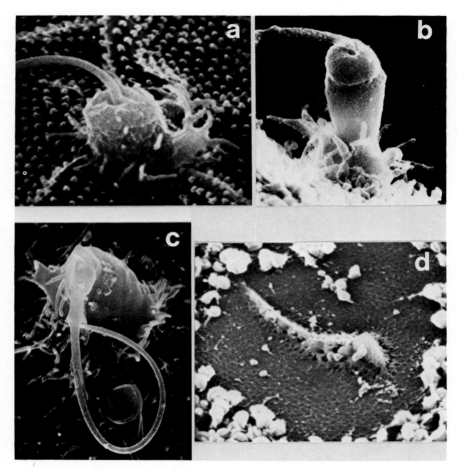

Fig. 1. Scanning electron microscopy of sperm incorporation. A. Egg microvilli project through the elevating vitelline layer as sperm–egg fusion proceeds. ×14,500. B. The clustering of egg microvilli around the successful sperm is quite apparent in eggs denuded of the vitelline layer. ×12,000. C. The forming fertilization cone with its elongating microvilli engulf the entire sperm. Short microvilli surround the sperm tail. ×8000. D. Cortical fibers cover the entering sperm in this image of an egg surface isolated seconds after insemination. ×5250. (a,d) *S. purpuratus.* Reprinted with permission from Schatten and Mazia (1976). (b.d.) *L. variegatus.* Reprinted with permission from Schatten and Schatten (1980).

glycine, 2 mM MgCl$_2$, 2 mM EGTA, pH 7.5 with NaHCO$_3$ and processed for scanning electron microscopy. Fig. 1d is an image of the inner face of the egg cortex, where the sperm can be observed during incorporation. The peripheral spherical granules are the undischarged cortical granules. The entering sperm head, midpiece, and tail lie on the egg surface and, importantly, are separated from the egg cytoplasm by a netting of cortical fibers. These fibers can be extracted with 0.6 M KI, and the extract contains a protein which comigrates with rabbit muscle actin during electrophoresis. The sequence inferred from these cortical images is that of the sperm rotating through the cortical netting to enter the egg cytoplasm, leaving a "scar" in the egg cortex through which it had penetrated.

B. Living Observations Recorded by Differential Interference, Time-Lapse Video Microscopy

While electron microscopy provides superb resolution, the investigator is forced to study fixed material and to attempt to order these static images into a reasonable sequence. To confirm the accuracy of these assembled sequences, time-lapse video recordings with differential interference optics have been employed using two types of microscopic chambers. In Fig. 2, a water immersion objective has been utilized to observe the earliest interactions between the gametes. In Fig. 3, the fertilized egg, which has been fixed to a polylysine-coated cover slip (Mazia *et al.*, 1975), is gently compressed on an oxygen-saturated fluorochemical layer; though the earliest sperm–egg interactions are not captured, the compression chamber provides remarkable clarity.

Shortly after insemination, sperm are found attached to the egg (Fig. 2B). Sperm adhesion to the egg surface is a rapid event, with the sperm sticking to the egg as a dart might to a target; bindin (Vacquier and Moy, 1977), an acrosomal adhesive protein, is presumably involved in this adhesion. Sperm gyrate about their attachment sites for varying times due to the beating of the sperm tail. Fusion (Fig. 2C) is characterized by the sudden immobilization of the erect sperm. A rapid contraction (see Table I for times and rates) radiates over the egg cortex and is followed by the slower elevation of the fertilization coat (Figs. 2D–F).

Sperm incorporation occurs in two stages: the formation of the fertilization cone and the gliding of the sperm along the egg cortex. The stationary sperm, held erect on the egg surface, is surrounded by egg microvilli (Fig. 1b) which elongate to form the fertilization cone. During this phase, the sperm head and midpiece do not move much from the fusion site; rather, the egg surface engulfs the sperm (Figs. 2D–H). Shortly afterward, the sperm head rotates from its erect position and the entire sperm glides along the subsurface region of the egg (Figs. 2H–L); this gliding motion results in a lateral displacement of the sperm head from the fusion site. The sperm tail begins to beat again (compare Figs. 2K and 2L), and this slow, erratic beating often brings the unincorporated portion of the tail into the "fertilization coat

Fig. 2. Water immersion optics. The acrosomal region of the spermotozoon attaches to the egg (A), and the sperm continues swimming (B) prior to fusion. As fusion occurs (C), the sperm is held erect and immotile, and the fertilization cone elevates from the fusion site. The sperm tail remains immotile as the fertilization cone forms to enlarge around the sperm head and midpiece (E–G). The sperm rotates as it penetrates the egg cytoplasm (H–L). The tail projects from the fertilization cone through the fully elevated fertilization coat (H–L). The

space" and later into the egg cytoplasm. The resumption of tail beating and the gliding along the egg cortex conclude sperm incorporation; later movements require the formation of the sperm aster.

III. PRONUCLEAR MOVEMENTS DURING FERTILIZATION

The earliest description of the movements of the pronuclei (reviewed by Wilson, 1925; Chambers, 1939; Allen, 1958) were so accurate that further information was available only after technical advances in microscopy. Transmission electron microscopy (Longo and Anderson, 1968) and, recently, differential interference microscopy (Hamaguchi and Hiramoto, 1980; Schatten, 1979, 1981a) and antitubulin immunofluorescence microscopy (Bestor and Schatten, 1980, 1981; Harris et al., 1979, 1980a, b) have added details to the process culminating in syngamy.

A. Living Records

The formation and enlargement of the sperm aster are central to the migrations of the pronuclei. This monastral structure forms at the base of the sperm head shortly after the incorporation of the sperm. The growth of the sperm aster moves the male pronucleus centripetally at a rate of 4.9 μm/min. (Figs. 2N–R); the chromatin of the male pronucleus decondenses as the sperm aster enlarges. The centripetal movement of the sperm aster is intercepted by the swift migration of the female pronucleus to the center of the sperm aster (Figs. 2S–Y, 3e–m). The migration of the female pronucleus occurs at a rate of 14.6 μm/min and is the fastest of the pronuclear movements. This migration begins after the sperm astral rays are observed to contact the female pronucleus (Fig. 3c). During the migration, the female pronucleus is distorted into an oblate spheriod (Fig. 3d). The now adjacent pronuclei move to the egg center by the continuing enlargement of the fibers of the sperm aster (Figs. 2EE–II). The centration of the pronuclei occurs at an average rate of 2.6 μm/min. Pronuclear fusion typically occurs at the egg center shortly after the sperm aster has been disassembled.

resumption of tail beating appears involved in moving the tail into the egg (M). The fertilization cone continues to elongate (M-V) as the sperm aster forms (P-U). The female pronucleus migrates along the fibers of the sperm aster to contact the male pronucleus (S-Z). The adjacent pronuclei are moved to the egg center by the continued growth of the sperm aster (AA-HH). The centers of the sperm aster, the sperm-derived centrioles, separate during pronuclear centration (EE-HH). The sperm aster disassembles (MM) prior to the formation of the interim apparatus (NN,OO). The mitotic apparatus forms (QQ) with an axis similar to the direction of centriole separation during centration. Cleavage is completed in UU. ×450. L. variegatus. Reprinted with permission from Schatten (1981b).

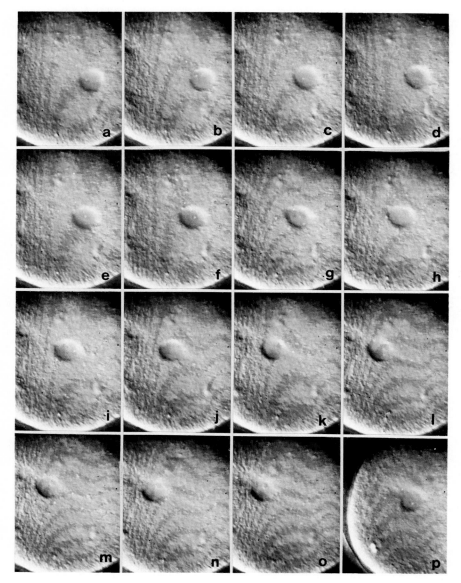

Fig. 3. Migration of the female pronucleus. The elongation of the sperm astral microtubules (a-c) establishes the initial contact between the pronuclei (d). The onset of movement is erratic, and it increases in velocity during the migration (e-j). The female pronucleus appears to slide along the sperm astral rays (g-i), and it is distorted into an oblate spheroid while moving. The fibers along which the female pronucleus moved appear to persist even

TABLE I

Movements during Fertilization[a]

A. Sperm incorporation	
0 min	Fusion between acrosome-reacted sperm and egg
0.25 min	Duration of egg cortical contraction (rate: 5.9 μm/sec)
0.5 min	Formation of fertilization cone (elongation rate: 2.6 μm/min)
1.0 min	Gliding of sperm along egg cortex (rate: 3.5 μm/min; average distance 12.4 μm) and resumption of sperm tail beating in egg cytoplasm
B. Formation of sperm aster	
4.4 min	Assembly of microtubules to form sperm aster
−6.8 min	Average rate: 4.9 μm/min
	Average distance traversed: 14.3 μm
C. Migration of female pronucleus	
6.8 min	Movement of female pronucleus to center of sperm aster
−7.8 min	Average rate: 14.6 μm/min
	Average distance traversed: 19.1 μm
D. Pronuclear centration	
7.8–14.1 min	Movement of adjacent pronuclei to egg center
	Average rate: 2.6 μm/min
	Average distance traversed: 12.3 μm
E. Syngamy	
14.7–15.2 min	Pronuclear fusion; male pronucleus coalesces into female pronucleus Rate: 14.2 μm/min

[a] Data from *L. variegatus* at 23°C.

B. Indirect Immunofluorescence with Tubulin Antibody

Since the studies of Zimmerman and Zimmerman (1967) demonstrated that colcemid will prevent syngamy, microtubules have been inferred to be of importance during the pronuclear migrations. To characterize the presence and especially the geometric configuration of the microtubules throughout fertilization, indirect immunofluorescence microscopy using column-purified pig brain tubulin antibody has been performed (Bestor and Schatten, 1980, 1981; Harris *et al.*, 1979, 1980a, b).

The unfertilized egg has no formed microtubules though the female pronucleus has numerous tubulin-staining punctate sources that are inferred to be "tubulin-containing structures" (Figs. 4a, c; Weisenberg, 1975). During sperm incorporation, the microtubules of the sperm axoneme are clearly apparent, and these are the only microtubules observable during this stage

after the migration (i–k). At the conclusion of the migration, the female pronucleus again assumes a spherical form (m). The continuing elongation of the sperm aster moves the adjacent pronuclei to the egg center (p). ×1000. *L. variegatus.* Reprinted with permission from Schatten (1981a).

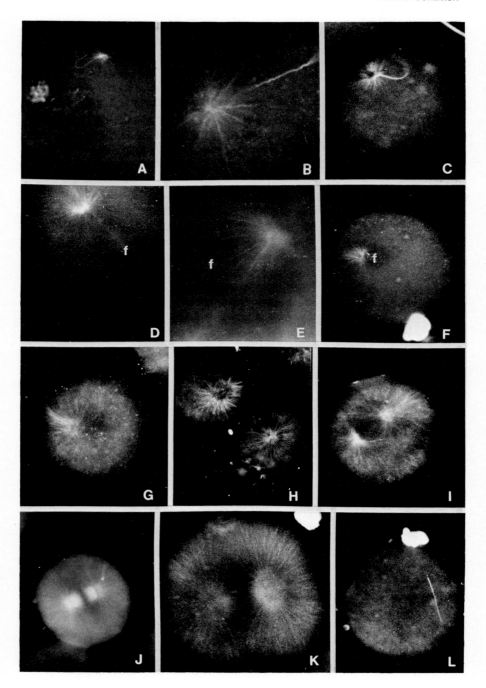

(Figs. 4a, c). The formation of the sperm aster results from microtubule polymerization at the base of the sperm head; the centrioles are the likely centers (Figs. 4b–c). The extension of these microtubules is closely coupled with the centripetal migration of the male pronucleus (Fig. 4d, e). Following contact of the female pronucleus by a few sperm astral microtubules, the migration of the female pronucleus occurs (Figs. 4f, g). Pronuclear centration is accomplished by the continued elongation of the astral microtubules, which are predominantly oriented between the egg cortex and the adjacent pronuclei (Figs. 4g–i). The sperm aster at this stage has two focal points, perhaps an indication of centriole separation (Figs. 4h, i). The sperm axoneme often remains associated with one of the centrioles throughout pronuclear migration and early development.

C. Microtubule Assembly Cycles

There are three cycles of microtubule assembly and disassembly during the first cell cycle and apparently two cycles during subsequent cell cycles, as observed by antitubulin immunofluorescence and differential interference microscopy. The first cell cycle is characterized by the formation and regression of the sperm aster, the interim apparatus, and the mitotic apparatus; the later cell cycles include the assembly and disassembly of only the interim apparatus and the mitotic apparatus.

Following the assembly of the sperm aster, with the associated pronuclear migrations, the sperm aster is disassembled (e. g., Fig. 2HH–JJ). Pronuclear fusion typically occurs after this disassembly. Toward the end of interphase, a second cycle of microtubule assembly and disassembly occurs; the formation and regression of the interim apparatus (Fig. 2LL, 4j). This structure has

---

Fig. 4. Indirect immunofluorescence with tubulin antibody. Following sperm incorporation, the sperm axoneme and assembling microtubules of the sperm aster are apparent (a). The punctate tubulin-staining sources associated with the female pronucleus (a,c) are the only fluorescent sources in the unfertilized egg and can be used as a marker for the location of the female pronucleus. The microtubule-organizing center for the sperm aster is located at the base of the sperm axoneme, near the sperm centrioles (b). The elongation of the microtubules comprising the sperm aster (b–d) results in the centripetal movement of the male pronucleus. Microtubules contact the female pronucleus (white f in Figs. d, e, f), and the swift migration of the female pronucleus ensues (f). The adjacent pronuclei are moved to the egg center by the continued elongation of the sperm astral microtubules (g). During centration, the sperm aster appears to have two focal points (h), perhaps an indication of the separation of the sperm centrioles. Following the disassembly of the sperm aster, the interim apparatus or "streak" forms (i). The interim apparatus is typically disassembled prior to the formation of the mitotic apparatus (j, k). Figure I is of a colchicine-treated egg, demonstrating that although the sperm aster will not form in the presence of this drug, the sperm axoneme is incorporated normally. All figures of *A. punctulata* except d and k, which are *L. variegatus*. a, c, f-j, I, k: 400 ×; d: 800 ×; b, e: 1,000×. Reprinted with permission from Bestor and Schatten (1981).

been described by Wilson (1925), and a variety of terms including "streak" and "interphase asters" (Harris *et al.*, 1980a) have been employed. The term "interim apparatus" is proposed here to describe the transitory nature and bipolarity of this structure. Possible roles for the interim apparatus might include the final positioning of the centrioles from the nuclear surface into the cytoplasm (see Section III, D), preparation of the nucleus for subsequent mitotic events, or even the recruitment of the available tubulin monomers which will be required for division. The final cycle of microtubule assembly and disassembly occurs during division (Figures 2QQ, RR, 4k).

D. Establishment of the First Cleavage Axis

The establishment of the first cleavage axis has been a problem of classical interest, and early investigations resulted in opposing views. Wilson and Mathews (1895), studying slightly compressed eggs, found cleavage to occur within 15° of the egg radius subtended through the sperm entry site. In contrast, Hörstadius (1928), investigating the pigmented *Paracentrotus lividus* egg, found that the sperm entry site had no bearing on the first cleavage axis. To determine whether the first embryonic axis is specified in the unfertilized egg or is established during sperm incorporation and the pronuclear migrations, time-lapse recordings of early development were analyzed (Schatten, 1981b).

In support of Wilson and Mathew's observations (1895), cleavage occurs within 13.5° of the egg radius subtended through the sperm entry site in compressed eggs; this angle increases to 32.6° in spherical eggs. Remarkably, though, the angle between cleavage and pronuclear centration is 5.8° in compressed eggs and 7.9° in spherical eggs. The antiparallel separation of the centrioles prior to syngamy, which occurs at right angles to the direction of pronuclear centration, appears responsible for the establishment of the axis of first division. The interim and mitotic apparatus are oriented parallel to the direction of centriolar separation; cleavage therefore is perpendicular and often similar to the egg radius subtended through the sperm entry site.

IV. MOTILITY DURING FERTILIZATION

A. Inhibitor Studies

Zimmerman and Zimmerman (1967) pioneered the study of aspects of fertilization using specific motility inhibitors with their now classical experiments demonstrating the dependence of pronuclear fusion on microtubule function by the use of colcemid. Aronson (1973), also using colcemid, Schatten (1977), using griseofulvin, and Schatten *et al.*, (1980), using the above compounds as well as maytansine, nocodazole, taxol, and vinblastine, have confirmed the requirement for functioning microtubules during the pronuclear migrations. The microfilament inhibitor cytochalasin B has been employed by Gould-Somero *et al.* (1977), Longo (1978, 1980), Byrd and Perry (1980), and Schatten and Schatten (1980), and it appears clear that cytochalasin B will prevent sperm incorporation, although sperm-induced egg activation occurs. The use of video microscopy in determining the precise effects of specific inhibitors has been crucial. Using the display of the video record on the monitor, one can get an instantaneous playback of results after one adds an inhibitor to the studied cell at any specified stage during the fertilization process. The results of these investigations are presented in Table II.

TABLE II

Effects of Motility Inhibitors during Fertilization[a]

Microfilament inhibitors	
Assembly inhibitors	
Cytochalasin D (10 μM)	
−15 min	Sperm incorporation blocked, no fertilization cone formed
At insemination	Sperm incorporation blocked, no fertilization cone formed
+15 sec	Formed fertilization cone resorbed: normal pronuclear migrations; cytokinesis blocked
Cytochalasin E (10 μM)	
−15 min	Sperm incorporation blocked; no fertilization cone formed
At insemination	Sperm incorporation blocked; no fertilization cone formed
+15 sec	Formed fertilization cone resorbed; normal pronuclear migrations; cytokinesis blocked
Cytochalasin B (25 μM)	
−15 min	Sperm incorporation blocked; no fertilization cone formed
At insemination	Sperm incorporation blocked; no fertilization cone formed
+15 sec	Formed fertilization cone resorbed; normal pronuclear migrations; cytokinesis blocked
Disassembly inhibitor	
Phalloidin (1 mM)	
−60 min	Huge fertilization cone formed; normal incorporation and pronuclear migrations; mitotic pole separation and cytokinesis arrested

(continued)

TABLE II (*Continued*)

Microtubule inhibitors	
Assembly inhibitors	
Colcemid (500 nm)	
−15 min	Normal incorporation; aster formation blocked
At insemination	Normal incorporation; sperm aster formation attempted and then regression
Colchicine (100 μM)	
−15 min	Normal incorporation; sperm aster formation blocked
At insemination	Normal incorporation; aster formation attempted, and then regression
Griseofulvin (50 μM)	
−15 min	Normal incorporation; unusually pronounced sperm gliding along cortex; aster formation blocked
At insemination	Normal incorporation; unusually pronounced sperm gliding along cortex; aster formation blocked
+5 min	Rapid resorption of sperm aster; migration of female pronucleus blocked
+7 min	Resorption of sperm aster; arrest of migrating female pronucleus
+10 min	Centration of pronuclei arrested
+14 min	Fusion of centered pronuclei arrested
Maytansine (50 nm)	
−15 min	Normal incorporation; pronounced sperm gliding; sperm aster arrested
At insemination	Normal incorporation; sperm aster formation attempted and then regression
+5 min	Lag of 3 min prior to regression of sperm aster; pronuclear centration blocked
10 nM @ −15 min	Incomplete centration; diminutive sperm aster
Nocodazole (50 nM)	
−5 min	Normal incorporation; sperm aster formation blocked
At insemination	Normal incorporation; sperm aster formation attempted and then regression
+5 min	Lag of 3 min prior to regression of sperm aster; pronuclear centration blocked
Podophyllotoxin (10 nM)	
−15 min	Normal incorporation; unusually pronounced sperm gliding along cortex; aster formation blocked
At insemination	Normal incorporation; sperm aster formation attempted and then regression
+10 min	Normal female pronuclear migration and centration; mitosis blocked

(*continued*)

TABLE II (*Continued*)

Vinblastine (1 μM)	
−15 min	Normal incorporation; unusually pronounced sperm gliding along cortex; aster formation blocked
At insemination	Normal incorporation; unusually pronounced sperm gliding along cortex; aster formation blocked
+5 min	Rapid resorption of sperm aster; migration of female pronucleus blocked
Photochemical reversal of assembly inhibitors	
Colchicine (100 μM)	
366 nm light at:	
−5 min	Normal incorporation; pronuclear movements and development
+3 min	Normal incorporation; pronuclear movements and development
+15 min	Pronuclear movements and syngamy arrested; two cytasters at mitosis associated with male pronucleus
Disassembly inhibitor	
Taxol (10 μM)	
−60 min	*De novo* aster formation
−30 min	Normal incorporation and cortical reaction; formation of huge persisting sperm aster; migration of female pronucleus inhibited
+15 min	Formation of huge persisting sperm aster after migration of female pronucleus
At insemination	Formation of huge persisting sperm aster after migration of female pronucleus
+15 min	Formation of enlarged mitotic apparatus; cleavage unsuccessful
+45 min	Formation of enlarged mitotic apparatus; cleavage unsuccessful

[a] Data from *L. variegatus*. Zero time = sperm–egg fusion.

1. Microfilament Inhibitors

Cytochalasin B, D, and E, inhibitors of microfilament elongation (Flanagan and Lin, 1980), and phalloidin, which inhibits the disassembly of formed microfilaments (Wieland, 1977; Wehland *et al.*, 1978), have been studied during fertilization. (The effect of cytochalasin D is greater than that of E, which is greater than that of B.) Cytochalasin B, D, and E prevent sperm incorporation and the formation of the fertilization cone. The cortical contraction and the secretion of the cortical granules occur, though slowly, and the egg surface develops an aberrant appearance within 45 minutes. Remarkably, the female pronucleus attempts to migrate in these treated eggs;

several minutes after fusion, the female pronucleus undergoes sudden and erratic motions, perhaps due to alterations in the cytoplasmic viscosity.

If cytochalasin B, D, or E is added 15 seconds after sperm–egg fusion is observed, the forming fertilization cone is rapidly (within 30 seconds) resorbed and the gliding of the sperm along the egg cortex is inhibited. However, formation of the sperm aster, migration of the female pronucleus, pronuclear centration, and syngamy proceed properly. Cytokinesis is inhibited. From these experiments, it is concluded that assembly of microfilaments is required for sperm incorporation. but not for the later pronuclear migrations (Schatten and Schatten, 1981).

Eggs incubated in phalloidine, which may not rapidly permeate the egg, produce huge, long-lasting fertilization cones at insemination. Sperm incorporation occurs normally and development proceeds properly until the time of cell division. Although the mitotic apparatus forms, the separation of the mitotic poles and the formation of the contractile ring do not occur. This may imply that while sperm incorporation does not require the sliding of microfilaments, the functioning of the contractile ring might. Conceivably, the separation of the poles during mitosis, which is involved in the separation of the fusing karyomeres, might possibly depend on the contractile ring.

2. Microtubule Inhibitors

As is apparent in Table II, inhibitors of microtubule assembly do not prevent sperm incorporation; the fertilization cone and fertilization coat elevate normally. Somewhat surprisingly, though, the gliding of the sperm along the egg cortex is far more pronounced in these treated eggs. This may well indicate that the gliding movement is terminated when the microtubules of the sperm aster assemble; this assembly might move the male pronucleus from the egg cortex into the cytoplasm as well as increase the drag of the male pronuclear complex to complete the incorporation sequence (Fig. 5i).

Colcemid, colchicine, griseofulvin, maytansine, nocodazole, podophyllotoxin, and vinblastine all prevent the formation of the sperm aster; each haploid nucleus undergoes the later cycles of chromatin condensation and decondensation, though syngamy never occurs. The migration of the female pronucleus is also dependent on microtubule assembly as determined by the use of griseofulvin and vinblastine. When added after the formation of the sperm aster, or even as the female pronucleus begins to migrate, this migration is terminated. Colcemid, colchicine, nocodazole, and maytansine require several minutes to permeate the egg, and these rapid addition experiments are not possible. The centration of the pronuclei is similarly dependent on microtubule assembly. Interestingly, when subthreshold concentrations of these inhibitors are employed, the pronuclear centration is terminated due to the inability fully to elongate the microtubules of the sperm aster.

To test if sperm incorporation is truly independent of microtubules, unfertilized eggs were incubated in colchicine, washed free of the excess drug, and inseminated. At 3 minutes postfusion, the colchicine was photochemically inactivated with light at 366 nm (Aronson and Inoué, 1970) and the normal sequence of pronuclear movements and later development ensued.

Indirect immunofluorescence of colcemid- and griseofulvin-inhibited cells reveals the presence of the sperm axoneme which entered during incorporation. The assembly of microtubules which form the sperm aster cannot occur with these drugs, as illustrated in Fig. 4l, a colchicine-treated egg.

B. Sperm Incorporation Requires Cortical Microfilaments

It has been proposed that microfilaments in the egg cortex may be active during sperm incorporation (Schatten and Mazia, 1976; Epel, 1978; Longo, 1980; Byrd and Perry, 1980; Schatten and Schatten, 1980), and the accumulating evidence supports this supposition. The localization of actin filaments in the egg cortex during fertilization (Burgess and Schroeder, 1977; Begg et al., 1978; Otto et al., 1980; Spudich and Spudich, 1979) and the finding that cytochalasin B interferes with sperm incorporation by preventing the formation of the fertilization cone (see Section II, A) argue for such a role. The initial phase of sperm incorporation, the enlargement of the fertilization cone up and around the sperm head, is likely to be the result of egg actin filaments forming and sliding along the actin fibers in the acrosomal process (Figs. 5d–f). The polarity of the actin in the acrosomal process (Tilney and Kallenbach, 1979) would permit actin filament sliding up and around the sperm head, but not in the opposite direction. Though cytochalasin B has been shown to prevent sperm incorporation, it is of interest that the bioelectric responses at fertilization occur normally with this inhibitor (Hülser and Schatten, 1980; Dale and DeSantis, 1981). The formation of actin filaments in the egg may well be responsible for sperm incorporation but does not appear necessary for gamete membrane fusion and the accompanying egg activation.

The second phase of sperm incorporation, the gliding of the sperm along the egg cortex, occurs after the formation of the fertilization cone. Perhaps the microfilaments in the acrosomal process slide along the newly formed microfilaments in the egg cortex to affect this gliding movement (Figs. 5g–i). Alternatively, the beating of the sperm tail within the egg cytoplasm might well be responsible for this motion.

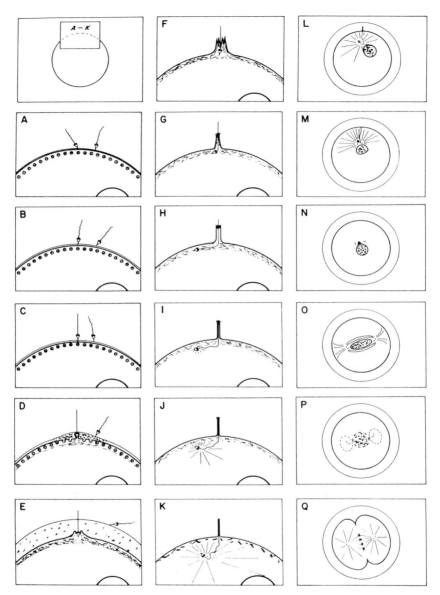

Fig. 5. The movements during fertilization. Sperm attach to the egg surface (a) and gyrate about their attachment sites (b) for varying times prior to fusion (c). Following a rapid cortical contraction radiating from the fusion site, the fertilization coat elevates (d-f). Sperm incorporation is characterized by the formation of the fertilization cone around the erect and stationary sperm; the sperm tail is immotile at this stage (d-f). The sperm glides along the

C. Pronuclear Centration Mediated by Microtubule Polymerization

The cytoplasmic movements of the pronuclei appear to be mediated by the formation of the microtubules around the sperm-derived centrioles. The evidence for the involvement of the microtubules during the pronuclear movements includes studies with microtubule inhibitors (Zimmerman and Zimmerman, 1967; Aronson, 1973; Schatten, 1977; Schatten et al., 1980) and direct observations of microtubules associated with the sperm aster both from transmission electron microscopy (Longo and Anderson, 1968) and indirect immunofluorescence with tubulin antibody (Harris et al., 1979, 1980a, b; Bestor and Schatten, 1980); microfilament inhibitors do not prevent the pronuclear movements (Schatten et al., 1980). The assembly of microtubules into the sperm aster results in the centripetal movement of the male pronucleus (Fig. 5j). The microtubules continue to elongate when they contact the female pronucleus (Fig. 5k). Due to the rapid migration of the female pronucleus, it is tempting to speculate that a mechanism other than microtubule assembly is responsible for the force generation (see Section IV, D). The adjacent pronuclei continue the centripetal migration due to the further extension of the sperm astral fibers (Fig. 5m). The idea that microtubule assembly itself directly results in the centripetal movements of the sperm aster and the later centration of the pronuclei is based on the following evidence: the rate of microtubule elongation, as observed by differential interference optics and antitubulin immunofluorescence, correlates precisely with the centrating motions, and inhibitors of microtubule assembly prevent both the formation of the sperm aster and the centripetal migrations.

D. Migration of the Female Pronucleus: Dynamic Equilibrium/Sliding Microtubules

The migration of the female pronucleus to the male pronucleus occurs at a rate three times that of the migration of the male pronucleus and six times that of

egg cortex during penetration (g–h). The formation of the sperm aster moves the male pronucleus centripetally (i–j). The migration of the female pronucleus occurs when the fibers of the sperm aster interconnect the pronuclei (k, l). The adjacent pronuclei are moved to the egg center by the continuing elongation of the sperm aster (m); the centrioles separate during this motion, and the sperm aster has two focal points. Syngamy typically occurs at the egg center after the disassembly of the sperm aster (n). The interim apparatus forms around and distorts the zygote nucleus (o). The axis of the interim apparatus (o) is usually identical to the direction of the sperm centriole separation (m) as well as the axis for the mitotic apparatus (q). The interim apparatus is disassembled prior to the nuclear breakdown at prophase (p). Cleavage is perpendicular to the axis of the mitotic apparatus and is usually parellel to the egg radius subtended through the sperm entry site (q). Reprinted with permission from Schatten (1981b).

the centering movement of the pronuclei. Though this migration is clearly dependent on assembled microtubules (see Table II), it is also inhibited by taxol, which stabilizes microtubules. This result argues that microtubule disassembly as well as microtubule assembly is required for the observed motion; if so, this is an example of motility in support of the dynamic equilibrium models proposed to account for chromosome movements (Inoué and Sato, 1967; Margolis et al., 1978). Should a dynamic equilibrium between microtubule assembly and disassembly be active during the migration of the female pronucleus, it is predicted that the attachment of the female pronuclear surface to the ends of the sperm aster microtubules, and the subsequent flow of the tubulin monomers through the interconnecting microtubules to the sperm aster center, would result in the observed motion.

Another viable model for microtubule motility which might well be active during the migration of the female pronucleus is that of dynein arms sliding along assembled microtubules, as in ciliary movements (Gibbons, 1977). Were this model to account for the female pronuclear migration, it is predicted that the pronuclear surface would possess dynein. Then the interaction between the sperm aster microtubules and the dynein on the female pronucleus would result in the observed motion. It should be noted, though, that it is not possible at this time to support or reject any model of microtubule motility.

E. Ionic Controls of Motility during Fertilization

The understanding of the ionic signals responsible for the program of egg activation (Epel, 1978) permits a correlation between the observed structural modifications and the altered ionic environment. Schackman et al. (1978) and Tilney et al. (1978) have implicated the influx of $[Ca^{2+}]$ and the efflux of $[H^+]$ during the acrosome reaction of the sperm. Perhaps the same ions regulate the assembly of the motile macromolecules active in the egg during fertilization, e.g., $[H^+]$ might control microfilament polymerization and $[Ca^{2+}]$ might control the assembly, and functioning of microtubules and of microfilaments. The regulation of motility during fertilization may prove to be a most promising model for the study of the factors controlling the cytoskeleton; candidates as likely regulatory agents, in addition to protons and calcium ions, include cyclic nucleotides, and their associated enzymes, and calmodulin. It is of interest that neither cytochalasin D nor colcemid affects the bioelectric responses at fertilization (Hülser and Schatten, 1980). Interestingly, Begg and Rebhun (1979) have shown that egg cortical actin assembly is pH dependent. Steinhardt et al. (1977) have demonstrated that the transient increase in cytoplasmic calcium following fertilization decreases at about 3 minutes after insemination. Since microtubules assemble in low-calcium environments (Borisy et al., 1972; Weisenberg, 1972), the cytoplasmic environment at 4 minutes would favor assembly; this is consistent with the finding that the sperm aster forms 4.4 minutes after sperm–egg fusion. Consistent with the results of Gibbons and Gibbons (1979) on isolated sperm tails, the incorporated sperm tail is quiescent during the burst in intracellular $[Ca^{2+}]$ following fusion. It is predicted that the cytoplasmic calcium concentration would remain at a relatively low level throughout the time of the pronuclear movements, until about 15 minutes when the sperm aster undergoes disassembly and the membranes of the pronuclear envelopes fuse at syngamy.

V. SUMMARY

The events during fertilization which lead to syngamy begin with the extension of the acrosomal process of the sperm, which might well occur near or even at the egg surface (Aketa and Ohta, 1977). The swimming spermatozoon rapidly adheres to the egg surface (Fig. 5a) and gyrates about its attachment site until fusion (Fig. 5b), which is characterized by the sudden immobilization of the erect sperm (Fig. 5c). The influx of $[Ca^{2+}]$ and the efflux of $[H^+]$ during the acrosome reaction may well be involved in triggering the rapid cortical contraction, the cortical reaction, and engulfment of the sperm by the fertilization cone following fusion. Sperm incorporation involves the assembly of microfilaments to form the fertilization cone which surround the entire sperm (Figs. 5d–f), and perhaps the sliding of the microfilaments of the acrosome along the newly formed ones in the egg cortex to effect the gliding of the sperm along the subsurface region of the egg (Figs. 5g–h). The elevating pH of the cytoplasm might regulate the formation of the fertilization cone.

The assembly of microtubules to form the sperm aster moves the male pronucleus centripetally and may well be responsible for terminating the lateral displacement of the sperm along the egg cortex (Figs. 5i–j). The prevailing $[Ca^{2+}]$ conditions at this stage should favor microtubule assembly. The elongating sperm astral microtubules contact the female pronucleus (Fig. 5k), and by a mechanism requiring both microtubule assembly and disassembly, the female pronucleus migrates to the center of the sperm aster (Fig. 5l). The continuing elongation of the sperm astral microtubules moves the now adjacent pronuclei to the egg center (Fig. 5m).

The pair of sperm-derived centrioles, which serve as the center for the sperm aster, separate during pronuclear centration. The sperm aster regresses prior to pronuclear fusion (Fig. 5n), perhaps due to a requirement for elevated $[Ca^{2+}]$ for the fusion of the nuclear membranes. The centrioles move to opposing poles on the zygote nucleus and serve to specify the first cleavage axis. The interim apparatus (streak) is assembled and disassembled prior to prophase (Fig. 5o, p); the axis of the interim apparatus is usually identical to that of the mitotic apparatus. Clear zones form at the regions which will be the mitotic poles during prophase (Fig. 5p), and the mitotic apparatus soon forms which leads to cytokinesis (Fig. 5q) and the completion of the first cell cycle.

ACKNOWLEDGMENTS

It is a pleasure to acknowledge the contributions of Dr. Heide Schatten, Dr. Timothy Bestor, and Dr. Dieter Hülser, the technical assistance of Mr. Calvin Simerly, Mr. Richard Roche for the illustration, and the generous donations of pharmaceuticals from Dr. J. Douros (Developmental Therapeutic Program, National Cancer Institute, Bethesda), Professor Th. Wieland (Max-Planck-Institut, Heidelberg), and Janssen Pharmaceutica (Beerse). The support of these investigations by the National Institutes of Health (Research Grant HD 12913; Research Career Development Award HD00363) is gratefully acknowledged.

REFERENCES

Aketa, K., and Ohta, J. (1977). When do sperm of the sea urchin *Pseudocentrotus* depressus undergo the acrosome reaction at fertilization? *Dev. Biol.* **61**, 366–372.

Allen, R. D. (1958). The initiation of development. *In* "The Chemical Basis of Development" (W. O. McElroy and B. Glass, eds.), Johns Hopkins Press, Baltimore, Maryland.

Aronson, J. F. (1973). Nuclear membrane fusion in fertilized *Lytechinus variegatus* eggs. *J. Cell Biol.* **58**, 126–134.

Aronson, J., and Inoué, S. (1970). Reversal by light of the action of N-methyl N-desacetyl colchicine on mitosis. *J. Cell Biol.* **45**, 470–477.

Austin, C. R. (1968). "The Ultrastructure of Fertilization." Holt, New York.

Bedford, J. M. (1970). Saga of mammalian sperm from ejaculation to syngamy. *In* "Mammalian Reproduction" (H. Gibian and E. J. Plotz, eds.), pp. 124–182. Springer-Verlag, Berlin and New York.

Begg, D. A., and Rebhun, I. I. (1979). pH regulates the polymerization of actin in the sea urchin egg cortex. *J. Cell Biol.* **79**, 846–852.

Begg, D. A., Rodewald, R., and Rebhun, L. I. (1978). The visualization of actin filament polarity. *J. Cell Biol.* **79**, 846–852.

Bestor, T. H., and Schatten, G. (1980). Immunofluorescence microscopy of microtubules involved in the pronuclear movements of sea urchin fertilization. *J. Cell Biol.* **87**, 136.

Bestor, T. H., and Schatten, G. (1981). Anti-tubulin immunofluorescence microscopy of microtubules present during the pronuclear movements of sea urchin fertilization. *Dev. Biol.* **87**, in press.

Borisy, G. G., and Olmsted, J. B. (1972). Nucleated assembly of microtubules in porcine brain extracts. *Science* **177**, 1196–1197.

Burgess, D. R., and Schroeder, T. E. (1977). Polarized bundles of actin filaments within microvilli of fertilized sea urchin eggs. *J. Cell Biol.* **74**, 1032–1037.

Byrd, W., and Perry, G. (1980). Cytochalasin B blocks sperm incorporation but allows activation of the sea urchin egg. *Exp. Cell Res.* **126**, 333–342.

Chambers, E. L. (1939). The movement of the egg nucleus in relation to the sperm aster in the echinoderm egg. *J. Exp. Biol.* **16**, 409–424.

Chambers, R. (1933). The manner of sperm entry in various marine ova. *J. Exp. Biol.* **10**, 130–141.

Dale, B., and DeSantis, A. (1981). The effects of cytochalasins B and D on the fertilization of sea urchins. *Dev. Biol.* **83**, 232–237.

Dan, J. C. (1950). Sperm entrance in echinoderms observed with phase contrast microscopy. *Biol. Bull.* (Woods Hole, Mass.) **99**, 399–411.

Eddy, E. M., and Shapiro, B. M. (1976). Changes in the topography of the sea urchin egg after fertilization. *J. Cell Biol.* **71**, 35–48.

Epel, D. (1978). Mechanisms of activation of sperm and egg during fertilization of sea urchin gametes. *Curr. Top. Dev. Biol.* **12**, 186–246.

Epel, D., Weaver, A. M., and Mazia, D. (1970). Methods for removal of the vitelline membrane of sea urchin eggs. I. Use of dithiothreitol (Cleland's Reagent). *Exp. Cell Res.* **61**, 64–68.

Flanagan, M. D., and Lin, S. (1980). Cytochalasins block actin filament elongation by binding to high affinity sites associated with F-actin. *J. Biol. Chem.* **255**, 835–838.

Gibbons, B. H., and Gibbons, I. R. (1979). Calcium-induced quiescence in reactivated sea urchin sperm. *J. Cell Biol.* **83**, 806a.

Gibbons, I. R. (1977). Structure and function of flagellar microtubules. In "International Cell Biology" (B. R. Brinkley and K. R. Porter, eds.), 348–357. Rockefeller Univ. Press, New York.

Gould-Somero, M., Holland, L., and Paul, M. (1977). Cytochalasin B inhibits sperm penetration into eggs of *Urechis caupo*. *Dev. Biol.* **58**, 11–22.

Hamaguchi, M. S., and Hiramoto, Y. (1980). Fertilization process in the heart urchin (*Clypeaster japonicus*, observed with differential interference microscopy. *Dev. Growth Diff.* **22**, 517–530.

Harris, P., Osborn, M., and Weber, K. (1979). A spiral array of microtubules in the fertilized egg cortex. *J. Cell. Biol.* **83**, 1813.

Harris, P., Osborn, M., and Weber, K. (1980a). Distribution of tubulin containing structures in the egg of the sea urchin *Strongylocentrotus purpuratus* from fertilization through first cleavage. *J. Cell Biol.* **84**, 668–679.

Harris, P., Osborn, M., and Weber, K. (1980b). A spiral array of microtubules in the fertilized sea urchin egg cortex examined by indirect immunofluorescence and electron microscopy. *Exp. Cell. Res.* **126**, 19–28.

Hörstadius, S. (1928). Über die Determination des Keimes bei Echinodermen. *Acta Zool.* **9**, 1–191.

Hülser, D., and Schatten, G. (1980). The bioelectric response of sperm-egg fusion and the cortical reaction in the sea urchin *Lytechinus variegatus*. *Eur. J. Cell Biol.* **22**, 253.

Inoué, S., and Sato, H. (1967). Cell motility by labile association of molecules. *J. Gen. Phyiol.* **50**, 259–292.

Longo, F. (1973). Fertilization: A comparative ultrastructural review. *Biol. Reprod.* **9**, 149–215.

Longo, F. (1978). Effects of cytochalasin B on sperm-egg interactions. *Dev. Biol.* **67**, 249–265.

Longo, F. (1980). Organization of microfilaments in sea urchin (*Arbacia punctulata*) eggs at fertilization: Effects of cytochalasin B. *Dev. Biol.* **74**, 422–433.

Longo, F., and Anderson, E. (1968). The fine structure of pronuclear development and fusion in the sea urchin *Arbacia punctulata*. *J. Cell Biol.* **66**, 198–200.

Margolis, R. L., Wilson, L., and Kiefer, B. (1978). Mitotic mechanism based on intrinsic microtubule behaviour. *Nature (London)* **272**, 450–452.

Mazia, D., Schatten, G., and Sale, W. (1975). Adhesion of cells to surfaces coated with polylysine. *J. Cell Biol.* **66**, 198–200.

Otto, J. J., Kane, R. E., and Bryan, J. (1980). Redistribution of actin and fascin in sea urchin eggs after fertilization. *Cell Motil.* **1**, 31–40.

Schackman, R. W., Eddy, E. M., and Shapiro, B. M. (1978). The acrosome reaction of *Strongylocentrotus purpuratus*. *Dev. Biol.* **65**, 483–495.

Schatten, G. (1979). Pronuclear movements and fusion at fertilization: Time lapse video observations. *J. Cell Biol.* **83**, 1001.

Schatten, G. (1981a). Movements and fusion of the pronuclei at fertilization of the sea urchin *Lytechinus variegatus:* Time lapse video microscopy. *J. Morphol*, **167**, 231–249.

Schatten, G. (1981b). Sperm incorporation, the pronuclear migrations and their relation to the establishment of the first embryonic axis: Time lapse video microscopy of the movements during fertilization of the sea urchin *Lytechinus variegatus*. *Dev. Biol.*, **85**, in press.

Schatten, G., and Mazia, D. (1976). The surface events at fertilization: The movements of the spermatozoon through the sea urchin egg surface and the roles of the surface layers. *J. Supramol. Struct.* **5**, 343–369.

Schatten, G., and Schatten, H. (1981). Effects of motility inhibitors during sea urchin fertilization: Microfilament inhibitors prevent sperm incorporation and restructuring of fertilized egg cortex, whereas microtubule inhibitors prevent pronuclear movements. *Exp. Cell Res.*, in press.

Schatten, H. (1977). *Untersuchungen über die Wirkung von Griseofulvin in Seeigeleiern und Mammalierzellen.* Dissertation, University of Heidelberg, West Germany.

Schatten, H. and Schatten, G. (1980). Surface activity at the egg plasma membrane during sperm incorporation and its Cytochalasin B sensitivity: Scanning electron microscopy and time lapse video microscopy of the sea urchin *Lytechinus variegatus*. *Dev. Biol.* **78**, 435–449.

Schatten, H., Bestor, T., and Schatten, G. (1980). Motility during fertilization: Sperm incorporation is prevented by microfilament inhibitors while pronuclear movements are prevented by microtubule inhibitors, and immunofluorescence of associated microtubules. *Eur. J. Cell Biol.* **22**, 356.

Spudich, A., and Spudich, J. A. (1979). Actin in triton-treated cortical preparations of unfertilized and fertilized sea urchin eggs. *J. Cell Biol.* **82**, 212–226.

Steinhardt, R., Zucker, R., and Schatten, G. (1977). Intracellular calcium release at fertilization in the sea urchin egg. *Dev. Biol.* **58**, 185–196.

Tilney, L. G., and Kallenbach, N. (1979). Polymerization of Actin VI. The polarity of the actin filaments in the acrosomal process and how it might be determined. *J. Cell Biol.* **81**, 608–623.

Tilney, L. G., Kiehart, D. P., Sardet, C., and Tilney, M. (1978). Polymerization of Actin. IV. Role of Ca^{++} and H^+ in the assembly of actin and in membrane fusion in the acrosomal reaction of echinoderm sperm. *J. Cell Biol.* **77**, 536–550.

Vacquier, V. D., and Moy, G. W. (1977). Isolation of bindin: The protein responsible for adhesion of sperm to sea urchin eggs. *Proc. Natl. Acad. Sci. U.S.A.* **74**, 2456–2460.

Wehland, J., Stockem, W., and Weber, K. (1978). Cytoplasmic streaming is inhibited by the actin-specific drug phalloidin. *Exp. Cell Res.* **115**, 451–454.

Weisenberg, R. C. (1972). Microtubule formation *in vitro* in solutions containing low calcium concentrations. *Science* **177**, 1104–1106.

Weisenberg, R. C. (1975). Role of intermediates in microtubule assembly *in vivo* and *in vitro*. *Ann. N.Y. Acad. Sci.* **253**, 78–89.

Wieland, Th. (1977). Modification of actin by phallotoxins. *Naturwissenschaften* **64**, 303–309.

Wilson, E. B. (1925). "The Cell in Development and Heredity". MacMillan, New York.

Wilson, E. B., and Mathews, A. P. (1895). Maturation, fertilization, and polarity in the echinoderm egg. New light on the "quadrille of the centers". *J. Morphol.* **10**, 319–342.

Zimmerman, A. M., and Zimmerman, S. (1967). Action of colcemid in sea urchin eggs. *J. Cell Biol.* **34**, 483–488.

4

The Architecture of and Chromosome Movements within the Premeiotic Interphase Nucleus

KATHLEEN CHURCH

I. INTRODUCTION: AN OVERVIEW

A. Presynaptic Homologous Chromosome Alignment

Renewed interest in the architecture of and chromosome movements within the premeiotic interphase nucleus occurred when Feldman (1966) proposed a mechanism for the action of a gene(s) located on chromosome 5 BL in hexaploid wheat. Varying numbers of chromosome 5 BL per cell alter the pattern of meiotic chromosome synapsis; in the absence of 5 BL (nullosomics), homeologous as well as homologous chromosomes may synapse. With the normal two doses synapsis is restricted to homologues, but with 6 doses (tri-isosomics) all synapsis is reduced (for a review, see Sears, 1976). Feldman (1966) proposed that the variations in synapsis could be explained

83

MITOSIS/CYTOKINESIS
Copyright © 1981 by Academic Press, Inc.
All rights of reproduction in any form reserved.
ISBN 0-12-781240-7

by spatial proximity of the pairing partners in cells entering meiosis and that this somatic nuclear architecture was influenced by 5 BL. Some evidence for the hypothesis was initially gained by the observation of a loose somatic association of homologues in wheat root tip cells (Feldman *et al.*, 1966), but Darvey and Driscoll (1971) were unable to confirm those observations. It should be emphasized, however, that the Feldman hypothesis does not necessarily apply to all somatic cells. What is important to the hypothesis is how much if any homologous association occurs in cells approaching meiosis (Stack and Brown, 1969; Sears, 1976).

Feldman certainly was not the first to propose that presynaptic homologous chromosome alignment is a regular feature of premeiotic cells. In achieving chiasma formation, meiocytes are remarkably organized in that chromosomes rarely interlock during synapsis. Thus, it has been difficult to accept that synapsis involves long, convoluted, randomly positioned chromosomes. Despite the conceptual attractiveness of an ordered interphase nucleus, the evidence for presynaptic homologous chromosome alignment as part of that order is mixed. Brown and Stack (1968), Comings (1968), and others in their reviews have concluded that premeiotic alignment is a necessary prerequisite to meiosis. The opposite point of view is expressed forcefully by John (1976) in his provocative article entitled "Myths and Mechanisms of Meiosis."

The problems in understanding the architecture of the premeiotic interphase nucleus stem from the near impossibility of observing interphase chromosomes directly. Recent evidence suggests that chromosomes do occupy individual domains within the interphase nucleus and under appropriate experimental conditions can be visualized (Stack *et al.*, 1977; Rao *et al.*, 1977; Brown *et al.*, 1979). However, the techniques have not yet been perfected to the point of allowing the investigator to determine the positional relationships of specific chromosomes, nor have they been applied to cells in premeiotic interphase. Thus evidence for (Smith, 1942; Brown and Stack, 1968; Maguire, 1972) and against (John, 1976) premeiotic interphase alignment has come from the technically difficult observations of chromosomes during mitotic divisions preceding meiosis or from observations of positional relationships of heterochromatic chromosome regions at premeiotic interphase. The observations and their interpretation have differed depending on the species and, in some cases, on the investigator (e.g., compare Maguire, 1967, and Palmer, 1971). Perhaps the strongest evidence against a significant role for premeiotic interphase alignment in meiotic success has come from analyses of complete chromosomal complements at early meiotic prophase by serial section reconstruction. Such analyses reveal that *Locusta* (Moens, 1969), maize (Gillies, 1975a), *Bombyx* (Rasmussen, 1976), and *Lilium* (Holm, 1977a) have in common a bouquet arrangement of the chromosomes (see Section I, B) but no obvious prealignment of homologous chromosomes within the bouquet. The obvious exception is *Drosophila melanogaster*, in which the presence of somatic pairing is undebated and conventional leptotene and zygotene of meiosis appear to have been eliminated (Carpenter, 1975).

B. Rabl and Bouquet Arrangements

Although there is controversy concerning the presence or absence of homologous chromosome alignment at premeiotic interphase, there is almost complete agreement that the telophase orientation of chromosomes persists into the interphase nucleus (Rabl, 1885, cited in Wilson, 1928; Fox, 1966; Fussell, 1975). This Rabl configuration is characterized by the pointing of the centromeres toward the spindle pole (centrosome) and the passive positioning of the telomeres depending on the length of the chromosomes and the position of the centromere on the chromosome. The telomeres but not the centromeres become anchored to the nuclear envelope (Church and Moens, 1976) and perhaps to each other as well (Wagenaar, 1969; Ashley, 1979). The Rabl configuration persists throughout mitotic interphase and prophase but in the meiocytes of many species, the Rabl configuration is transformed into the bouquet (Eisen, 1900, cited in Wilson, 1928). The transformation may occur during premeiotic interphase (Bowman and Rajhathy, 1977) or early meiotic prophase (Holm, 1977a). The bouquet differs from the Rabl configuration in that the nuclear envelope-anchored telomeres rather than the centromeres are clustered and pointed toward the spindle pole (Darlington, 1965; Moens, 1969). In animal meiocytes, the end result is a juxtaposition of the tightly clustered chromosome ends with the cytoplasmic region of the cell containing the golgi, the mitochondria, the centrioles, and the spindle microtubule remnants from the ultimate premeiotic division. This complex region of the cell is located near the mouths of the cytoplasmic channels (fusosomes) which connect the meiocytes (Virkki, 1974). The functional significance of the bouquet has remained a mystery, but historically the bouquet has been perceived as a facilitator for homologous chromosome synapsis (Rhodes, 1961).

C. Chromosome Movement at Premeiotic Interphase

Successful meiotic chromosome synapsis may also depend on characteristic chromosome movements that occur during premeiotic interphase. However, again there are widely different interpretations of experimental results due primarily to the necessarily indirect nature of the observations. Colchicine disrupts chiasma formation in plant meiocytes (see Bennett and Smith, 1979, for citations). More recent investigators have concluded that the colchicine-sensitive stage encompasses at least part of the premeiotic interphase interval and, depending on the species investigated, may or may not extend into early meiotic prophase. The stage of colchicine sensitivity is restricted to premeiotic interphase in *Triticum* (wheat) (Dover and Riley, 1973) and *Secale* (rye) (Bowman and Rajhathy, 1977) but extends from mid premeiotic interphase to early meiotic prophase in *Lilium* (Shepard *et al.*, 1974 and Bennett *et al.*, 1979b) and *Triticale* (Wheat-rye-hybrid) (Thomas and Kaltsikes, 1977). The well-established effect of colchicine on microtubular organization has often led to the conclusion that this same effect must carry over to its apparent action on presynaptic cells. Thus, colchicine-induced asynapsis in wheat has been attributed to an inhibition of chromosome movements involved in the initiation of homologous chromosome associations at premeiotic interphase (Sears, 1976) or the alteration of the ultimate premeiotic anaphase resulting in a disruption of normal chromosome associations at premeiotic interphase (Dover and Riley, 1973). Similar hypotheses have also been proposed to account for temperature-induced asynapsis in some genotypes of wheat (Bayless and Riley, 1972) and to explain the induction of interlocked bivalents by premeiotic interphase temperature treatment in locusts (Buss and Henderson, 1971). Bennett *et al.* (1979b) have suggested that the colchicine and heat are acting on separate premeiotic interphase events in wheat. Colchicine may be disturbing the orderly spatial distribution of centromeres at early premeiotic interphase, while heat may be disturbing the association of homologous telomeres at late premeiotic interphase. Thomas and Kaltsikes (1977) proposed that bouquet formation is the colchicine-sensitive stage wherein homologous chromosomes locate one and other. Finally, Levan (1939) and Shepard *et al.* (1974) suggest that colchicine affects ongoing meiotic synapsis in *Allium* and *Lilium*, respectively.

II. CENTROMERE AND NUCLEOLAR ORGANIZER POSITIONS AND MOVEMENTS

The approach my students (Mr. F. E. DelFosse and Ms. A. B. Hill) and I have taken to study the architecture of the premeiotic interphase nucleus comes from the original observation that the centromeres of plant chromosomes can be visualized in electron micrographs of interphase nuclei (Fig. 1)

Fig. 1. *Allium fistulosum* nucleus at premeiotic interphase. The centromere structure (arrow) is easily identified. The telomeric heterochromatin (T) is attached to the nuclear envelope. NU, nucleolus. Bar = 1 μm.

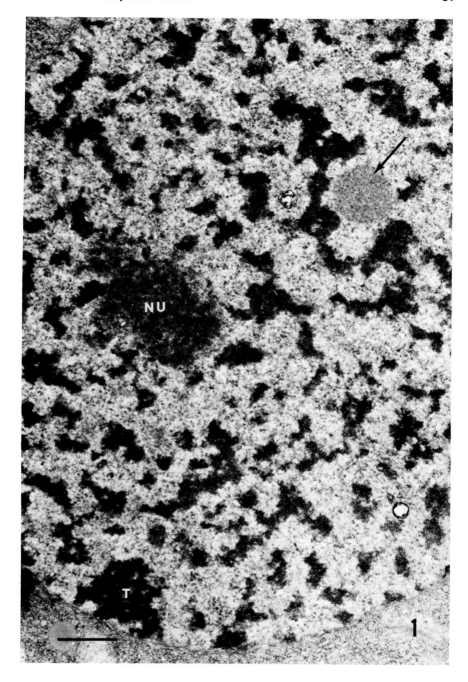

(Church and Moens, 1976). We observed that in the plant *Allium fistulosum*, premeiotic interphase centromeres aggregate into what we referred to as "centromere structures"; that the number of these structures varied from cell to cell; and that centromere structures were composed of one or more (up to seven) aggregated centromeres. The centromere structures were clustered and positioned opposite the telomeric heterochromatic masses, which in turn were attached to the nuclear envelope. Thus the premeiotic interphase nucleus appeared to be in the classic Rabl configuration. We could not determine which (indeed, if any) of the centromere associations were homologous, but we did demonstrate that centromeres could and often did associate nonhomologously (Moens and Church, 1977). Our observation that centromeres were closer together at premeiotic interphase than at early prophase suggested that centromeres were exhibiting movement and/or reorganization as the interphase nuclei entered either a meiotic (Church and Moens, 1976) or mitotic (Moens and Church, 1977) prophase.

These initial observations formed the foundation for an ambitious set of experiments, the long-term objectives of which are to determine (1) if the architecture of the interphase nucleus is dependent on chromosome homology, (2) if the centromere regions of chromosomes show characteristic movements throughout premeiotic interphase, and (3) if premeiotic interphase nuclear architecture or chromosome movements are disrupted by agents that decrease chiasma frequency when applied to premeiotic interphase. Three plant species, two in the genus *Lilium* and *Ornithogalum virens*, are being investigated.

Our technique is to thin-section serially nuclei in G_1 S, and G_2 of premeiotic interphase and to reconstruct the nuclei from electron micrographs of the serial sections. From the reconstructions, we determine the number and centromere composition of the centromere structures; digitize their positions within the nucleus; and, with the aid of a computer, determine the distances separating the individual centromere structures and the distance from each centromere structure to the nuclear envelope. From these measurements, we can determine if movement or reorganization of these structures occurs throughout premeiotic interphase. In addition to the centromere structures, other electron-dense chromosomal markers, including nucleolar-organizing regions and heterochromatic regions, are also being monitored.

To determine if the premeiotic interphase nuclear architecture is dependent on chromosome homology, *Lilium speciosum* cv. Rosemede, a chiasmate diploid lily ($2n = 24$), and Black Beauty, an achiasmate lily hybrid ($n + n = 24$) originating from a cross between *L. speciosum* and *L. henryi* (Uhring, 1968), are being investigated. Both species show a relatively good correlation between bud length and meiotic stage of development

(Bennett and Stern, 1975; Bennett *et al.*, 1979b). Cells from bud lengths spanning the entire premeiotic interphase interval have been completely examined. Although the correlation between bud/anther length and developmental stage may not be exact (Shull and Menzel, 1977), the relationship is accurate enough to place nuclei in approximate order (Table I). This is confirmed by the growth of nuclear volume which is known to accompany the passage of cells from G_1 to G_2. The number of centromere structures ranges from 15 to 23 in Rosemede (Table I and DelFosse and Church, 1981) and from 13 to 18 in Black Beauty (Table I and Hill, 1978). There are no

TABLE I

Centromere Structure Number and Composition throughout Premeiotic Interphase

Cell no.	Bud length (mm)	Nuclear volume (μm^3)	Total centromere structures	Singles	Aggregation of two or more centromeres
Rosemede					
1R	11.8	1800	18	13	5
2R	11.8	1800	19	14	5
3R	13.1	3000	23	22	1
4R	13.1	2600	18	13	5
5R	13.4	2400	17	10	7
6R	13.4	2600	19	14	5
7R	13.8	2300	16	10	6
8R	13.8	2700	15	8	7
9R	14.2	3400	19	15	4
10R	14.2	3000	21	19	2
Black Beauty					
1B	9.6	1750	18	13	5
2B	9.6	1850	13	6	5
3B	9.6	1800	17	10	7
4B	12.9	2300	16	12	4
5B	12.9	1850	17	12	5
6B	12.9	1950	18	12	6
7B	13.2	3000	16	11	5
8B	15.0	5500	16	10	6
9B	15.0	4900	15	8	7
10B	15.0	4200	17	10	7

	Avg. no. centromere structures	Avg. no. singles	Avg. no. aggregates
Rosemede	18.5 ± 2.3	13.8 ± 4.2	4.7 ± 1.9
Black Beauty	16.3 ± 1.5	10.4 ± 2.1	5.7 ± 1.1

obvious correlations between centromere structure number and bud length or nuclear volume in either cultivar; however, the average number of centromere structures per nucleus was slightly higher at all stages of premeiotic interphase in Rosemede than in Black Beauty. Since the number of centromere structures remains relatively constant throughout the premeiotic interphase interval, there is no evidence that new associations are occurring or, conversely, that old associations are being resolved as nuclei approach meiosis. From a volume analysis of the centromere structures, it is possible to determine whether each centromere structure is composed of one centromere or is an aggregate of two or more centromeres (Church and Moens, 1976). Such analysis reveals that the majority of centromere structures in both lily cultivars are composed of single centromeres but that Rosemede centromere structures are more likely to be singles than Black Beauty centromere structures (Table I). These observations suggest to us that centromere associations in lilies are not dependent on homology. They occur equally frequently if not more frequently in the achiasmate hybrid, in which no homologous centromeres are present compared to the chiasmate diploid lily, in which 12 pairs of homologous centromeres occur.

Observations on the behavior of nucleolar-organizing regions throughout premeiotic interphase led to a similar conclusion (Church *et al.*, in preparation). Rosemede contains three pairs of homologous nucleolar-organizing regions and Black Beauty contains five nucleolar-organizing regions. Four of the six Rosemede nucleolar-organizing regions are attached to large nucleoli and two to small nucleoli. These nucleoli may or may not fuse with each other, resulting in nuclei with different numbers (from one to five) and different sizes of nucleoli (Fig. 2). We assumed that homologous nucleolar-organizing regions organize similar-size nucleoli and that fused nucleoli result when nucleolar-organizing regions are positioned near each other in the interphase nucleus. From an analysis of nucleolar volumes, we could determine the pattern of nucleolar fusion. We concluded that nucleolar-organizing regions are not positioned with respect to each other by reason of homology but rather by chance events. Furthermore, the positions of the nucleolar-organizing regions appeared to be determined early in premeiotic interphase and to be maintained throughout the premeiotic interphase interval. We detected no differences in nucleolar-organizing region behavior between Black Beauty and Rosemede.

Other measurements in Rosemede nuclei support the conclusion from the nucleolar-organizing region observations that chromosomes move very little during premeiotic interphase in lilies (DelFosse and Church, 1981). Both the average distances separating centromere structures and the distance of the centromere structures from the nuclear envelope increased as cells progressed through premeiotic interphase (Table II). However, the magnitude

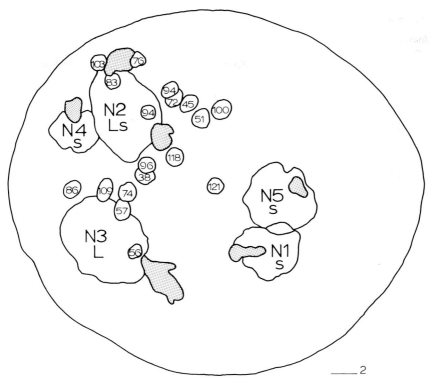

Fig. 2. Reconstruction of a premeiotic interphase nucleus from *Lilium speciosum*. Circles represent the 18 centromere structures. Numbers indicate the section in which the centromere structure was most prominent. Five nucleoli are present (N1–5). N1, N4, and N5 are small (s) nucleoli (based on volume analysis), and N3 is a large nucleolus (L). N2 is the result of an association between a large and small nucleolus (L). The six nucleolar-organizing regions are indicated by stippling. Bar = 1 μm.

of the increases was small and was correlated with the increase in nuclear volume. Furthermore, the coefficient of variation of the distances did not show any significant change throughout premeiotic interphase. These findings indicate that the centromere structures do not change their orientation with respect to each other, although they are moving farther apart as the nuclear volume increases. The centromere structures were found not to be positioned randomly within the nucleus but rather occupied a small percentage ($\simeq$ 20%) of the nuclear volume throughout premeiotic interphase.

Based on these observations on centromeres and nucleolar-organizing regions, we have come to the tentative conclusion that homology is either not involved or has minimal involvement in the architecture of the lily premeio-

TABLE II

Measurements from Reconstructed Rosemede Premeiotic Interphase Nuclei

Cell no.	Avg. distance between centromere structures (μm)	CV[b]	Avg. distance from centromere structure (μm)	CV
1R[a]	5.09	(36.9)	2.87	(61.5)
2R	5.69	(41.8)	2.05	(51.2)
3R	7.04	(41.6)	2.84	(53.8)
4R	5.19	(37.2)	3.30	(41.6)
5R	5.19	(39.1)	3.72	(49.0)
6R	6.91	(38.5)	3.22	(49.5)
7R	5.88	(39.8)	2.52	(56.8)
8R	6.10	(34.3)	3.59	(45.0)
9R	8.15	(38.7)	3.37	(52.2)
10R	8.03	(35.5)	3.46	(42.5)

[a] Cells are arranged in developmental sequence from early to late premeiotic interphase. Nuclear volumes for each cell can be obtained from Table I.
[b] Coefficient of variation.

tic interphase nucleus. We are aware that we are observing only limited regions of the chromosomes, but these observations do supplement those of Walters (1970) and Holm (1977b), who were unable to detect presynaptic homologous chromosome alignment in lilies at the preleptotene contraction stage or at early meiotic prophase, respectively. Our current view of the lily premeiotic interphase nucleus is that chromosomes are most likely positioned in the nucleus in the Rabl configuration as determined by the telophase position of the chromosomes at the ultimate premeiotic mitosis; that the positions of the chromosomes are determined by chance events rather than homology; and that they move very little prior to early meiotic prophase.

These observations make less attractive the hypothesis that the premeiotic interphase colchicine effect in lilies is to disrupt presynaptic chromosome alignment or characteristic premeiotic interphase chromosome movements. However, it is clear that considerable nuclear reorganization must occur in lilies at early meiotic prophase. For example, the nucleoli, along with the nucleolar-organizing regions, migrate to the nuclear envelope; they move on the envelope and subsequently fuse into a giant crescent-shaped nucleolus (Stern *et al.*, 1975). The ends are collected together in a limited region of the nuclear envelope in a bouquet arrangement (Holm, 1977b). It will therefore be of interest to determine if colchicine disrupts the nucleolar migration and the formation of the bouquet in *Lilium*, as it has been reported to do in both

Secale (Bowman and Rajhathy, 1977) and *Tritacale* (Thomas and Kaltsikes, 1977).

Unlike lilies, the majority of the observations on chromosome relationships in *Ornithogalum virens* $(2n = 6)$ indicate a high frequency of homologous chromosome association in root tip mitotic cells and an even higher frequency in premeiotic cells (Stack, 1971; Chauhan, 1973; Goden and Stack, 1976). Ashley (1979) has proposed a somatogram for diploid cells in this species that includes specific homologous and nonhomologous telomere attachments and predicts that homologous chromosomes lie close together and in particular relationships with nonhomologous chromosomes. Although the effect of colchicine on chiasma formation has not been determined, it is known that cold-temperature treatment during premeiotic interphase disrupts subsequent chiasma formation in *O. virens* (Church and Wimber, 1971).

We are now in the process of examining the effects of cold temperature and of colchicine on the premeiotic interphase nuclear architecture in this species. Although all of the data have not yet been analyzed, we have obtained intriguing preliminary results. It is clear that *O. virens* differs from *Lilium* in several ways. *O. virens* centromeres are almost twice as likely to enter into associations as are *Lilium* centromeres (Table III), and centromere associations in *O. virens* appear to be largely determined by homology rather than by chance events. One pair of homologous centromeres can be recognized in the interphase nucleus due to its position near the nucleolar-organizing region (Figs. 3, 4a,b). We have observed this pair of centromeres in 17 nuclei and of the 34 centromeres, 12 were unassociated, 18 were homologously associated, and 4 were nonhomologously associated. If the

TABLE III

Centromere Association Frequency

	No. of nuclei examined	No. of centromeres observed	No. of centromeres associated	Association frequency
Lilium				
Rosemede	10	240 ($2n = 24$)	102	0.43
Black Beauty	10	240 ($n + n = 24$)	136	0.57
Ornithogalum				
Untreated	12	72 ($2n = 6$)	59	0.82
Colchicine treatment	12	72 ($2n = 6$)	42	0.58
Low-temperature treatment	12	72 ($2n = 6$)	53	0.76

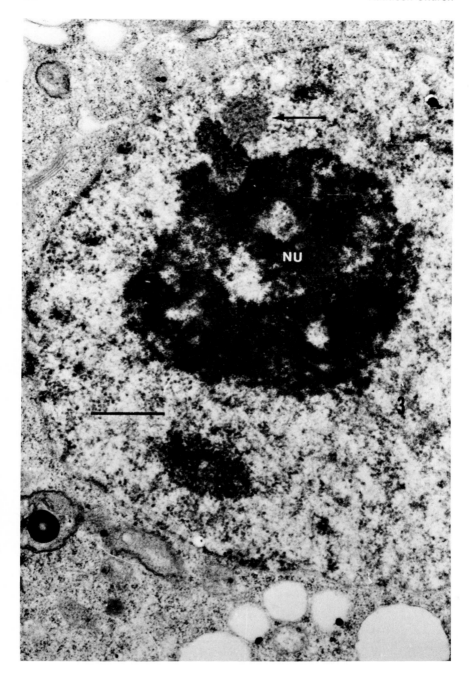

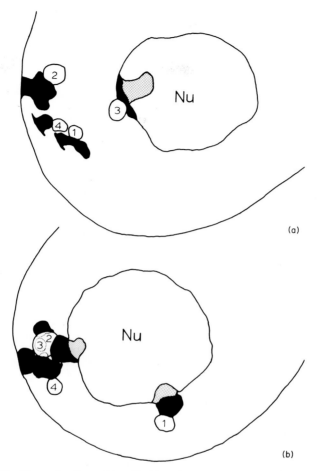

Fig. 4.(a,b)   Reconstructions of two *Ornithogalum virens* premeiotic interphase nuclei. In the first nucleus (a), centromere structures (circles) 1 and 4 are composed of single centromeres. Centromere structures 2 and 3 are doubles. The identifiable centromere structure (3) represents a homologous pair of associated centromeres. The pericentric heterochromatin of centromere structures 1, 2, and 4 is associated with the nuclear envelope. In the second nucleus (b), the identifiable homologous centromeres (1 and 2) are not associated; however, centromere structure 2 is composed of three aggregated centromeres. Thus at least one *nonhomologous* association is occurring. Stippled areas are nucleolar-organizing regions. Bar = 1 μm.

Fig. 3.   *Ornithogalum virens* nucleus at late premeiotic interphase. This particular centromere structure (arrow) can be identified in all cells by its close proximity to the nucleolar-organizing regions. The centromere structure is composed of two homologously associated centromeres. One nucleolus (NU) is found in each cell. Bar = 1 μm.

centromeres were associating randomly, we would expect only 20% of the associations to be homologous.

Another striking difference between *O. virens* and *Lilium* is reflected in the pattern of nucleolar fusion. *O. virens* possesses only one pair of nucleolar-organizing regions, and virtually all premeiotic interphase nuclei display only one nucleolus, indicating that homologous nucleolar-organizing regions are positioned near each other within the nucleus (Fig. 4a,b). Our preliminary data also indicate that colchicine treatment may be affecting centromere behavior. The centromere association frequency is reduced in the colchicine-treated nuclei compared to either untreated or cold-treated nuclei (Table III). Furthermore, of the 24 identifiable centromeres from 12 reconstructed colchicine-treated nuclei, the majority (13) were unassociated, 9 were nonhomologously associated, and only 2 were homologously associated. We are not yet in a position to draw definite conclusions, and until more data are forthcoming, I would prefer not to add more speculations to the literature on the colchicine effect. It is clear, however, that this approach is worth pursuing.

The observations on two different chiasmate species of lileacious plants (*L. speciosum* and *O. virens*) indicate that there are species-specific differences in premeiotic interphase nuclear architecture. This should not come as a surprise based on a literature perusal, but nevertheless it is most likely a reason for the often polemic discussions on the topic of presynaptic homologue alignment.

III. FROM THE RABL CONFIGURATION TO THE BOUQUET

Several mechanisms have been proposed for the transformation of the premeiotic interphase nucleus from the Rabl configuration to the bouquet. Janssens (in Wilson, 1928) proposed that nuclear rotation of 180° reversed the relationship of the centromeres and the centrosome. Virkki (1974) has observed that the fusomes (cytoplasmic channels connecting meiocytes) can migrate 180° on the cell surface and proposes that contact of the fusome mouth with the centriole and the chromosome ends could rotate a nucleus 180°. Virkki also points out that the bouquet could be achieved by simple migration of the centrosome 180° in the absence of nuclear rotation. Although nuclear rotation and/or centrosome migration may well be involved in the formation of the bouquet, these events alone will not account for all observations. Active chromosome movement must occur inside the nucleus at either premeiotic interphase or early meiotic prophase to transform the chromosomes to the bouquet arrangement, as shown by observations on grasshopper nuclei.

The karyotype of the male grasshopper, *Brachystola magna*, includes 11 acrocentric autosome pairs and a univalent X chromosome. All centromeres are flanked by pericentric heterochromatin which can be observed in electron micrographs of premeiotic interphase nuclei. Each premeiotic interphase nucleus contains 15 to 20 distinct heterochromatic chromocenters with 19 to 22 attachment sites to the nuclear envelope presumably representing the proximal autosomal ends. The chromocenters and nuclear envelope attachments are polarized in approximately one-half of the nucleus (Church, 1976). The architecture of the leptotene nucleus differs from the premeiotic interphase nucleus in that (1) the number of chromocenters is reduced from 15 to 20 small ones to 1 or 2 large ones, indicating that the heterochromatic regions are closer together, and (2) the proximal chromosome ends are clustered on the envelope more at leptotene than at premeiotic interphase. The distal chromosome ends can also be recognized at leptotene by the presence of electron-dense axial cores, and they too are clustered and attached to the envelope near the proximal ends (Church, 1977). Identical rearrangements of the chromosome ends have also been observed directly during the formation of the bouquet in rye (Bowman and Rajhathy, 1977). Thus relative chromosome positions change between premeiotic interphase and leptotene.

It has been recognized for many years that the centrosome may act as a focal point of attraction for the chromosome ends in the formation of the bouquet (Rhodes, 1961). Centriole-focused movements are well known in the literature (see Rickards, 1975, for a review). Whatever the mechanism, the nuclear envelope is strongly implicated as playing a substantial role in the movement. This, of course, is not a novel proposal. Membranes have been implicated in chromosome movement systems in prokaryotes, lower eukaryotes, and synapsis and postpachytene chromosome movements of higher eukaryotes (for reviews, see Moens, 1973; Virkki, 1974; Rickards, 1975). That the envelope is involved in the formation of the bouquet in *B. magna* is suggested by the intimate association of the chromosome ends with the nuclear envelope and by a correlation between changes in the architecture of the envelope and the migration of the ends (Church, 1976, 1977). Changes in the nuclear envelope during the interval between premeiotic interphase and early meiotic prophase include (1) a resolution of deep envelope invaginations that are characteristic of cells in premeiotic interphase in this species and (2) a dramatic rearrangement of the nuclear pores from a scattered arrangement over the entire envelope at premeiotic interphase to a clustered arrangement, with virtually all being located in the same general area of the nuclear envelope as the chromosome ends. Such pore rearrangements also occur in lilies during the formation of the bouquet (Holm, 1977a,b and personal observation) and during early meiotic prophase in

Psiotum nudum (Fabri and Bonzi, 1975). The significance of these observations is unclear, but they do demonstrate an active reorganization of the nuclear envelope at a time when the movement of chromosome ends within the framework of that envelope is occurring.

As should be evident from this review, the analysis of presynaptic chromosome behaviors has not yet advanced from the descriptive stage, and it is much too early to speculate on mechanisms or to present models. It is intriguing, however, that whatever is happening in the premeiotic interphase nucleus, it is strongly affected by colchicine. This has understandably led to the speculation that microtubules are involved (Dover and Riley, 1973). However, Hotta and Shepard (1973), in an attempt to identify the mechanism by which colchicine interferes with chromosome pairing in lilies, found that colchicine affects the nuclear envelope of meiotic cells by causing a reduction in DNA-binding proteins in the envelope. Furthermore, a colchicine-binding protein was found in the membranous fraction of isolated meiocyte nuclei, and preliminary characterization indicated that it was not tubulin. These results are intriguing but apparently have not been carried much further. Since the nuclear envelope seems to be implicated in both presynaptic and postsynaptic (Rickards, 1975) chromosome movement (see Chapter 5 in this volume), further studies on the molecular architecture of the meiotic nuclear envelope and the effects of premeiotic interphase-perturbing agents on the envelope may aid in the identification of a mechanism for moving chromosomes in the absence of microtubules.

IV. PROSPECTIVES

In our studies with lilies and *O. virens*, we are attempting a rather primitive genetic dissection of events occurring in premeiotic interphase by comparing the achiasmate hybrid to a chiasmate diploid lily and by using environmental agents to produce putative phenocopies of genetic blocks. However, these studies cannot be carried much further because of a lack of genetic variants and, perhaps more importantly, because of the limited number of chromosomal markers that can be monitored at premeiotic interphase. A recent discovery in cereals, however, may have opened up possibilities for such an analysis in a genetically well-defined plant system. Bennett *et al.* (1979a) have recently described intranuclear bundles of microfilaments in cereal pollen mother cells. This fibrillar material was present during premeiotic interphase, leptotene, and zygotene but was most abundant during premeiotic interphase or leptotene. Fibrillar material often linked masses of chromatin (most likely telomeres) and less often linked chromatin and the nuclear envelope. Although the function of fibrillar material is entirely unknown, it is speculated that it might have a dynamic role and function by moving chromosome segments in relation to each other and the nuclear envelope. Furthermore, the distribution of fibrillar material is altered by colchicine application (Bennett and Smith, 1979). Thus the stage is set for these investigators to determine the chemical nature and the genotypic and environmental behavior of this putative chromosome mover. However, it must be realized that fibrillar material thus far is unique to cereals. We have completely examined premeiotic interphase nuclei from three different plant species and two different animal species and have encountered nothing remotely resembling fibrillar material, again emphasizing that specific meiotic mechanisms may not be ubiquitous.

Our data on three different species representing two different genera of lileaceous plants demonstrate that different degrees of presynaptic homologous chromosome association occur depending on the species investigated. One might conclude, therefore, that although presynaptic homologous chromosome alignment may facilitate meiotic synapsis, it is not a necessary meiotic prerequisite. However, the significance of the premeiotic nuclear architecture to future meiotic success will probably not leave the realm of speculation until there is a greater understanding of the meiotic process itself. Despite the many studies, the process is poorly understood even at the descriptive level. The most convincing morphological evidence concerning the sequence of events in meiotic synapsis has come from studies of entire chromosomal complements during early meiotic prophase stages by using reconstructions of serial sections (Gillies, 1975b) or whole-mount surface spreads (Counce and Meyer, 1973). New techniques are also being developed for light microscopic observations of synaptic behavior in surface spread chromosomes (Moses, 1977). What is needed is an application of the above techniques to meiotic systems in which there is a body of genetic data with a catalogue of mutants that affect the pairing process. Meiotic mutants and their effects have been the subject of numerous reviews (Sears, 1976; Baker *et al.*, 1976; Golubovskaya, 1979), and it appears that the time is ripe for a genetic dissection of the pairing process at the fine structural level. Understanding these processes in well-defined systems could lay the foundation for future studies at the molecular level. Such an analysis in the wheat system would be especially illuminating. Much of the speculation for somatic pairing and premeiotic interphase chromosome movements has come from this system, yet little is known about the early meiotic events that these premeiotic interphase chromosome activities are proposed to facilitate.

ACKNOWLEDGMENTS

This work was supported by Grant BMS75-08685-A01 from the National Science Foundation. I thank Ms. Becky Payne and Ms. Rosalie Landers for their skill in preparing the manuscript.

REFERENCES

Ashley, T. (1979). Specific end to end attachment of chromosomes in *Ornithogalum virens*. *J. Cell Sci.* **38**, 357–367.

Baker, B. S., Carpenter, A. T. C., Esposito, M. S., Esposito, R. E., and Sandler, L. (1976). The genetic control of meiosis. *Annu. Rev. Genet.* **10**, 53–134.

Bayless, M. W., and Riley, R. (1972). An analysis of temperature-dependent asynapsis in *Triticum aestivum*. *Genet. Res.* **23**, 193–200.

Bennett M. D., and Smith, J. B. (1979). The effects of colchicine on fibrillar material in wheat meiocytes. *J. Cell Sci.* **38**, 33–47.

Bennett, M. D., and Stern, H. (1975). The time and duration of preleptotene chromosome condensation in the *Lilium* hybrid cv. Black Beauty. *Proc. Roy. Soc. London* **188**, 473–485.

Bennett, M. D., Smith, J. B., Simpson, S., and Wells, B. (1979a). Intranuclear fibrillar material in cereal pollen mother cells. *Chromosoma* **71**, 289–332.

Bennett, M. D., Toledo, L. A., and Stern, H. (1979b). The effect of colchicine on meiosis in *Lilium speciosum* cv. "Rosemede." *Chromosoma* **72**, 175–189.

Bowman, J. G., and Rajhathy, T. (1977). Fusion of chromocenters in premeiotic interphase of *Secale cereale* and its possible relationship to chromosome pairing. *Can. J. Genet. Cytol.* **19**, 313–321.

Brown, W. V., and Stack, S. M. (1968). Somatic pairing as a regular preliminary to meiosis. *Bull. Torrey Bot. Club* **95**, 369–378.

Brown, D. B., Stack, S. M., Mitchell, J. B., and Bedford, J. S. (1979). Visualization of the interphase chromosomes of *Ornithogalum virens* and *Muntiacus muntjak*. *Cytobiologie* **18**, 398–412.

Buss, M. E., and Henderson, S. A. (1971). Induced bivalent interlocking and the course of meiotic chromosome synapsis. *Nature (London), New Biol.* **234**, 243–246.

Carpenter, A. T. C. (1975). Electron microscopy of meiosis in *Drosophila melanogaster* females. I. Structure, arrangement, and temporal change of the synaptonemal complex in wild type. *Chromosoma* **51**, 157–182.

Chauhan, K. P. S. (1973). Association of homologous chromosomes in the somatic cells of *Ornithogalum virens*. *Cytologia* **38**, 29–33.

Church, K. (1976). Arrangement of Chromosome ends and axial core formation during early meiotic prophase in the male grasshopper *Brachystola magna* by 3D, EM reconstruction. *Chromosoma* **58**, 365–376.

Church, K. (1977). Chromosome ends and the nuclear envelope at premeiotic interphase in the male grasshopper *Brachystola magna* by 3D, EM recontruction. *Chromosoma* **64**, 143–154.

Church, K., and Moens, P. B. (1976). Centromere behavior during interphase and meiotic prophase in *Allium fistulosum* from 3D EM reconstruction. *Chromosoma* **56**, 249–263.

Church, K., and Wimber, D. E. (1971). Meiosis in *Ornithogalum virens* (Liliaceae). II. Univalent production by preprophase cold treatment. *Exp. Cell Res.* **64**, 119–124.

Comings, D. E. (1968). The rational for an ordered arrangement of chromatin in the interphase nucleus. *Am. J. Hum. Genet.* **20**, 440–460.

Counce, S. J., and Meyer, G. F. (1973). Differentiation of the synaptonemal complex and the kinetochore in Locusta spermatocytes studied by whole mount electron microscopy. *Chromosoma* **44**, 231–253.

Darlington, C. D. (1965). "Cytology." Little, Brown, Boston, Massachusetts.

Darvey, N. L., and Driscoll, C. J. (1971). Evidence against somatic association in wheat. *Chromosoma* **36**, 140–149.

DelFosse, F. E., and Church, K. (1981). Presynaptic behavior in *Lilium*. I. Centromere orientation and movement during premeiotic interphase (PMI) in *Lilium speciosum* var. (Rosemede) *Chromosoma* **81**, 701–716.

Dover, G. A., and Riley A. (1973). Effect of colchicine on meiosis in wheat. *Nature (London)* **216**, 687–688.

Fabri, F., and Bonzi, L. M. (1975). Observations on nuclear pores of freeze-etched spore mother cells of *Psitotum nudum* (L.) Beauv. during early stages of meiotic Prophase. I. Preliminary report. *Caryologia* **28**, 549–559.

Feldman, M. (1966). The effect of chromosomes 5B, 5D, and 5A on chromosome pairing in *Triticum aestivum*. *Proc. Natl. Acad. Sci. U.S.A.* **55**, 1447–53.

Feldman, M., Mello-Sampayo, T., and Sears, E. R. (1966). Somatic association in *Triticum aestivum*. *Proc. Natl. Acad. Sci. U.S.A.* **56**, 1192–99.

Fox, D. P. (1966). The effects of X-rays on chromosomes of locust embryos. II. Chromatid interchanges of the interphase nucleus. *Chromosoma* **20**, 173–194.

Fussell, C. P. (1975). The position of interphase chromosomes and late replicating DNA in centromere and telomere regions of *Allium cepa* L. *Chromosoma* **50**, 201–210.

Golubovskaya, I. N. (1979). The genetic control of meiosis. *Int. Rev. Cytol.* **58**, 247–290.

Gillies, C. B. (1975a). An ultrastructural analysis of chromosomal pairing in maize. *C. R. Trav. Lab. Carlsberg* **40**, 135–161.

Gillies, C. B. (1975b). Synaptonemal complex and chromosome structure. *Annu. Rev. Genet.* **9**, 91–109.

Goden, D. E., and Stack, S. M. (1976). Homologous and non-homologous chromosome associations by interchromosomal chromatin connectives in *Ornithogalum virens*. *Chromosoma* **57**, 309–318.

Hill, A. B. (1978). Centromere arrangements during premeiotic interphase. M.S. Thesis, Arizona State University, Tempe, Arizona.

Holm, P. B. (1977a). Three-dimensional reconstruction of chromosome pairing during the zygotene stage of meiosis in *Lilium longiflorum*. (Thunb.) *Carlsberg Res. Commun.* **42**, 103–151.

Holm, P. B. (1977b). The premeiotic DNA replication of euchromatin and heterochromatin in *Lilium longiflorum*. (Thunb.) *Carlsberg Res. Commun.* **42**, 249–281.

Hotta, Y., and Shepard, J. (1973). Biochemical aspects of colchicine action in meiotic cells. *Mol. Gen. Genet.* **122**, 243–260.

John, B. (1976). Myths and mechanisms of meiosis. *Chromosoma* **54**, 295–325.

Levan, A. (1939). The effect of colchicine on meiosis in *Allium*. *Hereditas* **35**, 9–26.

Maguire, M. P. (1967). Evidence for homologous pairing of chromosomes prior to meiotic prophase in maize. *Chromosoma* **21**, 221–231.

Maguire, M. P. (1972). Premeiotic mitosis in maize: Evidence for pairing of homologues. *Caryologia* **25**, 17–24.

Moens, P. B. (1969). The fine structure of meiotic chromosome polarization and pairing in *Locusta migratoria* spermatocytes. *Chromosoma.* **28**, 1–25.

Moens, P. B. (1973). Mechanisms of chromosome synapsis at meiotic prophase. *Int. Rev. Cytol.* **35**, 117–134.

Moens, P. B., and Church, K. (1977). Centromere sizes, positions, and movements in the interphase nucleus. *Chromosoma* **61**, 41–48.

Moses, M. J. (1977). Synaptonemal complex karyotyping in spermatocytes of the chinese hamster (*Cricetulus griseus*). I. Morphology of the autosomal complement in spread preparations. *Chromosoma* **60**, 99–125.

Palmer, R. G. (1971). Cytological studies of a meiotic mutant and normal maize with reference to premeiotic pairing. *Chromosoma* **35**, 233–246.

Rasmussen, S. W. (1976). The meiotic prophase in *Bombyx mori* females analyzed by three dimensional reconstructions of synaptonemal complexes. *Chromosoma* **54**, 245–293.

Rickards, G. K. (1975). Prophase chromosome movements in living house cricket spermatocytes and their relationship to prometaphase, anaphase, and granule movements. *Chromosoma* **49**, 407–455.

Rao, P., Wilson, B., and Puck, T. (1977). Premature chromosome condensation analysis. *J. Cell Physiol.* **91**, 131–142.

Rhodes, M. M. (1961). Meiosis. *In* "The Cell" (J. Bracket and A. E. Mirsky, eds.), Vol. 3, pp. 1–75. Academic Press, New York.

Sears, E. R. (1976). Genetic control of chromosome pairing in wheat. *Annu. Rev. Genet.* **10**, 31–51.

Shepard, J. Boothroyed, E. R., and Stern, H. (1974). The effects of colchicine on synapsis and chiasma formation in microsporocytes of *Lilium*. *Chromosoma* **44**, 423–437.

Shull, J. K., and Menzel, M. Y. (1977). A study of the reliability of synchrony in the development of pollen mother cells of *Lilium longiflorum* at first meiotic prophase. *Am. J. Bot.* **64**, 670–679.

Smith, S. G. (1942). Polarization and progression in pairing. II. Premeiotic orientation and the initiation of pairing. *Can. J. Res.* **20D**, 221–229.

Stack, S. M. (1971). Premeiotic changes in *Ornithogalum virens*. *Bull. Torrey Bot. Club* **98**, 207–214.

Stack, S. M., and Brown, W. V. (1969). Somatic pairing, reduction, and recombination: An evolutionary hypothesis of meiosis. *Nature (London)* **222**, 1275–76.

Stack, S. M., Brown, D. B., and Dewey, W. (1977). Visualization of interphase chromosomes. *J. Cell Sci.* **26**, 281–299.

Stern, H., Westergaard, M., and von Wettstein, D. (1975). Presynaptic events in meiocytes of *Lilium longiflorum* and their relation to crossing-over: A preselection hypothesis. *Proc. Natl. Acad. Sci. U.S.A.* **72**, 961–965.

Thomas, J. B., and Kaltsikes, P. J. (1977). The effect of colchicine on chromosome pairing. *Can. J. Genet. Cytol.* **19**, 231–249.

Uhring, J. (1968). Hybridizing experiments with *Lilium* x Black Beauty. *The Lily Yearbook* **21**, 44–52.

Virkki, N. (1974). The bouquet. *J. Agric. Univ. P.R.* **58**, 338–350.

Wagenaar, E. B. (1969). End to end attachments in mitotic interphase and their possible significance in meiotic chromosome pairing. *Chromosoma* **26**, 410–426.

Walters, M. S. (1970). Evidence on the time of chromosome pairing from the preleptotene spiral stage in *Lilium longiflorum* "Croft". *Chromosoma* **29**, 375–418.

Wilson, E. B. (1928). "The Cell in Development and Heredity." Macmillan, New York.

Chromosome Movements within Prophase Nuclei

GEOFFREY K. RICKARDS

I. INTRODUCTION

Chromosome movements during anaphase have traditionally been the focus of attention for students of mitosis and meiosis, largely because of their relative simplicity: the chromosomes move poleward without obvious pauses or reversals of direction and at a steady but low velocity. More recent attention has been directed at the more complex movements during prometaphase, in which, particularly in early prometaphase, chromosome movements are individually much more variable than those during anaphase. Statistically, however, they are predictable: they lead to the congression and precise orientation of chromosomes along the metaphase equator.

Forces for both anaphase and prometaphase movements are localized at kinetochores, normally have velocities of approximately 1 μm per minute, and both movements are inseparably associated with the spindle apparatus. The patterns of these movements are now well documented, chiefly through

MITOSIS/CYTOKINESIS
Copyright © 1981 by Academic Press, Inc.
ISBN 0-12-781240-7

studies on living cells (see Nicklas, 1971; Bajer and Molè-Bajer, 1972, for reviews). But the mechanism of chromosome movement, in terms of the force producers concerned and force production and its control, is still not understood. While most current and past views invoke microtubules of the spindle as force producers, it is not clear, in fact, that spindle microtubules do indeed function directly as force producers. They may function less directly as a skeleton on which another force producer (actin filaments, for example) is constructed and polarized (see McIntosh *et al.*, 1975; Nicklas, 1975; Forer 1978a, for reviews).

It has long been known, although often overlooked, that chromosome movements also occur in cells at prophase (see Schrader, 1953, for review). These movements are often more variable and unpredictable than even prometaphase movements, and it is perhaps partly for this reason that they are often ignored in attempts to explain the mechanism of chromosome movements on the spindle (Margolis *et al.*, 1978, for example). However, prophase movements are relevant to this mechanism, because I consider it unlikely that an efficient process capable of moving chromosomes within prophase nuclei would be dispensed with and an entirely new process suddenly adopted for the subsequent movements of prometaphase and anaphase. This chapter directs attention to these prophase chromosome movements and then to their relevance to the wider issue of the mechanism of chromosome movement on the spindle. In discussing this wider issue, I shall be referring solely to the poleward movement of chromosomes, not to the process in anaphase whereby chromosomes separate more widely from each other through separation of the poles (although I do not wish to imply that this latter process is unimportant).

II. SCOPE

The generally accepted definition of "prophase"—that it is said to terminate when the nuclear envelope breaks and the chromosomes first start movements associated directly with the spindle material (Mazia, 1961, pp. 165, 193; Bajer and Molè-Bajer, 1969, p. 463; Roos, 1973, p. 51)—will be used here. The emphasis is on the breakage of the nuclear envelope. While this definition creates unavoidable problems with those primitive eukaryotes in which mitosis takes place completely within an intact nuclear envelope, it does distinguish unambiguously prophase movements (before the breakage of the nuclear envelope) from prometaphase movements (which start after the nuclear envelope breaks). This definition is crucial to considerations of the role of microtubules in chromosome movements, since microtubules are not introduced to the environment of the chromosomes until the breakdown of the nuclear envelope; once the nuclear envelope breaks, microtubules quickly enter the nuclear area (Bajer and Molè-Bajer, 1969; Roos, 1973). The nuclear envelope can be seen in living cells under phase optics when viewed in optical transverse section, and its disappearance is easily monitored (Bajer and Molè-Bajer, 1969; Rickards, 1975). Hence the shift from prophase to prometaphase can be precisely delineated.

On the basis of the above definition, I exclude here full consideration of the interesting movements in certain mitotic cells that produce a polarization of chromosomes around the two centers, forming the so-called spurious anaphase (Fell and Hughes, 1949; Roos, 1976). This polarization is broadly similar to the one that is produced in prophase cells of some organisms (see later), but since the spurious anaphase movements take place *after* breakage of the nuclear envelope, they must be regarded as prometaphase (Roos, 1976). Paweletz (1974) described the polarization movements of chromosomes in HeLa cells and concluded that the chromosomes are attached to the nuclear membrane and are transported by polarized growth of the nuclear membrane and the formation of membrane invaginations at the centrioles; that is, the chromosomes move by a mechanism comparable to that proposed (but not established) for genome separation in bacteria (see Heath, 1974, for review). Such a conclusion, if sustainable, would have considerable significance to this chapter on prophase chromosome movements. However, scrutiny of the paper by Paweletz, and of the film on living cells on which his work was based (Mayer, 1955), shows clearly that these polarization movements (which are quite rapid) occur, in my view, during the very early stages of prometaphase, *after* initial breakage of the nuclear envelope; i.e., they are comparable to the spurious anaphase movements mentioned above. Thus, Fig. 4 of Paweletz's paper is of a prophase cell (the nuclear enevelope is intact); but although the chromosomes are arranged in the periphery of the nucleus, there are no obvious signs of polarization. Fig. 5 shows polarization of chromosomes around the centers, but then the nuclear envelope has already broken in a number of places. (Compare also Figs. 1g and i.) Only correlative light and electron microscopy can resolve the question of the timing of these movements relative to the breakdown of the nuclear envelope, and thus also the important question of whether the nuclear envelope or other elements (microtubules, for example) are involved in the movements (see Roos, 1976).

The polarization movements described later in this chapter occur long before the breakdown of the nuclear envelope.

III. CATEGORIES OF PROPHASE CHROMOSOME MOVEMENTS

One category of chromosome movements in prophase nuclei includes a wide range of movements which are more or less ubiquitous to mitotic and meiotic cells and in which the mitotic centers (centrioles, poles) probably play no causative role. These movements are summarized as follows. (1) The movements associated with the condensation (coiling) cycle of chromosomes during prophase (and later stages) have been documented from both fixed material (Vanderlyn, 1948) and living material (Bajer and Molè-Bajer, 1956). Unfortunately, little is known about the mechanics of this contraction process (see Swanson, 1960; Mazia, 1961; John and Lewis, 1965, 1969, for reviews), although the identification of actin and myosin in some nuclei has prompted the view that chromosome contraction may be mediated by interactions involving these "muscle contraction" proteins (Forer, 1974 and references therein). (2) During prophase, chromosomes move to peripheral positions, close to the nuclear envelope, so as to leave the center of the nucleus more or less hollow (see, for example, Comings and Okada, 1970; Paweletz, 1974). But since chromosomes are often attached to the nuclear envelope (Rickards, 1975 and references therein), attachment coupled with condensation will draw chromosomes toward the nuclear envelope, thereby emptying the center of the nucleus (Comings and Okada, 1970). However, in some organisms these peripheral movements occur in late prophase of meiosis, when chromosome contraction is largely complete (Schrader, 1953; Fig. 2), in which case they cannot occur solely in the manner proposed by Comings and Okada. (3) Finally, there are movements associated with synapsis of homologous chromosomes (see Chapter 4 by Church in this volume). These movements are still poorly understood (Schrader, 1953; Rhoades, 1961; Stern and Hotta, 1973); and, unfortunately, they occur when the chromosomes cannot be seen in living cells. However, it is noteworthy, in the light of later discussion, that synapsis has often been viewed as being initiated by the movement of chromosome ends along the nuclear envelope (Rasmussen, 1976, for example). Also, in some plants, chromosome movements associated with synapsis are sensitive to the "microtubule poison" colchicine (see Sears, 1976, for review); and both actin-like and myosin-like molecules have been identified in some nuclei (Jockusch *et al.*, 1971, 1973). The implication is that perhaps microtubules (tubulin), and actin and myosin, are involved in these chromosome movements, as they are in other cell movements (Pollard and Weihing, 1974). Much more work needs to be done on this most important area of chromosome movements in prophase nuclei.

The second category of chromosome movements differs from the first in that mitotic centers outside the nucleus play a dominant role. These are the movements of prophase nuclei that are associated with the formation of the bouquet stage and the late polarization of bivalents in meiotic cells. The theme to be developed in this chapter is that these movements are assuredly relevant to our understanding of the mechanisms of chromosome movements on the spindle.

IV. RABL ORIENTATION AND THE BOUQUET

In some cells there are *no* chromosome movements at prophase, other than those of the category I type. In 1885, Rabl noted that the centromeres of prophase chromosomes in mitotic cells usually lie grouped together on one side of the nucleus (the nuclear "pole"), and the chromosome arms lie roughly parallel with their ends basically aligned toward the opposite side of the nucleus. This "Rabl orientation" is a common feature in both plant and animal mitosis (Fig. 1a; see Roos, 1976 Fig. 1 at 0.00; Schrader, 1953, for review). The nuclear pole is close to the centriole, when one is present. However, rather than implying that chromosomes actively adopt this position during prophase, Rabl orientation is usually interpreted as reflecting the arrangement of chromosomes established during anaphase of the previous mitotic division (Schrader, 1953). Thus the orientation implies that chromosomes in some cells undergo no major movements during interphase and prophase. This conclusion is particularly clear in cells for which ciné records are available, in which substantial chromosome movements in prophase could not possibly have been missed had they been present (Bajer and Molè-Bajer, 1956; Bajer, 1959, for example).

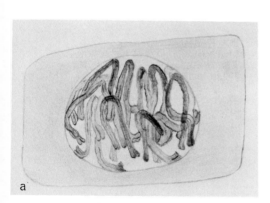
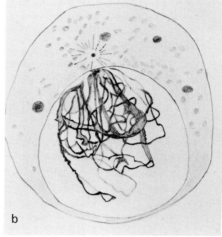

Fig. 1. (a) Camera-lucida drawing of Rabl orientation in prophase of mitosis in a root tip meristem cell of *Allium cepa*. The centromeres, marked by loops, are located at the nuclear "pole," while the ends lie loosely at the "anti-pole." Not all chromosomes are drawn, for clarity. (b) The pachytene bouquet from an oocyte of *Dendrocoelum lacteum*. The chromosome ends are grouped close together opposite the centrosome, the rest of the chromosomes forming loops into the body of the nucleus. Redrawn from Gelei (1921).

Rabl orientation is also characteristic of early stages of prophase of meiosis (Rhoades, 1961). However, in meiocytes of many animals and in pollen mother cells of at least some plants, Rabl orientation is replaced by a bouquet, in which the ends of the chromosomes are associated together on one side of the nucleus, close to the centrosome (centriole) when this is present, while the remaining parts of the chromosomes form loops into the body of the nucleus (Fig. 1b; see Hughes, 1952; Schrader, 1953; Rhoades, 1961, for reviews; Thomas and Kaltsikes, 1977 in rye). In some species the bouquet is formed during leptotene, in others during pachytene; and in some mantids it forms first in leptotene opposite the as yet undivided centrosome, lapses, and then re-forms as a double bouquet later during pachytene when the centrosomes have divided and separated (Hughes-Schrader, 1943). The bouquet involves an active interaction between the ends (not centromeres) of the chromosomes and the centrosome (when present), an interaction which must operate across the substantial width of the intact nuclear envelope (for some evidence of this intactness, see Jones, 1973). Unfortunately, formation of the bouquet has not been studied in living cells. Virrki (1974) concluded that the change from Rabl orientation to a bouquet is brought about by rotation of the nucleus relative to the centrosome, or by migration of the centrosome from close to the centromeres (in Rabl orientation) to the opposite side of the nucleus. Virrki implied that the chromosomes are not actively moved to produce the bouquet, though he did not rule out active migration of chromosome ends "alone the nuclear membrane towards the extranuclear polarizing agent" (1974, p. 339). However, in Rabl orientation the chromosome ends are but loosely associated on one side of the nucleus, and the outer areas of the nucleus are more or less entirely taken up by chromosomes. But in the bouquet, chromosome ends are closely associated; and often much of the periphery as well as the center of the nucleus is void of chromosomes (cf. Fig. 1a,b). These differences suggest that the bouquet is, in fact, the result of active chromosome movements. Also, the initial formation of a single bouquet, followed by a double bouquet in mantids, cannot possibly be accounted for without the active involvement of the chromosome ends; their migration along the membrane toward the center(s) is decidedly a more likely explanation for bouquet formation. Studies reported later in this chapter show quite clearly that chromosome ends *do* move along the inside of the nuclear membrane (see also Chapter 4 by Church in this volume).

In some plants, at least, bouquet formation is sensitive to colchicine (Thomas and Kaltsikes, 1977). Further study is very much needed.

The bouquet is a tantalizingly interesting phenomenon which deserves much more study. Correlative light and electron microscopy, for which very good techniques are now available (Nicklas *et al.*, 1979 and references therein), assuredly needs to be done in the first instance.

V. LATE PROPHASE POLARIZATIONS

The beautiful work of the Schraders (see Schrader, 1941; Hughes-Schrader, 1943; Schrader, 1953, for review) established that in spermatocytes of certain insects the bivalents during late prophase become polarized in a manner that is not entirely unlike the bouquet arrangement. Thus, in

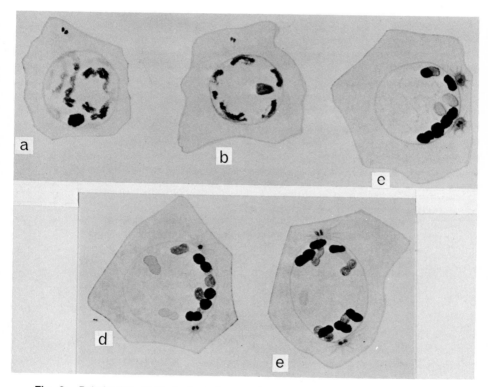

Fig. 2. Polarization of bivalents during prophase I in a spermotocyte of *Anisolabis maritima*. (a) Diplotene; note the double centriole in the cytoplasm. (b) Early diakinesis; chromosomes are peripherally oriented. (c–e) Successively later stages of diakinesis showing separation of the centrioles and polarization of chromosomes; the nuclear envelope is intact throughout. ×1800. From Schrader (1941).

the earwig *Anisolabis maritima* (Fig. 2) the bivalents in diakinesis are divided into two groups situated close to the two centers. Similar, although not well studied, examples of late prophase polarization are to be found in spermatocytes of the grasshopper *Psophus stridulus* (personal observations of Michel's 1944 commercially available film), in microsporocytes of some plants (Oksala and Therman, 1958), and in spermatocytes of *Nephrotoma suturalis* (La Fountain, 1980). Also, Heath (1978) showed by electron microscopy that mitotic prophase chromosomes of the fungus *Saprolegnia* are initially grouped together just within the nuclear envelope opposite the unseparated centrioles, but then later they are divided into two groups after (or as) the centrioles partially separate. Comparable prophase chromosome movements have now also been partially documented from living cells of the flagellate *Barbulanympha* (Ritter *et al.*, 1978). Also, the rapid, oscillatory movements of chromosomes during early stages of mitosis in the diatoms *Hantzschia* and *Nitzschia* described by Pickett-Heaps and Tippet (1980) may prove to be somewhat comparable to the similarly rapid and oscillatory prophase chromosome movements detailed below from cricket spermatocytes.

Polarization movements in prophase I have been observed in living spermatocytes of the house cricket, *Acheta domesticus* (Levine, 1966), in the Australian black cricket, *Teleogryllus commodus* (unpublished observations of A. Harris in my laboratory), and in certain spiders (Wise and Kubai, unpublished).

The movements in *Acheta* spermatocytes have now been studied in detail (Rickards, 1975). By the end of prophase I, the chromosomes lie in two groups close to the two centers (Fig. 3). This polarization develops gradually over the last hour or so of prophase. The movements involved are seen as soon as the chromosomes are dense enough to be clearly visible in living cells. They are exemplified in Fig. 4. In Fig. 4a, a bivalent lying in the optically upper (or lower) region of the nucleus rotated approximately 100° in 18 seconds. Analysis of this movement (Rickards, 1975) showed that force was applied to at least two, possibly three, parts of the bivalent. In Fig. 4b, a bivalent moved 9.5 μm around the nuclear envelope in 24 seconds. The movement was based at the initially upper end of the bivalent, and this end retained throughout an intimate association with the nuclear envelope, causing the bivalent to buckle and invert. The movement was directed toward the lower centriole. In Fig. 4c, the end of one bivalent moved 2.75 μm toward the lower right center in a series of in–out movements; then the end of another bivalent close by moved along a similar (although slightly convergent) track toward the same center.

The general characteristics of these movements are summarized as follows (see Rickards, 1975, for further details and illustrations).

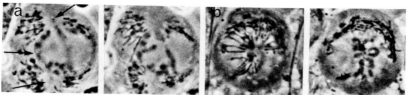

Figs. 3–6. From living spermotocytes in prophase I (diakinesis) of the house cricket, *Acheta domesticus;* phase contrast. Time in minutes (') and seconds (") at bottom right of prints. For culture technique and so on, see Rickards (1975). ×1000.

Fig. 3. Two levels of focus of each of two prophase cells, with the asters in the "side" view (a) and the "polar" view (b). In (a), the two asters, outlined by radially arranged, elongate mitochondria, lie in the same plane of focus; in (b) they are "top" and "bottom" (only the top one is illustrated). In (a), centrioles are visible in the aster center (upper and lower arrows), and the nuclear envelope is visible when viewed in optical transverse section (middle arrow). The bivalents are grouped loosely at the centers. In (b), print 1 is at the upper center and shows the radial array of mitochondria; print 2 is at the group of chromosomes just within the nuclear envelope at the center. The other aster (not shown), at which the remaining bivalents were grouped, was similar.

1. They occur during prophase (diakinesis) up to at least 1 hour before the breakdown of the nuclear envelope. The nuclear envelope can be seen clearly using phase-contrast microscopy (Fig. 3), and its disappearance (over approximately 1 minute) is easily monitored. The presence of an intact nuclear envelope in these prophase cells has been confirmed by electron microscopy (see Fig. 7).

2. Most frequently, it is the end of a chromosome that leads the movement, although sometimes it is the kinetochore; the rest of the chromosome follows. The chromosome regions at which force is applied (i.e., the leading regions) are always closely pressed to the inside of the nuclear envelope throughout a movement (Fig. 4b).

3. The movements are nearly always polarized radially to the centers (although they follow the curve of the nuclear envelope). The centrosomes (and sometimes the centrioles) can be seen at the focal point of radially arranged, elongate mitochondria (Fig. 3). As seen using the electron microscope, microtubules radiate from the centrioles, forming an aster (see Figs. 7 and 8).

4. The movements cover distances ranging from 0.5 to at least 9.5 μm (Fig. 4b,c). They occur both toward and away from the aster centers. Often a chromosome will execute in–out movements over many minutes. Movements away from a center are as frequent as those toward it, but the

average wristwatch, then the prophase chromosome movements often reach a velocity equal to that of the sweep second hand of the watch.

Four important features that also have emerged from these studies are as follows.

(i) Variability. The prophase movements are very variable. They vary as to the particular region, and the number of regions, of a chromosome at which force is applied; in the distance a chromosome travels; whether the movements are toward or away from a center; in their velocity; and in whether or not a chromosome moves. There is no obvious pattern to the movements. They give the distinct impression of being the inevitable consequence of the presence of force producing elements in the nucleus (presumably in preparation for prometaphase and anaphase).

(ii) Colcemid sensitivity. Rather surprisingly, perhaps, the movements are sensitive to colcemid (a milder derivative of colchicine). Thus, if spermatocytes are treated *in situ* for 30 minutes with 2×10^{-5} M colcemid, or if 10^{-3} M colcemid is microinjected into the immediate environment of a spermatocyte while it is being examined under the microscope, then, under both treatments, the asters disappear and chromosome movements cease. In the case of microinjection, these effects are seen within 2 or 3 minutes.

Cells treated with colcemid otherwise behave normally: chromosomes contract normally, and the nuclear envelope and nucleolus disappear. Treated cells can recover 1 hour or so after microinjection of colcemid and resume prophase chromosome movements. As a test for specificity, lumi-colcemid, which is devoid of antimitotic activity (Aronson and Inoué, 1970), has no visible effect on prophase cells: cells treated with lumi-colcemid form normal asters, show normal chromosome movements, and proceed through a normal prometaphase. Hence, in their sensitivity to colcemid and insensitivity to lumi-colcemid, these prophase movements are identical to those on the spindle in prometaphase and anaphase.

(iii) Granule movements. Within the asters of these prophase cells, there are granules of varying size and shape. These granules execute movements of a saltatory nature (see Rebhun, 1972), as illustrated in Fig. 5. What is remarkable is that the characteristics of these granule movements are identical to those of chromosome movements inside the nucleus in all essential respects: in their direction [radial to, and toward and away from, the center(s)]; in localization of force application (to only part of the granule mass); in distances covered; velocity (of the order of a micron per second); in the relationships between velocity and distances covered and motion toward and away from the center; and in their colcemid sensitivity. The only significant difference is that granule movements are not consistently and specifically associated with the nuclear envelope.

(iv) Prometaphase and anaphase movements. The two aster centers quickly become the poles of the developing spindle as the nuclear envelope disappears. During this short (1–2 minute) period, chromosome movements of the prophase type cease, while movements typical of prometaphase take over.

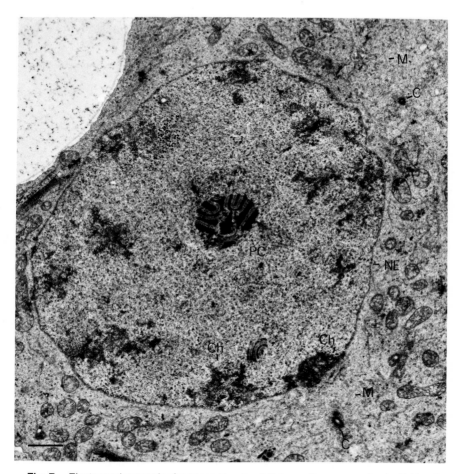

Fig. 7. Electron micrograph of a spermatocyte of *Teleogryllus commodus* in diakinesis. The two centrioles (c) each mark the center of a more or less radial array of microtubules (M) (and some membrane elements). The nuclear envelope (NE) is intact throughout, and the chromosomes (Ch) are closely appressed to the nuclear envelope; there are no microtubules in the nucleus. PC, polycomplex associated with the nucleolus. Glutaraldehyde/osmium fixation; uranyl acetate/lead citrate straining. Bar = 1.25 μm. From the work of A. Harris in my laboratory.

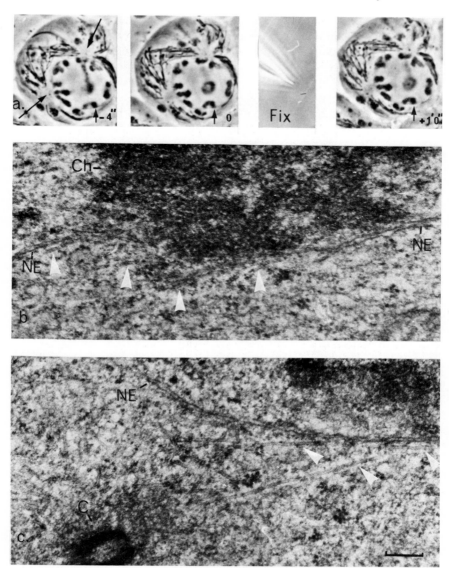

Fig. 8. Correlative living cell and electron microscopic observations of a spermatocyte of *Acheta domesticus* in diakinesis. (a) Light microscopic observations of the living cell and after its fixation. Prior to fixation, the two ends of the arrowed bivalent had shown numerous movements along the nuclear envelope toward and away from the lower left center (long arrow). The last of these movements is illustrated in the −4″ to zero time prints; the right-hand end (arrow) moved closer to the left-hand end. At zero time, microinjection of 6% glutaraldehyde began, with continuous observation of the cell under the oil-immersion objective. The

Prometaphase and anaphase chromosome movements on the spindle contrast sharply with those of prophase. First, in prophase the chromosomes are
intimately appressed to the inside of the nuclear envelope, but in prometaphase and anaphase they are intimately attached to the spindle by microtubules. Second, in prophase it is predominantly the ends of chromosomes, uncommonly the kinetochores, at which force is applied, and often
more than one part of a chromosome is active at one time. But on the
spindle, usually only the kinetochore is active (see Bajer and Östergren,
1961, for example, for activity of chromosome ends on the spindle). Third, the
relatively high velocity of prophase chromosome movements is reduced systematically when chromosomes associate with the spindle. Thus, some very
early prometaphase movements are (only) marginally slower (commonly
0.1–0.2 μm/sec, up to 0.5 μm/sec) than those of prophase; later prometaphase movements are distinctly slower (1.5–2.0 μm/min $= c.$ 0.03 μm/sec),
and during anaphase the movements are much slower still (0.25–0.5 μm/
min). Hence there is a systematic reduction in the velocity of chromosome
movement from prophase to anaphase. Fourth, the movements of chromosomes in prophase are saltatory, but on the spindle chromosomes regularly
move over extended distances without pause. Fifth, prophase movements
are bidirectional, but, on the spindle, chromosome movements are unidirectional in that they occur toward a center (pole) only; active movements away
from a center do not occur (see later dicussion of this point).

VI. RELEVANCE, EXPLANATIONS, AND IMPLICATIONS

The mechanism of poleward chromosome movement on the spindle unfortunately is still poorly understood (Sec. I). A prevalent view today is that
chromosomes move either through lengthening and shortening
(polymerization–depolymerization) of kinetochore microtubules or by relative sliding of kinetochore microtubules with continuous (interpolar) microtubules (see Nicklas, 1971, 1975; McIntoch *et al.*, 1975; Forer, 1978a, for

---

micropipette (0.5 μm internal diameter) and its light diffraction image are illustrated. The 1'0″
print shows the cell after microfixation (note the halo of glutaraldehyde around the cell).
Chromosome movement ceased almost immediately after the first injection of glutaraldehyde. After microfixation, the cell was processed for electron microscopy. See Nicklas *et
al.* (1979) for full details of the methods. (b, c) Electron micrographs of the same cell in (a),
showing the right-hand end of the relevant bivalent (b) and part of the lower left aster (c). In
(b), the nuclear envelope (NE) is indistinct where the chromosome (Ch) is closely appressed
to it. A number of obliquely sectioned (and thus somewhat indistinct) short lengths of microtubules (white arrowheads) lie close to the outside of the nuclear envelope. Their origin
could not be determined. In (c), many microtubules (white arrowheads) radiate from the
centriole region (C); some lie very close to the nuclear envelope (NE). Bar = 0.25 μm.

reviews; Margolis *et al.*, 1978). These hypotheses (and others; e.g., Bajer, 1973) imply that spindle microtubules are the force producers for chromosome movements and that chromosomes on the spindle move passively as a result of poleward transport of kinetochore microtubules. In other words, kinetochore microtubules both produce and transmit the force for chromosome movement. Moreover, the relatively low velocity of chromosome movement on the spindle, which implies low energy expenditure, has been viewed as indicative of a system of control aimed at limiting velocity in order to ensure precision (Nicklas, 1975 and others).

However, a hypothesis now receiving widespread attention is that microtubules are not themselves the force producing elements of the spindle, but that this role is played by proteins of the actin and myosin type. Spindle microtubules are then viewed as elements that have perhaps a number of supportive roles—orienting, polarizing, and controlling the force producers, for example. It is currently very much debated whether or not actin is a bona fide component of spindle fibers (see Forer, 1978a; Petzelt, 1979, for reviews; Aubin *et al.*, 1979). Further, while it is clear that force on the spindle is generated in kinetochore spindle fibers (see reviews cited above), it is not entirely clear that poleward transport of spindle fibers in itself moves chromosomes poleward. [Even the oft-appealed-to experiments of Inoué (1976) and Salmon (1976), and also those of Begg and Ellis (1979), are not decisive on this point.] Indeed, the motile behavior of kinetochores and chromosome ends in prophase suggests that in the spindle also the kinetochore itself might be actively and directly involved in interactions with the spindle force producers. Poleward transport of kinetochore microtubules in anaphase, for which there is some evidence (Hard and Allen, 1977 and references therein; see, however, Forer, 1976), might then simply be an associate of chromosome transport. The low velocity of chromosome movement on the spindle might then result from the use of most of the available spindle force in poleward transport of kinetochore microtubules, thereby leaving little force available for kinetochore (chromosome) transport. These alternative views are given further attention later.

It seems unlikely that the mechanism of poleward chromosome movement in anaphase is fundamentally different from the mechanism of movement in prometaphase. Temporally, the movements are very close; and both are inseparably associated with the spindle and have very similar characteristics (see Nicklas, 1971, for review). The same conclusion probably applies to chromosome movements in prophase, on the one hand, and prometaphase and anaphase, on the other (Sec. I; Schrader, 1953, p. 58). Although the characteristics of the prophase movements of cricket spermatocytes appear somewhat different from those of prometaphase and anaphase, these differences are probably indicating much that is relevant to our understanding of the spindle mechanism. Moreover, since the characteristics of granule movements in asters are identical to those of chromosome movements within the prophase nucleus, the mechanism of granule movement is probably identical basically to that of prophase chromosome movements. Hence the view adopted in the following discussion is that polarized movements of chromosomes in prophase, of cytoplasmic granules in asters, and of chromosomes (and granules) on the spindle are brought about by one and the same basic mechanism; and thus the characteristics of all three classes of movement should be examined together if we are to understand this basic mechanism.

The two major features of the late prophase chromosome movements in cricket spermatocytes to be explained are as follows (see Rickards, 1975, for additional points and references).

1. Their directionality and colcemid sensitivity. Chromosome movements in prophase are focused on the centrioles, as are both granule movements in the asters and chromosome movements on the spindle. There seems little doubt that in all these movements the centrioles are acting as centers on which the force-producing elements concerned are focused.

Movements toward and away from the centers in asters and in the spindle proceed along paths delineated by microtubules focused on the centriole (see, for example, Nicklas, 1971; Bajer and Molè-Bajer, 1972; Fuge, 1974 for chromosomes on the spindle; Rebhun, 1972, for granules in asters). These movements are sensitive to the "antimicrotubule" drug colchicine (and colcemid) but not to lumi-colchicine (see Margulis, 1973, for review), from which it is concluded that microtubules play a vital role in the movements by acting at least as structural elements along which force producers are oriented, if not as the force producers themselves. The same conclusion is drawn for chromosome movements in prophase nuclei: their directionality in relation to the centrioles and their colcemid (but not lumi-colcemid) sensitivity—and, indeed, also their similarities to granule movements in asters—lead to the conclusion that aster microtubules (Fig. 8) are in some way involved in producing these prophase chromosome movements.

2. Location and nature of the force producer(s). Theoretically, the force producer(s) involved in prophase chromosome movements might be located outside the nuclear envelope, or might be integrated with the envelope or be inside it. The directionality of the movements relative to the centers, their colcemid sensitivity, and their similarities to aster-associated granule movements might mean that microtubules of the asters are the force producers. This would require a dynamic molecular interaction across the substantial width of the nuclear envelope, which, although not entirely unlikely, is difficult to entertain seriously. Alternatively, a localized change in the nuclear envelope might occur such that the chromosomes can interact directly

with aster microtubules. Then the chromosomes might move laterally along
the aster microtubules (which might produce the force or orient another
force producer) in the same way that granules move along microtubules
(Rebhun, 1972; Murphy and Tilney, 1974). The localized change in the
nuclear envelope might be propagated with the movement along the membrane so as to maintain its intactness. This model, which is not very different
from the preferred model described later, would explain the colcemid sensitivity of the movements, their directionality, and their similarity to
movements of granules in asters. Electron microscopy has so far not revealed
associations of chromosomes with aster microtubules, but the nuclear envelope does lack clarity where chromosomes lie closely appressed to it (Fig.
8).

Force production as a property of the nuclear envelope *per se* warrants
close scrutiny, since the part of the chromosome actively involved in movement is closely appressed to the inside of the nuclear envelope. Thus,
force-producing elements with which chromosomes interact may be integral
components of the nuclear envelope. However, this would leave unex-
plained both the directionality and colcemid (but not lumi-colcemid) sensitivity of the movements. [Although in some cells colchicine binds to the
nuclear envelope, so also does lumi-colchicine (Stadler and Franke, 1974).]
Alternatively, the movements might be brought about by polarized transport
of the nuclear envelope, accompanied or even brought about by formation of
new membrane material "behind" a movement and withdrawal "in front"
(see Sec. V). This would transport chromosomes which are attached to the
membrane. But then one would expect in-concert movements of groups of
chromosomes to be common, not rare; and independent movements of the
type shown in Fig. 4c would hardly be expected. Also, the high velocity of
chromosome movements in prophase hardly suggests membrane transport
as a causative agent, particularly since there is no ruffling of the nuclear
envelope behind or in front of a movement.

Models that invoke force production as a property of the nuclear membrane *per se* imply that the mechanism of chromosome movement in prophase is different from that involved in granule movements in asters and
chromosome movements on the spindle. It is unlikely that membranes are
directly involved as force-producing elements in either of these cases. Moreover, prophase chromosome movements in *Acheta* occur despite the absence of an intact nuclear envelope (Fig. 6).

Since the prophase movements occur within the *Acheta* nucleus, the
model I currently favor predicts linear force producers within the nucleus
itself (but closely associated with the inside of the nuclear envelope). This
rules out the direct involvement of microtubules in force production, since
there are no microtubules within *Acheta* prophase nuclei, as in nuclei in

other organisms (Bajer and Molè-Bajer, 1969, 1972; Roos, 1973). This conclusion raises questions concerning the directionality of the movements and their colcemid sensitivity and has implications regarding the role of microtubules in granule movements in asters and chromosome movements on the spindle. The most plausible hypothesis is that for all these movements a two-component system operates in which microtubules play mainly a skeletal role and other elements, probably actin-like and myosin-like proteins, function as the force producers *per se*. The force producers ("muscles") are oriented along the skeletal elements ("bones"). Removal of the skeletal elements (by colchicine) would mean that directed movements could not occur despite the continued presence, in an unoriented form, of the force-producing components. A two-component system has been advocated from other lines of evidence for both granule movements in asters and chromosome movements on the spindle (see Rebhun, 1972; Forer, 1974, 1978a, for reviews).

VII. THE MOTILE MECHANISM: FACTS AND CURRENT PERSONAL VIEWS

I conclude, from the characteristics of prophase chromosome movements in crickets, that microtubules are not involved directly as force producers in chromosome movements. By implication, and in the light of additional evidence (see Forer, 1978a, for review), it is probable that actin–myosin interactions form the basis of force production in chromosome movements in prophase and in prometaphase and anaphase, as in other forms of intracellular motility (see Pollard and Weihing, 1974, for review). The basic mechanism of motility that these data collectively suggest is illustrated in its simplest form for granule movements in the aster. Thus, microtubules are polymerized at centrioles as nucleating centers (see Kirschner, 1978, for review). Along these skeletal elements, actin filaments are oriented and are then anchored in the cytoplasm. Myosin in particulate form (Pollard and Korn, 1972) is adsorbed onto the surface of granules; the myosin interacts with the anchored actin filaments to propel the "attached" granule along the filament surface (Schmitt, 1969). The motile mechanism is thus essentially the same as that in muscle contraction (Huxley, 1976): if the bipolar myosin filaments of skeletal muscle, which normally act against opposed actin filaments and remain stationary were to be bisected transversely, the myosin "half-filaments" would (presumably) slide toward the Z lines relative to the stationary actin filaments.

For chromosome movements in prophase nuclei the model predicts the presence of a static, structural (rather than dynamic) complex between microtubules of the asters, the nuclear envelope, and actin filaments within the inside of the nuclear envelope of the type known to exist in some organisms (see Franke and Scheer, 1974, for review). Myosin is adsorbed onto chromosome ends and kinetochores, which thereby interact directly with the actin filaments to be moved in the same manner as with granules in asters.

Finally, for chromosome movements on the spindle, the model suggests that microtubules are polymerized on kinetochores acting as nucleating centers (see Kirschner, 1978, for a review); and then actin filaments are oriented in relation to the kinetochore microtubules and anchored to the rest of the spindle. Myosin is adsorbed onto kinetochore microtubules, which are thereby transported poleward through myosin interactions with actin. Now, one possibility is that this poleward transport of kinetochore microtubules transports the attached chromosome. Indirect chromosome transport in this manner need not compromise basic principles of the model, because if the kinetochore microtubules are viewed as extensions of the kinetochore, then the mechanism of movement remains the same as for granules and prophase chromosomes. Alternatively, poleward transport of kinetochore microtubules might be associated with chromosome movement rather than a direct cause of it. In this case, the model predicts that myosin will also be adsorbed onto the kinetochore (as first suggested by Gawadi, 1974); and thence, chromosome movement proceeds in a manner more strictly identical to that for chromosome and granule movements in prophase, i.e., through direct interaction with the force producer.

Although the above model relates closely, in mechanistic terms, the movements of chromosomes in prophase nuclei with those of granules in asters and with chromosome movements on the spindle, it does not immediately explain the sharp changes in the characteristics of the movements as chromosomes become associated with the spindle. These changes concern (1) the chromosome region at which force is applied, (2) the polarity, and (3) the velocity of the movements. These changes are initiated at the breakdown of the nuclear envelope (which, the model implies, leads to dissociation of the chromosome from its prophase motile mechanism) and are followed by the development of spindle fibers at kinetochores (and, thus, association of the chromosome with the spindle motile mechanism). The differing characteristics of the movements might then be explained as follows.

1. From chromosome ends to kinetochores. In prophase, it is predominantly the chromosome ends that are actively involved in movement, while the kinetochores are almost exclusively involved in spindle movements. Now, while chromosome ends are, in general, frequently found associated with the nuclear envelope (Comings and Okada 1972, for example), in the spindle it is the kinetochore where chromosomal microtubules usually develop. In each case, this represents association of the chromosome with that

part of the cell (nuclear envelope or kinetochore microtubules, respectively) at which force production is confined. Thus, chromosome ends not attached to the nuclear envelope in prophase do not move; and directed movement of granules and chromosomes in the spindle does not occur outside the immediate environment of kinetochore spindle fibers (see Rickards, 1975, p. 445, for summary). Thus, the kinetochore microtubules are viewed not solely as structural elements but also as ones whereby force in the spindle is now available to only one part, and consistently the same part, of a chromosome (the kinetochore): the kinetochore microtubules in some way preempt an otherwise more generally available force-producing mechanism in the spindle. At the same time, the assured, firm attachment of chromosomes to the spindle is perhaps the mechanism whereby the saltatory characteristic of chromosome movement in prophase is eliminated: assured attachment guarantees force application to the kinetochore.

2. Polarities. The bipolarity of prophase chromosome movements, and of granules in the spindle (both toward and away from a center), might imply the presence of separate linear force producers with opposite polarities. However, the movements of both chromosomes and granules toward a center are different (in amplitude and velocity) from movements in the opposite direction (Sec. V) and movements away from a center often follow inward movements without detectable pauses or alterations in the angle of movement. These observations suggest that force production is, in fact, occurring in both directions along one and the same force producer (for a parallel in stretching of stimulated muscle, see Curtin and Davies, 1973). The implication is that the force producers are all polarized in one direction, probably in correspondence with inherent polarities of the aster microtubules: nucleation of microtubules at the centriole will ensure production of a microtubule array with a single polarity. The conclusion is that microtubules both orient and polarize the force producers.

Chromosome movements in both prometaphase and anaphase are best accounted for by force operating from a kinetochore toward the pole to which they are moving (Nicklas, 1971). Hence, movements of chromosomes away from a center in prophase are suppressed when chromosomes become associated with the spindle. (Congressional movements to the equator of a bivalent or mitotic chromosome during prometaphase are best explained by active movement of one kinetochore toward the more remote pole and passive movement of the other kinetochore away from the nearer pole. This important aspect of chromosome movement on the spindle, however, is very incompletely researched and understood.)

The unidirectionality of force production in the spindle implies that all the force producers in kinetochore fibers are polarized in one direction, in correspondence with inherent polarities of kinetochore microtubules: nucleation

of microtubules at the kinetochore (or kinetochore attachment to existing aster microtubules) will produce a microtubule array with a single
polarity. The suppression of active reverse movements in kinetochore
fibers might result from direct linkage of the kinetochore to spindle microtubules, which permit movement toward a pole but resist movement away
from it (see Nicklas, 1971; Begg and Ellis, 1979). Movements away from a pole
do occur during prometaphase, for example, in which one set of kinetochore
microtubules elongates while the other set shortens. Perhaps these reverse
movements are permitted because force and thus movement of one kinetochore
toward the one (remote) pole might set up special conditions allowing for
growth of kinetochore microtubules connected to the opposite pole.

   3. Velocity. The sequential decline in velocity of chromosome movements
from prophase to anaphase (from a micron per second to a micron per minute) may imply the purposeful sequential development of a regulatory system in the spindle such that velocity is now limited in order to achieve
precision. The rate of depolymerization of kinetochore microtubules might
regulate (limit) the rate of chromosome movement (irrespective of whether
or not kinetochore microtubules are the spindle force producers; Nicklas,
1971, 1975; Forer, 1974, 1978a and others). Although microtubule transport
and depolymerization probably take place during poleward movement of
chromosomes on the spindle, such transport and depolymerization are (presumably) unnecessary in chromosome movements in prophase nuclei (or
granule movements in asters). Hence, the absence of a requirement for
microtubule transport in prophase might mean that the force-producing
system is free of just the mechanism which, in the spindle, limits velocity.
Hence, chromosome movements in prophase have higher velocities.

   Alternatively, the reduced velocity in prometaphase and anaphase might
result from a reduction in that portion of the total energy available that is
manifest directly as chromosome motion *per se*. Thus, at prometaphase, and
then more so at anaphase, an increased proportion of the total energy available might be channeled into kinetochore microtubule transport that occurs
with but (in this view) is not the cause of chromosome transport. This implies
that in prophase a greater proportion of total energy is chaneled into actual
chromosome movement, because microtubule transport does not occur. The
spindle situation can then be likened to that of a person traveling along a
bush track which is being cleared by a working party ahead. The person's
energy expenditure and rate of progress ($\cong$ chromosome movement) are
limited by the high energy expenditure and slower progress of the working
party ($\cong$ microtubule transport and depolymerization). Whatever the truth,
the reduced variability of chromosome velocity on the spindle, compared to
that in prophase, is viewed as deriving from elimination of the higher velocities, which in prophase are possible because of greater energy input.

In light of the above, the behavior on the spindle of granules and chromosome fragments without kinetochores (akinetochoric bodies) is especially interesting. It is now well established that such bodies show polarized movements on the spindle with characteristics that are very similar to those of chromosome movements on the spindle (see Nicklas and Koch, 1972; Rickards, 1975, for reviews). Thus, movement is confined to the region between kinetochores and the pole; the movements are polarized and have velocities of about a micron per minute. There has been debate as to whether or not these movements disclose features relevant to an understanding of chromosome movement on the spindle (Nicklas, 1971). The conclusion that they are indeed relevant is now compelling. This is because the movements of chromosomes and granules that are not associated with the spindle (in prophase nuclei) have characteristics that are very similar to each other; and this similarity is preserved when chromosomes and granules associate with the spindle despite the changes in the characteristics of the movements.

The polarized movements of akinetochoric bodies on the spindle are to some extent saltatory in their initiation and their progress toward a pole (see data in Nicklas and Koch, 1972). Although the saltatory nature of the movements is not as obvious as that for granules (and chromosomes) associated with asters, this is at least partly illusory, because the movement is confined to kinetochore spindle fibers and because of their relative slowness. Moreover, while the limited data presently available suggest a preference for these movements to be poleward, reverse movements do occur (summary in Nicklas and Koch, 1972). Thus, although the movements of akinetochoric bodies on the spindle have certain characteristics in common with chromosome movements on the spindle, they retain some characteristics of granule movements in asters. This might reflect the fact that akinetochoric bodies are associated with but not attached to the spindle by microtubules.

VIII. CONCLUSIONS AND PROSPECTS

The discussions presented in this chapter suggest that the study of chromosome movements in prophase nuclei has contributed much information and insight relevant to an understanding of the mechanism of chromosome movement on the spindle. We now have detailed knowledge of the characteristics of one class of chromosome movements in prophase nuclei. But although this knowledge provides suggestions as to what is happening in prophase and on the spindle, they do not give definitive answers. Much has yet to be learned. Still needed are combined light and electron microscopic studies of the region where motile chromosome ends are associated with the nuclear envelope in chromosomes whose immediate history of movements has been documented. We now have good technical procedures for doing this (Fig. 8 and Nicklas *et al.*, 1979), although progress is still difficult. In this approach lies the distinct possibility that the model currently preferred can be refuted, since direct associations of chromosomes with microtubules through the nuclear membrane might be revealed—either end-on associations, as in some dinoflagellates (see Kubai, 1978, for a review), or lateral associations, as in granule movements in asters (Rebhun, 1972; Murphy and Tilney, 1974; see Sec. VI). And, on the positive side, the predicted structural complex between aster microtubules, the nuclear envelope, and force-producing fibers within the nuclear envelope must be documented. The problem of adequate combined fixation of microtubules, membranes and, particularly, actin (Maupin-Szamier and Pollard, 1978; Forer, 1978b) will need close attention. Also, a direct attack on the possible involvement of actin and myosin in prophase movements should be made through heavy meromyosin decoration and/or immunofluorescence studies (see Forer, 1978a, for review), although initial attempts at this have been technically discouraging.

Chromosome movements in prophase nuclei (and granule movements in asters) have called into question some long held but unproven views concerning the mechanism of chromosome movement in the spindle. They suggest that spindle (and aster) microtubules are probably not directly involved as force-producing elements, but instead play indirect (but equally vital) roles. Thus, kinetochore microtubules can be viewed as structural elements that orient and polarize the force producers; as elements that ensure the alignment and thus the confinement, and indeed the guarantee of steady force production in the spindle to one specific localized part of the chromosome; and as elements that suppress active chromosome movements away from a pole. These features are vital to the orderly orientation of chromosomes during prometaphase and thus their orderly separation during anaphase. Hence kinetochore microtubules are involved in the imposition of order on a set of chromosome movements that, in prophase nuclei, are saltatory, variable, and somewhat disordered and that, if transferred untamed, as it were, to the spindle would hardly be expected to promote orderly orientation and separation of chromosomes in mitosis and meiosis.

Some aspects of the model currently favored are more controversial than others, as, for example, the view that kinetochores might interact actively and directly with force-producing elements in spindle fibers. They are presented with the intention of promoting more consideration of them and of the characteristics of prophase chromosome movement from which they derive. The unfortunately meager data currently available on the polarity of actin fibers in the spindle (deduced from heavy meromyosin arrowhead directions; see Forer, 1978a, for review) are compatible with the model presented, as opposed to others; actin–microtubule associations have been observed in the spindle (see Forer, 1976, 1978a, for reviews); and there is documentation of direct linkage of actin filaments with kinetochores (Forer *et al.*, 1979).

The spindle model predicts that kinetochore microtubules themselves are transported poleward in response to their reaction with actin force producers. This transport is coupled with poleward chromosome movement in anaphase but might be dissociated from it in metaphase. Although microtubule transport during metaphase is not a vital component of the model, the constant poleward movement of material in spindle fibers (Hard and Allen, 1977, and references therein) suggests that such transport during metaphase does occur. Maintenance of kinetochore linkage to microtubules during poleward migration in metaphase would require that tubulin subunits be incorporated into kinetochore microtubules at the proximal (kinetochore) end, as in the model of Margolis *et al.* (1978). The exciting studies of Summers and Kirschner (1979) and of Bergen *et al.* (1980), indicating growth of microtubules by addition of subunits distal to the kinetochore nucleation site, might seem to challenge this requirement. But microtubule growth can occur at both ends under certain conditions (see above references); and therefore the possibility remains that addition of subunits at the kinetochore might take place *in vivo* during poleward transport of a fully grown microtubule, even though the initial production of the kinetochore microtubule had taken place by distal addition.

In mitosis of certain dinoflagellates, chromosome movements take place under conditions that are remarkably similar to those for prophase chromosome movements in cricket spermatocytes—that is, they take place within an intact nuclear envelope, without direct linkage to the extranuclear spindle (see Kubai, 1978, for review). Moreover, there is evidence (although this is incomplete) that the well-known and enigmatic anaphase I movements of paternal, as opposed to maternal, chromosomes in spermatocytes of *Sciara* (see Schrader, 1953, for review) are similar in many respects to the movements of prophase chromosomes in cricket spermatocytes (see the unpublished work of Zilz, 1970). Thus, chromosome movements which are not, or not directly, associated with a spindle are found in prokaryotes and in both primitive and more advanced eukaryotes; and they are found in the same cell in which the spindle does operate. This is intriguing from the point of view of the evolution of the mitotic mechanism (Kubai, 1978). Further study of these movements cannot fail to be rewarding.

ACKNOWLEDGMENTS

Thanks are due to Ken Ryan and Alfred Harris for their suggestions and stimulating discussion; and to Dr. N. Paweletz for a gratis copy of Mayer's film of mitosis in HeLa cells and for his helpful comments. This work was supported in part by research grant GM-13745 from the Division of Medical Sciences DHEW to Dr. R. B. Nicklas, Duke University, North Carolina.

REFERENCES

Aronson, J., and Inoué, S. (1970). Reversal by light of the action of N-methyl N-desacetyl colchicine on mitosis. *J. Cell Biol.* **45**, 470–477.

Aubin, J. E., Weber, K., and Osborn, M. (1979). Analysis of actin and microfilament-associated proteins in the mitotic spindle and cleavage furrow of PtK₂ cells by immunofluorescence. A critical note. *Exp. Cell Res.* **124**, 93–109.

Bajer, A. (1959). Changes of length and volume of mitotic chromosomes in living cells. *Hereditas* **45**, 579–596.

Bajer, A. (1973). Interaction of microtubules and the mechanism of chromosome movement (zipper hypothesis). I. General principle. *Cytobios* **8**, 139–160.

Bajer, B., and Molè-Bajer, J. (1956). Cine-micrographic studies on mitosis in endosperm. II. Chromosome, cytoplasmic and Brownian movements. *Chromosoma* **7**, 558–607.

Bajer, A., and Molè-Bajer, J. (1969). Formation of spindle fibres, kinetochore orientation and behavior of the nuclear envelope during mitosis in endosperm. Fine structural and *in vitro* studies. *Chromosoma* **27**, 448–484.

Bajer, A., and Molè-Bajer, J. (1972). Spindle dynamics and chromosome movements. *Int. Rev. Cytol. Suppl.* 3, **34**, 1–271.

Bajer, A., and Östergren, G. (1961). Centromere-like behaviour of non-centromere bodies. I. Neocentric activity in chromosome arms at mitosis. *Hereditas* **47**, 563–598.

Begg, D. A., and Ellis, G. W. (1979). Micromanipulation studies of chromosome movement. I. Chromosome-spindle attachment and the mechanical properties of chromosomal spindle fibres. *J. Cell Biol.* **82**, 528–541.

Bergen, L. G., Kuriyama, R., and Borisy, G. G. (1980). Polarity of microtubules nucleated by centrosomes and chromosomes of chinese hamster ovary cells *in vitro*. *J. Cell Biol.* **84**, 151–159.

Comings, D. E., and Okada, T. A. (1970). Condensation of chromosomes onto the nuclear membrane during prophase. *Exp. Cell Res.* **63**, 471–473.

Comings, D. E., and Okada, T. A. (1972). Architecture of meiotic cells and mechanisms of chromosome pairing. *Adv. Cell Mol. Biol.* **2**, 309–384.

Curtin, N. A., and Davies, R. E. (1973). Chemical and mechanical changes during stretching of activated frog skeletal muscle. *Cold Spring Harbor Symp. Quant. Biol.* **37**, 619–629.

Fell, H. B., and Hughes, A. F. (1949). Mitosis in the mouse: A study of living and fixed cells in tissue culture. *Q. J. Micros. Sci.* **90**, 355–380.

Forer, A. (1974). Possible roles of microtubules and actin-like filaments during cell division. *In* "Cell Cycle Controls" (J. M. Padilla, I. L. Cameron, and A. M. Zimmerman, eds.), pp. 319–335. Academic Press, New York.

Forer, A. (1976). Actin fibres and birefringent spindle fibres during chromosome movements. *In* "Cell Motility," Book C (R. Goldman, T. Pollard, and J. Rosenbaum, eds.), pp. 1273–1293. Cold Spring Harbor Lab., Cold Spring Harbor, New York.

Forer, A. (1978a). Chromosome movements during cell division: Possible involvement of actin filaments. *In* "Nuclear Division in the Fungi" (I. B. Heath, ed.), pp. 21–88. Academic Press, New York.

Forer, A. (1978b). Electron microscopy of actin. *In* "Principles and Techniques of Electron Microscopy" (M. A. Hyat, ed.), Vol. 9, pp. 126–174. Van Nostrand-Reinhold, Princeton, New Jersey.

Forer, A., Jackson, W. T., and Engberg, A. (1979). Actin in spindles of *Haemanthus katherinae* endosperm. II. Distribution of actin in chromosomal spindle fibres determined by analysis of serial sections. *J. Cell Sci.* **37**, 349–371.

Franke, W. W., and Scheer, U. (1974). Structure and function of the nuclear envelope. *In* "The Cell Nucleus" (H. Busch, ed.), Vol. 1, pp. 219–347. Academic Press, New York.

Fuge, H. (1974). Ultrastructure and function of the spingle apparatus, microtubules and chromosomes during cell division. *Protoplasma* **82**, 289–320.

Gawadi, N. (1974). Characterization and distribution of microfilaments in dividing locust testis cells. *Cytobios* **10**, 17–35.

Gelei, J. (1921). Weitere Studien über die Oögenese von *Dendrocoelum*, II. *Arch. Zellforsch.* **16**, 88.

Hard, R., and Allen, R. D. (1977). Behaviour of kinetochore fibres in *Haemanthus katherinae* during anaphase movements of chromosomes. *J. Cell Sci.* **27**, 47–56.

Heath, I. B. (1974). Genome separation mechanisms in prokaryotes, algae and fungi. *In* "The Cell Nucleus" (H. Busch, ed.), Vol. 2, pp. 487–515. Academic Press, New York.

Heath, I. B. (1978). Experimental studies of mitosis in the fungi. *In* "Nuclear Division in the Fungi" (I. B. Heath, ed.), pp. 89–176. Academic Press, New York.

Hughes, A. (1952). "The Mitotic Cycle: The cytoplasm and nucleus during interphase and mitosis." Butterworth, London.

Hughes-Schrader, S. (1943). Polarization, kinetochore movements, and bivalent structure in the meiotic chromosomes of male mantids. *Biol. Bull. (Woods Hole, Mass.).* **85**, 265–300.

Huxley, H. E. (1976). Introductory remarks: The relevance of studies on muscle to problems of cell motility. *In* "Cell Motility," Book A (R. Goldman, T. Pollard, and J. Rosenbaum, eds.), pp. 115–126. Cold Spring Harbor Lab., Cold Spring Harbor, New York.

Inoué, S. (1976). Chromosome movement by reversible assembly of microtubules. *In* "Cell Motility," Book C (R. Goldman, T. Pollard, and J. Rosenbaum, eds.), pp. 1317–1328. Cold Spring Harbor Lab., Cold Spring Harbor, New York.

Jockusch, B. M., Brown, D. F., and Rusch, H. P. (1971). Synthesis and some properties of an actin-like nuclear protein in the slime mould *Physarum polycephalum. J. Bact.* **108**, 705–714.

Jockusch, B. M., Ryser, U., and Behnke, O. (1973). Myosin-like protein in *Physarum* nuclei. *Exp. Cell Res.* **76**, 464–466.

John, B., and Lewis, K. R. (1965). The meiotic system. Protoplasmatologia Band VI F1, Springer-Verlag, Wien.

John, B., and Lewis, K. R. (1969). The chromosome cycle. Protoplasmatologia Band VI B, Springer-Verlag, Wien.

Jones, G. H. (1973). Light and electron microscope studies of chromosome pairing in relation to chiasma localization in *Stethophyma grossum* (Orthoptera; Acrididae). *Chromosoma* **42**, 145–162.

Kirschner, M. W. (1978). Microtubule assembly and nucleation. *Int. Rev. Cytol.* **54**, 1–71.

Kubai, D. F. (1978). Mitosis and fungal phylogeny. *In* "Nuclear Division in the Fungi" (I. B. Heath, ed.), pp. 177–229. Academic Press, New York.

LaFountain, J. R., Jr. (1980). Chromosome movements during meiotic prophase. *J. Cell Biol.* **87**, 237a.

Levine, L. (1966). *In vitro* male meiosis in the domestic cricket (*Achaeta domesticus*). *Cytologia* **31**, 438–451.

McIntosh, R., Cande, W. Z., and Snyder, J. A. (1975). Structure and physiology of the mammalian mitotic spindle. *In* "Molecules and Cell Movement" (S. Inoué and R. E. Stephens, eds.), pp. 31–76. Raven, New York.

Margolis, R. L., Wilson, L., and Kiefer, B. I. (1978). Mitotic mechanism based on intrinsic microtubule behaviour. *Nature (London)* **272**, 450–52.

Margulis, L. (1973). Colchicine-sensitive microtubules. *Int. Rev. Cytol.* **34**, 333–361.

Maupin-Szamier, P., and Pollard, T. D. (1978). Actin filament destruction by osmium tetroxide. *J. Cell Biol.* **77**, 837–852.

Mayer, A. (1955). Film des Institutes für experimentelle Krebsforschung der Universität Heidelberg.

Mazia, D. (1961). Mitosis and the physiology of cell division. *In* "The Cell" (J. Brachet and A. E. Mirsky, ed.), Vol. III, pp. 80–412. Academic Press, New York.

Michel, K. (1944). Hochschulfilm C443/1944 der Reichsanstalt für Film und Bild in Wissenschaft und Unterricht, Berlin.

Murphy, D. B., and Tilney, L. G. (1974). The role of microtubules in the movement of pigment granules in teleost melanophores. *J. Cell Biol.* **61**, 757–779.

Nicklas, R. B. (1971). Mitosis. *Adv. Cell Biol.* **2**, 225–294.

Nicklas (1972)

Nicklas, R. B. (1975). Chromosome movements: Current models and experiments on living cells. *In* "Molecules and Cell Movement" (S. Inoué and R. E. Stephens, eds.), pp. 97–117. Raven, New York.

Nicklas, R. B. (1977). Chromosome distribution: Experiments on cell hybrids and *in vitro*. *Philos. Trans. R. Soc. London, Ser. B* **277**, 267–276.

Nicklas, R. B., and Koch, C. A. (1972). Chromosome micromanipulations. IV. Polarized motions within the spindle and models for mitosis. *Chromosoma* **39**, 1–26.

Nicklas, R. B., Brinkley, B. R., Pepper, D. A., Kubai, D. F., and Rickards, G. K. (1979). Electron microscopy of spermatocytes previously studied in life: Methods and some observations on micromanipulated chromosomes. *J. Cell Sci.* **35**, 87–104.

Oksala, T., and Therman, E. (1958). The polarized stages in meiosis of liliaceous plants. *Chromosoma* **9**, 505–513.

Paweletz, N. (1974). Electronenmikroskopische Untersuchungen an frühen Studien de Mitose bei HeLa-Zellen. *Cytobiologie* **9**, 368–390.

Petzelt, C. (1979). Biochemistry of the mitotic spindle. *Int. Rev. Cytol.* **60**, 53–92.

Pickett-Heaps, J. D., and Tippit, D. H. (1980). Light and electron microscopic observations on cell division in two large diatoms, *Hantzschia* and *Nitzschia*. I. Mitosis *in vivo*. *Europ. J. Cell Biol.* **21**, 1–11.

Pollard, T. D., and Korn, E. D. (1972). The "contractile" proteins of *Acanthamoeba castellanii*. *Cold Spring Harbor Symp. Quant. Biol.* **37**, 573–583.

Pollard, T. D., and Weihing, R. R. (1974). Actin and myosin and cell movement. *C.R.C. Crit. Rev. Biochem.* **2**, 1–65.

Rabl, C. (1885). Über Zelltheilung. *Gegenbaurs Morphologisches Jahrbuch* **10**, 214–330.

Rasmussen, S. W. (1976). The meiotic prophase in *Bombyx mori* females analysed by three-dimensional reconstruction of synaptonemal complexes. *Chromosoma* **54**, 245–293.

Rebhun, L. I. (1972). Polarized intracellular particle transport: Saltatory movements and cytoplasmic streaming. *Int. Rev. Cytol.* **32**, 93–137.

Rhoades, M. M. (1961). Meiosis. *In* "The Cell" (J. Bracket and A. E. Mircky, eds.), Vol. III, pp. 1–75. Academic Press, New York.

Rickards, G. K. (1975). Prophase chromosome movements in living house cricket spermatocytes and their relationship to prometaphase, anaphase and granule movements. *Chromosoma* **49**, 407–455.

Ritter, H., Jr., Inoué, S., and Kubai, D. (1978). Mitosis in *Barbulanympha*. I. Spindle structure, formation and kinetochore engagement. *J. Cell Biol.* **77**, 638–654.

Roos, U.-P. (1973). Light and electron microscopy of rat kangaroo cells in mitosis. I. Formation and breakdown of the mitotic apparatus. *Chromosoma* **40**, 43–82.

Roos, U.-P. (1976). Light and electron microscopy of rat kangaroo cells in mitosis. III. Patterns of chromosome behavior during prometaphase. *Chromosoma* **54**, 363–385.

Salmon, E. D. (1976). Pressure-induced depolymerization of spindle microtubules. IV. Production and regulation of chromosome movements. *In* "Cell Motility," Book C (R. Goldman, T. Pollard, and J. Rosenbaum, eds.), pp. 1329–1341. Cold Spring Harbor Lab., Cold Spring Harbor, New York.

Schmitt, F. O. (1969). Fibrous proteins and neuronal dynamics. *Symp. Int. Soc. Cell Biol.* **8**, 95–111.

Schrader, F. (1941). The spermatogenesis of the earwig *Anisolabis maritima* Bon. with reference to the mechanism of chromosomal movement. *J. Morphol.* **68**, 123–148.

Schrader, F. (1953). "Mitosis: The Movement of Chromosomes in Cell Division." Columbia Univ. Press, New York.

Schrader (1963)

Sears, E. R. (1976). Genetic analysis of chromosome pairing in wheat. *Annu. Rev. Genet.* **10**, 31–51.

Stadler, J., and Franke, W. W. (1974). Characterization of the colchicine binding and membrane fractions from rat and mouse liver. *J. Cell Biol.* **60**, 297–303.

Stern, H., and Hotta, Y. (1973). Biochemical controls of meiosis. *Annu. Rev. Genet.* **7**, 37–66.

Summers, K., and Kirschner, M. W. (1979). Characteristics of the polar assembly and disassembly of microtubules observed *in vitro* by dark field microscopy. *J. Cell Biol.* **83**, 205–217.

Swanson, C. P. (1960). "Cytology and Cytogenetics." Macmillan, London.

Thomas, J. B., and Kaltsikes, P. J. (1977). The effects of colchicine on chromosome pairing. *Can. J. Genet. Cytol.* **19**, 231–249.

Vanderlyn, L. (1948). Somatic mitosis in the root tip of *Allium cepa:* A review and reorientation. *Bot. Rev.* **14**, 270–318.

Virrki, N. (1974). The bouquet. *J. Agric. Univ. P.R.* **58**, 338–349.

Zilz, M. L. (1970). *In vitro* male meiosis in the fungus gnat (*Sciara*) with analysis of chromosome movements during anaphase I and II. Ph.D. Dissertation, Wayne State University, Detroit, Michigan.

II

Mitotic Mechanisms and Approaches to the Study of Mitosis

6

Light Microscopic Studies of Chromosome Movements in Living Cells

ARTHUR FORER

I. INTRODUCTION

In this chapter, I concentrate on studies of living cells during cell division using the light microscope and on how these studies have contributed to our understanding of the mechanisms of chromosome movements. In my view, the study of living cells has contributed in two main ways. One contribution is descriptive: descriptions of chromosome movements as they occur in living cells verify those deduced from fixed and stained cells. Further, from studying living cells one obtains information generally unattainable from studies of fixed cells, such as speeds of movement and other dynamic aspects of the processes. The other contribution is experimental: cells are observed as they are perturbed with physical and/or chemical agents and from the observed effects one makes deductions about various aspects of the process, e.g., the nature of spindle fiber organization, where force is produced, or

135

MITOSIS/CYTOKINESIS
Copyright © 1981 by Academic Press, Inc.
All rights of reproduction in any form reserved.
ISBN 0-12-781240-7

the mechanisms that regulate force production or coordination between different chromosomes. This chapter presents what I think are some of the main descriptive and experimental contributions to our understanding of the mechanisms of chromosome movements. The limitations and potentials of the approach are discussed at the end of the chapter.

This chapter is not a comprehensive review that deals with all the data on chromosome movements that exist. Rather, it is limited to what I think are the main conclusions drawn from studies of living cells. The definition of "main" is, of course, a personal one, so perhaps it is relevant to state my biases at the outset. To me, the questions dealing with chromosome movement that are most likely to yield to experimental attack are those that deal with force production: what produces the force to move a chromosome? Where is the force applied on the chromosome? How much force is produced? What regulates whether the force production machinery is turned on or off? And so forth. Much of this chapter is concerned with these questions as studied in living cells.

II. DESCRIPTIVE STUDIES

Studies of chromosome movements on living cells have verified the picture that emerged from studying fixed and stained cells (e.g., Wilson, 1928; Schrader, 1953) and have added details on the dynamics of the process: speeds of movement at various stages, lengths of stages, changes in spindle fiber birefringences, etc. These dynamic aspects have allowed several important deductions to be made, one of which concerns the amount of force necessary to move a chromosome.

The force necessary to cause a chromosome to move poleward in anaphase has been estimated as 10^{-9}–10^{-8} dyne (Nicklas, 1965; Taylor, 1965; Gruzdev, 1972). This estimate derives from data on chromosome positions versus time (in anaphase in living cells), from which one calculates chromosome velocity. Chromosome velocities depend on temperature (review in Mazia, 1961; Schaap and Forer, 1979) and are constant throughout anaphase at about 1 μm/min. The force needed to move a chromosome is calculated from the chromosome velocities and from estimates of chromosome sizes and of cytoplasmic viscosities; the force produced by one myosin filament acting together with one actin filament is more than three orders of magnitude larger than that necessary to move a chromosome (discussion in Wolpert, 1965; Forer, 1969, 1974; and Nicklas, 1971, 1975). For this (and other) reasons, it has been suggested that the rate-limiting step for causing chromosome movement is other than force production (e.g., Nicklas, 1965; Taylor, 1965; McIntosh *et al.*, 1969; Forer, 1974, 1978). Whether or not this is true, very little force and energy are required to move chromosomes.

It should be emphasized that, descriptively, chromosome movement in anaphase is due either to chromosome-to-pole movement while the pole-to-pole distance remains constant, or to pole-to-pole elongation while the chromosome-to-pole distance remains fixed, or to combinations of both movements. The chromosome-to-pole movements might be the result of mechanisms that are quite different from those responsible for pole-to-pole elongation (e.g., Ris, 1943, 1949; Jacques and Biesele, 1954; Mazia, 1961; McIntosh *et al.*, 1975); thus when considering experimental results it is important to separate data that apply to chromosome-to-pole movements from data that apply to pole-to-pole movements. The discussion in this chapter concerns chromosome-to-pole movements unless specified otherwise.

Spindle fibers can be seen in living cells using polarization microscopy (Swann, 1951a,b; Inoué, 1952, 1953; reviews in Inoué, 1976; Inoué and Sato, 1967). The demonstrations of spindle fibers in living cells ended many years of debate about whether spindle fibers seen in fixed and stained preparations might be artifacts of fixation (reviews in Schrader, 1953; Mazia, 1961). It is relevant to point out, though, that with the polarizing microscope one is not looking at all of the material in a spindle fiber: the spindle fibers are seen because they have an optical property called "optical anisotropy," or "birefringence," but not all of the material in a spindle fiber is birefringent. One observes the birefringent component(s) using polarization microscopy, but the non-birefringent component(s) (and weakly birefringent components) may behave differently from those components primarily responsible for the birefringence. There are compelling arguments that a "traction fiber" causes chromosomes to move (e.g., Cornman, 1944), but the traction components of the spindle fiber need not be the same as the birefringent components (discussion in Forer, 1966, 1976, 1978). Nonetheless, that spindle fibers can be seen in living cells allows us to study directly the birefringent components of the fibers, knowing that spindle fibers exist and are not fixation artifacts.

Prometaphase chromosome movements have also been studied in living cells (review in Nicklas, 1971, 1975). In prometaphase the chromosomes move up and back in the spindle pole-to-pole direction. In early prometaphase the velocities are greater than in anaphase and movements tend to be toward the poles, while as prometaphase progresses chromosome movements become slower and slower and the chromosomes tend to be located more and more at the equator (see Chapter 5 by Rickards). "Mistakes" in orientation occur early in prometaphase but are usually corrected (e.g., Bauer *et al.*, 1961), so the question arises of how bipolar orientation is achieved. Nicklas and collaborators have shown that a crucial factor in obtaining bipolar orientation is *tension* on the chromosome (review in Nicklas, 1971; see also Chapter 7 by Ellis and Begg). Another question is how the metaphase positions are reached; though the data are not extensive, metaphase at least in part seems to be an equilibrium of forces pulling in opposite directions (review in Nicklas, 1971; McIntosh, 1979).

Finally, direct observation of living cells has shown that components found in the spindle that do not attach to spindle fibers (such as granules or akinetochoric chromosome fragments) are transported out of the spindle: granules move toward the nearest pole, at about the speed of chromosome movement, and then move out of the spindle. Such non-chromosomal movements might give clues about the mechanisms that move chromosomes (discussion in Nicklas, 1971; Nicklas and Koch, 1972; McIntosh, 1979), but less is known about these movements than about chromosome movements, and it is not at all clear that granule movements and chromosome movements are due to the same mechanisms.

III. EXPERIMENTAL STUDIES

A. Force Production and Spindle Organization

It is generally believed that the force that causes a chromosome to move to the pole is produced (or at least transmitted) by some component(s) present in the chromosomal spindle fiber—i.e., in the spindle fiber that is attached to the chromosome's kinetochore. One argument that leads to this belief is the negative one that all other hypotheses seem to be ruled out (e.g., discussions in Cornman, 1944; Schrader, 1953; Gruzdev, 1972). Positive arguments come from experimental treatment of living cells. These positive arguments are not foolproof, but combined with the negative arguments they lead to a reasonably strong conclusion that the force for chromosome movement arises from some component(s) associated with the chromosomal spindle fiber. One of the positive arguments arises from experiments using micromanipulation techniques that have demonstrated that chromosomes are mechanically attached to the spindle, and that the attachment is due to some component(s) associated with the chromosomal spindle fiber (see Chapter 7 by Ellis and Begg). The fact that there is a mechanical attachment suggests that there is a pulling (traction) force on the chromosome as it moves poleward in anaphase.

Other positive arguments arise from experiments using ultraviolet microbeam irradiations that have localized a force-producing element to the chromosomal spindle fiber: heterochromatic ultraviolet light focused to small portions (1–2 μm) of individual chromosomal spindle fibers in anaphase regularly stopped the associated chromosomes from moving. On the other hand, irradiations outside the spindle or between the separating chromosomes (the interzone), using equivalent doses, did *not* stop chromosome movement (Forer, 1966). Thus a uv-sensitive force-producing system seems to be associated with the spindle fiber but not the interzonal region. Further, the force-producing component of the spindle fiber seems to be associated with the complete length of the fiber, because irradiations at positions along the spindle fiber up to 80% of the distance from the chromosome to the pole, the maximum distance tested, were able to stop chromosome movement (Forer, 1969).

It could be, though, that components necessary for force production are indeed present in the interzone, as suggested by some hypotheses of motion (e.g., McIntosh *et al.*, 1969; Nicklas, 1971; Margolis and Wilson, 1978), but that the interzonal components just are not damaged by irradiations that damage the same components in the chromosomal spindle fiber. This might occur, for example, if there were components in the interzone that absorb the ultraviolet light and thereby "mask" the force-producing components. This possibility is ruled out by irradiations using monochromatic light and by measurements of absorption spectra. Interzonal irradiations with monochromatic light at three times the doses needed to stop chromosome movement when spindle fibers are irradiated still had no effect on chromosome movement (Sillers and Forer, 1981b). Furthermore, at the wavelengths in question (270 nm, 280 nm, 290 nm) the absorption of ultraviolet light by chromosomal spindle fibers is not much different from that of the interzone, certainly not enough to account for a dose three times higher having no effect (Sillers and Forer, 1981b). Thus one concludes that the uv-sensitive force-producing component is associated with the spindle fiber and does not extend into the interzone. Taken together with data from micromanipulation and other experiments, the net result is a strong presumption that the force for chromosome movement comes from the chromosomal spindle fiber.

It is relevant to point out that the force-producing component(s) associated with the chromosomal spindle fiber is probably different from the birefringent component(s). I make this interpretation because the two effects, (1) stopping movement and (2) reducing birefringence, occur independently of each other: chromosomes can stop moving with or without alteration of birefringence, and alteration of birefringence occurs with or without stopping movement (Forer, 1966).

Data on the chemical nature of the uv-sensitive force-producing spindle fiber component have been obtained by seeing which wavelengths are most effective in stopping chromosome movement, i.e., by obtaining an action spectrum for stopping chromosome movement. The action spectrum has two peaks, at 270 nm and 290 nm (Sillers and Forer, 1981b), so it would seem that movement can be stopped in at least two different ways. The action spectrum does not match the absorption spectrum of known spindle components such as tubulin, dynein, actin, myosin, and calmodulin, so it seems likely that one needs to compare the absorption of selected regions of the suspected molecules rather than the entire molecules (as in an absorption spectrum). To obtain further information on the uv-sensitive components, Sillers and Forer (1981b) compared the action spectrum for stopping chromosome movement with action spectra for stopping myofibril contraction and for stopping cilia from beating, since various hypotheses for chromosome movement invoke either dynein–microtubule interactions, as in cilia, or actin–myosin interactions, as in myofibrils (see discussions in Forer, 1978 and McIntosh, 1979). The action spectrum for blocking myofibril contraction is exactly parallel to that for stopping chromosome movement, with peaks at 270 nm and 290 nm and with a smaller dose required to block myofibril contraction than to stop chromosome movement. On the other hand, the action spectrum for stopping cilia from beating is quite different, with a single broad peak between 270 nm and 280 nm. Indeed, one can rule out the idea that the ciliary mechanochemical system is involved in the effect of 290-nm irradiation in stopping chromosome movement because the dose of 290-nm light needed to stop a cilium from beating is about four times higher than that needed to stop a chromosome from moving (Sillers and Forer, 1981b). My interpretation of these (and other) results is that an actin–myosin type system produces the force for chromosome movement while the microtubules have some other role, such as limiting the rates of chromosome movement (see discussions in Forer, 1974, 1978; Rickards, 1975; Forer *et al.*, 1979 and in Chapter 5 by Rickards; see also the counterarguments of Cande *et al.*, 1981).

I want to emphasize again that, unless otherwise stated, I am discussing the mechanisms of chromosome-to-pole movement. The preceding argument, then, as with most of the others, pertains to chromosome-to-pole movement.

The forces for anaphase chromosome movement are applied independently to each chromosome. One reaches this conclusion because the anaphase movements of different chromosomes moving to the same pole are independent of one another: the chromosomes may move at different times, or with different velocities. Also, one can move one chromosome (via micromanipulation) without altering the movement of others, and one can stop the movement of one chromosome (via ultraviolet microbeam irradiation) without stopping the movement of others (discussion in Forer, 1969, 1974; Nicklas, 1971). It is relevant to point out that the initial separation of chromosomes at the start of anaphase seems to be independent of sub-sequent anaphase movement: the initial separation of chromosomes occurs even though the spindle fibers have been damaged and poleward anaphase movements have been delayed. The initial separation of chromosomes oc-curs even in the complete absence of spindle fibers (discussion in Mazia, 1961; Forer, 1966, 1969).

Spindles seem to be labile structures: a spindle forms, remains in place for a relatively short time, and disappears again as the chromosomes move poleward in anaphase. Experimental treatments of living cells gave rise to the idea that the spindle is made up of organized structures in "dynamic equilibrium" with nonorganized subunits (e.g., Inoué, 1964, 1976; Inoué and Sato, 1967). For example, treatment with colchicine or lowered temper-atures causes the organized, birefringent (anisotropic) spindle to become disorganized and isotropic (and hence not visible with polarization micro-scopy). The spindles quickly reform after return to normal conditions. The interpretation (e.g., Inoué, 1976) is that the organized spindle is in dynamic equilibrium with unorganized subunits. Measurements of the amounts of spindle birefringence during experimental treatments of this kind have been used as a basis for a thermodynamic analysis of spindle organization (e.g., Inoué, 1959; Inoué and Ritter, 1975) and for the hypothesis that the motive force for pulling a chromosome poleward derives from depolymerization of the chromosomal spindle fiber (e.g., Inoué and Ritter, 1975; Inoué, 1976). There are several difficulties with the thermodynamic and mechanochemical interpretations of the data, though. One is the method for quantifying the birefringence (see Forer, 1976). For another, one does not know if the experimental agents (e.g., temperature) act *directly* on the postulated single step of the equilibrium or at some step(s) removed (e.g., see discussion in McIntosh, 1979). Further, the measurements are of single spots in whole spindles, including continuous spindle fibers as well as chromosomal spindle fibers. "Continuous" microtubules respond to experimental treatment dif-ferently from chromosomal (kinetochore) microtubules (e.g., Brinkley and Cartwright, 1975; Salmon et al., 1976; Lambert and Bajer, 1977), and con-tinuous fiber birefringence responds differently from chromosomal fiber birefringence (Izutsu et al., 1979; Salmon and Begg, 1980). Thus the data on portions of entire spindles are averages from a mixture of components with different responses to temperature, and the measured changes do not neces-sarily apply to chromosomal fibers. Finally, the data on birefringence are interpreted as measuring microtubules, yet, as far as I am concerned, there is still a question regarding how many components give rise to spindle bire-fringence (e.g., Forer, 1976, 1978). What is missing is non-subjective quan-titative data on birefringence in various regions of a spindle (e.g., along a single spindle fiber) and point-by-point comparison with microtubules. It is known that the coefficient of birefringence, the direct measure of concentration of oriented elements (see Forer, 1976), is different in different parts of a spindle (Swann, 1951a,b) as well as at different parts of a single spindle fiber (Forer, 1976; Salmon and Begg, 1980), so the term "spindle birefringence" has no real meaning except at individual regions. One therefore needs to compare the coefficient of birefringence with microtubules on a point-by-point basis.

Regardless of these criticisms of the interpretations of the data, there seems little doubt that the spindle is labile and in some kind of equilibrium with its surroundings.

B. Coordination between Chromosomes

Some experimental studies on mitosis have dealt not with force production but rather with "coordination" between chromosomes, by which I mean the influence of some chromosomes on the movements of other chromosomes in the same cell. As discussed above, the movements of chromosomes to poles in anaphase would seem to be due to traction forces that are applied independently to the different chromosomes; simultaneously with this *independence*, though, one finds *coordination* that is recognized in several ways (e.g., discussion in Forer, 1974, 1980). For example, the *timing* of anaphase movements is coordinated: chromosomes generally start anaphase movements at the same time. As another example, the *directions* that chromosomes move in at anaphase may be coordinated: non-random segregations occur in meiosis (e.g., Camenzind and Nicklas, 1968; Hughes-Schrader, 1969; Brown, 1969). Indeed, chromosomes will even reverse directions and move to the opposite poles after appropriate micromanipulation of *other* chromosomes (Forer and Koch, 1973). Discussions of various examples of these (and other) kinds of chromosome coordinations are given in Forer (1974, 1980), White (1973), and McIntosh (1979); a major question is, how are these coordinations achieved?

There are very few hypotheses that attempt to explain how chromosome movements are coordinated, and, indeed, there are very few experiments that deal with coordination mechanisms. Dietz (1969, 1972a,b) hypothesized that mechanisms of coordination are linked to mechanisms of movement. He argued that the force for chromosome movement arises from depolymerization of microtubules, and he suggested that the coordination in direction arises because individual spindle fibers share a common pool of subunits. On the other hand, others have suggested that coordination arises because of interactions between chromosomal spindle fibers (Camenzind and Nicklas, 1968; Forer and Koch, 1973). Recent experiments on coordination in crane fly spermatocytes have further implicated interactions between spindle fibers and have ruled out the idea that coordination occurs via a pool of sub-units. I now describe these experiments.

In anaphase I in crane fly spermatocytes three autosomal half-bivalents move toward each pole while the two unpaired sex chromosomes remain at the equator. The sex chromosomes each have a spindle fiber to *both* poles, and they move to opposite poles about 20–30 minutes after the autosomes begin to move poleward, i.e., they move poleward at about the time the autosomes reach the poles (e.g., Schaap and Forer, 1979). Sex-chromosome movements are themselves coordinated because the two unpaired sex chromosomes invariably go to opposite poles. Sex-chromosome movements also are coordinated with those of the autosomes: alterations in autosome segregation cause alterations in sex-chromosome movements (Dietz, 1969, 1972b; Forer and Koch, 1973). In particular, when autosomal segregation is altered such that four autosomal half-bivalents go to one pole and two autosomal half-bivalents go to the other, one sex chromosome moves to one pole (the one that has two autosomes) while the other sex chromosome remains at the equator and does not move. Dietz (1969, 1972a,b) suggested that this is due to interactions of sex-chromosomal spindle fibers with a pool of subunits, while Forer and Koch (1973) suggested that this is due to interactions between sex-chromosomal spindle fibers and autosomal spindle fibers. Experiments using ultraviolet microbeam irradiations verify the latter hypothesis, as follows.

Monochromatic ultraviolet light was focused to spots about 4 μm in diameter, and these spots were aimed at portions of individual autosomal spindle fibers; the autosomal spindle fibers were then irradiated with various energies of ultraviolet light of various wavelengths (Sillers and Forer, 1981a). We were surprised to find that irradiations of autosomal spindle fibers often altered sex-chromosome movement. Either one sex chromosome stayed at the equator while the other one moved to the pole opposite the half-spindle with the irradiated autosomal spindle fiber, or both sex chromosomes moved to the same pole (opposite the half-spindle with the irradiated autosomal spindle fiber). Because irradiations of autosomal spindle fibers altered sex-chromosome movements *only* when the autosomal spindle fiber was adjacent to at least one sex-chromosome spindle fiber, and because the two different effects on sex-chromosome movements were associated with two different arrangements of the irradiated autosomal spindle fiber vis-à-vis sex-chromosomal spindle fibers, we concluded that the coordination between autosomal and sex-chromosomal movements arises because of interactions between autosomal spindle fibers and sex-chromosomal spindle fibers. It seems likely that the irradiations alter a "control" system that turns on the sex-chromosome force production machinery rather than altering the force producers themselves (discussion in Sillers and Forer, 1981a). It is relevant to note that ultrastructural evidence in a different cell type also suggests that interchromosomal coordinations are mediated by physical connections (Kubai and Wise, 1981).

In addition to coordination between autosomes and sex chromosomes in crane fly spermatocytes, there is coordination between separating half-bivalents: when irradiation of a portion of an individual autosomal spindle fiber stops the associated half-bivalent from moving poleward, the partner half-bivalent (moving to the opposite pole) also stops moving (Forer, 1966; Sillers and Forer, 1981c). Since separating autosomal partner half-bivalents are mechanically linked, even when they are separated by 10–15 μm (Forer and Koch, 1973; Forer, unpublished; see also Chapter 7 by Ellis and Begg), I suggested that these mechanical linkages might be responsible for the coordination between separating partner half-bivalents (Forer, 1974). I tested this possibility using monochromatic ultraviolet light focused to spots 2 or 4 μm in diameter (Forer, in preparation). Irradiations of the interzonal region (between separating autosomal half-bivalents) have no effect on the autosomal movements toward the pole (Sillers and Forer, 1981c), but if the interzonal region between separating half-bivalents is irradiated *prior* to irradiation of the spindle fiber associated with one of the half-bivalents, then the coordination is uncoupled: only one half-bivalent stops moving (the one attached to the irradiated chromosomal spindle fiber). It remains to be seen whether interzonal irradiations that uncouple partner half-bivalents also destroy the mechanical linkage between partners.

In summary, more is known about force production for chromosome movement than about coordination between chromosomes, and I have summarized some of the studies from experimentally treated living cells that lead to the conclusions that the force is transmitted to chromosomes individually and independently by the attached chromosomal spindle fibers; that an actin–myosin system might be involved in force production; that the spindle is labile, in some kind of "dynamic equilibrium" with its surrounds; and that at least *some* examples of coordination between chromosomes are due to interactions between spindle fibers. I now discuss the limitations and potential of studying living cells in order to understand mitotic mechanisms.

IV. OVERVIEW: LIMITATIONS AND POTENTIAL

Light microscopic studies of living cells provide necessary descriptions of mitosis: any valid hypothesis of mitosis must explain the phenomena that occur in living cells (speeds of movements, interchromosomal coordinations, effects of experimental perturbations, etc.), and *in vitro* models must duplicate these phenomena. On the other hand, studies of living cells can provide only limited data on molecular mechanisms; understanding of the molecular events will probably come about by comparison of data from several approaches, such as biochemistry, as yet non-existent *in vitro* model systems, and quantitative electron microscopy. Our present understanding of the comparatively well-known contraction of skeletal muscle came primarily from combining biochemical studies of muscle proteins with light microscopic, x-ray diffraction, and electron microscopic studies that localized the muscle proteins both to various portions of the sarcomere and to various filaments in the sarcomere and that described the movements of filaments during muscle contraction (Needham, 1971). Light microscopic studies of living muscle cells *did* provide crucial evidence in favor of the sliding-filament hypothesis, however, which helped rule out other hypotheses (e.g., Edman, 1966; Gordon *et al.*, 1966), and I expect that light microscopic studies of mitosis will similarly contribute to our molecular understanding of mitosis. Specifically, I expect that studies of living cells will be important in testing the relevance of *in vitro* biochemical studies. Examples of such tests will now be discussed.

When microtubules are at steady state (i.e., when there is neither net assembly nor net disassembly of microtubules), microtubule subunits continually assemble onto one end of the microtubule while equal numbers of subunits disassemble from the other end of the microtubule. This opposite-end assembly–disassembly creates a "treadmill" in which added subunits gradually move from the assembly end to the disassembly end while the microtubule itself remains constant in length (Margolis and Wilson, 1977, 1978, 1979; Wilson and Margolis, 1978; Farrell *et al.*, 1979). One proposed molecular model of mitosis incorporates the idea that microtubules of constant length are treadmills (Margolis *et al.*, 1978; Margolis, 1978); the crucial point for my discussion is that, in this model, for microtubules that extend between kinetochore and pole, the kinetochore is the assembly end while the pole is the disassembly end. At metaphase, then, the microtubules are at steady state so that microtubule subunits continually add to microtubules at the kinetochore end, and the subunits move (treadmill) to the pole end and leave the microtubule at the pole end. During anaphase, the assembly is stopped at the kinetochore end of the microtubule while disassembly continues at the poleward end, thereby pulling the chromosome poleward.

On the other hand, other data lead to different conclusions. Under *polymerizing* conditions polymerization takes place at *both* ends of a microtubule, not just at one "assembly end," though the rate of polymerization at one end is much higher than at the other end (Allen and Borisy, 1974; Binder *et al.*, 1975; Rosenbaum *et al.*, 1975). Kinetic experiments on microtubules under *depolymerizing* conditions have suggested that association (assembly) and dissociation (disassembly) reactions *both* occur at *each* end of a microtubule and that treadmilling arises because one end has *net* assembly whereas the other end has *net* disassembly (Karr and Purich, 1979). Similar conclusions were reached from kinetic studies that showed that under polymerizing conditions one end of the microtubule polymerizes faster than the other and that under depolymerizing conditions the fast-polymerizing end depolymerizes faster than the slow-polymerizing end (Summers and Kirschner, 1979; Bergen and Borisy, 1980; Bergen *et al.*, 1980). From extrapolating the rates of polymerization (and depolymerization) at each end to steady state conditions one concludes that treadmilling would indeed exist (e.g., Bergen and Borisy, 1980; Kirschner, 1980) and that the *fast*-assembling end is equivalent to the *net*-assembly end of Margolis and Wilson (1978). Particularly relevant to the present discussion are experiments on microtubules organized by kinetochores and by centrioles, the conclusions of which are now summarized. When microtubules are organized (and caused to polymerize) by either centriole or kinetochore, that end associated with the organizing center *does not take part in association–dissociation reactions* (Summers and Kirschner, 1979; Bergen *et al.*, 1980; Kirschner, 1980). That is to say, if a microtubule is caused to grow from the kinetochore, then the poleward end of the microtubule is the site of all dissociation and association reactions while the kinetochore end is "dead" to association and dissociation. Furthermore, the poleward end of a microtubule grown from a kinetochore is the fast-polymerizing end (Summers and Kirschner, 1979; Bergen *et al.*, 1980). The two groups of workers therefore come to mutually contradictory conclusions. Margolis *et al.* (1978) conclude that the kinetochore end of a microtubule is *not* dead to association reactions and is in fact the *fast*-polymerizing (net-assembly) end, while Bergen *et al.* (1980) and Kirschner (1980) conclude that the kinetochore end of a microtubule *is* dead to association–dissociation reactions and is the *slow*-polymerizing (net-disassembly) end. Experimental studies of spindles *in vivo* are relevant to these conflicting interpretations, and, as I now argue, these data suggest that neither interpretation is correct without modification.

Local changes in spindle fiber birefringence can be produced by irradiation of a small portion of a spindle fiber with a focused beam of ultraviolet light: the birefringence in the irradiated region is greatly reduced (i.e., is near zero), while the birefringence on both sides of the irradiated region remains unchanged (Forer, 1964, 1965, 1966). Immediatedly after being formed the area of reduced birefringence moves poleward, in both metaphase and anaphase cells; both ends of the irradiated region move at the same speed, so the region moves poleward without changing shape. The irradiated area moves poleward in anaphase independently of whether the associated chromosome has stopped moving, and, in both metaphase and anaphase, once the irradiated region reaches the pole it disappears and a normal-looking chromosomal spindle fiber is formed. These results are directly relevant to the conflicting interpretations of mitotic spindle fiber organization described above. If spindle birefringence is due to microtubules, as most people believe (e.g., Inoué, 1976), the reduced birefringence is caused by depolymerizing the spindle fiber microtubules (e.g., see Begg and Ellis, 1979); the different interpretations of microtubule organization predict different behavior of such severed microtubules, as follows.

Margolis *et al.* (1978) argue that the kinetochore end of a microtubule (labeled K in Fig. 1) is the site of microtubule assembly while the pole end (labeled P in Fig. 1) is the site of disassembly, and that the microtubule is at steady state. If the irradiation severs the microtubule to produce the area of reduced birefringence, as in Fig. 1B, this is exactly equivalent to *shearing* the microtubules (Margolis and Wilson, 1978). One would thus predict that end 1 (Fig. 1B) is a disassembly end and end 2 (Fig. 1B) is an assembly end, and since this is at steady state the region of reduced birefringence (between 1 and 2) *would not move*. This prediction clearly does not fit the data. One could not argue that the microtubules are *not* in steady state, because points 1 and 2 move poleward at the same rate, and hence the rate of depolymerization equals the rate of polymerization. One might argue that when the irradiation severed the microtubule, it damaged the irradiated ends (i.e., ends 1 and 2) so that they no longer can take part in association–dissociation reactions. If so, end P would depolymerize and pull "dead" point 2 to the pole at the same time that end K would polymerize and push "dead" point 1 to the pole. This explanation is not possible, however, for two rea-

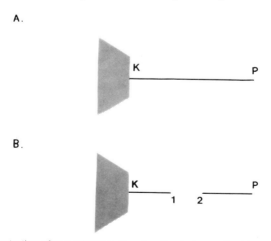

A.

B.

Fig. 1.   (A) Illustration of one microtubule extending between the kinetochore (K) and the spindle pole (P). (B) Illustration of the same microtubule after ultraviolet microbeam irradiation has severed it and created two free ends (labeled 1 and 2).

enough to answer such questions. It seems to me that progress is slowed if one tries to use as components of the force-producing mechanism those elements that really are part of the control systems, and that we will just have to wait to sort out such questions until we know more about force production or about control systems. But we should not try to force into hypotheses of force production data that really deal with control systems.

It is easy to get discouraged by our slow progress and by the difficulty of devising (and executing) critical experiments that will lead to greater under-standing of the basic mechanisms. This is especially so in our present un-stable world of inflating prices in which one has to spend large amounts of time away from productive science (let alone the lab bench) worry about the money that one needs to do science. In any event, detailed understanding of mitosis will probably result from a combination of different approaches, using different techniques. At present there seem to be a reasonable number of people doing experiments on mitosis with different approaches and differ-ent ideas, so I think that we can expect continued progress in understanding the molecular mechanisms of mitosis. As argued above, I think that light microscopic studies can contribute to this progress.

ACKNOWLEDGMENTS

It is a pleasure to acknowledge the continued, enjoyable collaboration and discussions with Arthur Zimmerman. It should also be clear from the text that graduate students working in my laboratory have contributed greatly. I thank Mrs. Anna Adhihetty and Mrs. Helen Spencer for excellent typing of my usual many versions. This work was supported by grants from the Natural Sciences and Engineering Research Council of Canada.

REFERENCES

Allen, C., and Borisy, G. G. (1974). Structural polarity and directional growth of microtubules of *Chlamydomonas* flagella. *J. Molec. Biol.* **90**, 381–402.

Aronson, J. F. (1971). Demonstration of a colcemid-sensitive attractive force acting between the nucleus and a center. *J. Cell Biol.* **51**, 579–583.

Aronson, J. F., and Inoué, S. (1970). Reversal by light of the action of N-methyl N-desacetyl colchicine on mitosis. *J. Cell Biol.* **45**, 470–477.

Bauer, H., Dietz, R., and Röbbelen, C. (1961). Die Spermatocytenteilungen der Tipuliden. III. Das Bewegungsverhalten der Chromosomen in Translokationsheterozygoten von *Tipula oleracea. Chromosoma* **12**, 116–189.

Begg, D. A., and Ellis, G. W. (1979). Micromanipulation studies on chromosome movement. II. Birefringent chromosomal fibres and the mechanical attachment of chromosomes to the spindle. *J. Cell Biol.* **82**, 542–554.

Bergen, L. G., and Borisy, G. G. (1980). Head-to-tail polymerization of microtubules *in vitro*. Electron microscope analysis of seeded assembly. *J. Cell Biol.* **84**, 141–150.

Bergen, L. G., Kuriyama, R., and Borisy, G. G. (1980). Polarity of microtubules nucleated by centrosomes and chromosomes of Chinese hamster ovary cells *in vitro*. *J. Cell Biol.* **84**, 151–159.

Binder, L. I., Dentler, W. L., and Rosenbaum, J. L. (1975). Assembly of chick brain tubulin onto flagellar microtubules from *Chlamydomonas* and sea urchin sperm. *Proc. Natl. Acad. Sci. U.S.A.* **72**, 1122–1126.

Brinkley, B. R., and Cartwright, J. (1975). Cold-labile and cold-stable microtubules in the mitotic spindle of mammalian cells. *Ann. N.Y. Acad. Sci.* **253**, 428–439.

Brown, S. W. (1969). Developmental control of heterochromatization in coccids. *Genetics, Suppl.* **61**, 191–198.

Camenzind, R., and Nicklas, R. B. (1968). The non-random chromosome segregation in spermocytes of *Gryllotalpa hexadactyla*. A micromanipulation analysis. *Chromosoma* **24**, 324–335.

Cande, W. Z., McDonald, K., and Meeusen, R. L. (1981). A permeabilized cell model for studying cell division: a comparison of anaphase chromosome movement and cleavage furrow constriction in lysed Pt K₁ cells. *J. Cell Biol.* **88**, 618–629.

Chalfie, M., and Thomson, J. N. (1979). Organization of neuronal microtubules in the nematode *Caenorhabditis elegans*. *J. Cell Biol.* **82**, 278–289.

Cornman, I. (1944). A summary of evidence in favor of the traction fiber in mitosis. *Am. Nat.* **78**, 410–422.

Dietz, R. (1969). Bau und Funktion des Spindelapparats. *Náturwissenschaften* **56**, 237–248.

Dietz, R. (1972a). Anaphase behaviour of inversions in living crane-fly spermatocytes. *Chromosomes Today* **3**, 70–85.

Dietz, R. (1972b). Die Assembly-Hypothese der Chromosomenbewegung und die Veränderungen der spindellänge während der Anaphase I in Spermatocyten von *Pales ferruginea (Tipulidae, Diptera)*. *Chromosoma* **38**, 11–76.

Edman, K. A. P. (1966). The relation between sarcomere length and active tension in isolated semitendinosus fibres of the frog. *J. Physiol. (London)* **183**, 407–417.

Farrell, K. W., Kassis, J. A., and Wilson, L. (1979). Outer doublet tubulin reassembly: evidence for opposite end assembly-disassembly at steady state and a disassembly end equilibrium. *Biochemistry* **12**, 2642–2647.

Forer, A. (1964). Evidence for two spindle fiber components: a study of chromosome movement in living crane fly (*Nephrotoma suturalis*) spermatocytes, using polarization microscopy and an ultraviolet microbeam. Ph.D. Thesis, Dartmouth College, Hanover, New Hampshire.

Forer, A. (1965). Local reduction of spindle fiber birefringence in living *Nephrotoma suturalis* (Loew) spermatocytes induced by ultraviolet microbeam irradiation. *J. Cell Biol.* **25** (No. 1, Pt. 2), 95–117.

Forer, A. (1966). Characterization of the mitotic traction system, and evidence that birefringent spindle fibers neither produce nor transmit force for chromosome movement. *Chromosoma* **19**, 44–98.

Forer, A. (1969). Chromosome movements during cell-division. *In* "Handbook of Molecular Cytology" (A. Lima-de-Faria, ed.), pp. 553–601. North-Holland Publ., Amsterdam.

Forer, A. (1974). Possible roles of microtubules and actin-like filaments during cell-division. *In* "Cell Cycle Controls" (G. M. Padilla, I. L. Cameron, and A. Zimmerman, eds.), pp. 319–336. Academic Press, New York.

Forer, A. (1976). Actin filaments and birefringent spindle fibers during chromosome movements. *In* "Cell Motility," Book C (R. Goldman, T. Pollard, and J. Rosenbaum, eds.), pp. 1273–1293. Cold Spring Harbor Lab., Cold Spring Harbor, New York.

Forer, A. (1978). Chromosome movements during cell-division: possible involvement of actin filaments. *In* "Nuclear Division in the Fungi" (I. B. Heath, ed.), pp. 21–88. Academic Press, New York.

Forer, A. (1980). Chromosome movements in the meiosis of insects, especially crane-fly spermatocytes. *In* "Insect Cytogenetics" (R. L. Blackman, G. M. Hewitt, and M. Ashburner, eds.), pp. 85–95. Blackwell Scientific Publications, Oxford, London, Edinburgh, Boston, Melbourne.

Forer, A., and Koch, C. (1973). Influence of autosome movements and of sex-chromosome movements on sex-chromosome segregation in crane-fly spermatocytes. *Chromosoma* **40**, 417–442.

Forer, A., Jackson, Wm. T., and Engberg, A. (1979). Actin in spindles of *Haemanthus katherinae* endosperm. II. Distribution of actin in chromosomal spindle fibres, determined by analysis of serial sections. *J. Cell Sci.* **37**, 349–371.

Fuge, H. (1974). The arrangement of microtubules and the attachment of chromosomes to the spindle during anaphase in tipulid spermatocytes. *Chromosoma* **45**, 245–260.

Fuge, H. (1977). Ultrastructure of the mitotic spindle. *Int. Rev. Cytol. Suppl.* **6**, 1–58.

Gordon, A. M., Huxley, A. F., and Julian, F. J. (1966). The variation in isometric tension with sarcomere length in vertebrate muscle fibres. *J. Physiol. (London)* **184**, 170–192.

Gruzdev, A. D. (1972). Critical review of some hypotheses concerning anaphase chromosome movements. *Tsitologiya* **14**, 141–149. (In Russian. English translation is: NRC Technical Translation 1758, National Research Council of Canada, Ottawa).

Hughes-Schrader, S. (1969). Distance segregation and compound sex chromosomes in mantispids (*Neuroptera: Mantispidae*). *Chromosoma* **27**, 109–129.

Inoué, S. (1952). The effect of colchicine on the microscopic and submicroscopic structure of the mitotic spindle. *Exp. Cell Res., Suppl.* **2**, 305–318.

Inoué, S. (1953). Polarization optical studies of the mitotic spindle. *Chromosoma* **5**, 487–500.

Inoué, S. (1959). Motility of cilia and the mechanism of mitosis. *Rev. Mod. Phys.* **31**, 402–408.

Inoué, S. (1964). Organization and function of the mitotic spindle. *In* "Primitive Motile Systems in Cell Biology" (R. D. Allen and N. Kamiya, eds.), pp. 549–598. Academic Press, New York.

Inoué, S. (1976). Chromosome movement by reversible assembly of microtubules. *In* "Cell Motility," Book C. (R. Goldman, T. Pollard, and J. Rosenbaum, eds.), pp. 1317–1328. Cold Spring Harbor Lab., Cold Spring Harbor, New York.

Inoué, S., and Ritter, H., Jr. (1975). Dynamics of mitotic spindle organization and function. *In* "Molecules and Cell Movement" (S. Inoué and R. E. Stephens, eds.), pp. 3–29. Raven, New York.

Inoué, S., and Sato, H. (1967). Cell motility by labile association of molecules: The nature of mitotic spindle fibers and their role in chromosome movement. *J. Gen. Physiol.* **50** (No. 6, Pt. 2), 259–292.

Izutsu, K., Sato, H., and Ohnuki, Y. (1979). Behavior of mitotic spindles in dividing cells—a description of a film. *In* "Cell Motility: Molecules and Organization" (S. Hatano, H. Ishikawa, and H. Sato, eds.), pp. 281–287. University of Tokyo Press, Tokyo.

Jacques, J. A., and Biesele, J. J. (1954). A study of Michel's film on meiosis in *Psophus stridulus* L. *Exp. Cell Res.* **6**, 17–29.

Karr, T. L., and Purich, D. L. (1979). A microtubule assembly/disassembly model based on drug effects and depolymerization kinetics after rapid dilution. *J. Biol. Chem.* **254**, 10885–10888.

Kirschner, M. W. (1980). Implications of treadmilling for the stability and polarity of actin and tubulin polymers *in vivo*. *J. Cell Biol.* **86**, 330–334.

Kubai, D. F., and Wise, D. (1981). Nonrandom chromosome segregation in *Neocurtilla (Gryllotalpa) hexadactyla*: an ultrastructural study. *J. Cell Biol.* **88**, 281–293.

Lambert, A.-M., and Bajer, A. S. (1977). Microtubule distribution and reversible arrest of chromosome movements induced by low temperature. *Cytobiologie* **15**, 1–23.

McIntosh, J. R. (1979). Cell division. *In* "Microtubules" (K. Roberts and J. S. Hyams, eds.), pp. 381–441. Academic Press, London.

McIntosh, J. R., Hepler, P. K., and Van Wie, D. G. (1969). Model for mitosis. *Nature (London)* **224**, 659–663.

McIntosh, J. R., Cande, W. Z., and Snyder, J. A. (1975). Structure and physiology of the mammalian mitotic spindle. *In* "Molecules and Cell Movement" (S. Inoué and R. E. Stephens, eds.), pp. 31–76. Raven, New York.

Margolis, R. L. (1978). A possible microtubule dependent mechanism for mitosis. *In* "Cell Reproduction: In honor of Daniel Mazia" (E. R. Dirksen, D. M. Prescott, and C. F. Fox, eds.), pp. 445–456. Academic Press, New York.

Margolis, R. L., and Wilson, L. (1977). Addition of colchicine-tubulin complex to microtubule ends: The mechanism of substoichiometric colchicine poisoning. *Proc. Natl. Acad. Sci. U.S.A.* **74**, 3466–3470.

Margolis, R. L., and Wilson, L. (1978). Opposite end assembly and disassembly of microtubules at steady state *in vitro. Cell* **13**, 1–8.

Margolis, R. L., and Wilson, L. (1979). Regulation of the microtubule steady state *in vitro* by ATP. *Cell* **18**, 673–679.

Margolis, R. L., Wilson, L., and Kiefer, B. I. (1978). Mitotic mechanism based on intrinsic microtubule behavior. *Nature (London)* **272**, 450–452.

Mazia, D. (1961). Mitosis and the physiology of cell division. *In* "The Cell" (J. Brachet and A. E. Mirsky, eds.), Vol. 3, pp. 77–412. Academic Press, New York.

Needham, D. M. (1971). "Machina Carnis. The biochemistry of muscular contraction in its historical development." Cambridge Univ. Press, London.

Nicklas, R. B. (1965). Chromosome velocity during mitosis as a function of chromosome size and position. *J. Cell Biol.* **25** (No. 1, Pt. 2), 119–135.

Nicklas, R. B. (1971). Mitosis. *Adv. Cell Biol.* **2**, 225–297.

Nicklas, R. B. (1975). Chromosome movement: current models and experiments on living cells. *In* "Molecules and Cell Movement" (S. Inoué and R. E. Stephens, eds.), pp. 97–117. Raven, New York.

Nicklas, R. B., and Koch, C. A. (1972). Chromosome micromanipulation. IV. Polarized motions within the spindle and models for mitosis. *Chromosoma* **39**, 1–26.

Rickards, G. K. (1975). Prophase chromosome movements in living house cricket spermatocytes and their relationship to prometaphase, anaphase and granule movements. *Chromosoma* **49**, 407–455.

Ris, H. (1943). A quantitative study of anaphase movement in the aphid Tamalia. *Biol. Bull. (Woods Hole, Mass.)* **85**, 164–178.

Ris, H. (1949). The anaphase movement of chromosomes in the spermatocytes of the grasshopper. *Biol. Bull. (Woods Hole, Mass.)* **96**, 90–106.

Rosenbaum, J. L., Binder, L. I., Granett, S., Dentler, W. L., Snell, W., Sloboda, R., and Haimo, L. (1975). Directionality and rate of assembly of chick brain tubulin onto pieces of neurotubules, flagellar axonemes, and basal bodies. *Ann. N.Y. Acad. Sci.* **253**, 147–177.

Salmon, E. D., and Begg, D. A. (1980). Functional implications of cold-stable microtubules in kinetochore fibers of insect spermatocytes during anaphase. *J. Cell Biol.* **85**, 853–865.

Salmon, E. D., Goode, D., Maugel, T. K., and Bonar, D. B. (1976). Pressure-induced depolymerization of spindle microtubules. III. Differential stability in HeLa cells. *J. Cell Biol.* **69**, 443–454.

Schaap, C. J., and Forer, A. (1979). Temperature effects on anaphase chromosome movement in the spermatocytes of two species of crane flies (*Nephrotoma suturalis* Loew and *Nephrotoma ferruginea* Fabricius). *J. Cell Sci.* **39**, 29–52.

Schrader, F. (1953). "Mitosis: The movements of chromosomes in cell division." Columbia Univ. Press, New York.

Sillers, P. J., and Forer, A. (1981a). Autosomal spindle fibres influence subsequent sex-chromosome movements in crane-fly spermatocytes. *J. Cell Sci.*, **49**, 51–67.

Sillers, P. J., and Forer, A. (1981b). Analysis of chromosome movement in crane-fly spermatocytes by ultraviolet microbeam irradiation of individual chromosomal spindle fibres. II. Action spectra for stopping chromosome movement and for stopping ciliary beating and myofibril contraction. *Can. J. Biochem.*, in press.

Sillers, P. J., and Forer, A. (1981c). Analysis of chromosome movement in crane-fly spermatocytes by ultraviolet microbeam irradiation of individual chromosomal spindle fibres. I. General effects of the irradiation. *Can. J. Biochem.*, in press.

Sluder, G. (1976). Experimental manipulation of the amount of tubulin available for assembly into the spindle of dividing sea urchin eggs. *J. Cell Biol.* **70**, 75–85.

Sluder, G. (1979). Role of spindle microtubules in the control of cell cycle timing. *J. Cell Biol.* **80**, 674–691.

Summers, K., and Kirschner, M. W. (1979). Characteristics of the polar assembly and disassembly of microtubules observed *in vitro* by darkfield light microscopy. *J. Cell Biol.* **83**, 205–217.

Swann, M. M. (1951a). Protoplasmic structure and mitosis. I. The birefringence of the metaphase spindle and asters of the living sea-urchin egg. *J. Exp. Biol.* **28**, 417–433.

Swann, M. M. (1951b). Protoplasmic structure and mitosis. II. The nature and cause of birefringence changes in the sea-urchin egg at anaphase. *J. Exp. Biol.* **28**, 434–444.

Taylor, E. W. (1965). Brownian and saltatory movements of cytoplasmic granules and the movement of anaphase chromosomes. *In* "Proc. Fourth Intern. Congress on Rheology, Brown University" (E. H. Lee, ed.), pp. 175–191. Wiley (Interscience), New York.

White, M. J. D. (1973). "Animal Cytology and Evolution." Cambridge Univ. Press, London and New York.

Wilson, E. B. (1928). "The Cell in Development and Heredity." Macmillan, New York.

Wilson, L., and Margolis, R. L. (1978). Opposite end assembly and disassembly of microtubules: A steady state mechanism. *In* "Cell Reproduction: In Honor of Daniel Mazia" (E. R. Dirksen, D. M. Prescott, and C. F. Fox, eds.), pp. 241–258. Academic Press, New York.

Wolpert, L. (1965). Cytoplasmic streaming and amoeboid movement. *Symp. Soc. gen. Microbiol.* **15**, 270–293.

7

Chromosome Micromanipulation Studies

GORDON W. ELLIS and DAVID A. BEGG

I. INTRACELLULAR CHROMOSOME MICRURGY

A. Introduction

Micromanipulation (micrurgy) is a powerful tool for studying the morphological and mechanical aspects of chromosome movement. In addition, it has proved to be useful for studying chromosome-orienting mechanisms. Such experiments have established a set of facts which, although they themselves do not explain the phenomena in chromosome movement, supply tests which have served to dispense with inappropriate theoretical models.

155

B. Background

Just over 100 years after Flemming's (1879, 1880) discovery of mitosis, the mechanism of its operation is still the subject of experiment and controversy. Early workers, whose studies are summarized by Wilson (1925), had thought, on the basis of fixed and stained material, that chromosomes must be pulled to the poles by "traction" fibers; but by the time of this third edition of Wilson's monumental treatise, more work had been done on living cells and the outlook had changed. In particular, Chambers's micrurgical investigations of cell structure had been influential in framing the views that prevailed half a century after Flemming. Kite and Chambers (1912), in the first paper of its kind, described dissecting out, from a grasshopper spermatocyte, a single chromosome with spindle fiber attached and described the fiber as a "slightly refractive elastic gel." Later, Chambers (1914) described the spindle in similar cells as being a hyaline gel of higher viscosity than the general cytoplasm but having in life no detectable fibrous structure. Similar reports (Chambers, 1915a, 1915b, 1923, 1925) reinforced the concept of the spindle as a weak, non-fibrous gel in which the chromosomes were embedded with little mechanical constraint. The following quotation from Wilson's introduction to "General Cytology," edited by E. V. Cowdry (1924) (and containing a chapter by Chambers on the physical structure of the cytoplasm), typifies the attitude toward cell structure that had developed after a period in which living cells were studied in the absence of phase contrast or sensitive polarizing microscopes. Referring to the earlier views, Wilson said:

> Among the most familiar was the concept of protoplasm as a netlike or reticular structure (Leydig, Klein, Heitzmann, VanBeneden, Carnoy) or as consisting of separate fibrillae suspended in a transparent ground substance (Flemming, Heidenhain, Ballowitz). These views were unduly influenced by the study of fixed or coagulated cells; and the reticular theory in particular has proved extremely improbable, if not wholly untenable. On the other hand, more modern studies have constantly added weight to the view of the earlier investigators that living protoplasm considered as a whole has the properties of a more or less viscid liquid and constitutes what is now considered a colloidal system.

This was, perhaps, a majority view, but it was not unanimous. A minority persisted in the opinion that spindle fibers were more than fixation artifacts. Wada (1935), from micrurgical punctures of *Tradescantia* stamen hair cells, obtained evidence that in anaphase the chromosomes were mechanically attached to the apparently fibrous half-spindle. Others (Schrader, 1935; Beams and King, 1936; Shimamura, 1940) used centrifugation to gain evidence that chromosomes were mechanically attached to and supported by the spindle. Finally, in the early 1950s, three papers removed the mitotic spindle from the realm of artifact and started its functional and structural characterization. Mazia and Dan (1952) reported the mass isolation and preliminary biochemical characterization of the mitotic apparatus; Carlson

(1952) described the careful, non-destructive micrurgical exploration of the spindle in living grasshopper neuroblasts; and Inoué (1953) showed the existence and activities of birefringent spindle fibers in living *Chaetopterus* oocytes and lily pollen-mother cells.

Carlson's report contradicted Chambers regarding the attachment of chromosomes to the spindle and showed the half-spindle between chromosomes and pole to be mechanically anisotropic, with much lower viscosity in the direction of the spindle axis than across it. In anaphase, he found the interzone to be much more fluid than the half-spindles. He also showed that the spindle could be dislodged from its anchorage to the cell cortex and moved about in the cell without interrupting either chromosome movement to the pole or pole-to-pole elongation. His study of the relationship of spindle position to cleavage furrow location was later extended by Kawamura (1960). (See Chapter 16 by Conrad and Rappaport.) That Carlson did not explore the connections of the individual chromosomes to the spindle in any detail was, in retrospect, probably in part a consequence of the amphiorientation of the mitotic chromosomes of the neuroblasts, which gives them a much tighter coupling to the spindle than is found in the bivalents of grasshopper spermatocytes, and in part a technological limitation due to the micromanipulator he used. The piezoelectric micromanipulator introduced in 1962 (Ellis, 1962) greatly simplified the task of controlling microneedle movement and freed the operator to try to cope with the formidable logistics of intracellular chromosome manipulation. Using this instrument, Nicklas launched a program of chromosome micromanipulation that has been both productive and surprising in its results. The next section describes the current state of knowledge regarding chromosome–spindle interactions as found through chromosome micrurgy, largely by Nicklas and associates but including contributions from our laboratory and by others.

C. Findings from Recent Chromosome Micromanipulation Studies

This section is an attempt to list conclusions that can be drawn from chromosome micrurgy experiments performed to date. Following each statement is a summary of the evidence supporting it. When previously unpublished data are included, the materials and methods employed are those described in Begg and Ellis (1979a). This list is inevitably a biased one and may not call attention to all of the features that other investigators feel can be concluded from the data. In particular, we have made no attempt here to relate these findings to specific theories of chromosome movement.

1. Chromosomes are attached to the spindle at the kinetochore. Even Chambers reported that chromosomes are weakly attached to the spindle

(1951; and Chambers and Chambers, 1961), although he did not report attempts to localize the site of attachment. Carlson (1952), working with grasshopper neuroblasts, did note that the chromosome arms were free, particularly in the case of the larger chromosomes that were located at the kinetochore (centromere). Nicklas and Staehly (1967), working with grasshopper spermatocytes, and Begg and Ellis (1979a), using both grasshopper and crane fly spermatocytes, have clearly demonstrated that tugging with the tip of a microneedle at any point in the chromosome, other than at a kinetochore, will deform the chromosome. The pattern of deformation that results clearly shows that kinetochores are the points at which the chromosome is mechanically linked to the spindle.

2. The kinetochore is connected to the pole by a relatively inextensible fiber. Nicklas and Staehly (1967) found, from prometaphase through anaphase in grasshopper spermatocytes, that stretching a bivalent with a microneedle greatly extended the chromosome but did not significantly change the distance between the kinetochores and the poles. Begg and Ellis (1979a,b) have confirmed this for both grasshopper and cranefly spermatocytes and have added the qualifications described in No. 4 below.

3. The mechanical connection between chromosome and spindle is coextensive with the birefringent chromosomal fiber. Both colcemid and vinblastine treatment of grasshopper and crane fly spermatocytes abolish spindle birefringence; concurrently, chromosomes lose their mechanical connection to the poles and become freely displaceable within the cytoplasm (Begg and Ellis, 1979b). Inactivation of colcemid by ultraviolet light (366 nm) restores both the spindle birefringence and the mechanical connection. Moving a birefringent chromosomal fiber with a microneedle selectively moves the chromosome to which the fiber is attached. In this experiment, the pull by the microneedle can bend the fiber sharply (ca. 90°) without damaging it; at the same time, the chromosome attached to the fiber is clearly stretched (Begg, 1975b).

4. The extensibility of the kinetochore fiber decreases as the fiber birefringence increases. Generally, as grasshopper or crane fly spermatocytes progress from prometaphase to anaphase, the birefringence associated with kinetochoric fibers increases and the birefringence of the interpolar spindle decreases. As these cells progress to anaphase, the distance a kinetochore can be displaced from the pole toward which it is oriented, by a single smooth pull from the microneedle, decreases. While the same generalization could be made regarding fiber extensibility and meiotic stage (i.e., the chromosomal fiber becomes progressively less extensible as the cell proceeds from prometaphase to anaphase), the observation of an exception makes the relationship between birefringence and fiber strength seem more basic than that between stage and fiber strength. In this case, a chromosome

in a prometaphase crane fly spermatocyte in meiosis I was found to be closely associated with a strongly birefringent interpolar fiber. A pull on this chromosome produced much less displacement than was typical for other chromosomes at a similar stage (Begg, 1975b; Begg and Ellis, 1979b).

5. The chromosomal fiber birefringence and the mechanical properties of the fiber are probably both due to microtubules. Microtubule-depolymerizing agents abolish both the mechanical connection between kinetochore and spindle and the fiber birefringence (No. 3 above).

Chromosomes detached from the spindle by micromanipulation (see also No. 12, et seq., below) are, for a short time, free in the cytoplasm and have no birefringent chromosomal fiber. However, after being released, they reestablish their connection to the spindle by means of a kinetochoric fiber that is at first invisible under the light microscope and that gradually gains strength and birefringence until it eventually resembles the original kinetochoric fiber in both appearance and function (Begg, 1975b). Brinkley and Nicklas (1968) and Nicklas *et al.* (1979a) have found, through electron microscopic examination of kinetochores of detached chromosomes, that those kinetochores which showed some sign of movement prior to fixation have microtubules attached to their kinetochores and running in the direction of movement, while those kinetochores that had not moved before fixation have no microtubules (Brinkley and Nicklas, 1968; Nicklas, 1971), or no more than two microtubules (Nicklas *et al.*, 1979a), associated with them. Because we find that chromosome movement in these circumstances invariably indicates mechanical reattachment, these observations provide a *prima facie* case for microtubules as the mechanical link to the chromosome.

Sato *et al.* (1975) have demonstrated the relationship between birefringence and microtubules as studied in isolated spindles. Several others have compared the birefringence of spindles and chromosomal fibers in intact cells with the results of electron microscopic examination and have agreed or disagreed that the relationship was valid for intact cells. Most recently, Marek (1978) has compared the volume birefringence in spindles of grasshopper spermatocytes with the microtubule count in electron microscopic sections and has found the number of microtubules to be proportional to the birefringence but lower in number than called for by the Wiener equation. While this could indicate the existence of a significant quantity of non-microtubule elements parallel to the microtubules in life but lost on fixation, a more economical explanation is, as Sato *et al.* (1975) reported, that microtubules are themselves less well preserved in intact cells than in isolated spindles. This explanation is strengthened by work in progress by Nicklas (personal communication, 1980), who after a thoughtful comparison of methods and results between La Fountain (1976) and Forer and Brinkley (1977) has devised a method of fixation that gives, in preliminary tests, consistently increased numbers of microtubules in spindles fixed in intact spermatocytes, including the species Marek used.

6. Chromosomal movement occurs at the rate at which the chromosomal spindle fiber shortens. During a normal anaphase, both the birefringent chromosomal fiber *in vivo* (Salmon and Begg, 1980) and the kinetochoric microtubules in electron microscopic preparations (McIntosh *et al.*, 1975; LaFountain, 1976) occupy only the space between the kinetochore and pole (pericentriolar region in electron microscopy) as the chromosome moves. That is, the fiber extends only from kinetochore to pole, and it runs fairly directly between the two. While this statement has no meaning for fibers other than those currently associated with a particular kinetochore, the point is that the chromosome does not move along the fiber; it is always terminal relative to the kinetochoric end of the fiber, and the polar end of the fiber never extends past the pole. While the chromosome is moving, the fiber is never longer than the distance between kinetochore and pole (not considering the slight extra length due to the arch of some peripheral chromosomal fibers); yet when the fiber is made longer than this distance experimentally by pushing the chromosome toward the pole, the chromosome stops moving until the "slack" is taken up (see Nos. 8 and 9 below).

7. Anaphase shortening of the chromosomal spindle fiber is independent of the position of the chromosome relative to the interpolar spindle. During anaphase movement, the half-bivalents in both *Trimeratropis* and *Nephrotoma* spermatocytes can be swung freely about their polar connections in an arc that can carry them well beyond the spindle boundary (Begg, 1975b; Begg and Ellis, 1979a). If left in any of these displaced positions, the chromosome continues its movement toward the pole at a normal velocity. This is similar to the condition of grasshopper spermatocytes (Nicklas and Staehly, 1967).

8. The chromosomal fiber is stiff in tension, but it is relatively flexible in bending or in compression. This is difficult to test in prometaphase through metaphase because one cannot load a chromosomal fiber in compression without also loading in tension the fiber to the opposite pole, but in anaphase the situation is less confusing. During anaphase, one can readily push (Nicklas and Staehly, 1967; Begg and Ellis, 1979a; also see No. 9 below) a chromosome toward the pole it is approaching without significantly distorting the chromosome in reaction to the push. If released in its new position, closer to the pole, the chromosome will remain for some time in that position without noticeable rebound. If the chromosome is returned toward its original position, it will move with little applied force (little or no distortion) until it reaches a position near the original distance from the pole, at which point it will move no farther without severe distortion. (See also No. 3 above.)

9. During anaphase, the chromosomal fiber appears to be composed of non-crosslinked filaments for most of its length. In both grasshopper and crane fly spermatocytes, when one pushes toward the pole a chromosome having a prominently birefringent chromosomal fiber, the birefringent fiber disappears and the space between the kinetochore and the pole shows no trace of a coherent folded fiber. Careful examination shows only a diffuse birefringence with little or no orientation. This result is most economically interpreted as showing that the fiber substrands have splayed out as the fiber was compressed. The continued presence of the now invisible fiber is readily demonstrated, as in No. 8 above. If one leaves the chromosome in its displaced position, it remains where it is placed until, when the other autosomes have traveled a comparable distance toward the pole, it resumes its movement toward the pole. Careful measurement shows this movement to be a resumption of travel at its original rate (Begg and Ellis, 1979a). This result is most simply explained by supposing that the fiber has continued to shorten, *even though splayed*, at the same rate as before pushing, and when the slack is taken up, the chromosome continues on its way. The fibers of polewardly displaced chromosomes have not yet been examined under electron microscopy; consequently one cannot say with assurance that the microtubules of the fiber splay as entirely independent units. Considering the sensitivity of the polarizing microscope used (Begg, 1975b), the maximum undetected subunit size for crosslinked microtubules is probably five microtubules or less.

10. While the primary connection of the chromosome to the spindle is via the kinetochore, in anaphase mechanical links may connect the arms of partner half-bivalents. Carlson (1952) reported that he found mechanical linkage between the arms of separating sister chromatids of anaphase grasshopper neuroblasts. These links correspond to the interzonal fibers found in fixed and stained material (for example, see Schrader, 1944). However, in most grasshopper spermatocyte preparations (Nicklas and Staehly, 1967; Begg and Ellis, 1979a), no significant mechanical linkage is found between partner half-bivalents. In two species of crane fly spermatocytes, Forer and Koch (1973) report that partner half-bivalents separating in anaphase show linkage, as manifested by the fact that pushing one half-bivalent perpendicular to the interpolar axis resulted in movement of the partner half-bivalent as well. Begg and Ellis (1979a) found in spermatocytes of one of the same species of crane fly (*Nephrotoma ferruginea* Fabricius) used by Forer and Koch (1973) that anaphase partner half-bivalents in freshly prepared cultures showed no mechanical linkage, while in cells that entered anaphase after being in culture for an hour or more, mechanical linkage was detected between partner half-bivalents. Both the frequency and the degree

of bridging increased with time. In all cases where mechanical linkage was detected, what appeared to be chromosomal bridges were found between the linked half-bivalents. Similar linkage could be induced in grasshopper spermatocytes by lowering the pH of the Ringer solution below 7.2 (Begg, 1975b). This apparent contradiction has not been resolved. Forer has assured us that careful examination of his photographic data reveals no chromosomal bridges. We could see what appeared to be chromosomal bridges under the conditions described, but we did not fix and stain the material to verify the chromosomal nature of these links. Conceivably, what happens is that an undefined amorphous matrix, as yet undetected in electron microscopic examination, invests the chromosomes and has mechanical properties that vary according to physiological conditions. Thus in one case, the chromosomes separate without significant linkage, while in the other the matrix is more viscous and "strings" are pulled out between separating half-bivalents. In extreme cases, the viscosity could become so high that the chromosomes would be stretched out in bridges. Schrader (1944) clearly distinguishes between interzonal fibers and other spindle elements, particularly continuous or interpolar spindle fibers, in terms of their staining, response to centrifugation, and locale. He also cites some reports describing interzonal fibers as varying in their presence or absence in some cells. His explanation is essentially that given above. Today not as much care is given to distinguishing between interzonals and interpolar fibers, and some accounts equate interzonal fibers with interpolar microtubules revealed by the separating chromosomes (McIntosh *et al.*, 1975).

11. During prometaphase and metaphase the kinetochores may have, in addition to their normal links to the pole to which they are oriented, other weaker connections; but in anaphase these connections are not seen and, in any event, are not essential for chromosome movement.

The micrurgical evidence for these connections in prometaphase through metaphase (Begg and Ellis, 1979a) comes in part from the observation that during this period it is difficult to move grasshopper spermatocyte kinetochores beyond the periphery of the spindle by tugging on the chromosome. This could have other explanations and by itself is not very convincing. Other evidence is found (Nicklas and Koch, 1969) from experiments in which reorientation is induced without first detaching the chromosome. Here, after tilting the chromosome by pushing one half-bivalent toward the interpolar axis, the chromosome was bent into a J shape by pushing the middle of the bivalent toward one pole. The kinetochore originally oriented toward this pole was now seen to be rotated to a new position in which it faces the equator (and the opposite pole). After a time in this position, the kinetochore was found to have transferred its connection to the opposite pole. This is, of course, an important observation in itself; but it also suggests to us the presence of a constraint on the rotated kinetochore from a direction other than that of its known link to the nearer pole. For the chromosome to bend in this fashion in response to the needle, either the chromosomal fiber must be thought of as being stiff in compression but having a flexible "swivel" at the kinetochore or the kinetochore must have a link to something else located in the spindle between it and the other pole. Nicklas *et al.* (1979c) have shown serial reconstructions of electron micrographs of spindles that had a manipulated metaphase bivalent pushed as far as possible from the rest of the spindle before fixation. Here, in addition to the expected kinetochoric microtubules directed toward the pole, they found a few kinetochoric microtubules running toward the axis of the spindle. The distal anchorage of these microtubules has not yet been traced. Although the presence of these microtubules could be invoked to explain the difficulty in pulling the kinetochores free of the spindle, the facts that swinging the chromosome around the periphery of the spindle is not inhibited, and that no reaction to the pull is shown by other chromosomes in the spindle, leave the explanation in doubt. Of course, no single explanation is required and perhaps, as others have suggested, the polarization of kinetochoric links toward a single pole in anaphase is a gradual process, beginning with bipolar links in prometaphase which become completely unipolar by anaphase. In any event, no trace of this restriction remains in anaphase.

12. The connection between the kinetochore and the kinetochore–fiber can be disrupted experimentally by mechanical agitation. Nicklas's (1967) original finding has been repeated on numerous occasions (Brinkley and Nicklas, 1968; Camenzind and Nicklas, 1968; Nicklas and Koch 1969; Henderson and Koch, 1970; Nicklas, 1971; Nicklas and Koch, 1972; Forer and Koch, 1973; Begg, 1975b; Nicklas, 1977a; Marek, 1978; Nicklas *et al.*, 1979a; D. A. Begg and G. W. Ellis, manuscript in preparation), and the reality of the phenomenon and its consequences are well known. However, the mechanics of the detachment process itself remain a mystery that is all the more intriguing because no microtubule stubs are left on the kinetochore, and the kinetochore shows neither functional nor structural damage (Brinkley and Nicklas, 1968; Nicklas *et al.*, 1979a).

13. The ease of chromosome detachment decreases from prometaphase through metaphase as the chromosomal birefringence increases. In grasshopper spermatocytes, in early prometaphase I a single pull with the microneedle can be sufficient to detach a chromosome from the spindle (Begg, 1975a,b; D. A. Begg and G. W. Ellis, manuscript in preparation). In later prometaphase, the kinetochore : chromosomal fiber junction must be repeatedly prodded with the microneedle, alternating with tugs on the chromosome, before the chromosome suddenly, usually during a tug, becomes freely movable in the cytoplasm. In metaphase, when the kinetochores

have prominent birefringent fibers, the chromosomes become extremely difficult to detach. This firm connection persists most of the way through anaphase. The increase in strength of attachment parallels the increase in birefringence and the decrease in extensibility of the fiber, as noted in No. 4 above. Late metaphase detachment is so difficult that it is rarely achieved. When it is achieved, anaphase is *not* blocked, contrary to Nicklas's (1967) speculation.

In early to mid-anaphase, detachment of bivalents is extremely difficult, if not impossible (at least it has not yet been accomplished), in spermatocytes from at least two species of grasshopper (Nicklas, 1967; Begg, 1975b), while in crane fly spermatocytes it is relatively easy (Forer and Koch, 1973). We have not explored chromosome detachment in crane fly spermatocytes, and Forer and Koch did not discuss the relative strength of attachment as a function of either stage of birefringence. They did find—unlike grasshopper spermatocytes, in which only the X chromosome is detachable in most of anaphase (Nicklas, 1967)—that all the autosomes were detachable throughout anaphase.

14. Mechanically isolated chromosomes will spontaneously reestablish connections to the spindle. Chromosomes detached from the spindle may be pushed around freely in the cytoplasm while showing none of the stretching that reveals mechanical attachment (Nicklas, 1967). Shortly after (ca. 1 to 2 minutes) such a chromosome is released from the microneedle, it will be found to stretch if moved again and, if undisturbed, it will spontaneously move to orient on the spindle. The chromosomal site of attachment, as revealed by this stretching, is always the kinetochore. The motion of chromosomes returning to the spindle after detachment is as much as three to four times faster than anaphase chromosome velocity (Nicklas, 1967; Begg, 1975a,b; D. A. Begg and G. W. Ellis, manuscript in preparation).

15. The reestablished kinetochoric connections may form initially between one kinetochore and either pole, between one kinetochore and the spindle somewhere other than the pole, or between both kinetochores and a single pole. Nicklas (1967) reported that the kinetochore reestablished a connection to the pole it was facing regardless of its location with respect to the poles. This has been confirmed for chromosomes released for reattachment from within the spindle boundaries; but for chromosomes released well outside the spindle boundaries, the initial reattachment is to some point near the equator or within the half-spindle (Begg, 1975a,b; D. A. Begg and G. W. Ellis, manuscript in preparation). In such cases, the point of reattachment can be located by gently swinging the chromosome about its point of attachment. The center of the arc described by the kinetochore's path locates the attachment site. More recently, Nicklas *et al.* (1979a) have confirmed this finding and have found, in electron microscopic examination of such chromo-

somes, that microtubules directed toward the equator are the sole microtubular link to the spindle. They also report finding cases in which microtubules run toward both poles.

In grasshopper spermatocytes, when anaphase onset occurs while a chromosome is still detached (Begg, 1975b), the detached bivalent separates into two half-bivalents synchronously with the chromosomes still on the spindle. But the two half-bivalents do not move beyond the initial separation until, after a variable lag period, they both move back to the spindle periphery and then proceed at normal anaphase velocity to the *same* pole (Fig. 1).

16. Newly established connections require mechanical tension for stability. This assertion may not be a valid generalization in that it is based on events following malorientation when both half-bivalents orient to the same pole (Nicklas, 1967; Nicklas and Koch, 1969; Henderson and Koch, 1970) following detachment. Here, in the absence of artificially imposed tension in the chromosomal fibers, the chromosome will spontaneously reorient about 16 minutes following reattachment (Nicklas and Koch, 1969). Chromosomes that form normal reattachments reestablish co-orientation within a short enough period following initial reattachment so the need for tension is not really tested. Attempts to induce instability in a previously unoperated chromosomal fiber by pushing the chromosome toward the pole in order to relax the fiber (Nicklas and Koch, 1969) have produced equivocal results. In some cases, the chromosome reorients without detachment. In other cases, the kinetochore that is displaced toward the pole remains offset when released and then slowly returns to its normal position without reorienting.

17. The strength of newly established connections increases with time, as does their birefringence. When a detached chromosome shows reattachment to the spindle by movement (Nicklas, 1967) or by the reappearance of kinetochores protruding from the body of the bivalent (Begg, 1975b; D. A. Begg and G. W. Ellis, manuscript in preparation), tugging on the chromosome with the microneedle will stretch the chromosome, confirming reattachment; but a hard tug will detach the chromosome again. With the passage of time following the first sign of reattachment, the chromosome becomes more difficult to detach again (Begg, 1975b) (see also No. 5 above).

18. Detached chromosomes will orient on spindles other than the one from which they were detached. Nicklas (1971, 1977a) found that although he had been unsuccessful in transferring chromosomes from one cell to another, he could transfer chromosomes from one spindle to another in fused cells. Marek (1978) has confirmed and extended this finding.

19. The difference between meiosis I chromosomal orientation and mitotic chromosomal orientation lies in the kinetochore arrangement rather than in the spindles. Meiosis I and meiosis II spermatocytes can be fused without

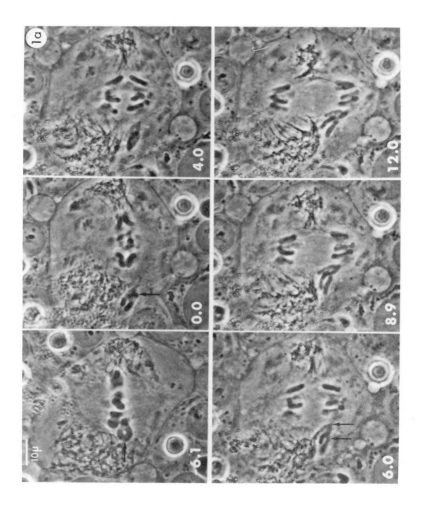

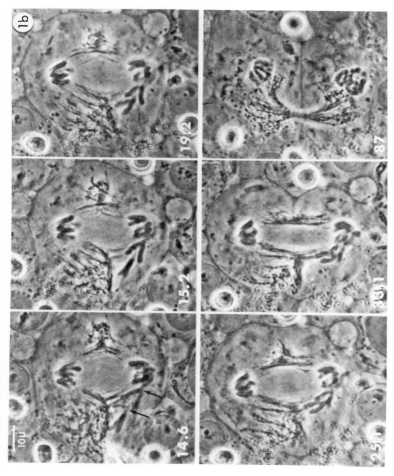

Fig. 1. Behavior of a detached bivalent in anaphase. The cell is a primary spermatocyte of *Trimeratropis maritima*. (a) −6.1 min: Before detachment. The cell is in metaphase. The arrow indicates the bivalent which will be detached. 0.0 min: Detached bivalent (arrow) appears doubled. The cell has entered anaphase. 4.0–12.0 min: Both half-bivalents (arrows) move directly to the lower spindle pole. (b) 14.6 and 15.2 min: The reattached half-bivalent is extensively stretched without significant displacement of its kinetochore away from the spindle pole. 19.2–33.1 min: Manipulated chromosomes continue to move toward the same spindle pole. 87 min: Cleavage.

disturbing the organization of their division apparatus (Nicklas, 1977a). In such cases, one can detach a meiosis I bivalent and transfer it to the meiosis II spindle. The reverse transfer is also possible, although more difficult. In both cases, the chromosomes orient on the new spindle in the same manner as on their original spindle. The result is that one finds in the same spindle meiosis I bivalents orienting syntelically alongside second-division chromosomes, with each chromatid orienting to opposite poles. As Nicklas reports, these observations have provided a *prima facie* case in favor of Östergren's (1951) view that differences in the kinetochores govern the mode of chromosome orientation, in opposition to Lima de Faria's (1958) view.

20. The strength of the chromosomal fiber is greatly in excess of the strength needed to withstand the tension required to move the chromosome. If we regard the chromosome as an elastic body that deforms reversibly as a function of the force applied to it (Nicklas, 1963), then it can be used as a spring balance to compare an experimental force with forces applied to the kinetochores during the normal course of division. Nicklas has done this on a number of occasions (e.g., Nicklas and Koch, 1969). If we collect observations from various sources (Nicklas and Staehly, 1967; Nicklas, 1967; Nicklas and Koch, 1969; Henderson and Koch, 1970; Henderson *et al.*, 1970; Begg, 1975b; Begg and Ellis, 1979a), we find that even in early prometaphase the chromosome can be stretched more with a microneedle, without breaking the connection, than it would be deformed normally during anaphase movement. At any subsequent time through anaphase, the chromosomal fiber is stronger than in prometaphase. In testing for chromosomal reattachment to the spindle, chromosome deformation in response to attempts to move it with a microneedle shows that even the earliest tests that are positive regarding attachment impose on the new connection a force in excess of that involved in moving the chromosome. Nicklas and associates (Brinkley and Nicklas, 1968; Nicklas, 1971; Nicklas *et al.*, 1979a) have shown that the fiber making this connection may consist of as few as three to five microtubules.

21. The astral connections to the cell cortex are weaker in ensemble than a single metaphase or anaphase chromosomal fiber. During metaphase and anaphase, attempts to pull a chromosome away from the pole toward which it has oriented result in displacement of the entire spindle within the cell by a distance much greater than the change in distance, if any, between chromosome and pole (Begg and Ellis, 1979a).

22. In some cells, micrurgical changes in chromosome orientation or position can influence the orientation and/or segregation of other chromosomes in the same spindle to which they have no detectable mechanical connection (Camenzind and Nicklas, 1968; Forer and Koch, 1973).

II. METHODOLOGY FOR CHROMOSOME MICRURGY

For background on micromanipulation in general, see Chambers and Kopac (1950), Kopac (1959, 1964), and El Badry (1963). Also, see articles on particular techniques, other than chromosome micrurgy, in "Methods in Cell Physiology," edited by David M. Prescott.

A. Cell Selection

Micromanipulation, including microinjection, has been used to explore other mechanical features in a wide variety of cell types. However, the number of cell types used for chromosome micromanipulation experiments is much more limited. This is a consequence of the multiple requirements placed on cells as candidates for this kind of experimental approach. The cell must be of reasonably large size so that the manipulation and its result can be seen clearly; the cell must be accessible to the microneedle without serious damage to the cell or undue restriction on the movement of the microtool; it must undergo mitosis or meiosis in the operating chamber reliably and in a resonable period of time; it must have a relatively small number of chromosomes that are individually distinguishable and clearly discernible *in vivo;* the chromosomes should be relatively large so that their topography is easily resolved in a light microscope; and the cell must be readily available and easy to immobilize in the operating chamber. Mammalian tissue cells in culture are readily available but fall short on a number of other requirements: they tend to be too small for good visibility; they have too many chromosomes which are also too small; and they require sterile conditions and elevated temperatures, which complicate access and risk stress birefringence in the optics. Amphibian tissue cells, such as newt lung epithelial cells, are interesting from the standpoint of chromosome size and morphology, but these cells have not yet been placed successfully into continuous culture and their use in primary culture would add handicapping complications to an already complex procedure. Plant cells are generally poor prospects because of their tough cell walls, but endosperm cells lack this barrier, and those of *Haemanthus kathrinae* and related species are also very attractive on the basis of chromosome size. Unfortunately, these cells are very fastidious in terms of their culture conditions and have not yet been immobilized in viable condition in an operating chamber. Consequently all of the interesting chromosomal micrurgy to date has been done on insect cells, particularly spermatocytes of grasshoppers, crickets, and crane flies. Perhaps it is time to test some of the conclusions drawn from insect cells on some other, more distantly related, cell types.

B. Operating Chambers

The purpose of the operating chamber is to present the cell to the optics of the microscope while making it accessible to the microtool and maintaining it in good condition physiologically. Early workers used hanging drop (or, for inverted microscopes, lying drop) cultures in moist chambers (see Chambers and Kopac, 1950; El Badry, 1963). These are rarely satisfactory for long-term observation and impose unacceptable limitations on the microscope optics. DeFonbrune (1949) used a thinner chamber in which evaporation was retarded by surrounding with oil the droplet of medium containing the specimen. Today, most investigators use an oil chamber of some sort. Nicklas and associates use an open top chamber which consists of a glass slide (Nicklas and Staehly, 1967) or a metal slide (Nicklas *et al.*, 1979a) that has been drilled through to form a well, the bottom of which is made of a cover slip attached to the slide. This is necessarily used in conjunction with an inverted microscope. The specimen is placed on the cover slip in a droplet of medium and covered with a layer of inert fluorocarbon oil. This system provides excellent access to the cell, and Nicklas and associates have been quite successful using it. It does complicate the optical situation somewhat because one must either immerse the condenser in the fluorocarbon oil or contend with suboptimal illuminating conditions caused by projecting the beam from the condenser through a distorting interface.

We use a chamber bounded at the top and bottom with parallel cover slips, with a 1-mm space between them. For either the upright or inverted microscope, the specimen is on the top or the bottom cover slip, in a droplet of medium, and the rest of the space is filled with fluorocarbon oil. This combination is optically equivalent to a normal microscope slide and cover slip and permits use of the standard optical components used for any type of light microscopy. The usual form of the chamber (Begg and Ellis, 1979a) is a metal or plastic plate with a ¾-inch (19-mm) square hole cut out of one side. An alternate form, which provides access from nearly all sides, can be seen on the stage of the microscope shown in Fig. 4a. This form is helpful for use on rotating stages when specimen orientation must be changed over more than 90°.

C. Microscope Selection

A microscope for use in micromanipulation should have at least its fine focusing mechanism on the objective carrier rather than on the stage support. Otherwise, focusing motion produces relative movement between the specimen and the microtool, with potentially disastrous results. Only a few of the upright microscope stands that are presently manufactured satisfy this requirement, and those that do are often quite limited in their optical capabilities. Fortunately, several manufacturers have recently introduced inverted microscopes that both meet this requirement and can be equipped with a wide variety of optics. For use with polarization microscopy or with differential interference contrast (DIC) microscopy, the microscope should have a rotatable stage; not all of the new microscopes offer this option.

D. A Second-Generation Piezoelectric Micromanipulator

The original piezoelectric micromanipulator (Ellis, 1962) employed transducers fabricated from two X-cut slices of single-crystal rochelle salt (sodium–potassium tartrate) cemented together to form a "twister Bimorph" (Sawyer, 1931). These transducers had the advantage of being true piezoelectric devices with a linear voltage versus displacement function, but they were fragile and required restricted temperature and humidity conditions. Today, these transducers have been superseded by piezoelectric ceramic devices, and rochelle salt transducers are no longer available. Ceramic transducers offer superior durability and will tolerate a wide range of environmental conditions. They do, however, present two problems for their use as micromanipulator transducers. The first is a simple geometrical problem. Ceramic Bimorphs are inherently "benders" and cannot be constructed as "twisters." Therefore, the linkage geometry must be altered to accommodate them. Generally, this can be done for any of the linkage patterns previously described (Ellis, 1962) by substituting a bender trans-

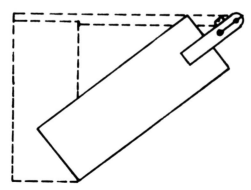

Fig. 2. A bender Bimorph transducer (solid line) can replace a twister Bimorph transducer (dashed line) and its torque arm, as shown here. The diagonal placement allows the maximum-length bender in the space available and, hence, the maximum range of generated movement.

3a

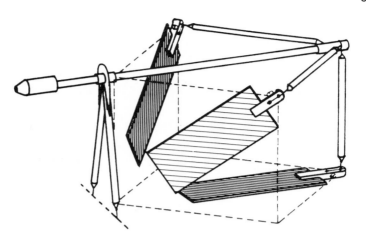

ducer for each twister transducer plus its torque arm (Fig. 2). Fig. 3 shows the linkage of the principal operating head adapted for ceramic transducers. The other problem, potentially more serious, is that ceramic transducers are polarized electrostrictive polycrystalline materials rather than single crystal piezoelectrics and tend to depolarize with use. This leads to a progressively reduced range of movement and a shift in zero-signal position. If the transducer is energized only in the direction of its original polarization, it will remain stable (Germano, 1969). However, in order to have ample range of movement with reasonably sized transducers, one risks dielectric breakdown if only one of the elements of the Bimorph is energized. The solution we have found is the use of diode switching to direct the energizing voltage to the half of the transducer whose poling is in phase with the applied voltage. In practice, with the series Bimorph which has its two elements oppositely polarized, one element is short-circuited and the entire signal is applied to the other element in appropriate polarity regardless of the sign of the applied voltage (circuit shown in Fig. 5).

For the coarse positioning mechanism, we have followed Nicklas's lead and used a commercially available mechanical micropositioner, in our case one manufactured by Kulicke and Soffa. We have added a stable base of adjustable height to allow use of a variety of microscopes. The assembled micromanipulator is shown in Fig. 4. The new operating head can be used with the elaborate control unit of the original description (Ellis, 1962), or a minimal control unit providing only manual control can be economically constructed using one of the ubiquitous joystick controls originally intended for video game control but now available from a variety of electronic bargain houses (see Fig. 4b). For this application, one should pick one with dual 100,000-ohm potentiometers on each axis. An appropriate circuit is shown in Fig. 5.

Fig. 3. (a) Dimetric projection of the transducer linkage geometry of the new micromanipulator. The three ceramic transducers (crosshatched) lie diagonally in three orthogonal faces of a cube. The wire-hinged links between each transducer and the tool shaft lie in three orthogonal edges of the cube, with each perpendicular to the plane of the transducer it links to the shaft. (b) Operating head with top and side covers removed to show the horizontal motion transducers and their links to the tool shaft. The transducers are each 2 inches (50 mm) by 0.8 inch (20 mm) PZT-5H Bimorphs. Except for the substitution of ceramic bender transducers for the rochelle salt twisters of the original and the switching diodes (not visible), this operating head is similar to the principal operating head described previously (Ellis, 1962). The diode switching circuit, needed for stable operation of the ceramic transducers, is shown in Fig. 5.

Fig. 4. 4a. A view of the completed micromanipulator showing the coarse positioning mechanism (Kulicke and Soffa Manufacturing Co., 135 Commerce Drive, Industrial Park, Fort Washington, PA 19034) mounted on an adjustable-height stand. The micrometer spin-

(continued)

dles providing coarse position control are marked. V, vertical; L, longitudinal; T, transverse. The stand is an L bracket fabricated of 1-inch (25-mm) aluminum alloy plate stock. A series of threaded mounting holes in the upright allows the height of the coarse manipulator and the operating head to be varied for use with different microscopes. The microscope shown is not optimal for micrurgy because only the coarse focus moves the objective. The new operating chamber, shown on the microscope stage, has the upper cover slip mounted on a ring cantilevered over the lower cover slip by a single support. This ensures good optical conditions (see text), while access for microtools is unobstructed for over 350° around the chamber. (b) Control station. The new operating head will operate without modification with the control station originally described (Ellis, 1962), or it may be used with less elaborate controllers such as the one shown here. It uses a two-axis gimballed joystick to control the horizontal-movement potentiometers (voltage-dividing resistors). Vertical movement is controlled by a Knobpot (Bourns, model 36005-1, 100,000 ohms) mounted on the joystick. The circuit is electrically similar to that described in the earlier paper for use with the satellite operating heads. The unit shown has a specially fabricated joystick, but that has proved to have little significant advantage over the inexpensive commercial unit shown in the left foreground. These joysticks, manufactured in enormous numbers for use in video game controllers, are available in several forms at extremely low cost from a number of electronic bargain houses. The one shown is the best that we have found for micromanipulator use. It has two 100,000-ohm potentiometers on each horizontal axis and is usable with the circuit shown in Fig. 5. Ideally, one would modify this unit by increasing the length of the control handle and adding a vertical motion potentiometer to it, but it can be used (with some handicap) with the vertical control mounted separately. Caution: Many of the game-controller joysticks have potentiometers with nonlinear resistance elements. Be sure to use one with linear taper potentiometers.

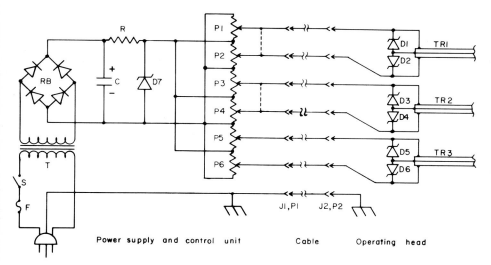

Fig. 5. Minimal micromanipulator circuit. This circuit shows all the electronic components necessary to operate a piezoelectric micromanipulator for chromosome micrurgy. It does not include provision for high speed or automatic movement, but these are not necessary for this

(continued)

ACKNOWLEDGMENTS

We would like to thank Mr. Edward Horn for his superlative craftsmanship in constructing the new micromanipulators. Part of the work reported was supported by NIH Grant GM 23475 and NSF Grant PCM 76-81451.

REFERENCES

Beams, H. W., and King, R. L. (1936). The effects of ultracentrifuging upon chick embryonic cells, with special reference to the "resting" nucleus and the mitotic spindle. *Biol. Bull. (Woods Hole, Mass.)* **71**, 188–198.

Begg, D. A. (1975a). Chromosome movement following detachment from the meiotic spindle by micromanipulation. *J. Cell Biol.* **67**, 25a. (Abstr.)

Begg, D. A. (1975b). "Micromanipulation studies of chromosome movement: The mechanical attachment of chromosomes to the spindle by birefringent chromosomal spindle fibers and the behavior of chromosomes following detachment from the spindle." Ph.D. Thesis, University of Pennsylvania, Philadelphia.

Begg, D. A., and Ellis, G. W. (1979a). Micromanipulation studies of chromosome movement. I. Chromosome-spindle attachment and the mechanical properties of chromosomal spindle fibers. *J. Cell Biol.* **82**, 528–541.

Begg, D. A., and Ellis, G. W. (1979b). Micromanipulation studies of chromosome movement. II. Birefringent chromosomal fibers and the mechanical attachment of chromosomes to the spindle. *J. Cell Biol.* **83**, 542–554.

Brinkley, B. R., and Nicklas, R. B. (1968). Ultrastructure of the meiotic spindle of grasshopper spermatocytes after chromosome micromanipulation. *J. Cell Biol.* **39**, 16a–17a. (Abstr.)

---

use. C, filter capacitor, 40 microfarads, 150WVDC; D1–D7, zener diode, 100 volts, 1 watt (1N4764 or equal); F, fuse, 0.1 amp; J1,P1 and J2,P2, miniature 7 pin jack and plug; P1–P4, 100,000-ohm potentiometers (voltage-dividing resistors) on joystick horizontal axes. P5, 100,000-ohm Knobpot (Bourns, model 3600S-1, or other potentiometer for vertical movement control). P6, vertical range control potentiometer, 100,000 ohms, 1 watt (may be replaced with two ¼-watt fixed resistors if a vertical movement range equal to half of the horizontal range is acceptable); R, fixed resistor, 7500 ohms, 1 watt; RB, rectifier bridge, 125VRMS at 1 amp (e.g., International Rectifier Co. type 1KAB40); S, power switch; T, isolation transformer, 115 volts primary : 115 volts secondary at 15VA (Triad type N48X or equal); TR1–TR3, PZT-5H ceramic Bimorph transducers (Vernitron Piezoelectric Division, 242 Forbes Road, Bedford, Ohio 44146). These transducers are special-order items. Those used in the operating heads described here are 2 inches long (the maximum available) by 0.8 inch wide (to fit this design), but other sizes are available. Although the control circuit shown is the minimum necessary, and more elaborate versions are desirable for some purposes, the same operating head circuit is used for all. The diodes, D1–D6, ensure that only that half of the transducer that is poled in the same direction as the incoming signal is energized at any given time. The use of zener diodes for this purpose also protects the transducer from accidental damage from prolonged overvoltage conditions, although this not a likely risk with the circuit shown.

Camenzind, R., and Nicklas, R. B. (1968). The non-random chromosome segregation in spermatocytes of *Gryllotalpa hexadactyla:* A micromanipulation analysis. *Chromosoma* **24**, 324–335.

Carlson, G. J. (1952). Microdissection studies of the dividing neuroblast of the grasshopper *Chortophaga viridifasciata* (de Geer). *Chromosoma* **5**, 199–220.

Chambers, R., Jr. (1914). Some physical properties of the cell nucleus. *Science* **40**, 824–827.

Chambers, R., Jr. (1915a). Microdissection studies on the germ cell. *Science* **41**, 290–293.

Chambers, R., Jr. (1915b). Microdissection studies of the physical properties of protoplasm. *Lancet-Clinic,* **113**, 363–365.

Chambers, R., Jr. (1925). Etudes de microdissection. IV. Les structures mitochrondriales et nucleaires dans les cellules germinales males chez la sauterelle. *La Cellule* **35**, 107–124.

Chambers, R. (1951). Micrurgical studies on the kinetic aspects of cell division. *Ann. N.Y. Acad. Sci.* **51**, 1311–1326.

Chambers, R., and Chambers, E. L. (1961). "Explorations into the Nature of the Living Cell." Harvard Univ. Press, Cambridge, Massachusetts.

Chambers, R., and Kopac, M. J. (1950). Micrurgical technique for the study of cellular phenomena. *In* "Handbook of Microscopical Technique" (R. McClung Jones, ed.), 3rd ed. pp. 492–543. Paul B. Hoeber, New York.

Chambers, R. Jr., and Sands, H. C. (1923). A dissection of the chromosomes in the pollen mother cells of *Tradescantia virginica* L. *J. Gen. Physiol.* **5**, 815–819.

Cowdry, E. V. (1924). "General Cytology." Univ. of Chicago Press, Chicago, Illinois.

El Badry, H. M. (1963). "Micromanipulators and Micromanipulation." Academic Press, New York, and Springer-Verlag, Vienna.

Ellis, G. W. (1962). Piezoelectric micromanipulators. *Science* **138**, 84–91.

Flemming, W. (1879). Beitrage zur Kenntnis der Zelle und ihrer Lebenserscheinungen. I. *Arch. Mikrosk. Anat. Entwicklungsmech.* **16**, 302–436.

Flemming, W. (1880). Beitrage zur Kenntnis der Zelle und ihrer Lebenserscheinungen. II. *Arch. Mikrosk. Anat. Entwicklungsmech.* **18**, 151–259.

de Fonbrune, P. (1949). "Technique de Micromanipulation." Monographies de l'Institut Pasteur, Masson, Paris.

Forer, A., and Brinkley, B. R. (1977). Microtubule distribution in the anaphase spindle of primary spermatocytes of a crane fly (*Nephrotoma suturalis*). *Can. J. Genet. Cytol.* **19**, 503–519.

Forer, A., and Koch, C. (1973). Influence of autosome movements and of sex-chromosome movements on sex-chromosome segregation in crane fly spermatocytes. *Chromosoma* **40**, 417–442.

Germano, C. P. (1969). Some design considerations in the use of Bimorphs as motor transducers. Engineering report. Vernitron Piezoelectric Division, Bedford Ohio.

Henderson, S. A., and Koch, C. A. (1970). Co-orientation stability by physical tension: A demonstration with experimentally interlocked bivalents. *Chromosoma* **29**, 207–216.

Henderson, S. A., Nicklas, R. B., and Koch, C. A. (1970). Temperature-induced orientation instability during meiosis: An experimental analysis. *J. Cell Sci.* **6**, 323–350.

Inoué, S. (1953). Polarization optical studies of the mitotic spindle. I. The demonstration of spindle fibers in living cells. *Chromosoma* **5**, 487–500.

Kawamura, K. (1960). Studies on cytokinesis in neuroblasts of the grasshopper, *Chortophaga viridifasciata* (De Geer). *Exp. Cell Res* **21**, 1–18.

Kite, G. L., and Chambers, R., Jr. (1912). Vital staining of chromosomes and the function and structure of the nucleus. *Science* **36**, 639–641.

Kopac, M. J. (1959). Micrurgical studies on living cells. *In* "The Cell" (J. Brachet and A. E. Mirskey, eds.), Vol. I, pp. 161–191. Academic Press, New York.

Kopac, M. J. (1964). Micromanipulators: Principals of design, operation, and application. *In* "Physical Techniques in Biological Research" (Oster *et al.*, eds.), Vol. V, pp. 191–233. Academic Press, New York.

LaFountain, J. R., Jr. (1976). Analysis of birefringence and ultrastructure of spindles in primary spermatocytes of *Nephrotoma sutulalis* during anaphase. *J. Ultrastruct. Res.* **54**, 333–346.

Lima de Faria, A. (1958). Recent advances in the study of the kinetochore. *Int. Rev. Cytol.* **7**, 123–157.

Marek, L. F. (1978). Control of spindle form and function in grasshopper spermatocytes. *Chromosoma* **68**, 367–398.

McIntosh, J., Richard, W., Cande, Z., and Snyder, J. A. (1975). Structure and physiology of the mammalian mitotic spindle. *In* "Molecules and Cell Movement" (S. Inoué and R. E. Stephens, eds.), pp. 31–76. Raven, New York.

Mazia, D., and Dan, K. (1952). The isolation and biochemical characterization of the mitotic apparatus of dividing cells. *Proc. Natl. Acad. Sci. U.S.A.* **38**, 826–838.

Nicklas, R. B. (1963). A quantitative study of chromosomal elasticity and its influence on chromosome movement. *Chromsoma* **14**, 276–295.

Nicklas, R. B. (1967). Chromosome micromanipulation. II. Induced reorientation and the experimental control of segregation in meiosis. *Chromosoma* **21**, 17–50.

Nicklas, R. B. (1971). Mitosis. *Adv. Cell Biol.* **2**, 225–297.

Nicklas, R. B. (1974). Chromosome segregation mechanisms. *Genetics* **78**, 205–213.

Nicklas, R. B. (1975). Chromosome movement: Current models and experiments on living cells. *In* "Molecules and Cell Movement" (S. Inoué and R. Stephens, eds.), pp. 97–117. Raven, Press, New York.

Nicklas, R. B. (1977a). Chromosome distribution: Experiments on cell hybrids and *in vitro*. *Phil. Trans. R. Soc. London, Ser. B* **277**, 267–276.

Nicklas, R. B. (1977b). Chromosome movement: Facts and hypotheses. *In* "Mitosis. Facts and Questions." (M. Little, N. Paweletz, C. Petzelt, H. Ponstingl, D. Schroeter, and H.-P. Zimmermann, eds.), pp. 150–166. Springer-Verlag, Berlin and New York.

Nicklas, R. B. (1979). Chromosome movement and spindle birefringence in locally heated cells: Interaction versus local control. *Chromosoma* **74**, 1–37.

Nicklas, R. B., and Koch, C. A. (1969). Chromosome Micromanipulation. III. Spindle fiber tension and the reorientation of mal-oriented chromosomes. *J. Cell Biol.* **43**, 40–50.

Nicklas, R. B., and Koch, C. A. (1972). Chromosome micromanipulation. IV. Polarized motions within the spindle and models for mitosis. *Chromosoma* **39**, 1–26.

Nicklas, R. B., and Staehly, C. A. (1967). Chromosome micromanipulation. I. The mechanics of chromosome attachment to the spindle. *Chromosoma* **21**, 1–16.

Nicklas, R. B., Brinkley, B. R., Pepper, D. A., Kubai, D. F., and Rickards, G. K. (1979a). Electron microscopy of spermatocytes previously studied in life: Methods and some observations on micromanipulated chromosomes. *J. Cell Sci.* **35**, 87–104.

Nicklas, R. B., Kubai, D. F., and Ris, H. (1979b). Electron microscopy of the spindle in locally heated cells. *Chromosoma* **74**, 39–50.

Nicklas, R. B., Hays, T. S., and Kubai, D. F. (1979c). A new approach to the organization of the mitotic spindle. *J. Cell Biol.* **83**, 375a. (Abstr.)

Östergren, G. (1951). The mechanism of co-orientation in bivalents: The theory of co-orientation by pulling. *Hereditas* **37**, 85–156.

Salmon, E. D., and Begg, D. A. (1980). Functional implications of cold-stable microtubules in kinetochore fibers of insect spermatocytes during anaphase. *J. Cell Biol.* **85**, 853–865.

Sato, H., Ellis, G. W., and Inoué, S. (1975). Microtubular origin of mitotic spindle form birefringence. *J. Cell Biol.* **67**, 501–517.

Sawyer, C. B. (1931). The use of rochelle salt crystals for electrical reproducers in microphones. *Proc. Inst. Radio Eng.* **19**, 220–229.

Schrader, F. (1935). On the reality of spindle fibers. *Biol. Bull. (Woods Hole, Mass.)* **67**, 519–553.

Schrader, F. (1944). "Mitosis." Columbia Univ. Press, New York.

Shimamura, T. (1940). The mechanism of nuclear division and chromosome arrangement. VI. Studies on the effect of the centrifugal force upon nuclear division. *Cytologia* **11**, 186–216.

Wise, D. (1978). On the mechanism of prophase congression: Chromosome velocity as a function of position on the spindle. *Chromosoma* **69**, 231–241.

Wada, B. (1935). Mikrurgische Untersuchungen lebender Zellen in der Teilung. II. Das Verhaltten der Spindelfigure und einige ihrer physikalishen Eigenshaften in den somatischen Zellen. *Cytologia* **6**, 381–406.

Wilson, E. B. (1925). "The Cell in Development and Heredity," 3rd ed. Macmillan, New York.

8

Mitotic Mutants

BERL R. OAKLEY

I. INTRODUCTION

Although the major morphological events of mitosis have been known for 100 years (Flemming, 1878, 1880), the mechanisms of chromosomal movement remain unknown. Substantial progress has been made, however, in several areas of mitotic study. Mitosis has been observed extensively under the light microscope, and the patterns of chromosomal movement have been studied in detail (reviewed by Forer in Chapter 6, this volume; Nicklas, 1971). Micromanipulation and microbeam experiments have provided a great deal of information on where forces are produced in the spindle and how equal distribution of daughter chromatids is insured (Forer, Chapter 6, this volume; Ellis and Begg, Chapter 7, this volume; Nicklas, 1971). Electron microscopic studies have described the general features of spindle morphology in a number of organisms, and more precise studies have analyzed the three-dimensional spacing of spindle microtubules (Oakley and Heath, 1978; Pickett-Heaps and Tippit, 1978 and earlier; McIntosh et al., 1979; reviewed by Heath in Chapter 11, this volume).

While much additional work remains to be done in each of these areas, the aspect of chromosomal movement which is least completely understood is the biochemistry of the spindle. The chemical identities of the mitotic motors remain unknown and, in fact, only two proteins, the microtubule proteins α- and β-tubulin, have been shown to function in chromosomal movement.

In this chapter, I will discuss genetics as a methodology for studying mitotic mechanisms. I will discuss specific ways in which mitotic mutants can be isolated and used in identifying and determining the function of the proteins and other gene products involved in chromosomal movement and other forms of microtubule-mediated cell motility. Although other aspects of mitosis will not be discussed in detail, such as the mechanisms of chromosomal condensation, most of the techniques for studying chromosomal movement genetically are applicable to other problems in mitotic research. I will discuss previous work in detail as it relates to the methodology, but readers are directed to the reviews of Hartwell (1978) and Morris (1980) for more detailed discussions of earlier work.

II. THE ADVANTAGES OF GENETIC APPROACHES

The major advantages of genetics are that phenotypic effects can be as-cribed unambiguously to particular mutations and that the effects of muta-tions can be observed *in vivo*, thus eliminating the ambiguity often encoun-tered when one tries to relate *in vitro* results to the situation *in vivo*. Genetics is not a replacement for biochemical or morphological studies of the mitotic apparatus because it is impossible to understand mitotic mutants without biochemical and morphological study. In concert with biochemical and morphological studies, however, genetics should reveal a great deal more about mitosis than would be possible with any of these approaches individually.

One of the most important steps in understanding chromosomal move-ment is identifying the functional components of the spindle. Biochemical methods which have been used to identify spindle components have gener-ally yielded ambiguous data. Several spindle isolation procedures have been developed (Mazia and Dan, 1952; Mazia et al., 1961; Kane, 1965; Sisken et al., 1967; Forer and Zimmerman, 1974; Salmon and Segall, 1981). With these procedures, however, it has not been possible to demonstrate that all the components of the isolated mitotic apparatus were originally in the mito-tic apparatus or that components of the mitotic apparatus were not lost during the purification procedures. Many of the problems of identifying spindle components by mitotic apparatus isolation will be difficult to overcome. Although there seem to be some elements surrounding many mitotic apparatuses, most spindles are not membrane bound and there is no barrier preventing cytoplasmic substances from lodging in the spindle during spindle isolation. Moreover, we do not know if all the substances which function in chromosomal movement are bound to the spindle or if all substances present in the spindle *in vivo* are necessary for spindle function.

Fluorescent antibody studies (reviewed by Aubin, Chapter 10, this volume) have provided evidence that actin (Sanger, 1975; Cande *et al.*, 1977), calmodulin (Marcum *et al.*, 1978), microtubule-associated proteins (Connolly *et al.*, 1977, 1978; Sherline and Schiavone, 1977, 1978), and myosin (Fujiwara and Pollard, 1976) are present in the spindle. Wang and Taylor (1979), however, have demonstrated that even fluorescently labeled ovalbumin will appear to be concentrated in the mitotic spindle, so that fluorescent antibody demonstrations that proteins are present in spindles must be considered suspect. Electron microscopic studies using heavy meromyosin or heavy meromyosin subfragment-1 have also suggested that actin is present in the spindle (Gawadi, 1974; Forer *et al.*, 1979 and earlier). These data are in some ways more convincing than antibody studies but do not rule out the possibility that actin is translocated from other points of the cell to the spindle during preparation. Indeed, this ambiguity is inherent in nearly all fluorescent antibody or heavy meromyosin labeling procedures. If the cell is permeable enough to allow antibodies or heavy meromyosin to enter the cell and associate with the spindle, the cell is also permeable enough to allow other proteins or even protein aggregates to translocate.

There are many important unanswered questions in the much studied area of microtubule chemistry, including the central questions of how microtubule assembly is regulated *in vivo* and how microtubules function in the generation of motion. Here again, one of the major difficulties is relating results obtained *in vitro* to the living cell. In the area of the regulation of microtubule assembly, for example, it has been shown that many substances, including various microtubule-associated proteins (Borisy *et al.*, 1975; Weingarten *et al.*, 1975), polycations (Erickson and Voter, 1976), and even dimethyl sulfoxide (Himes *et al.*, 1977), stimulate microtubule assembly *in vitro*, while other substances, including Ca^{2+} (Borisy *et al.*, 1975) and RNA (Bryan *et al.*, 1975), inhibit microtubule assembly. It is not clear which, if any, of these substances are involved in the regulation of microtubule assembly *in vivo*, but it is unlikely to me that all of them are. It is clear, however, that the spatial and temporal regulation of microtubule assembly *in vivo* is much finer than has been achieved *in vitro*.

These examples are cited to point out that we need ways of relating *in vitro* results to *in vivo* phenomena: genetics can potentially provide us with this capability. The special utility of genetics stems, first, from the fact that one can isolate mutations with a particular phenotype and map the mutations which produce these phenotypes to specific genes, and, second, from the fact that the phenotypic effects of the mutations are observable *in vivo*. If, for example, a mutation in an actin gene caused a blockage of mitosis, one would have very good evidence that actin is involved in mitosis. As another example, it has been difficult to determine whether microtubule-associated proteins which stimulate microtubule assembly *in vitro* have a similar function *in vivo;* if, however, one isolates a mutation which interferes with microtubule assembly under some conditions, and determines that this mutation is in a microtubule-associated protein, one can be confident that the microtubule-associated protein has a role in microtubule assembly *in vivo*.

To use genetics to discover the functional components of the spindle, one must (1) isolate mutants in genes whose products function in mitosis, (2) isolate the products of the defective genes, and (3) determine the function of the defective gene products.

III. ISOLATING MITOTIC MUTANTS

Two basic approaches for isolating mitotic mutants are presently feasible. One is to isolate conditional mutants and examine them microscopically to determine which mutants are blocked in mitosis under restrictive conditions. The other is to isolate conditional mutations in known mitotic proteins (e.g., tubulin) and then to isolate mutations in other genes essential for mitosis as extragenic suppressors of these mutations.

The procedure which has been used most often to isolate mitotic mutants has been to isolate a large number of mutants which are conditional for growth (heat sensitivity has been the condition most often used) and to test these mutants microscopically for blockage in mitosis under restrictive conditions. The rationale behind this approach is that if a mutation causes a blockage of mitosis, it must be in a gene whose product is essential for the completion of mitosis. It is important at this point to distinguish between cell cycle mutants, which are mutants blocked at any point in the nuclear division cycle, including interphase, and mitotic mutants, which are specifically blocked in mitosis. Cell cycle mutants have been isolated in a number of organisms, including *Aspergillus nidulans* (Morris, 1976a,b; Orr and Rosenberger, 1976), *Chlamydomonas* (Howell and Naliboff, 1973), *Saccharomyces cerevisiae* (reviewed by Hartwell, 1978), *Schizosaccharomyces pombe* (Nurse *et al.*, 1976; Thuriaux *et al.*, 1978), and *Tetrahymena* (Bruns and Sanford, 1978). Mitotic mutants have been isolated in *Saccharomyces* (reviewed by Hartwell, 1978), *Aspergillus* (Orr and Rosenberger, 1976; Morris, 1976a,b, 1980), and the mammalian (HM-1) cell line (Wang, 1974,

1976). I will confine subsequent discussion to mutants which are blocked in mitosis under restrictive conditions because these mutants are most likely to be of use in understanding the mechanisms of mitosis. For reviews of cell cycle mutants, see Hartwell (1978), Morris (1980), and Pringle (Chapter 1, this volume).

Although mutations which cause blockage of mitosis at restrictive temperatures must be in genes whose products are essential for the completion of mitosis, these products need not be part of the machinery which moves chromosomes. Some of the yeast mitotic mutants, for example, are defective in DNA ligase (Johnston and Nasmyth, 1978). While these mutants are interesting for other reasons, they do not reveal anything about the mechanisms of mitosis except by confirming the fact that mitosis is blocked in *Saccharomyces cerevisiae* if DNA replication is blocked. Whether DNA synthesis mutants would cause mitotic blockage in other organisms has not been determined. Since the spindle microtubules are present during all of the cell cycle in *Saccharomyces* (B. Byers, personal communication; see also Byers and Goetsch, 1974), it is possible that many mitotic mutants in this organism are merely cell cycle mutants which happen to be blocked at a point in the cell cycle at which the spindle is present. In other organisms in which the spindle is present for a much smaller percentage of the cell cycle, this would be a minor problem. More generally, there must be many controls and feedback mechanisms involved in the regulation of mitotic events, such as the initiation of anaphase. Mutations which cause a mitotic blockage could be found in genes involved in the regulation of mitosis as well as in genes whose products are directly involved in the movement of chromosomes.

Despite these complications, many of the mutations which block cells in mitosis should be in genes whose products function directly in chromosomal movement. Moreover, the control of mitosis is as interesting and important as the mechanisms of chromosomal movement, and mutations which affect the control systems are well worth studying in their own right.

IV. MITOTIC MUTANTS DEFECTIVE IN β-TUBULIN

While there are advantages to studying randomly isolated mitotic mutants, there is another approach which is more certain to give useful results in the short term. This approach involves isolating mutations in genes which code for known mitotic apparatus proteins and then isolating mutations in genes which code for other mitotic proteins as extragenic suppressor mutations. This approach is based on the pioneering work of Jarvik and Botstein (1975), who showed in phage P22 that (1) revertants of missense mutations were often conditional [heat sensitive $(hs-)$ or cold sensitive $(cs-)$], (2) the reversion was often due to a mutation in a gene other than the gene which carried the original defect (i.e., it was an extragenic suppressor mutation), (3) extragenic suppressor mutations were almost always in genes whose products interacted physically with the original gene products, and (4) $hs-$ or $cs-$ extragenic suppressors could be reverted to produce additional mutations. Extrapolation of these results to the microtubule system suggests that if a conditional mutation in the gene for a known microtubule protein could be isolated, it should be possible to isolate mutations in the genes for other microtubule proteins as revertants of the original mutation.

A key step in using this approach to study microtubule structure and function is to isolate conditional mutants "defective" in a gene whose product is a known microtubule protein. While no direct selection for such mutants is currently available, the work of Van Tuyl (1977), Sheir-Neiss *et al.* (1978), and Morris *et al.* (1979) has demonstrated an indirect procedure which can be used to isolate such mutants. Van Tuyl (1977) isolated a large number of mutations in *Aspergillus nidulans* which conferred resistance to the anti-microtubule agent benomyl, and he mapped the majority of these mutations to a locus he named *benA*. Sheir-Neiss *et al.* (1978) showed that *benA* was a structural gene for the microtubule protein β-tubulin, and Morris *et al.* (1979) showed that three alleles of *benA* were heat sensitive. Morris *et al.* (1979) isolated a number of $hs+$ revertants of $hs-$ *benA* alleles and analyzed them genetically. Some of the reversions were due to mutations in the *benA* gene (back mutations), and others were due to mutations in other genes (extragenic suppressor mutations). One of the extragenic suppressor mutations was in a gene for α-tubulin (Morris *et al.*, 1979), and the other suppressor mutations have been mapped to at least three additional loci (N. R. Morris and C. E. Oakley, unpublished). This work shows that reversion analysis can be used to isolate mutations not only in a gene for α-tubulin but in other genes whose products are part of, or interact with, microtubules.

Revertants are maximally useful if they are conditional. This is because one can determine their phenotypes and thus gain insight into the role their products play in mitosis, and because one can use them to isolate additional revertants. We have isolated a cold-sensitive *benA* back mutation which is blocked in mitosis at restrictive temperatures (D. R. Kirsch, C. F. Roberts, B. R. Oakley, and N. R. Morris, unpublished) and cold-sensitive extragenic suppressors in other as yet unidentified genes (C. E. Oakley, C. F. Roberts, N. R. Morris, and B. R. Oakley, unpublished).

This approach is particularly exciting because one knows that one's starting point is β-tubulin and thus that mutations which are isolated as suppressors of $hs-$ or $cs-$ *benA* mutations must code for products which are part of microtubules or otherwise interact with β-tubulin. We should, for example, be able to isolate mutations in genes which code for microtubule-associated proteins in this way and eventually determine the role of these proteins in microtubule assembly and chromosomal movement.

Sato (1976) and White *et al.* (1979) have also isolated antimocrotubule-drug-resistant mutants in *Chlamydomonas* and *Dictyostelium*, respectively. Some of these mutants are heat sensitive for growth, but these mutations have not yet been demonstrated to be in microtubule proteins and it has not been determined that they are mitotic mutants. Cabral *et al.* (1980) isolated drug-resistant mutants defective in a gene for β-tubulin in a mammalian (CHO) cell line, but these mutants are not heat sensitive or cold sensitive in the heterozygous diploid state in which they were isolated. Ling *et al.* (1979) have also isolated antimicrotubule-drug-resistant mutants in a CHO cell line but have not determined the defective gene(s) or reported any conditionality for these mutants (see Chapter 9 by Ling in this volume).

Kemphues *et al.* (1979) have used a third approach to find mutants of *Drosophila melanogaster* which are defective for microtubule function. Isolating mutants defective for microtubule function is difficult in diploid organisms because recessive mutations are not expressed. Kemphues and co-workers, however, identified electrophoretically a β-tubulin which appears only in testes. They reasoned that this β-tubulin might be involved in meiosis, spermiogenesis, or sperm motility and that mutations in the gene which codes for this β-tubulin might confer male sterility. They then screened male sterile mutants and isolated a dominant mutation {[*ms(3)KK^D*] since renamed *B2t^D*} in the structural gene which codes for the testes-specific β-tubulin. This mutation causes morphological abnormalities in sperm flagella and in meiosis (Kemphues *et al.*, 1979, 1980).

V. THE PHENOTYPES OF MITOTIC MUTANTS

One may obtain important clues to the roles of the products of the defective genes by determining the precise morphological stages at which the mutants are blocked. In cells which contain small spindles that cannot be studied *in vivo*, this type of analysis requires tedious, detailed ultrastructural study using serial sections; this has been performed on two sets of mutants, the *Saccharomyces cdc* (cell division cycle) mutants (Byers and Goetsch, 1974) and the *Aspergillus bim* (blocked in mitosis) mutants (Morris, 1980; B. R. Oakley and N. R. Morris, in preparation). Byers and Goetsch found that at the restrictive temperatures, each mutant was blocked at a characteristic stage in the mitotic cycle (which they called the "termination phenotype") and that among the mutants some were blocked in early, some in medial, and some in late mitotic stages (termination phenotypes are summarized in Fig. 1). These mutants give important clues to the functions of the products of the defective genes because these gene products must be involved in the transition from the stage at which blockage occurs to later mitotic stages. None of the *Saccharomyces* mutants had obvious spindle defects such as the absence of a particular type of microtubule, which would

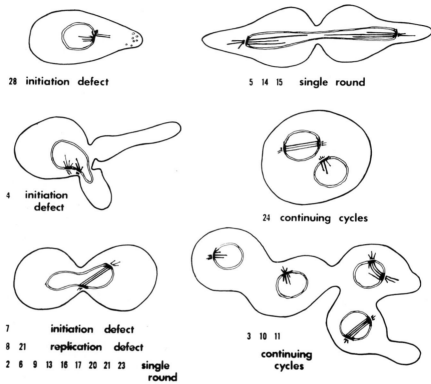

28 initiation defect

5 14 15 single round

4 initiation defect

24 continuing cycles

7 initiation defect

8 21 replication defect

2 6 9 13 16 17 20 21 23 single round

3 10 11 continuing cycles

Fig. 1. Termination phenotypes and DNA synthetic behavior of *Saccharomyces cerevisiae cdc* mutants. When each temperature-sensitive *cdc* mutant is shifted from permissive to restrictive temperatures, growth gradually comes to a halt, with the cells in a characteristic morphology known as the "termination" or "terminal phenotype." The diagram shows schematically the termination phenotypes for the *Saccharomyces cdc* mutants. In each figure, the outer solid line represents the nuclear envelope and the solid lines abutting the nuclear envelope represent microtubules. Below each figure are listed the *cdc* genes which are blocked with each phenotype and the corresponding DNA synthetic behavior. *Cdc* 4 is blocked in early nuclear division, *cdc* 7, 8, 21, 2, 6, 9, 13, 16, 17, 20, 21, and 23 are blocked in medial nuclear division, and *cdc* 5, 14, and 15 are blocked in late nuclear division. Although *cdc* 24, 3, 10, and 11 are blocked with a characteristic cellular morphology, their nuclear morphology is variable. From Byers and Goetsch (1974). Reproduced by permission of the Cold Spring Harbor Laboratory, copyright 1974.

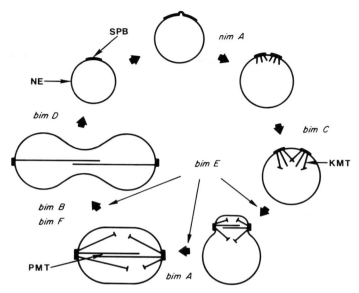

Fig. 2. Nuclear division cycle of *Aspergillus nidulans* and terminal phenotypes of *bimA–F* and *nimA* mutants. *NimA* is blocked very late in interphase, and the *bim* mutants are blocked in mitosis. All the *bim* mutants are blocked with characteristic terminal phenotypes except for *bimE,* which is found with various phenotypes and thus appears to progress slowly through mitosis. *BimC* is defective in spindle formation, and *bimD* apparently lacks kinetochore microtubules. The spindles of the other *bim* mutants show no apparent structural defects. Abbreviations: NE, nuclear envelope; SPB, spindle pole body; KMT, kinetochore microtubule; PMT, polar microtubule. Adapted from Morris (1980), Fig. 1. Reproduced by permission of Cambridge University Press, copyright 1980.

give further clues to the function of the defective gene product and consequently its identity.

In *Aspergillus,* as in *Saccharomyces,* we have found that most mutants are blocked with a characteristic morphology or terminal phenotype (Fig. 2). Several mutants are blocked in configurations which are apparently identical to wild-type stages, but some mutants have more interesting phenotypes. The spindle is incompletely formed in *bimC,* for example, which suggests that the defective gene product is involved in microtubule assembly. Another mutant, *bimD,* apparently lacks kinetochore microtubules, so the defective gene product may be necessary for the formation of this type of microtubule. Another mutant, *bimE,* is found in several mitotic stages at restrictive temperatures so it appears to progress slowly through mitosis.

The blockage of growth at restrictive temperatures in the original *hs⁻ benA* mutants was not complete enough to allow one to determine their ultrastructural phenotype with confidence. In collaboration with N. R. Morris, I have isolated two new tightly *hs*− *benA* mutants, *benA31* and *benA33*. At the light microscopic level these alleles are blocked in mitosis at restrictive temperatures, and at the ultrastructural level they have interesting phenotypes. One of the two mutants, *benA31*, is apparently defective for microtubule assembly at restrictive temperatures because only a few short spindle microtubules are found in *benA31* blocked in mitosis. This mutant in effect allows one to regulate microtubule assembly *in vivo* merely by shifting temperatures and is potentially very useful for studying microtubule assembly. The other mutant, *benA33*, is not defective for microtubule assembly because abundant, fully formed spindles are found in *benA33* blocked in mitosis (Oakley and Morris, 1981). Since chromosomal movement is blocked in this mutant but microtubule assembly is not, this mutant probably is defective for the functioning of assembled microtubules. If microtubule disassembly is necessary for chromosomal movement, the defect in function could be due to a reduced rate of microtubule disassembly at restrictive temperatures. If intermicrotubule sliding is necessary for chromosomal movement, the defect in function could be due to a failure of the microtubules to interact normally with microtubule-associated proteins or other substances which function in chromosomal movement. Indirect evidence suggests that the former possibility is more likely (Oakley and Morris, 1981). The biochemical resolution between these possibilities should reveal a great deal about the mechanisms of mitosis.

In addition to their use in studying mitosis, these mutants are also helpful in determining which cellular functions are microtubule mediated. Until recently, the determination of whether a particular cellular function was microtubule mediated was dependent on microtubule inhibitor studies. Microtubule inhibitors are not, however, demonstrably specific (see Oakley and Morris, 1980 for a discussion and further references). In collaboration with N. R. Morris, I have shown that the specificity of antimicrotubule drugs can be demonstrated with drug-resistant β-tubulin mutants (Oakley and Morris, 1980). Since the microtubules of the *hs*− *benA* mutants are not functional at restrictive temperatures, one can determine those cellular functions which use the β-tubulin encoded by the *benA* gene simply by determining those functions which are blocked at restrictive temperatures in a strain with an *hs*− β-tubulin and are not blocked in a wild strain at the same temperature. Both nuclear division and nuclear movement are blocked at restrictive temperatures in *benA31* and *benA33*, so the microtubules involved in each process must contain β-tubulin encoded by the same gene, the *benA* gene. In *Drosophila*, however, Kemphues *et al.* (1979) have shown that a specific β-tubulin gene is expressed only in testes, so in this organism different tubulin genes are expressed in different tissues. These data are interesting in themselves and also because they show that genetics should allow us ultimately to determine the role of tubulin in development and differentiation.

VI. ISOLATING THE PRODUCTS OF DEFECTIVE GENES

If mitotic mutants are to aid in understanding mitosis, one must know the chemical identities of the products of the defective genes. Isolating the products of the defective genes is more difficult than determining the phenotypes of mitotic mutants, but there are several promising approaches. One approach involves the use of two-dimensional gel electrophoresis to detect electrophoretically altered gene products. Many mutations involve the replacement of one amino acid by another amino acid with a different charge. Thus there is a reasonable chance that a mutation will alter the isoelectric point of a protein and consequently that the position of the protein will be altered in the isoelectric-focusing dimension of a two-dimensional gel. Sheir-Neiss et al. (1978) have shown, for example, that many β-tubulin mutants have altered isoelectric points. Even if a protein does not have an altered charge, one can isolate revertants in the same gene and many of these will carry charge alterations (Morris et al., 1979). It should thus be possible to detect many defective proteins in this way. This approach should be particularly useful in concert with an initial screening to reduce the number of proteins to be examined on the two-dimensional gels. If one were able to isolate spindles from cell types studied genetically, the number of proteins in the two-dimensional gels would be reduced and the task of finding a protein with an altered isoelectric point would be greatly simplified. The identification of a defective gene product on a two-dimensional gel will not, of itself, reveal the biochemical function unless the gene product is easily identifiable on gels—as, for example, actin would be. Two-dimensional gel electrophoresis can, however, reveal the isoelectric points and molecular weights of defective gene products, and these data should provide an assay, albeit cumbersome, that can be used in purifying the gene product. Two-dimensional gels will not necessarily reveal all defective gene products, however. Some proteins essential for mitosis may be present in small quantities which will not allow their visualization on two-dimensional gels, and some gene products necessary for mitosis may not be proteins.

A second, and intrinsically more powerful, approach to finding defective gene products is to develop an assay for the missing function. Assays can be of various kinds, but the most obvious is an *in vitro* mitotic system. If such a system were available, one could isolate mutant spindles blocked at restrictive temperatures and add cell-free extracts, hoping to restore spindle function. The fractions which promoted spindle function would contain the wild-type version of the defective gene product, and this substance could be purified by further fractionation. The missing element in this approach is a simple, reliable *in vitro* mitotic system. Several groups have worked on such systems (Cande *et al.*, 1974; Sakai *et al.*, 1976; Nicklas, 1977; see also Snyder, Chapter 13, this volume) but whether these systems work easily and consistently enough to be useful as an assay is questionable. Another problem with present *in vitro* systems is that they do not use organisms which are amenable to genetic analysis. The development of *in vitro* mitotic systems in genetically useful organisms such as *Aspergillus* and *Saccharomyces*, although possibly difficult, should lead to substantial progress in understanding spindle biochemistry.

Another type of assay involves not the isolation of the defective gene product but the defective gene itself. Most mutations which cause blockage in mitosis are recessive (Morris, 1976b; Hartwell, 1978); that is, heterozygous diploids undergo mitosis and grow normally under restrictive conditions. Thus introduction of a wild-type copy of the defective gene into the cell by transformation can be expected to restore normal division and growth. Clark and Carbon (1980) and Nasmyth and Reed (1980) have used this fact to develop a procedure for isolation of wild-type copies of the *Saccharomyces* cell division cycle genes *cdc10* and *cdc28*, respectively. The procedure they used was, first, to create a library of wild-type DNA cloned into appropriate shuttle vectors. Shuttle vectors are plasmids which transform at high frequencies and have origins of replication for yeast and for *Echerichia coli*, so that they can replicate in either host and can thus be recovered intact from transformed yeast (Beggs, 1978; Struhl *et al.*, 1979). Heat-sensitive ($hs-$) *cdc* mutants were treated with these libraries, and $hs+$ transformants were selected. The plasmids containing the wild-type alleles of the *cdc* genes were recovered from the transformants and used to transform *E. coli*. The plasmid was then grown in and purified from *E. coli*. With this procedure, it should be comparatively easy to isolate the wild-type allele of the defective gene for any of the mitotic mutants in *Saccharomyces*, and it is hoped that shuttle vectors will be developed for other species as well.

Isolating the defective gene is not as useful as isolating its product because one needs to isolate the product to determine its biochemical function. Nevertheless, isolating the gene should be a very useful step toward identifying the gene product. For example, one should be able to use the DNA containing the wild-type allele to isolate the mRNA encoded by the gene. The mRNA could be translated *in vitro* to obtain small quantities of the gene product. These small quantities should give clues which would allow one to purify large quantities. If there are no posttranslational modifications of the gene product, one should be able to determine its isoelectric point and molecular weight and use these data to facilitate large-scale purification.

One faces basically the same problems in isolating the products of mitotic mutants isolated as extragenic suppressors of *benA*. One advantage of these mutants, however, is that one knows that their products interact with β-tubulin. One might, for example, examine proteins which co-polymerize with the tubulins in the mutants to see if any of them showed isoelectric shifts on two-dimensional gels, as many of the *benA* mutants do.

VII. CONCLUSION

The use of genetics to study mitosis is in many ways in its infancy. Up to now, genetics has not contributed greatly to our understanding of mitotic mechanisms, but I hope I have demonstrated that it is an extremely promising tool for studying mitosis. Moreover, it is a methodology which is still developing rapidly. This is especially true because eukaryote genetics is in the throes of a revolution spawned by recombinant DNA technology. For example, in the short period between the initial and final preparations of this manuscript, recombinant DNA technology has permitted the isolation of centromeric DNA in *Saccharomyces* (Clark and Carlson, 1980) and the sequencing of *Drosophila* α- and β-tubulin genes (Valenzuela *et al.*, 1981). This technology will undoubtedly facilitate genetic studies of mitosis in many ways. The cloning of actin and tubulin genes from *Drosophila* (Cleveland *et al.*, 1980), for example, will allow the mapping of these genes and eventually the construction of deletion strains which are haploid for these genes. These strains should, in turn, greatly facilitate the isolation of mutations in these genes. What is perhaps most exciting is that recombinant DNA technology is developing so rapidly that it will facilitate the study of mitosis in many ways which are as yet unforeseeable.

ACKNOWLEDGMENTS

I would like to acknowledge many fruitful discussions with Dr. N. R. Morris, Dr. C. F. Roberts, and Dr. D. R. Kirsch and to thank Dr. Morris and Dr. B. McL. Breckenridge for critically reading the manuscript. Supported by NIH grant GM23060.

REFERENCES

Beggs, J. D. (1978). Transformation of yeast by a replicating hybrid plasma. *Nature (London)* **275**, 104–109.

Borisy, G. G., Marcum, J. M., Olmstead, J. B., Murphy, D. B., and Johnson, K. A. (1975). Purification of tubulin and associated high molecular weight proteins from porcine brain and characterization of microtubule assembly *in vitro*. *Ann. N.Y. Acad. Sci.* **253**, 107–132.

Bruns, P. J., and Sanford, Y. M. (1978). Mass isolation and fertility testing of temperature-sensitive mutants in *Tetrahymena*. *Proc. Natl. Acad. Sci. U.S.A.* **75**, 3355–3358.

Bryan, J., Nagle, B. W., and Doenges, K. H. (1975). Inhibition of tubulin assembly by RNA and other polyanions: Evidence for a required protein. *Proc. Natl. Acad. Sci. U.S.A.* **72**, 3570–3574.

Byers, B., and Goetsch, L. (1974). Duplication of spindle plaques and integration of the yeast cell cycle. *Cold Spring Harbor Symp. Quant. Biol.* **38**, 123–137.

Cabral, F., Sobel, M. E., and Gottesman, M. M. (1980). CHO mutants resistant to colchicine, colcemid or griseofulvin have an altered β-tubulin. *Cell* **20**, 29–36.

Cande, W. Z., Snyder, J., Smith, D., Summers, K., and McIntosh, J. R. (1974). A functional mammalian spindle prepared from mammalian cells in culture. *Proc. Natl. Acad. Sci. U.S.A.* **71**, 1559–1563.

Cande, W. Z., Lazarides, E., and McIntosh, J. R. (1977). A comparison of the distribution of actin and tubulin in the mammalian mitotic spindle as seen by indirect immunofluorescence. *J. Cell Biol.* **72**, 552–567.

Clarke, L., and Carbon, J. (1980). Isolation of the centromere-linked *CDC*10 gene by complementation in yeast. *Proc. Natl. Acad. Sci. U.S.A.* **77**, 2173–2177.

Clark, L., and Carbon, J. (1980). Isolation of a yeast centromere and construction of functional small circular chromosomes. *Nature* **287**, 504–509.

Cleveland, D. W., Lopata, M. A., MacDonald, R. J., Cowan, N. J., Rutter, W. J., and Kirschner, M. W. (1980). Number and evolutionary conservation of α- and β-tubulin and cytoplasmic β- and γ-actin genes using specific cloned cDNA probes. *Cell* **20**, 95–105.

Connolly, J. A., Kalnins, V. I., Cleveland, D. W., and Kirshner, M. W. (1977). Immunofluorescent staining of cytoplasmic and spindle microtubules in mouse fibroblasts with antibody to τ protein. *Proc. Natl. Acad. Sci. U.S.A.* **74**, 2437–2440.

Connolly, J. A., Kalnins, V. I., Cleveland, D. W., and Kirshner, M. W. (1978). Intracellular localization of the high molecular weight microtubule accessory protein by indirect immunofluorescence. *J. Cell Biol.* **76**, 781–786.

Erickson, H. P., and Voter, W. A. (1976). Polycation-induced assembly of purified tubulin. *Proc. Natl. Acad. Sci. U.S.A.* **73**, 2813–2817.

Fleming, W. (1878, 1880). Beitrage zur Kenntniss der Zelle und ihrer hebensersheinungen. I. *Arch. Mikrosk. Anat. Entwicklungsmech.* **16**, 302–436; **18**, 151–259. Reprinted in English in *J. Cell Biol. Suppl.* 1, **25**, 1–69.

Forer, A., and Zimmerman, A. M. (1974). Characteristics of sea-urchin mitotic apparatus isolated using a dimethylsulphoxide/glycerol medium. *J. Cell Sci.* **16**, 481–497.

Forer, A., Jackson, W. T., and Engberg, A. (1979). Actin in spindles of *Haemanthus katherinae* endosperm II. Distribution of actin in chromosomal spindle fibres, determined by analysis of serial sections. *J. Cell Sci.* **37**, 349–371.

Fujiwara, K., and Pollard, T. D. (1976). Fluorescent antibody localization of myosin in the cytoplasm, cleavage furrow, and mitotic spindle of human cells. *J. Cell Biol.* **71**, 848–875.

Gawadi, N. (1974). Characterization and distribution of microfilaments in dividing locust testis cells. *Cytobios* **10**, 17–35.

Hartwell, L. H. (1978). Cell division from a genetic perspective. *J. Cell Biol.* **77**, 627–637.

Hartwell, L. H., Mortimer, R. K., Culotti, J., and Culotti, M. (1973). Genetic control of the cell division cycle in yeast: V. Genetic analysis of *cdc* mutants. *Genetics* **74**, 267–286.

Himes, R. H., Burton, P. R., and Gaito, J. M. (1977). Dimethyl sulfoxide-induced self-assembly of tubulin lacking associated proteins. *J. Biol. Chem.* **252**, 6222–6228.

Howell, S. H., and Naliboff, J. A. (1973). Conditional mutants in *Chlamydomonas reinhardtii* blocked in the vegetative cell cycle. *J. Cell Biol.* **57**, 760–772.

Jarvik, J., and Botstein, D. (1975). Conditional-lethal mutations that suppress genetic defects in morphogenesis by altering structural proteins. *Proc. Natl. Acad. Sci. U.S.A.* **72**, 2738–2742.

Johnston, L. H., and Nasmyth, K. A. (1978). *Saccharomyces cerevisiae* cell cycle mutant *cdc9* is defective in DNA ligase. *Nature (London)* **274**, 891–893.

Kane, R. E. (1965). The mitotic apparatus: Physical chemical factors controlling stability. *J. Cell Biol.* **25**, 137–144.

Kemphues, K. J., Raff, E. C., Raff, R. A., and Kaufman, T. C. (1980). Mutation in a testis-specific β-tubulin in *Drosophila:* Analysis of its effects on meiosis and map location of the gene. *Cell* **21**, 445–451.

Kemphues, K. J., Raff, R. A., Kaufman, T. C., and Raff, E. C. (1979). Mutation in a structural gene for a β-tubulin specific to testis in *Drosophila melanogaster. Proc. Natl. Acad. Sci. U.S.A.* **76**, 3991–3995.

Ling, V., Aubin, J. E., Chase, A., and Sarangi, F. (1979). Mutants of chinese hamster ovary (CHO) cells with altered colcemid-binding activity. *Cell* **18**, 423–430.

McIntosh, J. R., McDonald, K. L., Edwards, M. K., and Ross, B. M. (1979). Three-dimensional structure of the central mitotic spindle of *Diatoma vulgare. J. Cell Biol.* **83**, 428–442.

Marcum, J. M., Dedman, J. R., Brinkley, B. R., and Means, A. (1978). Control of microtubule assembly-disassembly by calcium-dependent regulator protein. *Proc. Natl. Acad. Sci. U.S.A.* **75**, 3771–3775.

Mazia, D., and Dan, K. (1952). The isolation and biochemical characterization of the mitotic apparatus of dividing cells. *Proc. Natl. Acad. Sci. U.S.A.* **38**, 826–838.

Mazia, D., Mitchison, J. M., Medina, H., and Harris, P. (1961). The direct isolation of the mitotic apparatus. *J. Biophys. Biochem. Cytol.* **10**, 467–474.

Morris, N. R. (1976a). A temperature-sensitive mutant of *Aspergillus nidulans* reversibly blocked in nuclear division. *Exp. Cell Res.* **98**, 204–210.

Morris, N. R. (1976b). Mitotic mutants of *Aspergillus nidulans. Genet. Res.* **26**, 237–254.

Morris, N. R. (1980). Chromosome structure and the molecular biology of mitosis in eukaryotic micro-organisms. *In* "The Eukaryotic Microbial Cell" (G. W. Gooday, D. Loyd, and A. P. J. Trinci, eds.), pp. 41–76. Cambridge Univ. Press, London and New York.

Morris, N. R., Lai, M. H., and Oakley, C. E. (1979). Identification of a gene for α-tubulin in *Aspergillus nidulans. Cell* **16**, 437–442.

Nasmyth, K. A., and Reed, S. I. (1980). Isolation of genes by complementation in yeast: Molecular cloning of a cell-cycle gene. *Proc. Natl. Acad. Sci. U.S.A.* **77**, 2119–2123.

Nicklas, R. B. (1971). Mitosis. *Adv. Cell Biol.* **2**, 225–297.

Nicklas, R. B. (1977). Chromosome distribution experiments on cell hybrids and *in vitro. Phil. Trans. R. Soc. London, Ser. B.* **277**, 267–276.

Nurse, P., Thuriaux, P., and Nasmyth, K. (1976). Genetic control of the cell division cycle in the fission yeast *Schizosaccharomyces pombe. Mol. Gen. Genet.* **146**, 167–178.

Oakley, B. R., and Heath, I. B. (1978). The arrangement of microtubules in serially sectioned spindles of the alga *Cryptomonas. J. Cell Sci.* **31**, 53–70.

Oakley, B. R., and Morris, N. R. (1980). Nuclear movement is β-tubulin dependent in *Aspergillus nidulans. Cell* **19**, 255–262.

Oakley, B. R., and Morris, N. R. (1981). A β-tubulin mutation in *Aspergillus nidulans* that blocks microtubule function without blocking assembly. *Cell* **24**, 837–845.

Orr, E., and Rosenberger, R. F. (1976). Initial characterization of *Aspergillus nidulans* mutants blocked in the nuclear replication cycle. *J. Bacteriol.* **126**, 895–902.

Pickett-Heaps, J. D., and Tippit, D. H. (1978). The diatom spindle in perspective. *Cell* **14**, 455–467.

Sakai, H., Mabuchi, I., Schimoda, S., Kuriyama, R., Ogawa, K., and Mohri, H. (1976). Induction of chromosome motion in the glycerol isolated mitotic apparatus: Nucleotide specificity and effects of anti-dynein and myosin sera on the motion. *Dev., Growth Differ.* **18**, 211–219.

Salmon, C. D., and Segall, R. R. (1980). Calcium-labile mitotic spindles isolated from sea urchin eggs (*Lytechinus variegatus*). *J. Cell Biol.* **86**, 355–365.

Sanger, J. W. (1975). The presence of actin during chromosomal movement. *Proc. Natl. Acad. Sci. U.S.A.* **72**, 2451–2455.

Sato, C. (1976). A conditional cell division mutant of *Chlamydomonas reinhardtii* having an increased level of colchicine resistance. *Exp. Cell Res.* **101**, 251–259.

Sheir-Neiss, G., Lai, M. H., and Morris, N. R. (1978). Identification of a gene for β-tubulin in *Aspergillus nidulans*. *Cell* **15**, 639–647.

Sherline, P., and Schiavone, K. (1977). Immunofluorescence localization of proteins of high molecular weight along intracellular microtubules. *Science* **198**, 1038–1040.

Sherline, P., and Schiavone, K. (1978). High molecular weight MAP's are part of the mitotic spindle. *J. Cell Biol.* **77**, R9–R12.

Sisken, J. E., Wilkes, E., Donnelly, G. M., and Kakefuda, T. (1967). The isolation of the mitotic apparatus from mammalian cells in culture. *J. Cell Biol.* **32**, 212–216.

Struhl, K., Stinchcomb, D. T., Scherer, S., and Davis, R. W. (1979). High frequency transformation of yeast: Autonomous replication of hybrid DNA molecules. *Proc. Natl. Acad. Sci. U.S.A.* **76**, 1035–1039.

Thuriaux, P., Nurse, P., and Carter, B. (1978). Mutants altered in the control coordinating cell division with cell growth in the fission yeast *Schizosaccharomyces pombe*. *Mol. Gen. Genet.* **161**, 215–220.

Van Tuyl, J. M. (1977). Genetics of fungal resistance to systemic fungicides. Thesis, Agricultural University, Wageningen, The Netherlands.

Valenzuela, P., Querogu, M., Zaldivar, J., Rutter, W. J., Kirschner, M. W., and Cleveland, D. W. (1981). Nucleotide and corresponding amino acid sequences encoded by α- and β-tubulin on RNAc. *Nature* **289**, 650–655.

Wang, R. J. (1974). Temperature-sensitive mammalian cell line blocked in mitosis. *Nature (London)* **248**, 76–78.

Wang, R. J. (1976). A novel temperature-sensitive mammalian cell line exhibiting defective prophase progression. *Cell* **8**, 257–261.

Wang, Y.-L., and Taylor, D. L. (1979). Distribution of fluorescently labeled actin in sea urchin eggs during early development. *J. Cell Biol.* **82**, 672–679.

Weingarten, M. D., Lockwood, A. H., How, S.-Y., and Kirschner, M. W. (1975). A protein factor essential for microtubule assembly. *Proc. Natl. Acad. Sci. U.S.A.* **72**, 1858–1862.

White, E., Scandella, D., and Katz, E. R. (1979). CIPC resistant mutants of *Dictyostelium discoideum*. *J. Cell Biol.* **83**, 341a.

9

Mutants as an Investigative Tool in Mammalian Cells

V. LING

I. INTRODUCTION

The utility of the genetic approach for elucidating aspects of the cell division process is the theme of this chapter. Essential to this approach is the *accessibility of appropriate mutants*, since a powerful strategy is to analyze the relationships between functional and structural changes that result from specific mutational changes. Within this context, I will deal mainly with mammalian cells in culture and describe some of the mutants isolated to date.

Two somewhat different approaches have emerged with respect to isolating cell division mutants. One is to screen temperature-sensitive (*ts*) conditional growth mutants for lines that can be recognized as being defective in the cell division process, for example, blocked at mitosis at a restrictive temperature. As will be described, such lines have been obtained in a number of laboratories. The challenge of this approach, however, is in the identification of the structural alteration(s) associated with the *ts* phenotype,

197

MITOSIS/CYTOKINESIS
Copyright © 1981 by Academic Press, Inc.
All rights of reproduction in any form reserved.
ISBN 0-12-781240-7

since in the past this has proven extremely difficult (Stanners, 1978). The other approach is to isolate mutants resistant to well-defined drugs known to affect specific events in cell division, for example, antimitotic compounds such as colchicine, griseofulvin, and vinblastine (Dustin, 1978). One advantage of this approach is that mutants resistant to such compounds stand a good chance of being defective in the drug target and, consequently, the structural alteration(s) can be more easily localized. How such alterations affect the cell division process can then be investigated.

No attempts are made here to discuss in detail experimental strategies and procedures associated with selections of mutants of mammalian cells. The interested reader might refer to previous reviews of this area (Thompson and Baker, 1973; Baker and Ling, 1978).

II. TEMPERATURE-SENSITIVE (*TS*) CELL DIVISION MUTANTS

Since we are dealing with a vital function, it is possible that mutants with major defects in the cell division process may be isolated only if they carry conditional mutations, for example, temperature sensitivity (*ts*) in the expression of the altered phenotype. While there is as yet no direct selection for cell division mutants, procedures for isolating *ts* mutants for growth (DNA synthesis) are well established for cells in culture (c.f. Thompson and Baker, 1973). Such *ts* mutants are obtained by an enrichment procedure whereby drugs are used to kill cells able to synthesize DNA at the "nonpermissive" temperature (say, 39°C); the culture is then shifted to the "permissive" temperature (say, 34°C), and colonies formed at this temperature are isolated for analysis. Usually only a small percentage of isolated colonies display a stable *ts* phenotype on later testing, since cells by chance not in the DNA-synthesizing phase of the cell cycle during incubation at the nonpermissive temperature will also survive the selection.

Screening of stable *ts* growth mutants isolated in this manner indicates that cell division mutants can be obtained (Table I) and that two classes have been identified. One accumulates bi- or multinucleated cells on incubation at the nonpermissive temperature and is apparently affected in some step of cytokinesis. The other appears to be conditionally blocked at mitosis. One phenotype shared by the cell division mutants listed in Table I is that mutant cells round up at the nonpermissive temperature as they accumulate at their temperature sensitive step. This may be a potentially useful marker to exploit for selecting cell division mutants.

A. *ts* Cytokinesis Mutants

One of the putative cytokinesis mutants, *ts111*, isolated by Hatzfeld and Buttin (1975) from Chinese hamster cells, stops increasing in cell number

TABLE I

Temperature-Sensitive (ts) Cell Division Mutants in Mammalian Cells

Cell line	ts phenotype	Frequency per ts growth mutants examined	Reference
NW1 from Syrian hamster cells	Binucleated cells	Few per 20	Smith and Wigglesworth (1972)
ts111 from Chinese hamster cell line CCL39	Multinucleated cells	1/3	Hatzfeld and Buttin (1975)
MS1–1 from Chinese hamster ovary cells	Multinucleated cells	—	Thompson and Lindl (1976)
ts546 from hamster cell line HM-1	Scattered metaphase chromosomes	1/50	Wang (1974) Wang and Yin (1976)
ts2 from murine leukemic cell line L5178Y	Scattered metaphase chromosomes; multinucleated cells	1/10	Shiomi and Sato (1976)
ts655 and ts654 from hamster cell line HM-1	Condensation of chromatin and loss of nuclear boundary	2/8	Wang (1976)

within 24 hours after shifting to the nonpermissive temperature (39°C), after which the cell number remains relatively constant. Macromolecular synthesis, however, is not arrested, and giant cells accumulate such that by 48 hours at the restrictive temperature, they represent the majority of the population. The giant cells are heterogeneous, containing either one large nucleus with several nucleoli or multiple (e.g., 15) distinct nuclei. Similarly, the same diversity is observed in mitotic cells, some containing one large metaphase figure with over 100 chromosomes while others contain several metaphase figures. Even after 96 hours at the restrictive temperature, the viability of *ts111* is still largely maintained since its colony-forming ability on shifting back to the permissive temperature is still about 25% that of the parental line.

On shifting back to the permissive temperature after a time at the restrictive temperature, *ts111* cells resume division almost immediately. Many "small cells" appear, some with a karyoplast-like appearance, while others are anucleated (cytoplasts?). This observation raises the possibility that these small cells may have been generated from an extrusion-like process similar to that generally observed on application of cytochalasin B to mammalian cells. Investigating this possibility further, Hatzfeld and Buttin (1975) compared the effect of cytochalasin B on *ts111* and the parental line. They observed that at the permissive temperature of 34°C a lower concentration of drug (0.4 μg/ml) was required to cause more than 50% of the *ts111* cells to accumulate as multinucleates, while 0.8 μg/ml of cytochalasin B was required to produce the same effect in the parental cells. Whether or not the *ts111* phenotype results from an alteration in the same cellular component(s) affected by cytochalasin B is not known at present, but it is an intriguing possibility.

Another cytokinesis mutant, *(MS1-1)*, isolated by Thompson and Lindl (1976) from Chinese hamster ovary cells, displays features similar to those of *ts111*. This mutant was obtained by several cycles of enrichment for cells displaying differential detachment and attachment to the growth surface at the nonpermissive (38.5°C) and permissive (34°C) temperatures, respectively. The design of the selection scheme was intended originally to select for a mitotic mutant based on the observation that mitotic cells detach from the growth surface (Terasima and Tolmach, 1963). In a number of different isolations, Thompson and Lindl (1976) obtained only mutants of the MS1-1 phenotype, and no mitotic mutants were observed with this selection procedure. Like the *ts111* mutant of Hatzfeld and Buttin (1975), *MS1-1* accumulates of multinuculeate cells at the nonpermissive temperature (39.5°C) and, after several days, the increase in cell number ceases. Examination of the sensitivity of the cells to drugs such as colcemid and cytochalasin B indicates that they are similar to parental cells with respect to inhibition of growth at the permissive temperature of 34°C. Ultrastructural examination of MS1-1 cells synchronized with colcemid and then allowed to progress through mitosis at the restrictive temperature indicates that telophase doublets appear not to have the normal midbody characteristic of the completion of cytokinesis. In the mutant, the cleavage furrow has not proceeded to completion, although constriction has occurred. Nevertheless, none of the major ultrastructural components (e.g., microfilaments and microtubules) observed in parental cells prepared under the same conditions were missing in the mutant cells, and no gross abnormality could be detected.

Cell–cell hybrid studies crossing MS1-1 with another non-ts CHO cell line indicated that the ts phenotype of MS1-1 is recessive. This result is of some significance, since it points clearly toward an approach whereby the question of whether the molecular defects of two putative cytokinesis mutants such as ts111 and MS1-1 are different can be answered genetically by complementation analysis.

B. ts Mitotic Mutants

Wang (1974) has isolated a line (ts546) from Chinese hamster cells, which displays properties of a ts mitotic mutant. The colony-forming ability of ts546 at 39°C is about 5×10^{-7}, while at 33°C it is close to 50%. On shifting up to the nonpermissive temperature, no generalized macromolecular synthesis defect was observed; however, cells became detached from the growth surface after rounding up, so that by the twentieth hour 80% of the cells became detached (Wang and Yin, 1976). Cytological examination of the detached cells indicates that mitotic cells accumulate with time after shifting up to a transient maximum of 23% by the eighth hour and then decline to less than 1% by the twenty-fourth hour. A large proportion of the mitotic cells blocked at metaphase possess scattered metaphase chromosomes similar in appearance to colchicine-induced metaphase chromosomes. The subsequent decline in the mitotic index appears to result from the chromosomes of the blocked cells fusing into aggregates, and from the nuclear membrane re-forming around these aggregates, yielding interphase-like cells containing one or more nuclei (Wang and Yin, 1976). In summary, properties displayed by ts546 at the nonpermissive temperature are quite similar to those of colchicine-treated mammalian cells which also round up, arrest at metaphase, and subsequently form mono- or multinucleated interphase-like cells (Terasima and Tolmach, 1963; Dustin, 1978). Thus it is possible that the defect in ts546 resides on the colchicine-binding target, the microtubules. It has not been reported whether or not a complete mitotic spindle is formed but is nonfunctional at the nonpermissive temperature in ts546 cells, but this is obviously a question of interest.

A mutant line (*ts2*) similar to *ts546* has been isolated from murine leukemic cells by Shiomi and Sato (1976). The authors raise the possibility that this mutant may be defective in both mitosis and cytokinesis; however, the description of the ts phenotype suggests that its behavior is almost identical to that of *ts546* and resembles colchicine-treated cells. Thus this line has been classified as a mitotic mutant (Table I). It would be of considerable interest to determine whether the phenotypes of *ts546* (Wang, 1974) and *ts2* (Shiomi and Sato, 1976) are genetically complementary in cell–cell hybrids.

Characterization of the ts phenotype of *ts2* employing a synchronously growing population indicates that cells shifted to the restrictive temperature at G_1 are blocked in metaphase and binucleate cells accumulate. On the other hand, cells shifted to the restrictive temperature during G_2 are not blocked during the first mitotic cycle, but the second one is affected. These results are compatible with a model which suggests that certain events prior to G_2 need to be successfully completed for normal mitosis to take place, and that the mutant *ts2* is conditionally defective in one of these events.

Wang (1976) has isolated another class of mitotic mutants that appears to be conditionally blocked during prophase of the mitotic cycle. Again, by screening *ts* growth mutants, he has identified two mutants, *ts655* and *ts654*, with the following properties. At the nonpermissive temperature, interphase cells accumulate at a stage whereby the cells are rounded and dark-staining clumps of nuclear material are seen but no nuclear boundary is visible. This appearance is compatible with the blocking of these mutants at prophase of the mitotic cycle. The isolation of mutants *ts655* and *ts654* is of some significance since it indicates that mutations can be obtained resulting in conditional blocks at different stages of the mitotic cycle. Moreover, these mutants provide a system for investigating the biochemistry of a large population of cells blocked at this particular stage of mitosis. Previously, biochemical studies of mitotic cells were carried out on metaphase cells accumulated by blocking drugs.

III. DRUG-RESISTANT MUTANTS

Colchicine and cytochalasin B are well-characterized compounds used extensively as specific agents to block mitosis and cytokinesis in eukaryotic cells. In principle, isolation of cells resistant to such drugs could yield mutants altered in a cell component likely to be important in cell division. Studies on the fungus *Aspergillus nidulans* have clearly demonstrated the feasibility of this approach (Davidse and Flach, 1977; Sheir-Neiss *et al.*, 1978; Morris *et al.*, 1979). Mutations in the α- and β-tubulin genes have been identified in mutant strains isolated for resistance to the antimitotic compound benomyl. Using the same strategy, we have initiated a program to isolate mutants in Chinese hamster ovary (CHO) tissue culture cells resistant to antimitotic drugs and other compounds. Two major classes of mutants have been observed: one has altered membrane (reduced drug permeability), and the other is altered in some property located inside the cell. In this section, I will concentrate mainly on the colcemid-resistant, microtubule-altered mutants obtained in our laboratory (Ling *et al.*, 1979) and in that of Cabral *et al.* (1980). Lines resistant to cytochalasin B have been obtained, but they appear to be membrane-altered mutants (V. Ling, unpublished result).

Colchicine and antimitotic compounds such as vinblastine, podophyllotoxin, and griseofulvin apparently bind to microtubule proteins and inhibit the formation of the mitotic spindle and the cytoskeletal network of microtubules (Wilson *et al.*, 1974; Dustin, 1978; Weber and Osborn, 1979). The actual binding site of colchicine and its analogues has been identified as being on the tubulin subunits. The binding sites of vinblastine and griseofulvin are demonstrated to be different from that of colchicine and from each other. In this context, it seems possible that cells with different microtubule alterations could be obtained by isolating mutants resistant to different antimitotic drugs, and that characterization of such putative mutants could provide further insights into the role(s) of microtubules in the cell division process.

Colcemid-Resistant CHO Cells

Independent CHO cell clones resistant to colcemid have been isolated in our laboratory (Ling *et al.*, 1979). A selection scheme involving sequential clonal selections of mutants for increased drug resistance was employed so that genetically related lines with different degrees of resistance (from approximately 2- to 10-fold) were available for analysis (Fig. 1 and Table II).

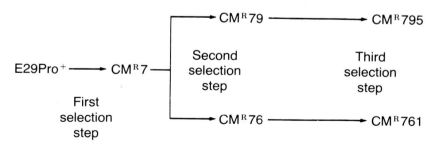

Fig. 1.   Interrelationship of one series of CHO cell clones selected stepwise for increased colcemid resistance (see Ling *et al.*, 1979). Some of their properties are described in Table II.

TABLE II

Cross-Resistance Pattern of Different Colcemid-Resistant CHO Cells[a,c]

	Relative drug resistance[b]		
Cell lines	Colcemid	Colchicine	Vinblastine
E29Pro+ (parental line to the CM^R mutants)	1	1	1
CM^R 7	2	2	1
CM^R 79	~4	2.5	1.5
CM^R 76	~4	2.5	1
CM^R 795	9	5	2
CM^R 761	~7	3	1
Cmd4	2.5	4	1.5

[a] Data for this table were compiled from unpublished results from Ling *et al.*, 1979, and from Cabral *et al.*, 1980. Line Cmd4 was selected by Cabral *et al.* The rest were obtained in our laboratory (cf. Fig. 1).

[b] Relative resistance was calculated from D_{10} mutant cells divided by D_{10} of parental cells where D_{10} is the concentration of drug required to reduce the relative colony-forming ability of a cell line to 10%.

[c] No cross resistance to puromycin, daunomycin, and griseofulvin was observed in these lines.

The defect(s) in these mutants appears to reside in an altered ability to bind colcemid in the tubulin from mutant cells since the more resistant cells display a greater degree of reduced drug-binding affinity. Cabral *et al.* (1980) also have isolated a colcemid-resistant CHO cell clone (Cmd-4). It was obtained in a single step, and the relative resistance of this line was about 2.5-fold (Table II). The colcemid-resistant mutants obtained from both laboratories display an incompletely dominant phenotype with respect to drug resistance in cell:cell hybrids. Thus complementation studies to reveal different classes of mutants are not possible.

Whether or not the colcemid-resistant clones are in fact different, however, can be examined by analyzing their resistance to different drugs. We have found that analysis of a cell's cross-resistant pattern is useful for classifying different mutants (e.g., membrane-altered versus microtubule-altered; c.f. Baker and Ling, 1978; Ling *et al.*, 1979). As can be seen in Table II, the colcemid-resistant clones are resistant only to compounds known to bind tubulin—for example, colchicine, colcemid, and vinblastine—but not to compounds such as puromycin and daunomycin. Previously, we found that colchicine-resistant, membrane-altered mutants are cross-resistant to these latter compounds (Ling and Thompson, 1974; Baker and Ling, 1978). Thus at this level of analysis, the data are consistent with the idea that colcemid-resistant lines (Table II) contain altered tubulin.

Examining in greater detail the cross-resistant patterns of the CM^R lines isolated in our laboratory (Table II), two classes can be identified. One is resistant to vinblastine (e.g., CM^R795), whereas the other is not; the latter behaves like the parental line (e.g., CM^R761) with respect to vinblastine resistance. These results raise the possibility that mutant CM^R761 possess a tubulin alteration localized at the colcemid-binding domain, while the alteration in CM^R795 is located in a different domain, possibly one in which both the colcemid- and vinblastine-binding sites are pleiotropically affected. The colcemid-resistant mutant Cmd-4 isolated by Cabral *et al.* (1980) is also different from the CM clones (Table II) in that its resistance to colchicine is higher than it is to colcemid. Thus it seems likely that at least three different classes of microtubule-altered, colcemid-resistant mutants have been identified in CHO cells.

It is perhaps noteworthy that none of the colcemid-resistant clones display cross resistance to the antimitotic compound griseofulvin (Table II). The binding target of griseofulvin has not yet been well defined, but it may be located on microtubule-associated proteins (Roobol *et al.*, 1977). Mutants isolated for resistance to griseofulvin, however, do display cross resistance to colcemid and colchicine (Cabral *et al.*, 1980; V. Ling, unpublished observation). Further investigation of these mutants may provide information concerning the griseofulvin-binding site.

Structural alterations in the tubulin of the colcemid-resistant lines have been observed (Keates *et al.*, 1979; Cabral *et al.*, 1980; Keates *et al.*, 1981). In the Cmd-4 mutant Cabral *et al.* (1980) found an additional β-tubulin spot migrating to a more basic isoelectric point when the cytoskeletal proteins of that line were examined by two-dimensional gel electrophoresis. We have observed, however, a more basic α-tubulin subunit component in CM^R795 (Keates *et al.*, 1981). Since in our case the tubulin protein was purified by cycles of *in vitro* assembly and disassembly prior to analysis by gel electrophoresis, it is likely that the variant α-tubulin is functional at least in the *in vitro* system. Investigating this assumption in greater detail, we found that in CM^R795 the *in vitro* assembly of microtubules showed more resistance to colcemid inhibition than that observed for the parental line, and further, that the critical concentration for assembly was significantly lowered (Keates *et al.*, 1981). These results strongly suggest that the altered α-tubulin subunit (and possibly other altered components in the CM^R795 line) do participate in forming microtubules *in vivo*.

To determine whether or not the *in vivo* microtubules of mutant CM^R795 and other colcemid-resistant clones contain altered components, studies using immunofluorescence technique with specific antitubulin antisera were undertaken (Connolly *et al.*, 1981). The results indicate that both the cyto-skeletal network of microtubules and the spindle in a number of mutant lines are apparently normal in the absence of drugs. In the presence of colcemid, however, a 6-fold higher concentration of drug was required to break down microtubules in both the spindle and cytoskeleton of CM^R795 compared to that of the parental line. No difference was observed between these two lines when griseofulvin (a drug to which both lines are equally sensitive) was used to break down microtubules. It is interesting, also, that the shape and size of the microtubule-containing paracrystals formed in the presence of vinblastine are different in the mutant cells compared with the parental cells (Connolly *et al.*, 1981). The results from these *in vivo* studies clearly indicate that altered microtubule components (including the altered α-tubulin subunit, plus possible others not yet identified) do form an intergral part of the microtubule system in mutant cells. Further, this study provides evidence that it is the actual interaction of colcemid with tubulin proteins which results in microtubule breakdown *in vivo*.

Taken together, the results of the work described in this section indicate that different mutations in the microtubule proteins can be obtained, that the mutant proteins form part of the *in vivo* microtubule structure, and that the altered components do affect aspects of the microtubule properties both *in vitro* and *in vivo*. Thus such mutants could be very useful tools for analyzing the role of microtubules in the cell division process.

IV. CONCLUDING REMARKS

It is evident from the work reviewed in this chapter that the prospect of obtaining a wide range of useful cell division mutants in mammalian cells looks promising. The two approaches taken in this field have both yielded appropriate mutants and provided new insights. When *ts* mutants for growth were screened, *ts* "cell division mutants" were found to comprise a relatively high proportion of the observed mutants (2–30%; see Table I). One possible explanation for this observed high percentage is that the products from many genes participate in cell division and that mutations in any one of those genes result in the phenotypic expression of a defect in the overall process. *A priori*, it can be imagined that the cell division process must involve in a coordinated manner a large number of different biochemical pathways and cell organelles. Thus this model predicts that the mutational target at risk is very large. Other explanations are of course possible, but the above hypothesis is testable. The results of the cell–cell hybrid analysis performed by Thompson and Lindl (1976) with their *ts* cytokinesis mutant indicate that complementation studies with different *ts* cell division mutants are feasible. Such studies with a wide range of appropriate phenotypically different mutants should provide information on the minimum number of complementation groups or genes that participate in the cell division process. It is crucial to know whether this number is large or small. Information of this nature is essential for shaping future investigative approaches.

Isolation of drug-resistant mutants to the antimitotic compounds colcemid and griseofulvin has yielded mutants possessing altered microtubule components. Such mutants are useful for determining whether or not microtubules participate in particular functions. For example, we have used the CM[R] mutants to demonstrate the involvement of microtubules in the movement of cell surface components in CHO cells (Aubin *et al.*, 1980). In a similar manner, the involvement of β-tubulin in nuclear movement has been demonstrated in mutants of *A. nidulans* (Oakley and Morris, 1980). Thus appropriately characterized drug-resistant mutants can be used to investigate aspects of the cell division process.

It is interesting that the altered phenotype of the *ts* cytokinesis mutants (Hatzfeld and Buttin, 1975; Thompson and Lindl, 1976) mimics the effect of cytochalasin B treatment, and that of the *ts* mitotic mutants (Wang, 1974; Shiomi and Sato, 1976) mimics the effect of colchicine treatment. It is possible that these mutants are defective in the drug target proteins (microfilament and microtubule proteins, respectively). In this respect, it is significant that Sato (1976) has isolated a *ts* cell division mutant in *Chlamydomonas reinhardii* that also has increased resistance to colchicine. Reversion analysis indicates that both the *ts* and drug-resistant phenotypes are the consequence of the same mutation. Thus in that mutant, at least, it appears that an alteration in a microtubule component, probably in the vicinity of the colchicine-binding site, results in a pleiotropic expression of a *ts* phenotype for cell division.

Mutants are most useful when their altered gene products are identified and characterized. Though this task admittedly is often a difficult one, it seems clear that this area deserves continued attention and innovation. The inability to identify the structural defects associated with the *ts* phenotype of the conditional cell division mutants (Table I) has limited the usefulness of those mutants. It is anticipated that with more current techniques of characterization, some of which are described in this book, this deficiency can be overcome. In this context, recent rapid advances in mammalian cell genetics with respect to manipulation of genetic material may also be productively applied. The prospect of isolating, amplifying, and transferring particular genes (Wigler *et al.*, 1979) into a variety of cells to analyze their effects on the cell division process is an exciting one.

ACKNOWLEDGMENTS

I thank my colleagues J. E. Aubin, A. Chase, J. Connolly, V. Kalnins, R. Keates, M. Naik, and F. Sarangi for their collaboration and helpful discussions. This work was supported by the National Cancer Institute of Canada, the Medical Research Council of Canada, and the Ontario Cancer Treatment and Research Foundation.

REFERENCES

Aubin, J. E., Tolson, N., and Ling, V. (1980). The redistribution of fluoresceinated concanavalin A in Chinese hamster ovary cells and their colcemid-resistant mutants. *Exp. Cell Res.* **126**, 75–85.

Baker, R. M., and Ling, V. (1978). Membrane mutants of mammalian cells in culture. In "Methods in Membrane Biology" (E. Korn, ed.), Vol. 9, pp. 337–384. Plenum, New York.

Cabral, F., Sobel, M. E., and Gottesman, M. M. (1980). CHO mutants resistant to colchicine, colcemid, or griseofulvin have an altered β-tubulin. *Cell* **20**, 29–36.

Connolly, J. A., Kalnins, V. I., and Ling, V. (1981). Microtubules in colcemid resistant mutants of CHO cells. *Exp. Cell Res.* **132**, 147–156.

Davidse, L. C., and Flach, W. (1977). Differential binding of methyl benzimidazol-2-yl carbamate to fungal tubulin as a mechanism of resistance to this antimitotic agent in mutant strains of *Aspergillus nidulans*. *J. Cell Biol.* **72**, 174–193.

Dustin, P. (1978) "Microtubules." Springer-Verlag, Berlin and New York.

Hatzfeld, J., and Buttin, G. (1975). Temperature-sensitive cell cycle mutants: A Chinese hamster cell line with a reversible block in cytokinesis. *Cell* **5**, 123–129.

Keates, R. A. B., Sarangi, F., and Ling, V. (1979). Microtubule proteins isolated from colcemid resistant mutants of CHB cells show altered sensibility to colcemid *in vitro*. *Proc. 11th Int. Congr. Biochem.* p. 580.

Keates, R. A. B., Sarangi, F., and Ling, V. (1981). Structural and functional alterations in microtubule protein from Chinese hamster ovary cell mutants. *Proc. Natl. Acad. Sci. (USA)* **78**, (in press).

Ling, V., and Thompson, L. H. (1974). Reduced permeability in CHO cells as a mechanism of resistance to colchicine. *J. Cell. Physiol.* **83**, 103–116.

Ling, V., Aubin, J. E., Chase, A., and Sarangi, F. (1979). Mutants of Chinese hamster ovary (CHO) cells with altered colcemid-binding affinity. *Cell* **18**, 423–430.

Morris, N. R., Lai, M. H., and Oakley, C. E. (1979). Identification of a gene for α-tubulin in Aspergillus nidulans. *Cell* **16**, 437–442.

Oakley, B. R., and Morris, R. N. (1980). Nuclear movement is β-tubulin-dependent in Aspergillus nidulans. *Cell* **19**, 255–262.

Roobol, A., Gull, K., and Pogson, C. I. (1977). Evidence that griseofulvin binds to a microtubule associated protein. *FEBS Lett.* **75**, 149–153.

Sato, Ch. (1976). A conditional cell division mutant of *Chlamydomonas reinhardii* having an increased level of colchicine resistance. *Exp. Cell Res.* **101**, 251–259.

Sheir-Neiss, G., Lai, M. H., and Morris, N. R. (1978). Identification of a gene for β-tubulin in Aspergillus nidulans. *Cell* **15**, 639–647.

Shiomi, T., and Sato, K. (1976). A temperature-sensitive mutant defective in mitosis and cytokinesis. *Exp. Cell Res.* **100**, 297–302.

Smith, B. J., and Wigglesworth, N. M. (1972). A cell line which is temperature-sensitive for cytokinesis. *J. Cell. Physiol.* **80,** 253–260.

Stanners, C. P. (1978). Characterization of temperature-sensitive mutants of animal cells. *J. Cell. Physiol.* **95,** 407–416.

Terasima, T., and Tolmach, L. J. (1963). Growth and nucleic acid synthesis in synchronously dividing populations of HeLa cells. *Exp. Cell Res.* **30,** 344–362.

Thompson, L. H., and Baker, R. M. (1973). Isolation of mutants of cultured mammalian cells. *Methods Cell Biol.* **6,** 209–281.

Thompson, L. H., and Lindl, P. A. (1976). A CHO cell mutant with a defect in cytokinesis. *Somat. Cell Genet.* **2,** 387–400.

Wang, R. J. (1974). Temperature-sensitive mammalian cell line blocked in mitosis. Nature (*London*) **248,** 76–78.

Wang, R. J. (1976). A novel temperature-sensitive mammalian cell line exhibiting defective prophase progression. Cell **8,** 257–261.

Wang, R. J., and Yin, L. (1976). Further studies on a mutant mammalian cell line defective in mitosis. *Exp. Cell Res.* **101,** 331–336.

Weber, K., and Osborn, M. (1979). Intracellular display of microtubular structure revealed by indirect immunofluorescence microscopy. *In* "Microtubules" (K. Roberts, and J. S. Hyams, eds.), pp. 279–313. Academic Press, New York.

Wigler, M., Sweet, R., Sim, G. K., Wold, B., Pellicer, A., Lacy, E., Maniatis, T., Silverstein, S., and Axel, R. (1979). Transformation of mammalian cells with genes from prokaryotes and eukaryotes. *Cell* **16,** 777–785.

Wilson, L., Bamburg, J. R., Mizel, S. B., Grisham, L. M., and Creswell, K. M. (1974). Interaction of drugs with microtubule proteins. *Fed. Proc. Fed. Am. Soc. Exp. Biol.* **33,** 158–166.

10

Immunofluorescence Studies of Cytoskeletal Proteins during Cell Division

J. E. AUBIN

I. INTRODUCTION

The immunofluorescence approach to studies of the cytoskeleton* is by now familiar to most biologists. Thus, the rudiments of the technique will

*The word "cytoskeleton" is used here as a general term to encompass components of the cellular filament systems, including microtubules, microfilaments, intermediate filaments, and certain of their accessory proteins. Except where specifically noted in the text, the term is not meant to imply a strictly static skeletal function for any of these systems, and indeed, it is quite clear that all three major filament systems are strikingly dynamic during various cellular phenomena, including mitosis.

211

MITOSIS/CYTOKINESIS
Copyright © 1981 by Academic Press, Inc.
All rights of reproduction in any form reserved.
ISBN 0-12-781240-7

not be reviewed here. Instead, the reader is referred to several detailed immunology sources for descriptions of antibody production, fluorochrome conjugation, and details of suitable controls (for example, see Sternberger, 1979; Johnson *et al.*, 1978). For aspects of immunofluorescence as applied to components of the cytoskeleton, two reviews are useful (Kalnins and Connolly, 1981; Weber and Osborn, 1981). In this chapter, I would like to outline the immunofluorescence approach to studies of mitosis, with a discussion of the particular problems encountered in applying the technique to mitotic cells.

For more detailed models and reviews of mitosis with emphasis on particular fibrous systems, the reader is referred to other chapters in this volume as well as to recent volumes on the subject (for example, see Inoué and Stephens, 1975; Little *et al.*, 1977). This chapter will not attempt to explain definitively how chromosomes move or how the cell divides but rather will suggest possible mechanisms which deserve attention by pointing out what we have learned from immunofluorescence about the locations of various cytoskeletal proteins and their accessory and perhaps control proteins during mitosis and cytokinesis. Whether or not it reveals the mitotic mechanism, immunofluorescence does point out the strikingly dynamic rearrangements of which cytoskeletal proteins are capable. In the mitotic cell, new structures form, function, and then disappear, and the cell experiences extensive changes in shape and motile function. On a molecular basis, none of these events is understood. It is known from many approaches that cytoskeletal proteins are involved, sometimes actively, perhaps at other times passively. In mitosis, the main advantage of immunofluorescence over other localization methods is that it can detect the whereabouts of the proteins of interest in many cells, and in the entire cell, including some proteins which cannot as yet be seen or identified definitively at the electron or light microscopic level. Thus, although basically a static approach, immunofluorescence does allow the definition of structural aspects of the cytoskeleton at each step of the division process. By looking at enough steps, that is, at temporally close rearrangements, the static approach may help to reveal the dynamics. Its primary disadvantage arises out of the very nature of immunofluorescence: one must attempt to determine by appropriate tests and controls that the antibodies employed are indeed specific for a particular protein, that cross reactivity is minimal, and that the fluorescent molecules used to tag first and/or second antibodies do not cause nonspecific binding to cellular components other than those of interest. The technique does, ultimately, fail to provide the resolution of an ultrastructural approach (though combination antibody–electron microscopy may help to bridge this gap), the mechanistic descriptions possible of biochemistry, or the definitiveness of a functional assay (though combined antibody–functional tests may now be feasible; see

Section V). However, it does often point us in the right direction by suggesting possible mechanisms worth testing at the molecular level.

II. HISTORICAL VIEW OF IMMUNOFLUORESCENCE OF MITOTIC CELLS

There are at least three classes of fibrous elements comprising an extensive cytoskeleton in eukaryotic cells: (1) microtubules, (2) microfilaments, and (3) intermediate (100 Å, 10-nm filaments (for review, see Kalnins and Connolly, 1981; Weber and Osborn, 1981). Each will be dealt with in turn, first with a brief review of its possible role in mitosis and then with a presentation of new data which have been accumulated.

A. Microtubules

Extensive electron microscopic analyses, including serial sectioning, have revealed that the major fibrous components of the spindle visible by present fixation techniques are microtubules (for reviews, see McIntosh et al., 1976; Fuge, 1977; Heath, this volume). Bright spindle and intracellular bridge microtubule immunofluorescence was first demonstrated by Dales (1972) and Dales et al. (1973) using antiserum raised against vinblastine-induced paracrystals (Nagayama and Dales, 1970). The specific antigenic determinant(s) of this antiserum was uncertain, however, because the antigen used was not demonstrated to contain only tubulin. Vinblastine was known to precipitate other cellular constituents in vitro (Wilson et al., 1970; Bryan, 1972), although in vivo it was thought to be more specific for the tubulin dimer (Bryan, 1972). In any case, no cytoplasmic microtubule network was detected when cells were stained in interphase with this antiserum. After production of antisera specific for the microtubule subunit protein, tubulin, the immunofluorescence visualization of brightly staining cytoplasmic microtubules in interphase was clear. An intensely fluorescent mitotic spindle containing a high concentration of microtubules, against a very weakly staining cytoplasm containing few or no microtubules, was also evident. The stages of mitosis could be accurately determined from the fluorescent microtubule images (Fuller et al., 1975; Weber et al., 1975; Aubin et al., 1976; Brinkley et al., 1976, Sato et al., 1976; Weber, 1976).

It is well to point out at this time that when a direct comparison has been made (see especially Osborn et al., 1978), the immunofluorescent localization of microtubules agrees with the electron microscopic analysis of microtubule distributions (with one exception: for discussion of midbody staining with tubulin antibodies, see Weber et al., 1975; Connolly, 1979). This is an important consideration in anticipation of discrepancies between the two techniques which have become evident in discussion of microfilaments. Although one early paper reported immunological differences found with antibodies against tubulin from *Arbacia punctulata* when this serum was used in immunodiffusion with tubulins from different lower organisms (Fulton *et al.*, 1971), the general finding in immunofluorescence studies has been that antibodies raised against tubulin from diverse sources have been widely cross-reactive on cells of various tissues and diverse species of both animals and plants (for a review of relevant literature, see Weber *et al.*, 1977; Kalnins and Connolly, 1981; Weber and Osborn, 1981). The discrepancy between Fulton *et al.* (1971) and other results found later in numerous laboratories may reflect the fact that immunofluorescence as a probe for antigen detection is far more sensitive than immunodiffusion (Sternberger, 1979), although it may not resolve differences in antigenic determinants of various tubulins to the extent of a more quantitative assay such as a radioimmune assay (Morgan *et al.*, 1977; Gozes *et al.*, 1977; Joniau *et al.*, 1977; Hiller and Weber, 1978). It should also be stressed that despite early worries to the contrary, and despite the fact that microtubules may be sensitive to the fixation procedures employed (Osborn and Weber, 1977; Webster *et al.*, 1978a), it is routinely possible, using at least some standard fixation procedures, to demonstrate at the electron microscopic level that tubulin antibodies decorate cytoplasmic microtubules (DeMey *et al.*, 1976) and that recognizable microtubules are specifically decorated after some procedures used for immunofluorescence (Weber *et al.*, 1978; Webster *et al.*, 1978a,b). It is beyond the scope of this chapter to go systematically through the volumes of data demonstrating the specificity of numerous purified tubulin antibodies for microtubules and tubulin-containing structures. Since this has been done, and since the antibodies are recognizing microtubules in whole cells under a variety of conditions, then it is fair to say that up to now, immunofluorescence of mitotic cells with tubulin antibodies has agreed with the findings of light and electron microscopy and has given impressive overviews of tubulin-containing spindle structures.

B. Microfilaments

It is now well established that actin and other microfilament-associated proteins are present in non-muscle cells (Pollard and Weihing, 1974; Korn, 1978). The display of microfilaments has been documented both by immunofluorescence microscopy with specific antibodies (e.g., Lazarides and Weber, 1974; Weber and Groeschel-Stewart, 1974; Goldman *et al.*, 1975; Lazarides and Burridge, 1975; Wang *et al.*, 1975; Lazarides, 1976; Webster *et al.*, 1978b; Osborn and Weber, 1979) and by electron microscopy (e.g., Brown *et al.* 1976; Lenk *et al.*, 1977; Schloss *et al.*, 1977; Small and Celis,

1978a; Webster *et al.*, 1978a,b; Isenberg and Small, 1978; Henderson and Weber, 1979). Specific decoration with heavy meromyosin (HMM) has also been used to identify microfilaments at both the light microscopic level with fluoresceinated HMM* and by the distinctive arrowhead pattern visualized in the electron microscope (Ishikawa *et al.*, 1969; for reviews, see Forer, 1978; Sanger and Sanger, 1979).

The presence of these so-called contractile proteins in nonmuscle cells suggested that there might be a general physicochemical basis for contraction and motile function, including mitosis, in muscle and non-muscle cells (Huxley, 1973, 1976; Korn, 1978). A number of observations have been made suggesting that actin might be located in the spindle. As early as 1965, isolated sea urchin egg spindles were shown to bind fluorescent HMM (Aronson, 1965). Subsequently, a number of investigators showed decorated filaments in the spindles of a variety of cell types at the electron microscopic level (Behnke *et al.*, 1971; Gawadi, 1971, 1974; Forer and Behnke, 1972; Forer and Jackson, 1976; Hinkley and Telser, 1974; Schloss *et al.*, 1977; Forer *et al.*, 1979; reviewed in McIntosh *et al.*, 1975; Sanger, 1977; Sanger and Sanger, 1979; Petzelt, 1979). These electron microscopic studies are not without reservations, however, because they have been extensively discussed (references as above) I will not review them here. Less controversial, and more consistently found in the electron microscope, are the belts of aligned microfilaments in the cleavage furrow and contractile ring described in numerous cells during cytokinesis (Schroeder, 1975, 1976; Sanger and Sanger, 1979, 1980). While molecular aspects of their mechanism, regulation, and association with membrane are not understood, it does seem clear that microfilaments are present in this region in a high local concentration and with clear orientation. As for microfilament-associated proteins, although myosin can be isolated from various non-muscle cells, almost no ultrastructural data for its distribution in dividing cells have been found. Schroeder (1975, 1976) suggested that thick filaments (putative myosin filaments) might occur in close association (and even as cross-bridges) with contractile ring microfilaments. More recently, Sanger and Sanger (1980) have also documented thick, possibly myosin, filaments in this region. Spindle myosin has not been identified ultrastructurally. However, even if actin is present in the spindle, it may be unnecessary to have associated myosin, since actin may function in some motile and/or cytoskeletal activity without associated myosin, for example, by sol–gel transformations (see, Tilney *et al.*, 1973; Tilney, 1976; and recent reviews: Stossel, 1978; Taylor and Condeelis, 1979; Hellewell and Taylor, 1979; Weber and Osborn, 1981). The

---

*Although not strictly an *immunofluorescent* probe, fluoresceinated HMM will be discussed here, as it has been used frequently in mitotic studies, and fixation and control experiments similar to those for antibodies must be done for it.

TABLE I

Fluorescence Analysis of Microfilaments in Dividing Cells Using Antibodies and HMM

Probe	Fixation/extraction	Metaphase				Anaphase		Telophase	Reference
		Kinetochores	Poles	Spindle fibers[b]	Spindle diffuse[c]	Chromosome to pole	Interzone	Contractile ring and cleavage furrow	
Actin antibodies									
1. IgG fraction Second antibody: fl[d]-IgGs	"Lysis" buffer (tubulin, pipes, GTP, ATP, MgSO₄, Triton X-100); formalin	−[e]	+[f]	+	−	+	−	+	Cande et al., 1977
	Formalin; "Lysis" buffer	−	+	−	+	ND[g]	ND	ND	Cande et al., 1977
2. Antigen affinity-purified IgGs Second antibody: fl-IgGs	Formalin; acetone	−	+	−	+	+	+	−	Herman and Pollard, 1979
3. Antigen affinity-purified IgGs	8–10 different procedures, including,[h] "Lysis" buffer; formalin								
a. Antiactin I	Methanol	−	−	−	−	−	−	+	Aubin et al., 1979a,b
b. Antiactin II	Formalin; acetone	−	−	−	−	−	−	+	Aubin et al., 1979a,b
Second antibody: preabsorbed fl-IgGs	Glycerination								
4. IgG fraction Second antibody: fl-IgGs	Formalin; acetone	−	±[i]	−	± – +[i]	ND	ND	ND	Izutsu et al., 1979
HMM									
1. Fluorescein–HMM	Glycerination	+	+	+	−	+	−	+	Sanger, 1975, 1977 Sanger and Sanger, 1979
2. Rhodamine–HMM	Formalin; acetone	−	+	+	+	+	+	−	Herman and Pollard, 1978
3. Rhodamine or fluorescein–HMM	Formalin; acetone	−	+	+	−	+	+	−	Herman and Pollard, 1979
4. Fluorescein–S1	Glycerination; formalin	−	+	+	−	+	−	+	Schloss et al., 1977

[a] Localization

5. Fluorescein-HMM	Formalin; methanol[j] Methanol Formalin; Triton X-100 Glycerination	–	–	–	–	–	–	+	Aubin *et al.*, 1979a,b
Myosin antibodies									
1. a. IgG fraction	Acetone Acetone; formalin	–	–	–	+	+	–	+	Fujiwara and Pollard, 1976, 1978
b. Antigen affinity-purified IgGs Direct immunofluorescence	Acetone	–	–	–	–	ND	–	+	Fujiwara *et al.*, 1978
2. Antigen affinity-purified IgGs Second antibody: preabsorbed fl-IgGs	Formalin; methanol[j] Formalin; acetone Methanol Formalin; Triton X-100	–	–	–	–	–	–	+	Aubin *et al.*, 1979a,b
Tropomyosin antibodies									
1. Antigen affinity-purified IgGs Second antibody: preabsorbed fl-IgGs	Formalin; Methanol[j] Formalin; Acetone Formalin; Triton X-100	–	–	–	–	–	–	+	Aubin *et al.*, 1979a,b
α-Actinin antibodies									
1. Antigen affinity-purified IgGs or IgG fraction Direct or second antibody: fl-IgGs	Acetone; formalin	–	–	–	–	–	–	+	Fujiwara *et al.*, 1978

[a] These data are compiled from the pictures, statements, and interpretations of the authors of the appropriate references.
[b] Fluorescence staining of the spindle was visualized as discrete fibers.
[c] Fluorescence staining of the spindle was visualized as a diffuse spindle shape rather than as fibrous.
[d] fl, fluorescent-labeled second antibody; fluorescein or rhodamine.
[e] Fluorescence intensity is equivalent to that of the general cytoplasmic fluorescence = –.
[f] Fluorescence intensity is increased over the general cytoplasmic fluorescence = +.
[g] ND, Not done or not reported in figures or text of reference.
[h] All fixations were done for each of the two actin antibodies; all gave the same result as shown.
[i] Believed to be non-specific by the authors since nonimmune IgGs gave comparable staining.
[j] All fixations gave identical results as shown.

possibility of microtubule–actin filament interactions as functional units should not be ruled out either. Indeed, in addition to preliminary electron microscopic data outlined in the references above, one report suggests that actin filaments and microtubules may interact *in vitro* (Griffith and Pollard, 1978).

Because ultrastructural analysis of spindle actin and actin-associated proteins may be difficult at the present time, a number of investigators have turned to the light-fluorescence microscope. Sanger (1975) and, subsequently, others (see Schloss *et al.*, 1977 and Herman and Pollard, 1978, 1979) repeated and refined the fluorescein–HMM approach of Aronson (1965) to study actin during cell division (for a review, see Sanger, 1977; Sanger and Sanger, 1979). Rather than to detail the observations again, it is more useful to stress that by using fluorescein–HMM as a probe, many discrepancies in results have been documented, including such important considerations as whether or not both chromosome pole fibers and interzone fibers stain, and whether or not the contractile ring stains in the same cell type (see, for example, Table I and specific references therein).

A second fluorescence approach using the IgG fractions of actin antisera was presented first by Cande *et al.* (1977). A later analysis used specific affinity-purified antibodies (Herman and Pollard, 1979). Table I summarizes the obvious inconsistencies in the localization of actin as determined by fluorescein–HMM, and by these actin antibodies, under a variety of fixation–extraction procedures. The immunofluorescence and fluorescence approaches described above have suggested that actin may be present in the spindle region at concentrations higher than those in the surrounding cytoplasm. However, *some, but not all*, results have indicated that it may also be present at increased levels in the polar regions and in the interzone; *some, but not all*, results have suggested that it may be found in a fibrous versus a diffuse configuration; and *some, but not all*, results have supported the high concentration in the cleavage furrow and contractile ring as seen in the electron microscope.

Some microfilament-associated proteins have been studied in the dividing cell. Using IgGs isolated from myosin antisera, Fujiwara and Pollard (1976, 1978) demonstrated bright but diffuse cytoplasmic staining in mitotic cells with more intense staining over the poles and from chromosomes to pole, in the metaphase and very early anaphase spindle, and in the cleavage furrow during cytokinesis. However, when the specific myosin antibodies were purified by affinity chromatography, the cleavage furrow continued to stain more intensely than the general cytoplasm, but polar and spindle areas stained with the same intensity as the rest of the cytoplasm (see Table I). Affinity-purified α-actinin antibodies stained the spindle region with no greater intensity than bright diffuse cytoplasmic fluorescence, although more intense staining of the cleavage furrow was evident in cytokinesis

(Fujiwara *et al.*, 1978). One preliminary report suggested that tropomyosin antibody stained the spindle poles and bundles perpendicular to the cleavage furrow (Sanger, 1977), but the unusual interphase staining pattern (appearance of cytokeratin-like filaments rather than microfilament bundles) makes interpretation of these results difficult.

C. Intermediate Filaments

It is now clear that while intermediate filaments may share some similar ultrastructural and possibly some common biochemical features in a wide variety of cells, they display a high degree of polymorphism as detected by immunological and gel electrophoretic analysis (see Davison *et al.*, 1977; Bennett *et al.*, 1978; Franke *et al.*, 1978; Lazarides and Balzer, 1978; Starger *et al.*, 1978; Lazarides, 1980; Weber and Osborn, 1981). Almost nothing is known about the function(s) of intermediate filaments, though they have been postulated to be structural and/or "mechanical integrators of various cytoplasmic organelles" (for a review, see Lazarides, 1980). Their relative stability and insolubility in physiological buffers makes it of interest to study these filaments in more detail during cell division. However, while a number of studies have outlined ultrastructural aspects of intermediate filaments in interphase (Ishikawa *et al.*, 1968; Goldman and Knipe, 1972; for review, see Lazarides, 1980; Kalnins and Connolly, 1981; Weber and Osborn, 1981), relatively little is known about the detailed ultrastructure of these filaments during mitosis (Ishikawa *et al.*, 1968; Goldman and Follet, 1970; Blose and Chacko, 1976; Starger *et al.*, 1978). During interphase in many types of mesenchymal cells and some (if not all) non mesenchymal cells grown *in vitro*, both ultrastructure and immunofluorescence studies have shown that the vimentin-type intermediate filaments form extensive cytoplasmic arrays of wavy filaments (Franke *et al.*, 1978; Hynes and Destree, 1978, Starger *et al.*, 1978; Gordon *et al.*, 1978; Bennett *et al.*, 1978; Lehto *et al.*, 1978; Small and Celis, 1978b). Starger *et al.* (1978) reported in an electron microscopic study that these vimentin filaments largely disaggregated in BHK-21 cells during prometaphase and reappeared in filamentous caps in the perinuclear area during spreading of the daughter cells. Hynes and Destree (1978) prepared antibodies against the subunit protein of the 10-nm filaments of NIL-8 hamster cells; this protein, with an apparent molecular weight of 58 K, is presumably closely related to vimentin. In cells labeled with the anti-58 K serum, the 10-nm filaments were detectable throughout mitosis. At metaphase and anaphase the filaments coursed through the cytoplasm, excluding the spindle region, which was easily detectable in comparable cells stained with tubulin antibodies. In telophase, tubulin antibody, but not the anti-58 K serum, stained the intercellular bridge. In dividing gerbil fibroma cells, the vimentin-type filaments detected by some rabbit autoimmune sera (designated CS24 serum; Gordon *et al.*, 1978) behaved similarly to those described above, with microtubule staining and intermediate filament staining patterns mutually exclusive in metaphase and telophase. In vascular endothelial cells, 10-nm vimentin-type filaments form a thick birefringent perinuclear whorl in interphase (Blose and Chacko, 1976; Blose, 1979). Ultrastructural analysis showed that these filaments were present through several stages of mitosis (Blose and Chacko, 1976). With CS24 serum (Gordon *et al.*, 1978), Blose (1979) described by immunofluorescence that the perinuclear bundle seemed to remain a relatively intact and closed structure during prometaphase and metaphase, becoming wavy as the cell rounded up. In anaphase, the bundle elongated to a rectangle which contained the spindle apparatus and chromosomes before separating into perinuclear crescents around the daughter cell nuclei. From these observations it has been argued that the vimentin-type filaments are not functional elements of the spindle itself, but that they may form a scaffolding to provide topological information to other cytoskeletal proteins and cellular organelles such as mitochondria during mitosis and cytokinesis (Blose, 1979).

III. THE PROBLEMS

Some of the major problems encountered by those attempting to use immunofluorescence to study mitosis should be clear by now. I will briefly enumerate them here, with mention of the technical improvements used to circumvent some of the difficulties. First, it has been evident to many workers that a cell type which remains relatively flat during mitosis (e.g., PtK1 and PtK2) facilitates the immunofluorescence analysis. Not only are the stages of mitosis more easily and more precisely defined, but the display of cytoskeletal proteins is better visualized in the well-spread cells, where the limitations of depth of focus imposed by the fluorescence microscope become less of a problem (for a discussion, see Aubin *et al.*, 1979; Sanger and Sanger, 1979). This is especially advantageous when staining with probes such as antibodies to actin and other microfilament-associated proteins, since the cytoplasms of mitotic cells labeled with such probes remain very brightly stained throughout the division cycle sometimes making detection of particular substructures, e.g., the spindle, very difficult (compare tubulin antibody staining). The advantage of this flat cell is lessened somewhat by the fact that many clones of PtK2 express very low mitotic indices. This low mitotic index has recently been greatly enhanced by synchronization using double thymidine block, ensuring large numbers of mitotic cells for detailed analyses (Aubin *et al.*, 1980b). Second, immunofluorescence allows visualization of the entire cell and all cells in the population. Unfortunately, many immunofluorescence studies of mitosis have not taken advantage of this feature, though to determine unambiguously the possible mitotic roles of various proteins, it is essential to know their locations at all stages of the division process. Third, many workers have recognized the importance of attempting to prove not only the specificity of their antibodies for particular proteins but also the absence of any nonspecific interactions of these antibodies with other cellular components. One must also control for nonspecific binding of labeled second antibodies employed in the indirect immunofluorescence method and for the difficulty in performing unambiguous blocking experiments in the direct fluorescence approach. Although antigen affinity purification of cytoskeletal antibodies was shown earlier to be useful in preparing specific antibodies and lowering nonspecific staining (for example, see Fuller *et al.*, 1975; Weber *et al.*, 1976; Fujiwara and Pollard, 1976; Osborn *et al.*, 1978), such highly purified antibodies have not always been used in studies on mitosis (see Section IIA,B,C and Table I). Moreover, a number of animal autoimmune sera have been described as having reactivity to various components of the cytoskeleton [(for example, rabbit autoantibody to cytokeratin filaments (Osborn *et al.*, 1977), centrioles (Connolly and Kalnins, 1979), vimentin-like filaments (Gordon *et al.*, 1978), and tubulin (Karsenti *et al.*, 1977); smooth muscle autoantibodies, including antiactin IgGs in human patients with chronic active hepatitis (Gabbiani *et al.*, 1973); and autoantibody to kinetochores in human scleroderma sera (Moroi *et al.*, 1980)]. Because autoantibodies may be more widespread than previously thought, it has also been suggested that not only first but also second antibodies should be affinity purified or preabsorbed for immunofluorescence studies to minimize any artifact and background which may be introduced by such molecules (Aubin *et al.*, 1979, 1980a,b). Fourth, fixation artifacts may be detected by comparing the staining patterns observed under a variety of fixation conditions (see, for example, Cande *et al.*, 1977; Herman and Pollard, 1978; Aubin *et al.*, 1979; Section IV,B). Finally, with the availability of new and precise fluorescence filter combinations and highly purified antibodies, double-label experiments have become quite feasible and are now frequently used. They allow the possibility of localizing unambiguously two antigens relative to one another, to help define true locations and some fixation artifacts.

IV. RECENT ADVANCES

A. Microtubules: Centriole Migrations and Spiral Arrays

Much is still unknown about the timing of the breakdown of cytoplasmic microtubules versus the timing of spindle formation and the location and migration of centriole duplexes. Use of the light and electron microscopes has allowed investigation of centriolar ultrastructure and simultaneously the state of the nuclear membrane, chromosomes, and microtubules. Although much has been learned, these techniques have been limited to relatively small numbers of cells, and often it has been difficult even from serial sections to get a good detailed overview of the entire cell (see, for example, Roos, 1973; Pickett-Heaps, 1975; Berns *et al.*, 1977; Zeligs and Wollman, 1979). The advantages of using immunofluorescence microscopy to study centriole locations in large numbers of cells have been suggested in studies employing both tubulin antibody (Osborn and Weber, 1976) and some autoimmune rabbit sera (Connolly and Kalnins, 1979). Specific tubulin antibodies have been used to quantify the percentages of cells showing particular and variable locations of centriole duplexes in interphase and early in mitosis from prophase to metaphase (Aubin *et al.*, 1980a). Besides the expected migration of duplexes around the nucleus, centriole duplexes sharing no apparent association with the nucleus also formed presumptive spindle poles with the establishment of prometaphase spindles. Cytoplasmic microtubules were monitored at the same time to ascertain the timing of various stages of centriole movement with the appearance of astral microtubules and the disappearance of cytoplasmic microtubules. Such a rapid technique for visualization of both the centrioles and their milieu among the depolymerizing cytoplasmic microtubules is excellent for quantitation and for the general definition of centriole migration, and it suggests a number of further detailed studies to understand better various molecular events accompanying mitosis.

Another use of specific tubulin antibodies was reported by Harris *et al.* (1980a,b; see also Chapter 2). Studying the first-division cycle of sea urchin eggs (*Stronglycentrotus purpuratus*), Harris and co-workers looked at all stages of the cycle, from sperm entry and fusion of the pronuclei up to prophase and through the first division. Microtubule-stabilizing fixation buffers were used as well as affinity-purified first and second antibodies to reduce the nonspecific fluorescent background. A beautiful transient spiral cortical array of microtubules was identified, the dynamics of which were followed up to the streak stage (Harris *et al.*, 1980a). The breakdown of the monaster, the interphase asters, and the asters of the mitotic spindle proceeded from the cell center to the cell periphery and was followed by the growth of new asters, also proceeding outward (Harris *et al.*, 1980b). Harris and co-workers, analyzing this pattern, have suggested that there exists a transient, wave-like movement of some condition that favors microtubule depolymerization (see Harris, Chapter 2, this volume). These studies reflect the still exciting discoveries to be made using tubulin antibodies (and, it is hoped, other antibodies as well) to reveal dynamic rearrangements which despite years of light and electron microscopy had previously been unknown and/or inaccessible by traditional approaches.

B. Microfilaments: Lack of Evidence for Increased Staining in the Spindle Region

The distributions of actin, myosin, and tropomyosin were studied again (Aubin *et al.*, 1979a) in an attempt to clarify some of the discrepancies outlined above. PtK2 cells were labeled with affinity-purified first antibodies and fluorescent second antibodies preabsorbed on fixed PtK2 cells. Eight different commonly used fixation and extraction procedures, including glycerination and cell lysis conditions, were compared. A number of observations were made which had not been reported previously. Microfilament bundles clearly disappeared as prophase progressed, but rather than the uniformly diffuse staining reported previously, some microfilamentous staining was detectable in the diffuse cytoplasmic background during mitosis. Fluorescein–HMM, two different actin antibodies, and myosin and tropomyosin antibodies, while not excluded from the spindle region, gave no evidence of increased labeling of the spindle region over cytoplasmic background (Fig. 1). All these antibodies and fluorescein–HMM did stain the cleavage furrow and contractile rings very brightly (Fig. 2), in agreement with all other reports (Cande *et al.*, 1977; Schloss *et al.*, 1977; Sanger and Sanger, 1979), with one exception (Herman and Pollard, 1978, 1979). The reasons spindle actin or myosin were not found at increased levels in the spindle in these studies, in contrast to previous reports (see Table I and Section IIB), are discussed in detail in Aubin *et al.* (1979a; see also Section V). It should be noted that Izutsu *et al.* (1979) have reported that they too find no evidence for actin in the spindle at levels greater than the cytoplasmic levels. It is possible that at least some of the staining observed previously may have been the result of non-specific binding of IgGs in only partially purified antibody preparations. Similarly, prior to purification on F-actin, some preparations of fluorescein–HMM, even after removal of overconjugated HMM on DEAE–cellulose, have been found to bind nonspecifically (Aubin *et al.*, 1979). It is still possible that the small numbers of microfilaments seen in the mitotic spindle in some electron microscopic studies (see earlier references) would not be identified by immunofluorescence techniques over the general cytoplasmic fluorescence. However, these more recent results do argue against models in which, on the basis of fluorescence alone, microfilaments are considered to be major structural components of the spindle.

A further cautionary note arose from this study. Several previous immunofluorescence and electron microscopic studies using HMM or S1 have

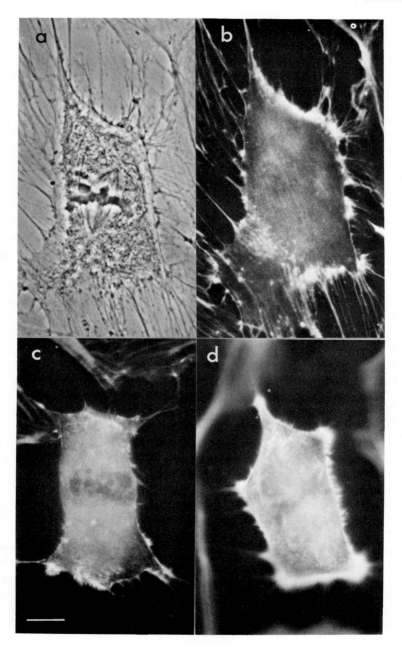

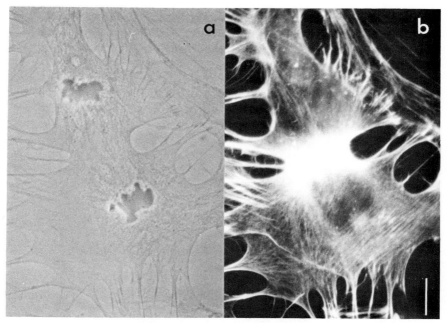

Fig. 2. (a) Phase and (b) fluorescence micrographs of a PtK2 cel in telophase. The cell was fixed in −10°C methanol for 5 minutes and stained with affinity-purified actin antibodies, as described in Fig. 1. The contractile ring region stains intensely over the general cytoplasmic fluorescence. Bar = 9 μm.

relied on glycerination to render cells permeable to the macromolecules. It has been suspected that this procedure might allow loss and rearrangement of cytoskeletal proteins, and glycerination procedures have been shortened, where possible, to minimize this action. By comparing the staining patterns observed under a number of fixations, Aubin *et al.* (1979a) found that glycerination did indeed seem to alter the distributions of myosin (for exam-

---

Fig. 1. PtK2 cells in metaphase. (a) Phase and (b) fluorescence micrographs of a PtK2 cell in metaphase. The cell was fixed in 3.7% formalin, followed by −10°C methanol, and was stained with affinity-purified rabbit antibody to chicken gizzard actin. The second antibody was fluorescent goat-antirabbit IgGs preabsorbed on PtK2 cells. Note that no increase over general cytoplasmic fluorescence is detected in the spindle region, which is clearly visible in phase. (c,d) Fluorescence micrographs of PtK2 cells in metaphase. The cells were fixed as above; (c) was stained with affinity-purified rabbit antibody to chicken gizzard myosin, and (d) was stained with affinity-purified rabbit antibody to bovine brain tropomyosin. The second antibodies were as above. Note: no increased intensity of staining in the spindle region in either case. Further details in Aubin *et al.* (1979). Bar = 10 μm.

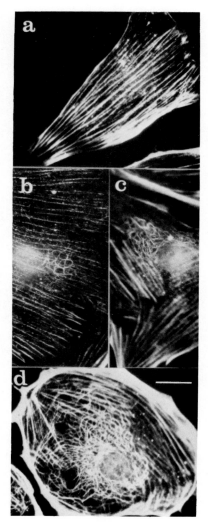

Fig. 3. Fluorescence micrographs of PtK2 cells glycerinated in 50% v/v glycerol/ standard salt solution at 4°C for the times indicated, followed by fixation in 3.7% formalin and staining with rabbit antibody to tropomyosin and fluorescent goat-antirabbit IgGs, as in Fig. 1d. (a) Glycerination for 8 hours. The cytoplasmic microfilament bundles are stained brightly in a pattern characteristic of tropomyosin. (b,c) Glycerination for 12 hours. Small areas of the cytoplasm reveal altered tropomyosin staining, with whorls and wavy filaments visible. (d) Glycerination for 24 hours. Large areas of the cytoplasm reveal tropomyosin antibody staining in thick, wavy filaments. See Aubin *et al.* (1979) for further discussion. Bar = 15 μm. Reprinted with permission from Aubin *et al.* (1979).

ple, it could no longer be found as a major component in the cleavage furrow), tropomyosin (normal stress fiber staining became aberrant, revealing wavy filaments quite unlike stress fibers) (Fig. 3), and nonimmune IgGs (which could be detected in increased amounts in the spindle region).

C. Intermediate Filaments: Different Filaments, Different Functions

Synchronized PtK2 cells were used in one study to look in more detail at the distribution of intermediate filaments during mitosis (Aubin *et al.*, 1979b; 1980b). Since this cell type expresses simultaneously both the vimentin type and the cytokeratin type of intermediate filaments (Franke *et al.*, 1978; Osborn *et al.*, 1980), it was thought that a direct comparison might elucidate possible divergent behavior and different functions of the two filament systems during mitosis. Furthermore, unlike the perinuclear bundle of vimentin filaments present even before mitosis in vascular endothelial cells, the vimentin filaments of PtK2 and numerous other cells form an extensive cytoplasmic array in interphase. We wanted to describe the sequence of rearrangements leading to the closed circular bundle evident by metaphase in these latter cell types. With respect to the latter objective, several new and striking features were noted for the vimentin-type filaments, including gradual formation of vimentin bundles around the nucleus and the migrating centrioles during prophase. The most unexpected feature occurred in prometaphase when a transient cage-like structure of vimentin fibers formed to surround the developing spindle (Fig. 4). The spindle poles formed the foci for this cage, with vimentin fibers swirling around the polar regions and chromosomes and the spindle microtubules contained within the body of the cage.

Zieve *et al.* (1980) subsequently also reported a filamentous cage around the mitotic apparatus in detergent-extracted CHO and HeLa cells examined ultrastructurally. When these investigators used the intermediate filament antiserum of Starger *et al.* (1978) and the anti-58K serum of Hynes and Destree (1978) (see also Section IIC) in indirect immunofluorescence studies, the detergent-extracted mitotic cells appeared to fluoresce. However no filamentous staining or details of a fibrous cage could be seen. A possible association of vimentin filaments with centriolar regions as detected in these mitotic cells by Aubin *et al.* (1979b; 1980b) is of interest. Such an association had been noted once at the ultrastructural level (Sandborn, 1970), and centrioles were found to be enriched in fractions of purified BHK cell intermediate filaments (Starger *et al.*, 1978). However, the general association of vimentin filaments and centrioles had not been described previously at the immunofluorescence level or as a general feature in large

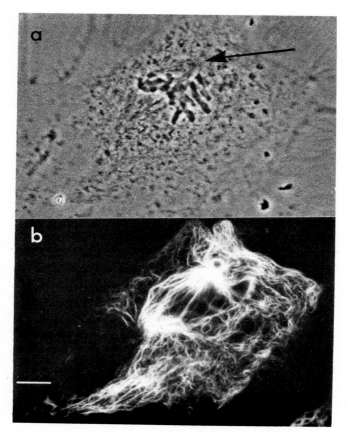

Fig. 4. (a) Phase and (b) fluorescence micrographs of a PtK2 cell in prometaphase. The cell was fixed in −10°C methanol and stained with affinity-purified guinea pig antibody to vimentin followed by preabsorbed fluorescent-goat-antiguinea pig IgGs. The "cage" surrounding the chromosomes and developing spindle is clearly visible. Note also the swirls of vimentin bundles around the centriolar region (e.g., upper pole marked by arrow). Bar = 7.5 μm. Reprinted with permission from Aubin *et al.* (1980b).

numbers of cells and may be interesting and important to investigate further. Cytokeratin filaments, which had not been described previously during mitosis, appeared to enjoy a more "passive" mitotic role and did not rearrange so drastically; they did not form the very thick perinuclear bundles characteristic of the vimentin-type filaments, and remained as an extensive array on the substrate-attached side of the cell and in the periphery of the cell body (Aubin *et al.*, 1979b; 1980b). Neither filament was found in the spindle itself. That both the vimentin-type and cytokeratin-type filaments did rearrange during mitosis, but did so in very different manners, suggests that the two systems are capable of forming different structures and may play different roles in mitosis. Our data have suggested that the cytokeratin filaments might continue to perform a skeletal function during mitosis, while the vimentin filaments probably do not. That vimentin filaments may perform another role, such as providing structural information to microtubules and chromosomes, is a possibility consistent with their localization (Blose, 1979; Aubin *et al.*, 1980b).

Much more can be learned about these filament systems by immuno-fluorescence. For example, a number of double-label experiments with specific vimentin and actin antibodies were done to attempt to define a mechanism for the intriguing manner by which the vimentin filaments might cleave to distribute into daughter cells (Aubin *et al.*, 1980b). Although the "break" of the vimentin filaments was found to coincide frequently with contractile ring and cleavage furrow formation, they would seem not to be causally related phenomena, since intact vimentin filament bundles were frequently seen after contractile ring formation. Consistent with this, cytochalasin B did not always inhibit the partition of these filaments in PtK2 cells, as had been reported previously (Blose, 1979). A partition mechanism based on enzymatic digestion, for example, employing a specific Ca^{2+}-activated protease as described in one report (Nelson and Traub, 1980), is an interesting possibility.

D. Accessory and Regulatory Proteins

I will discuss only the three classes of accessory and/or regulatory proteins which have been studied by immunofluorescence. The explosion of information on other regulatory proteins (e.g., sol–gel regulators of F-actin) makes it important to study these proteins in mitosis and cytokinesis as well.

1. Microtubule-Associated Proteins

The dyneins are a series of high-molecular-weight (HMW) proteins (Fronk *et al.*, 1973) with Mg^{2+}-dependent ATPase activities (Gibbons and Fronk, 1972) localized at the arms of the outer doublet microtubules of cilia and flagella (see Gibbons, 1975). The identification of the dynein–microtubule system as a mechanism for cilia and flagella movement (for example, see Gibbons, and Gibbons, 1974; Gibbons, 1975; Warner and Mitchell, 1980) lends support to speculation regarding similar mechanisms for chromosome movement. Dynein-like arms were postulated by McIntosh *et al.* (1969) in the sliding filament model of mitosis. Arm-like projections have been described on spindle microtubules (for example, see McIntosh, 1974; McIntosh *et al.*, 1975; Fuge, 1977). Antibodies prepared to a tryptic peptide (fragment

A) of dynein inhibit dynein ATPase activity (Ogawa and Mohri, 1975) and brightly stain the mitotic spindles in sea urchin eggs (Mohri et al., 1976) and a variety of other cell types (Izutsu et al., 1979).

Many microtubules other than those in cilia and flagella appear to be organized by intertubular bridges and/or to have projecting side arms (for a review, see Murphy and Borisy, 1975; Dentler et al., 1975; Snyder and McIntosh, 1976; Amos, 1977). HMW (ca. 300,000 molecular weight) microtubule-associated proteins (MAPs; MAP1 and MAP2) copurify with tubulin in in vitro assembly–disassembly procedures and stimulate the in vitro polymerization of tubulin (Murphy and Borisy, 1975; Sloboda et al., 1976). These MAPs, in particular MAP2, appear to form periodic side arms on microtubules assembled in vitro (Murphy and Borisy, 1975; Dentler et al., 1975; Amos, 1977; Herzog and Weber, 1978; Kim et al., 1979; Zingsheim et al., 1979). Although their function in vivo is not understood, including any role for them in in vivo tubulin assembly, they appear likely candidates for possible cross-linking proteins (for example, see Goldman and Knipe, 1972; Smith et al., 1977; Heggeness et al., 1978). Immunofluorescence localization of MAPs in vivo has supported the in vitro ultrastructural and biochemical suggestion that MAPs are found in association with microtubules in interphase cells (Sherline and Schiavone, 1977; Connolly et al., 1978). When used on mitotic cells, these antibodies to HMW MAPs were shown to stain cells brightly in a manner analogous to tubulin antibody staining from prophase through telophase (Sherline and Schiavone, 1978; Connolly et al., 1978).

Other microtubule accessory proteins, found to copurify with tubulin and to stimulate the polymerization of purified tubulin in vitro, are a set of four polypeptides of lower molecular weight (ca. 70,000 molecular weight) called "tau" (Weingarten et al., 1975; Cleveland et al., 1977). As with HMW MAPs, the in vivo function of tau is not known, but immunofluorescence studies have indicated that both antiserum to electrophoretically purified peptides (Connolly et al., 1977) and affinity-purified antibodies to one of the peptides (Lockwood, 1978) stain cytoplasmic microtubules in a manner similar to tubulin antibody staining. Connolly et al. (1977) found bright staining of the metaphase spindle with tau antiserum and, in a more extensive analysis of mitosis, Lockwood (1978) saw labeling with tau antibody which appeared analogous to that seen with tubulin antibody from early prophase through telophase.

It remains to be seen if HMW MAPs and tau function as cross-bridges and/or microtubule assembly regulators in vivo. That they are both found associated with mitotic spindle microtubules has led to further speculation regarding models for microtubule–microtubule interactions or microtubule interactions with other fibrous elements in the movement of chromosomes.

More precise definition of which MAPs and tau peptides are important is necessary, however.

2. Calcium- Dependent Regulator Protein

Calcium regulation of cell cycle events will be covered in detail in other chapters in this volume (see Harris, Chapter 2, and Nagle and Egrie, Chapter 15). To introduce the mitosis data accumulated from immunofluorescence studies, however, it is useful to note that a Ca^{2+}-dependent regulatory protein of cyclic nucleotide phosphodiesterase (calcium-dependent regulator or CDR protein, hereafter termed "calmodulin") has been implicated in a variety of cellular functions, including motility (for reviews, see Cheung, 1980, Klee et al., 1980; specifically in relation to cytoskeleton, see Dedman et al., 1979; Means and Dedman, 1980). Affinity-purified antibodies to calmodulin have been prepared and used to localize calmodulin in interphase and mitotic cells. Anderson et al. (1978) and Dedman et al. (1978, 1979) studied the distribution of calmodulin antibody fluorescence through mitosis. Calmodulin localized in polar regions and along the part of the spindle fibers close to the pole, as well as in parts of the cytoplasm close to the intercellular bridge. Labeling of similar cells with tubulin antibodies showed that staining by antibodies to calmodulin and to tubulin were similar but not identical.

These reports are exciting, but they do not yet support unambiguously any particular mechanism of spindle dynamics or chromosome movement. That calmodulin may be so specifically localized in high concentrations in the mitotic spindle region suggests that it may be performing a specific Ca^{2+}-binding function there, to control the Ca^{2+} concentration and regulate other agents such as cyclic nucleotide phosphodiesterase. Calmodulin shares an extensive homology with troponin-C (Amphlett ét al., 1976; Dedman et al., 1977; Cheung, 1980; Means and Dedman, 1980), has been shown to regulate myosin-linked ATPase of muscle and non muscle cells by activating the myosin light chain kinase (Dabrowska et al., 1978; Hathaway et al., 1978), and has been reported to localize along microfilament bundles in interphase cells (Dedman et al., 1978), suggesting that it could act on the actomyosin microfilament system. Alternatively, it has been proposed that calmodulin might regulate microtubule assembly–disassembly, as some in vitro polymerization experiments have suggested (Marcum et al., 1978; Kumagai and Nishida, 1979).

3. Cyclic Nucleotide-Dependent Protein Kinases

Cyclic nucleotides have been suggested to be involved in the regulation of microtubule assembly and in some rearrangements of cytoskeletal components in certain morphological changes occurring in fibroblasts (Willingham and Pastan, 1975; Pastan and Willingham, 1978). There is considerable evidence indicating that cyclic nucleotide levels fluctuate during the cell cycle. For instance, in Novikoff hepatoma cells, cGMP levels rise while cAMP levels fall as the cells enter mitosis and the pattern reverses as the cells progress out of metaphase (Zeilig and Goldberg, 1977). The only known mechanism of cyclic nucleotides involves the activation of specific protein kinases, but the specific target proteins of these kinases are not known (Greengard, 1978). However, it has been shown that certain of the microtubule-associated proteins are phosphorylated specifically *in vitro* by cAMP kinases (Sloboda *et al.* 1975; Sandoval and Cuatrecasus, 1976; Cleveland *et al.* 1977) and that a cAMP kinase copurifies with tubulin from rat brain (Sandoval and Cuatrecasus, 1976). Antibodies to a number of cyclic nucleotides and cyclic nucleotide-dependent protein kinases have been analysed during mitosis. Using immunofluorescence, Browne *et al.* (1980) reported that the R^{II} of cAMP-kinase and the cGMP-kinase holoenzyme are associated with the mitotic spindle of PtK_1 cells with a staining pattern closely similar to tubulin. By following the distributions of these molecules during the stages of mitosis, these authors suggested their association with the microtubules rather than the actin microfilaments especially since they associated with the intercellular bridge but not the contractile ring. Antibodies to cAMP, cGMP, and the R^I and catalytic subunits of the cAMP-dependent protein kinase did not label either the spindle or the intercellular bridge area. Browne *et al.* are careful to point out, however, that these latter molecules might be associated with the spindle also, but may be lost during fixation since many of these molecules may be soluble and extracted during the preparative stages. Alternatively, their antigenic determinants may be unavailable to the antibodies when they are bound to their substrates. Interestingly, only a fraction of the metaphase and anaphase spindles stained with the R^{II} and cGMP-kinase antibodies suggesting either again a difficulty with the accessibility of the antigens or perhaps a very transitory association of the molecules with spindle components. The presence of cGMP-kinase and the R^{II} of cGMP kinase on the spindle suggests that a specific protein phosphorylation step may be important in spindle function. The simultaneous localization of calmodulin in the spindle region suggests that the two systems may interact in the control of mitosis (see Browne *et al.*, 1980). What the particular target proteins are, i.e., whether they are specifically microtubules, awaits further evidence from a variety of sources.

V. CONCLUSION

Final mechanisms for chromosome movement and cell division will be known only when the results of all the techniques are well understood. I have not been able to include here data from other techniques which lend support to immunofluorescence results; other chapters in this book will be helpful in this regard. To conclude here, I would like to point out some of the technical problems which remain and some current directions in which I believe immunofluorescence approaches are going.

By comparing simultaneously the localization of more than one component of the cytoskeleton, one may find similarities or differences which may suggest functional differences and/or dynamic and regulatory mechanisms. Examples of this were given earlier; a number of other interesting examples have been documented. For example, in Aubin *et al.* (1979a,b), the distributions of actin, myosin, and tropomyosin were found to be distinctly different in the intercellular bridge in late telophase–early G_1. Similarly, Fujiwara *et al.* (1978) compared myosin and α-actinin antibody staining and found that these molecules also are visualized differently in that region. Yet all these microfilament-associated proteins have been found in the cleavage furrow and contractile rings (see references in Sections II,B, and IV,B). These observations lead to intriguing speculations about which of these proteins may occur together at certain times, for certain functions, and may imply separate kinetic or regulatory mechanisms for their distributions. Such double-label and comparative studies have also been useful in dissecting out what we believe are functional differences in two simultaneously expressed intermediate filament systems. Without other *in vivo* data on these filament systems, e.g., without specific inhibitors (see, however, a report that phorbol-12-myristate-13-acetate may have a selective effect on intermediate filaments: Croop *et al.*, 1980), we could, by comparing the two systems, document clear differences between them during mitosis, implying separate regulatory and functional control. Data such as these are not yet accessible by other techniques.

It has become evident that several fixation–extraction procedures may have to be employed to ascertain the true distributions of various proteins at either the light or electron microscopic level. Results with glycerination protocols have suggested that under some conditions gross rearrangements of some cytoskeletal proteins may occur (Aubin *et al.*, 1979a,b). Connolly and Kalnins (1980) have reported a loss of immunofluorescence staining with antibodies to HMW MAPs, but not with antibodies to tau or tubulin, when triton X-100 lysis of cells precedes formaldehyde fixation. Parallel electron microscopic studies indicated that the side arms associated with the microtubules seemed to be lost by this procedure. Herman and Pollard (1978) and Sanger and Sanger (1980) attempted to ascertain the percentages of cellular actin extracted under various fixation–extraction protocols, and Maupin-Szamier and Pollard (1978) have reported a loss during some fixations of "unstabilized" microfilaments. It is likely that as further studies of this sort are conducted, other examples of preferential loss or rearrangement of particular antigens may be found, so that only by a comparison of several procedures can one hope to define the true localization of certain proteins and possibly classify them into subgroups.

From immunofluorescence studies, it would seem that the question of actin and microfilament-associated proteins in the spindle may still be open. Earlier immunofluorescence data had suggested that actin and myosin were present in high concentrations and possibly were participating in the spindle fibers in the mitotic spindle (see references, Section II,B). More recent analyses have detected no increased levels of these contractile proteins in the spindle, although since the spindle region was not darker than the surrounding cytoplasm, microfilament-associated proteins were not preferentially excluded from that region (see Table I and Section IV,B). Various explanations can be given for these results, but perhaps the most difficult parameter to assess in these fluorescence studies is the possibility that different actins and myosins may exist in the cell and participate in a compartmentalized fashion in specific functions. That more than one actin gene is expressed in cells of various tissues is now well established (for example, see Elzinga and Lu, 1976; Vandekerckhove and Weber, 1978, 1980a for reviews). Firtel *et al.* (1979) have isolated several recombinant plasmids containing sequences of *Dictyostelium discoideum* DNA complementary to actin messenger RNA and have suggested that this unicellular slime mold may have 17 actin genes potentially giving rise to several different actins. Cleveland *et al.* (1980) have similarly estimated a minimum of six to seven actin genes in mammalian cells. Vandekerckhove and Weber (1980b) have found that in vegetative *Dictyostelium* cells, apparently only one amino acid sequence is expressed (minor actins would have to account for less than 10% of the isolated actins). Other minor actins might be expressed in these cells, or at some other specific time in differentiation, which leaves the possibility that there are physiologically different actins for various functions. It is clear that more than one myosin gene exists and different myosins are expressed in different tissues (see Korn, 1978 for a review). Two different myosins appeared to be expressed simultaneously in one cell, *Acanthamoeba* (Pollard *et al.*, 1977), and more recently, evidence has been presented for differential intracellular localizations of the three *Acanthamoeba* myosin isoenzymes (Gadasi and Korn, 1980). Thus, it might be speculated that more specific microfilament antibody reagents might pick out particular functional subclasses of actin or myosin in different cellular locations. Up to now, most antibodies to actin and many to myosin have expressed quite broad species and tissue cross reactivity. However, at the level of a quantitative radioimmunoassay, it has been shown that certain actin antibodies may express tissue specificities, e.g., discriminating skeletal α-actin and cytoplasmic β-

and γ-actins (Morgan *et al.*, 1980). Such specificity is not usually seen in immuno-fluorescence studies (see however, Groeschel-Stewart, 1980). One way to approach this sort of analysis more rigorously would involve the production of monoclonal antibodies (Kohler and Milstein, 1975) to very specific regions of actin, myosin, or other cytoskeletal molecules, some perhaps as yet unidentified, and to use such antibodies in immunofluorescence and other studies.

Finally, the immunofluorescence approach is being expanded by the application of microinjection and image intensification analysis for both antibodies and labeled proteins (for reviews, see Taylor and Wang, 1980). The microinjection of cytoskeletal antibodies as a way of analyzing and/or altering cytoskeleton function has been used a number of times in the past. I will summarize here only studies associated with mitosis. Mabuchi and Okuno (1977) microinjected antimyosin into living starfish blastomeres; nuclear division was normally not inhibited, although in some experiments the size of the spindle appeared to be reduced. However, cytokinesis was inhibited, consistent with the immunofluorescence results which have localized myosin in the cleavage furrow and contractile ring. Rungger *et al.* (1979) have reported that actin antibodies microinjected into *Xenopus* oocytes block chromosome condensation. Labeled proteins are also being introduced into living cells. Fluorescently labeled α-actinin was microinjected into fibroblasts (Feramisco, 1979) and was found to accumulate to high concentrations in the cleavage furrow, among other locations, as traditional immunofluorescence had shown earlier (Fujiwara *et al.*, 1978). Wang and Taylor (1979) have injected labeled actin into living sea urchin eggs during early development. Specific increases in actin-associated fluorescence appeared to occur transiently in the cortex of the fertilized egg; no increase over general cytoplasmic fluorescence was found in the cleavage furrow, although brightly fluorescent filaments running perpendicular to the furrows were seen. The spindle region stained brightly with labeled actin; however, it also fluoresced brightly with the control fluorescently labeled ovalbumin. Wang and Taylor (1979) attempted to explain some of these findings by analyzing cell volumes and the varying path lengths over which fluorescence is measured in different parts of the cell; these technicalities may be less problematic and interpretation easier in flatter cell types.

Clearly, much more work and analyses remain to be done on these new and rather exciting techniques. They do offer the potential of combining visual analysis, as in traditional immunofluorescence, with dynamic and functional analysis, as in inhibition of particular steps of mitosis with highly specific antibodies. Five to six years ago, immunofluorescence gave us a new and stimulating glimpse of cytoskeletal architecture in non-muscle cells. Since then, it has greatly increased our understanding of localization and mechanism by allowing us to visualize the cytoskeletal components during various cellular functions. This approach, refined and extended in new directions, is bound to enhance our knowledge of mitosis in the future.

ACKNOWLEDGMENTS

I would like to thank Dr. A. O. Jorgensen for her helpful suggestions and discussions regarding this manuscript and Dr. V. I. Kalnins for sharing information on tau and MAP antibodies before its publication. Very special thanks are extended to Drs. K. Weber and M. Osborn, in whose laboratories I spent a year as a Postdoctoral Fellow of the Medical Research Council of Canada, and with whom I enjoyed many lively and exciting discussions. I thank Elisa Krissilas for cheerful and excellent typing assistance.

REFERENCES

Amos, L. A. (1977). Arrangement of HMW-associated proteins on purified mammalian brain microtubules. *J. Cell Biol.* **72**, 642–654.

Amphlett, G. W., Vanaman, T. C., and Perry, S. V. (1976). Effect of the troponin C-like protein from bovine brain (brain modulator protein) on the MG^{2+}-stimulated ATPase of skeletal muscle actomyosin. *FEBS Lett.* **72**, 163–168.

Anderson, B., Osborn, M., and Weber, K. (1978). Specific visualization of the distribution of the calcium dependent regulatory protein of cyclic nucleotide phosphodiesterase (modulator protein) in tissue culture cells by immunofluorescence microscopy mitosis and intracellular bridge. *Cytobiologie* **17**, 354–364.

Aronson, J. F. (1965). The use of fluorescein heavy meromyosin for the cytological demonstration of actin. *J. Cell Biol.* **26**, 293–298.

Aubin, J. E., Subrahmanyan, L., Kalnins, V. I., and Ling, V. (1976). Antisera against electrophoretically purified tubulin stimulate colchicine-binding activity. *Proc. Natl. Acad. Sci. U.S.A.* **73**, 1246–1249.

Aubin, J. E., Osborn, M., and Weber, K. (1979a). Analysis of actin and microfilament-associated proteins in the mitotic spindle and cleavage furrow of PtK2 cells by immunofluorescence microscopy. *Exp. Cell Res.* **124**, 93–109.

Aubin, J. E., Osborn, M., and Weber, K. (1979b). The distribution of microfilament-associated proteins and intermediate filament proteins in mitotic PtK2 cells. *J. Cell Biol.* **83**, 370a.

Aubin, J. E., Osborn, M., and Weber K. (1980a). Variations in the distribution and migration of centriole duplexes in mitotic PtK2 cells studied by immunofluorescence microscopy. *J. Cell Sci.* **43**, 177–194.

Aubin, J. E., Osborn, M., Franke, W. W., and Weber, K. (1980b). Intermediate filaments of the vimentin-type and the cytokeratin-type are distributed differently during mitosis. *Exp. Cell Res.* **129**, 149–165.

Behnke, O., Forer, A., and Emmerson, J. (1971). Actin in sperm tails and meiotic spindles. *Nature (London)* **234**, 408–410.

Bennett, G. S., Fellini, S. A., Croop, J., Otto, J. J., Bryan, J., and Holtzer, H. (1978). Differences among 100-A filament subunits from different cell types. *Proc. Natl. Acad. Sci. U.S.A.* **75**, 4364–4368.

Berns, M. W., Rattner, J. B., Brener, S., and Meredith, S. (1977). The role of the centriolar region in animal cell mitosis: A laser microbeam study. *J. Cell Biol.* **79**, 526–532.

Blose, S. H., and Chacko, S. (1976). Rings of intermediate (100 A) filament bundles in the perinuclear region of vascular endothelial cells: Their mobilization by colcemid and mitosis. *J. Cell Biol.* **70,** 459–466.

Blose, S. H. (1979). Sen-nanometer filaments and mitosis: Maintenance of structural continuity in dividing endothelial cells. *Proc. Natl. Acad. Sci. U.S.A.* **76,** 3372–3376.

Brinkley, B. R., Fuller, G. M., and Highfield, D. P. (1976). Tubulin antibodies as probes for microtubules in dividing and nondividing mammalian cells. *In* "Cell Motility" (R. Goldman, T. Pollard, and J. Rosenbaum, eds.), pp. 435–456. Cold Spring Harbor Lab., Cold Spring Harbor, New York.

Brown, S., Levinson, W. and Spudich, J. A. (1976). Cytoskeletal elements of chick embryo fibroblasts revealed by detergent extraction. *J. Supramol. Structure* **5,** 119–128.

Browne, C. L., Lockwood, A. H., Su, J.-L., Blano, J. A. and Steiner, A. L. (1980). Immunofluorescent localization of cyclic nucleotide-dependent protein kinases on the mitotic apparatus of cultured cells. *J. Cell Biol.* **87,** 336–345.

Bryan, J. (1972). Vinblastine and microtubules. II. Characterization of two protein subunits from the isolated crystals. *J. Mol. Biol.* **66,** 157–168.

Cande, W. F., Lazarides, E., and McIntosh, J. R. (1977). A comparison of the distribution of actin and tubulin in the mammalian spindle as seen by indirect immunofluorescence. *J. Cell Biol.* **72,** 552–567.

Cheung, W. Y. (1980). Calmodulin plays a pivotal role in cellular regulation. *Science* **207,** 19–27.

Cleveland, D. W., Hwo, S.-Y., and Kirschner, M. W. (1977). Purification of tau: A microtubule-associated protein that induces assembly of microtubules from purified tubulin. *J. Mol. Biol.* **116,** 207–226.

Cleveland, D. W., Lopota, M. A., MacDonald, R. J., Cowan, N. J., Rutter, W. J., and Kirschner, M. W. (1980). Number and evolutionary conservation of α- and β-tubulin and cytoplasmic β- and α-actin genes using specific cloned cDNA probes. *Cell* **20,** 95–105.

Connolly, J. A. (1979). Microtubule-associated proteins and microtubule assembly *in vivo.* Ph.D. Thesis, University of Toronto, Toronto, Canada.

Connolly, J. A., and Kalnins, V. I. (1979). Visualization of centrioles and basal bodies by fluorescent staining with nonimmune rabbit sera. *J. Cell Biol.* **79,** 526–532.

Connolly, J. A., and Kalnins, V. I. (1980). Effect of triton X-100 on the distribution of microtubule-associated proteins *in vivo. Eur. J. Cell Biol.* **21,** 296–300.

Connolly, J. A., Kalnins, V. I., Cleveland, D. W., and Kirschner, M. W. (1977). Immunofluorescent staining of cytoplasmic and spindle microtubules in mouse fibroblasts with antibody to tau protein. *Proc. Natl. Acad. Sci. U.S.A.* **74,** 2347–2440.

Connolly, J. A., Kalnins, V. I., Cleveland, D. W., and Kirschner, M. W. (1978). Intracellular localization of high molecular weight microtubule accessory protein by indirect immunofluorescence. *J. Cell Biol.* **76,** 781–786.

Croop, J., Toyamar, Y., Dlugosz, A. A., and Holtzer, H. (1980). Selective effects of phorbol 12-myristate 13-acetate on myofibrils and 10-nm filaments. *Proc. Natl. Acad. Sci. U.S.A.* **77,** 5273–5277.

Dabrowska, R., Sherry, J. M. F., Armatorio, D. K., and Hartshorne, D. J. (1978). Modulator protein as a component of the myosin light chain kinase from chicken gizzard. *Biochemistry* **17,** 253–258.

Dales, S. (1972). Concerning the universality of a microtubule antigen in animal cells. *J. Cell Biol.* **52,** 748–754.

Dales, S., Hsu, K. C., and Nagayama, A. (1973). The fine structure and immunological labelling of the achromatic mitotic apparatus after disruption of cell membranes. *J. Cell Biol.* **59,** 643–660.

Davidson, P. F., Hong, B.-S., and Cooke, P. (1977). Classes of distinguishable 10 nm cytoplasmic filaments. *Exp. Cell Res.* **109**, 471–474.

Dedman, J. R., Potter, J. D., and Means, A. R. (1977). Biological cross-reactivity of rat testis phosphodiesterase activator protein and rabbit skeletal troponin C. *J. Biol. Chem.* **252**, 2437–2440.

Dedman, J. R., Welsh, M. J. and Means, A. R. (1978). Ca^{2+}-dependent regulator: Production and characterization of a monospecific antibody. *J. Biol. Chem.* **253**, 7515–7521.

Dedman, J. R., Brinkley, B. R., and Means, A. R. (1979). Regulation of microfilaments and microtubules by calcium and cAMP. *Adv. Cyclic Nucleotide Res.* **11**, 131–174.

DeMey, J., Hoebeke, I., de Brabander, M., Geuens, G., and Joniau, M. (1976). Immunoperoxidase visualization of microtubules and microtubular proteins. *Nature (London)* **264**, 273–275.

Dentler, W. L., Granett, S., and Rosenbaum, J. L. (1975). Ultrastructural localization of the high molecular weight proteins associated with *in vitro*-assembled brain microtubules. *J. Cell Biol.* **65**, 237–241.

Elzinga, M., and Lu, R. C. (1976). Comparative amino acid sequence studies on actins. *In* "Contractile Systems in Non-Muscle Tissues" (S. V. Perry, A. Margreth, and R. Adelstein, eds.), pp. 29–37. Elsevier, Amsterdam.

Feramisco, J. R. (1979). Microinjection of fluorescently labeled α-actinin into living fibroblasts. *Proc. Natl. Acad. Sci. U.S.A.* **76**, 3967–3971.

Firtel, R. A., Timm, R., Kimmel, A. R., and McKeown, M. (1979). Unusual nucleotide sequences at the 5' end of actin genes in *Dictyostelium discoideum. Proc. Natl. Acad. Sci. U.S.A.* **76**, 6206–6210.

Forer, A. (1978). Chromosome movements during cell division: Possible involvement of actin filaments. *In* "Nuclear-Division in Fungi" (I. B. Heath, ed.), pp. 21–88. Academic Press, New York.

Forer, A., and Behnke, O. (1972). An actin-like component in spermatocytes of a crane fly (*Nephratoma suturalis* Loew). I. The spindle. *Chromosoma* **39**, 145–173.

Forer, A., and Jackson, W. T. (1976). Actin filaments in the endosperm mitotic spindle in a higher plant, *Haemanthus katherinae* Baker). *Cytobiologie Z. Exp. Zellforsch.* **12**, 199–214.

Forer, A., Jackson, W. T., and Engberg, A. (1979). Actin in spindles of *Haemanthus katherinae* endosperm. II. Distribution of actin in chromosomal spindle fibres, determined by analysis of serial sections. *J. Cell Sci.* **37**, 349–371.

Franke, W. W., Schmid, E., Osborn, M., and Weber, K. (1978). Different intermediate-sized filaments distinguished by immunofluorescence microscopy. *Proc. Natl. Acad. Sci. U.S.A.* **75**, 5034–5038.

Fronk, E., Gibbons, I. R., and Ogawa, K. (1973). Multiple forms of dynein associated with flagellar axonemes from sea urchin sperm. *J. Cell Biol.* **67**, 125a.

Fuge, H. (1977). Ultrastructure of mitotic cells. *In* "Mitosis: Facts and Questions" (M. Little, N. Paweletz, C. Petzelt, H. Ponstingl, D. Schroeter, and H. P. Zimmerman, eds.), pp. 51–77. Springer-Verlag, Berlin and New York.

Fujiwara, K., and Pollard, T. D. (1976). Fluorescent antibody localization of myosin in the cytoplasm, cleavage furrow, and mitotic spindle of human cells. *J. Cell Biol.* **71**, 848–875.

Fujiwara, K., and Pollard, T. D. (1978). Simultaneous localization of myosin and tubulin in human tissue culture cells by double antibody staining. *J. Cell Biol.* **77**, 182–195.

Fujiwara, K., Porter, M. E., and Pollard, T. D. (1978). α-actinin localization in the cleavage furrow during cytokinesis. *J. Cell Biol.* **79**, 268–276.

Fuller, G. M., Brinkley, B. R., and Boughter, J. M. (1975). Immunofluorescence of mitotic spindles by using monospecific antibody against bovine brain tubulin. *Science* **187**, 948–950.

Fulton, C., Kane, R. E., and Stephens, R. E. (1971). Serological similarity of flagellar and mitotic microtubules. *J. Cell Biol.* **50**, 762–773.

Gabbiani, G., Ryan, G. B., Lamelin, J.-P., Vassalli, P., Cruchard, A., and Luscher, E. F. (1973). Human smooth muscle autoantibody. *Am. J. Pathol.* **72**, 473–478.

Gadasi, H., and Korn, E. D. (1980). Evidence for differential intracellular localization of the *Acanthamocha* myosin isoenzymes. *Nature (London)* **286**, 452–456.

Gawadi, N. (1971). Actin in the mitotic spindle. *Nature (London)* **234**, 410.

Gawadi, N. (1974). Characterization and distribution of microfilaments in dividing locus testes cells. *Cytobios* **10**, 17–35.

Gibbons, I. R. (1975). The molecular basis of flagellar motility in sea urchin spermatozoa. *In* "Molecules and Cell Movement" (S. Inone and R. E. Stephens, eds.), pp. 207–232. Raven, New York.

Gibbons, I. R., and Fronk, E. (1972). Some properties of bound and soluble dynein from sea urchin sperm flagella. *J. Cell Biol.* **54**, 365–381.

Gibbons, B. H., and Gibbons, I. R. (1974). Properties of flagellar "rigor waves" produced by abrupt removal of adenosine triphosphate from actively swimming sea urchin sperm. *J. Cell Biol.* **63**, 970–985.

Goldman, R., and Follet, E. (1970). Birefringent filamentous organelle in BHK-21 cells and its possible role in cell spreading and motility. *Science* **169**, 286–288.

Goldman, R. D., and Knipe, D. M. (1972). Functions of cytoplasmic fibers in non-muscle cell motility. *Cold Spring Harbor Symp. Quant. Biol.* **37**, 523–534.

Goldman, R. D., Lazarides, E., Pollack, R., and Weber, K. (1975). The distribution of actin in non-muscle cells. The use of actin antibody in the localization of actin within the microfilament bundles of mouse 3T3 cells. *Exp. Cell Res.* **90**, 333–344.

Gordon, W. E., Bushnell, A., and K. Burridge (1978). Characterization of the intermediate (10 nm) filaments of cultured cells using an autoimmune rabbit antiserum. *Cell* **13**, 249–261.

Gozes, I., Geiger, B., Fuchs, S., and Littauer, U. Z. (1977). Immunochemical determination of tubulin. *FEBS Lett.* **73**, 109–114.

Griffith, L. M., and Pollard, T. D. (1978). Evidence for actin filament-microtubule interaction mediated by microtubule-associated proteins. *J. Cell Biol.* **78**, 958–965.

Groeschel-Stewart, U. (1980). Immunochemistry of Cytoplasmic Contractile Proteins. *Int. Rev. Cytol.* **65**, 193–255.

Harris, P., Osborn, M., and Weber, K. (1980a). A spiral array of microtubules in the fertilized sea urchin egg cortex examined by indirect immunofluorescence and electron microscopy. *Exp. Cell Res.* **126**, 227–236.

Harris, P., Osborn, M., and Weber, K. (1980b). Distribution of tubulin-containing structures in the egg of the sea urchin *Strongylocentrotus purpuratus* from fertilization through first cleavage. *J. Cell Biol.* **84**, 668–679.

Hathaway, D. R., Sobieszek, A., Eaton, C. R., Adelstein, R. S. (1978). Regulation of platelet and smooth muscle myosin kinase activity. *Fed. Proc. Fed. Am. Soc. Exp. Biol.* **37**, 1328.

Heggeness, M. H., Simon, M., and Singer, S. J. (1978). Association of mitochondria with microtubules in cultured cells. *Proc. Natl. Acad. Sci. U.S.A.* **75**, 3863–3866.

Hellewell, S. B., and Taylor, D. L. (1979). The contractile basis of amoeboid movement. VI. The solation-gelation coupling hypothesis. *J. Cell Biol.* **83**, 633–648.

Henderson, D., and Weber, K. (1979). Three-dimensional organization of microfilaments and microtubules in the cytoskeleton. Immunoperoxidase labelling and stereo-electron microscopy of detergent-extracted cells. *Exp. Cell Res.* **124**, 301–316.

Herman, I. M., and Pollard, T. D. (1978). Actin localization in fixed dividing cells stained with fluorescent heavy meromyosin. *Exp. Cell Res.* **114**, 15–25.

Herman, I. M., and Pollard, T. D. (1979). Comparison of purified actin-actin and fluorescent heavy micromyosin staining patterns in dividing cells. *J. Cell Biol.* **80**, 509–520.

Herzog, W., and Weber, K. (1978). Fractionation of brain microtubule-associated proteins. Isolation of two different proteins which stimulate tubulin polymerization *in vitro. Eur. J. Biochem.* **92**, 1–8.

Hiller, G., and Weber, K. (1978). Radioimmunoassay for tubulin: A quantitative comparison of the tubulin content of different established tissue culture cells and tissues. *Cell* **14**, 795–804.

Hinkley, R., and Telser, A. (1974). Heavy meromyosin-binding filaments in the mitotic apparatus of mammalian cells. *Exp. Cell Res.* **86**, 161–164.

Huxley, H. E. (1973). Muscular contraction and cell motility. *Nature (London)* **243**, 445–449.

Huxley, H. E. (1976). Introductory remarks: The relevance of studies on muscle to problems of cell motility. *In* "Cell Motility" (R. Goldman, T. Pollard, and J. Rosenbaum, eds.), pp. 115–126. Cold Spring Harbor Lab., Cold Spring Harbor, New York.

Haynes, R. D., and A. T. Destree (1978). 10 nm filaments in normal and transformed cells. *Cell* **13**, 151–163.

Inoué, S., and Stephens, R. E. (1975). "Molecules and Cell Movement." Raven Press, New York.

Isenberg, G., and Small, J. V. (1978). Filamentous actin, 100 A filaments and microtubules in neuroblastoma cells: Their distribution in relation to sites of movement and neuronal transport. *Cytobiologie Z. Exp. Zellforsch.* **16**, 326–344.

Ishikawa, H., Bischoff, R., and Holtzer, H. (1968). Mitosis and intermediate-sized filaments in developing skeletal muscle. *J. Cell Biol* **38**, 538–555.

Ishikawa, H., Bischoff, R., and Holtzer, H. (1969). Formation of arrowhead complexes with heavy meromyosin in a variety of cell types. *J. Cell Biol* **43**, 312–328.

Izutsu, K., Owaribe, K., Hatano, S., Ogawa, K., Komada, H., and Mohri, H. (1979). Immunofluorescent studies on actin and dynein distributions in mitotic cells. *In* "Cell·Motility: Molecules and Organization" (S. Hatano, H. Ishikawa, and H. Sato, eds.), pp. 621–638. Univ. of Tokyo Press, Tokyo.

Johnson, G. D., Holborow, E. J., and Dorling, J. (1978). Immunofluorescence and immunoenzyme techniques. *In* "Immunocytochemistry". (D. M. Weir, ed.), pp. 15.1–15.30. Blackwell, Oxford.

Joniau, M., de Brabander, M., de Mey, J., and Hoebeke, J. (1977). Quantitative determination of tubulin by radioimmunoassay. *FEBS Lett.* **78**, 307–312.

Kalnins, V. I., and Connolly, J. A. (1981). Application of immunofluorescence to studies of cytoskeletal antigens. *In* "Advances in Cellular Neurobiology" (S. Federoff and L. Hertz, eds.), Vol. 2. Academic Press, New York (in press).

Karsenti, E., Guilbert, B., Bornens, M., and Avrameas, S. (1977). Antibodies to tubulin in normal nonimmunized animals. *Proc. Natl. Acad. Sci. U.S.A.* **74**, 3997–4001.

Kim, H., Binder, L. I., and Rosenbaum, J. L. (1979). The periodic association of MAP$_2$ with brain microtubules *in vitro. J. Cell Biol.* **80**, 266–276.

Klee, C. B., Crouch, T. H., and Richman, P. G. (1980). Calmodulin. *Annu. Rev. Biochem.* **49**, 489–516.

Kohler, G., and Milstein, C. (1975). Continuous cultures of fused cells secreting antibody of predefined specificity. *Nature (London)* **256**, 495–497.

Korn, E. (1978). Biochemistry of actomyosin-dependent cell motility: A Review. *Proc. Natl. Acad. Sci. U.S.A.* **75**, 588–599.

Kumagai, H., and Nishida, E. (1979). The interactions between calcium-dependent regulator protein of cyclic nucleotide phosphodiesterase and microtubule proteins. II. Association of calcium-dependent regulator protein with tubulin dimer. *J. Biochem.* **85**, 1267–1274.

Lazarides, E. (1976). Actin, alpha-actinin and tropomyosin interaction in the structural organization of actin filaments in non-muscle cells. *J. Cell Biol.* **68**, 202–219.

Lazarides, E. (1980). Intermediate filaments as mechanical integrators of cellular space. *Nature (London)* **283**, 249–256.

Lazarides, E., and Balzer, D. R. (1978). Specificity of desmin to avian and mammalian muscle cells. *Cell* **14**, 429–438.

Lazarides, E., and Burridge, K. (1975). Alpha-actinin immunofluorescent localization of a muscle structural protein in non-muscle cells. *Cell* **6**, 289–298.

Lazarides, E., and Weber, K. (1974). Actin antibody: The specific visualization of actin filaments in non-muscle cells. *Proc. Natl. Acad. Sci. U.S.A.* **71**, 2268–2272.

Lehto, V.-P., Virtanen, I. and Kurki, P. (1978). Intermediate filaments anchor the nuclei in nuclear monolayers of cultured human fibroblasts. *Nature (London)* **272**, 175–177.

Lenk, R., Ransom, L., Kaufman, J., and Penman, S. (1977). A cytoskeletal structure with associated polyribosomes obtained from HeLa cells. *Cell* **10**, 67–78.

Little, N., Paweletz, N., Petzelt, C., Ponstingl, H., Schroeter, D., and Zimmerman, H.-P. (1977). "Mitosis Facts and Questions." Springer-Verlag, Berlin and New York.

Lockwood, A. H. (1978). Tubulin assembly protein: Immunochemical and immunofluorescent studies on its function and distribution in microtubules and cultured cells. *Cell* **13**, 613–627.

Mabuchi, I., and Okuno, M. (1977). The effect of myosin antibody on the division of starfish blastomeres. *J. Cell Biol.* **74**, 251–263.

McIntosh, J. R. (1974). Bridges between microtubules. *J. Cell Biol.* **61**, 166–187.

McIntosh, J. R., Hepler, P. K., and Van Wie, D. G. (1969). Model for mitosis. *Nature (London)* **224**, 659–663.

McIntosh, J. R., Cande, W. Z., and Snyder, J. A. (1975). Structure and physiology of the mammalian mitotic spindle. *In* "Molecules and Cell Movement" (S. Inoué and R. E. Stephens, eds.), pp. 31–76. Raven, New York.

McIntosh, J. R., Cande, W. Z., Lazarides, E., McDonald, K., and Snyder, J. A. (1976). Fibrous elements of the mitotic spindle. *In* "Cell Motility" (R. Goldman, T. Pollard, and J. Rosenbaum, eds.), pp. 1261–1272. Cold Spring Harbor Lab., Cold Spring Harbor, New York.

Marcum, J. M., Dedman, J. R., Brinkley, B. R., and Means, A. R. (1978). Control of microtubule assembly-disassembly by calcium-dependent regulatory protein. *Proc. Natl. Acad. Sci. U.S.A.* **75**, 3771–3775.

Maupin-Szamier, P., and Pollard, T. D. (1978). Actin filament destruction by osmium tetroxide. *J. Cell Biol.* **77**, 837–852.

Means, A. R., and Dedman, J. R. (1980). Calmodulin: An intracellular calcium receptor. *Nature (London)* **285**, 73–77.

Mohri, H., Mohri, T., Mabuchi, I., Yazaki, I., Sakai, H., and Ogawa, K. (1976). Localization of dynein in sea urchin eggs during cleavage. *Dev., Growth Differ.* **18**, 391–397.

Morgan, J. L., Rodkey, L. S., and Spooner, B. S. (1977). Quantitation of cytoplasmic tubulin by radioimmunoassay. *Science* **197**, 578–580.

Morgan, J. L., Holladay, C. R., and Spooner, B. R. (1980). Immunological differences between actins from cardiac muscle, skeletal muscle, and brain. *Proc. Natl. Acad. Sci. U.S.A.* **77**, 2069–2073.

Moroi, Y., Peebles, C., Fritzler, M. J., Steigerwald, J., and Tan, E. M. (1980). Autoantibody to centromere (kinetochore) in scleroderma sera. *Proc. Natl. Acad. Sci. U.S.A.* **77**, 1627–1631.

Murphy, D. B., and Borisy, G. G. (1975). Association of high molecular weight proteins with microtubules and their role in microtubule assembly *in vitro*. *Proc. Natl. Acad. Sci. U.S.A.* **72**, 2696–2700.

Nagayama, A., and Dales, S. (1970). Rapid purification and immunological specificity of mammalian microtubular paracrystals posessing an ATPase activity. *Proc. Natl. Acad. Sci. U.S.A.* **66**, 464–471.

Nelson, W. J., and Traub, P. (1980). Characterization of a Ca^{2+}-activated protease specific for intermediate-sized filament protein. *Eur. J. Cell Biol.* **22**, 373.

Ogawa, K., and Mohri, H. (1975). Preparation of antiserum against a tryptic fragment (Fragment A) of dynein and an immunological approach to the subunit composition of dynein. *J. Biol. Chem.* **250**, 6476–6483.

Osborn, M., and Weber, K. (1976). Cytoplasmic microtubules in tissue culture cells appear to grow from an organizing structure towards the plasma membrane. *Proc. Natl. Acad. Sci. U.S.A.* **73**, 867–871.

Osborn, M., and Weber, K. (1977). The display of microtubules in transformed cells. *Cell* **12**, 561–574.

Osborn, M., and Weber, K. (1979). Microfilament-associated proteins in tissue culture cells viewed by stereo immunofluorescence microscopy. *Eur. J. Cell Biol.* **20**, 28–36.

Osborn, M., Franke, W. W., and Weber, K. (1977). Visualization of a system of filaments of 7-10 nm thick in cultured cells of an epithelioid line (PtK2) by immunofluorescence microscopy. *Proc. Natl. Acad. Sci. U.S.A.* **74**, 2490–2494.

Osborn, M., Born, T., Koitsch, H.-J., and Weber, K. (1978). Stereo immunofluorescence microscopy. I. Three-dimensional arrangement of microfilaments, microtubules, and tonofilaments. *Cell* **14**, 477–488.

Osborn, M., Franke, W. W., and Weber, K. (1980). Direct demonstration of the presence of two immunologically distinct intermediate-sized filament systems in the same cell by double immunofluorescence microscopy. Vimentin and cytokeratin fibers in cultured epithelial cells. *Exp. Cell Res.* **125**, 37–46.

Pastan, I., and Willingham, M. (1978). Cellular transformation and the morphological phenotype of transformed cells. *Nature* **274**, 645–650.

Petzelt, C. (1979). Biochemistry of the mitotic spindle. *Int. Rev. Cytol.* **60**, 53–92.

Pickett-Heaps, J. D., (1975). Aspects of spindle evolution. *Ann. N.Y. Acad. Sci.* **253**, 352–361.

Pollard, T. D., Porter, M. E., and Stafford, W. (1977). Characterization of a second form of myosin from *Acanthamoeba. J. Cell Biol.* **75**, 262a.

Pollard, T. D., and Weihing, R. R. (1974). Actin and myosin and cell movement. *Crit. Rev. Biochem.* **2**, 1–65.

Roos, U.-P. (1973). Light and electron microscopy of rat kangaroo cells in mitosis: Formation and breakdown of the mitotic apparatus. *Chromosoma* **40**, 43–82.

Rungger, D., Rungger-Brandle, E., Chaponnier, C., and Gabbiani, G. (1979). Intranuclear injection of anti-actin antibodies into *xenopus* oocytes blocks chromosome condensation. *Nature (London)* **282**, 320–321.

Sandborn, E. (1970). Cells and tissues by light and electron microscopy. p. 62. Academic Press, New York.

Sandoval, I. V., and Cuatrecasus, P. (1976). Protein kinase associated with tubulin: affinity chromatography and properties. *Biochemistry* **15**, 3424–3432.

Sanger, J. W. (1975). Changing patterns of actin localization during cell division. *Proc. Natl. Acad. Sci. U.S.A.* **72**, 1913–1916.

Sanger, J. W. (1977). Nontubulin molecules in the spindle. *In* "Mitosis: Facts and Questions" (M. Little, N. Paweletz, C. Petzelt, H. Ponstingh, D. Schrocter, and H.-P. Zimmerman, eds.), pp. 98–1131. Springer-Verlag, Berlin and New York.

Sanger, J. W., and Sanger, J. M. (1979). The cytoskeleton and cell division. *In* "Methods and Achievements in Experimental Pathology" (G. Gabbiani, ed.), Vol. 8, pp. 110–142. Karger, Basel.

Sanger, J. M., and Sanger, J. W. (1980). Banding and polarity of actin filaments in interphase and cleaving cells. *J. Cell Biol.* **86**, 568–575.

Sato, H., Ohnuki, Y., and Fujiwara, K. (1976). Immunofluorescent anti-tubulin staining of spindle microtubules and critique for the technique. *In* "Cell Motility" (R. Goldman, T. Pollard, and J. Rosenbaum, eds.), pp. 419–433. Cold Spring Harbor Lab., Cold Spring Harbor, New York.

Schloss, J. A., Milsted, A., and Goldman, R. D. (1977). Myosin subfragment binding for the localization of actin-like microfilaments in cultured cells: A light and electron microscope study. *J. Cell Biol.* **74**, 794–815.

Schroeder, T. E. (1975). Dynamics of the contractile ring. *In* "Molecules and Cell Movement" (S. Inoue and R. E. Stephens, eds.), pp. 305–334. Raven, New York.

Schroeder, T. E. (1976). Actin in dividing cells: Evidence for its role in cleavage but not mitosis. *In* "Cell Motility" (R. Goldman, T. Pollard, and J. Rosenbaum, eds.), pp. 265–278. Cold Spring Harbor Lab., Cold Spring Harbor, New York.

Sherline, P., and Schiavone, K. (1977). Immunofluorescence localization of proteins of high molecular weight along intracellular microtubules. *Science* **198**, 1038–1040.

Sherline, P., and Schiavone, K. (1978). High molecular weight MAPs are part of the mitotic spindle. *J. Cell Biol.* **77**, R9–R12.

Sloboda, R. D., Dentler, W. L., and Rosenbaum, J. L. (1976). Microtubule-associated proteins and the stimulation of tubulin assembly *in vitro*. *Biochemistry* **15**, 4497–4505.

Small, J. V., and Celis, J. E. (1978a). Filament arrangements in negatively stained cultured cells: The organization of actin. *Cytobiologie Z. Exp. Zellforsch.* **16**, 308–325.

Small, J. V., and Celis, J. E. (1978b). Direct visualization of the 10-nm (100 A)-filament network in whole and enucleated cultured cells. *J. Cell Sci.* **31**, 393–409.

Smith, D. S., Jarlfors, U., and Cayer, M. L. (1977). Structural cross-bridges between microtubules and mitochondria in central axons of an insect (*Periplaneta americana*). *J. Cell Sci.* **27**, 255–272.

Snyder, J. A., and McIntosh, J. R. (1976). Biochemistry and physiology of microtubules. *Annu. Rev. Biochem.* **45**, 699–720.

Starger, J. M., Brown, W. E., Goldman, A. E., and Goldman, R. D. (1978). Biochemical and immunological analysis of rapidly purified 10-nm filaments from baby hamster kidney (BHK-21) cells. *J. Cell Biol.* **78**, 93–109.

Sternberger, L. A. (1979). Immunofluorescence. *In* "Immunocytochemistry" (Sternberger, L. A., ed.), 2nd ed. pp. 24–58. Wiley, New York.

Stossel, T. P. (1978). Contractile proteins in cell structure and function. *Annu. Rev. Med.* **29**, 427–457.

Taylor, D. L., and Condeelis, J. S. (1979). Cytoplasmic structure and contractility in amoeboid cells. *Int. Rev. Cytol.* **56**, 57–144.

Taylor, D. L., and Wang, Y. L. (1980). Fluorescently labelled molecules as probes of the structure and function of living cells. *Nature (London)* **284**, 405–410.

Tilney, L. G. (1976). Polymerization of actin. III. Aggregates of nonfilamentous actin and its associated proteins: A storage form of actin. *J. Cell Biol.* **69**, 73–89.

Tilney, L. G., Hatano, S. H., Ishikawa, H., and Mooseker, M. S. (1973). The polymerization of actin: Its role in the generation of the acrosomal process of certain echinoderm sperm. *J. Cell Biol.* **59**, 109–126.

Vandekerckhove, J., and Weber, K. (1978). At least six different actins are expressed in a higher mammal: an analysis based on the amino acid sequence of the amino-terminal tryptic peptide. *J. Mol. Biol.* **126**, 783–802.

Vandekerckhove, J., Leavitt, J., Kakunaga, T. and Weber, K. (1980a). Coexpression of a mutant β-actin and the two normal β- and γ-cytoplasmic actins in a stably transformed human cell line. *Cell.* **22**, 893–899.

Vandekerckhove, J., and Weber, K. (1980b). Vegetative *Dictyostelium* cells containing 17 actin genes express a single major actin. *Nature (London)*. **284**, 475–477.

Wang, Y.-L., and Taylor, D. L. (1979). Distribution of fluorescently labelled actin in living sea urchin eggs during early development. *J. Cell Biol.* **82**, 672–679.

Wang, K., Ash, J. F., and Singer, S. J. (1975). Filamin: a new high molecular weight protein found in smooth muscle and non-muscle cells. *Proc. Natl. Acad. Sci. U.S.A.* **72**, 4483–4487.

Warner, F. D., and Mitchell, D. R. (1980). Dynein-the mechanochemical couping adenosine triphosphatase of microtubule-based sliding filament mechanisms. *Int. Rev. Cytol.* **66**, 1–44.

Weber, K. (1976). Visualization of tubulin-containing structures by immunofluorescence microscopy: Cytoplasmic microtubules, mitotic figures, and vinblastine-induced paracrystals. *In* "Cell Motility" (R. Goldman, T. Pollard, and J. Rosenbaum, eds.) pp. 403–417. Cold Spring Harbor Lab., Cold Spring Harbor, New York.

Weber, K., and Groeschel-Stewart, W. (1974). Antibody to myosin: The specific visualization of myosin-containing filaments in nonmuscle cells. *Proc. Natl. Acad. Sci. U.S.A.* **71**, 4561–4564.

Weber, K., and Osborn, M. (1981). The cytoskeleton. *In* "Muscle and Non-Muscle Motility" (A. Stracher, ed.), Academic Press, New York (in press).

Weber, K., Bibring, Th., and Osborn, M. (1975). Specific visualization of tubulin-containing structures in tissue culture cells by immunofluorescence. Cytoplasmic microtubules, vinblastine-induced paracrystals, and mitotic figures. *Exp. Cell Res.* **95**, 111–120.

Weber, K., Wehland, H., and Herzog, W. (1976). Griseofulvin interacts with microtubules both *in vivo* and *in vitro*. *J. Mol. Biol.* **102**, 817–829.

Weber, K., Osborn, M., Franke, W. W., Seib, E., and Scheer, U. (1977). Identification of microtubular structures in diverse plant and animal cells by immunological cross-reaction revealed in immunofluorescence microscopy using antibody against tubulin from porcine brain. *Cytobiologie Z. Exp. Zellforsch.* **15**, 284–302.

Weber, K., Rathke, P. C., and Osborn, M. (1978). Cytoplasmic microtubular images in glutaraldehyde-fixed tissue culture cells by electron microscopy and by immunofluorescence microscopy. *Proc. Natl. Acad. Sci. U.S.A.* **75**, 1820–1824.

Webster, R. E., Osborn, M., and Weber, K. (1978a). Visualization of the same PtK2 cytoskeletons by both immunofluorescence and low power electron microscopy. *Exp. Cell Res.* **117**, 47–61.

Webster, R. E., Henderson, D., Osborn, M., and Weber, K. (1978b). Three-dimensional electron microscopical visualization of the cytoskeleton of animal cells: Immunoferritin identification of actin- and tubulin-containing structures. *Proc. Natl. Acad. Sci. U.S.A.* **75**, 5511–5515.

Weingarten, M. D., Lockwood, A. H., Hwo, S.-Y., and Kirschner, M. W. (1975). A protein factor essential for microtubule assembly. *Proc. Natl. Acad. Sci. U.S.A.* **72**, 1858–1862.

Willingham, M., and Pastan, I. (1975). Cyclic AMP and cell morphology in cultured fibroblasts, effects on cell shape, microfilament and microtubule distribution, and orientation to substratum. *J. Cell Biol.* **67**, 146–159.

Wilson, L., Bryan, J., Ruby, A. and Mazia, D. (1970). Precipitation of proteins by vinblastine and calcium-ions. *Proc. Nat. Acad. Sci. U.S.A.* **66**, 807–812.

Zeilig, C. E., and Goldberg, N. D. 1977. Cell-cyclic related changes of 3′ : 5′-cyclic AMP in Norikoff hepatoma cells. *Proc. Nat. Acad. Sci. U.S.A.* **74**, 1052–1056.

Zieve, G. W., Heidemann, S. R., and McIntosh, J. R. (1980). Isolation and partial characterization of a cage of filaments that surrounds the mammalian mitotic spindle. *J. Cell Biol.* **87**, 160–169.

Zeligs, J. D., and Wollman, S. H. (1979). Mitosis in rat thyroid epithelial cells *in vivo*. II. Centriole and pericentriolar material. *J. Ultrastruct. Res.* **66**, 97–108.

Zingsheim, H.-P., Herzog, W., and Weber, K. (1979). Differences in surface morphology of microtubules reconstituted from pure brain tubulin using two different microtubule-associated proteins: The high molecular weight MAP2 proteins and tau proteins. *Eur. J. Cell Biol.* **19**, 175–183.

11

Mitosis through the Electron Microscope

I. BRENT HEATH

I. INTRODUCTION

The analysis of mitosis has always relied heavily on various microscopic techniques. Thus it was only natural that the introduction of routine electron microscopic techniques should lead to a new burst of work on mitosis. However, it was not until the introduction of glutaraldehyde as a fixative less than 20 years ago that the electron microscope began to yield significant information. The superior resolution of the electron microscope has substantially expanded our knowledge of mitosis and has provided much of the informa-

MITOSIS/CYTOKINESIS

tion upon which most current hypothetical models are based. Unfortunately the techniques used have their limitations, many of which are too rarely acknowledged. The purpose of this chapter is to concentrate on both the unique achievements and the currently recognizable limitations of electron microscopy as applied to the study of mitosis. The chapter is limited in scope to the author's perceptions of the most significant or convincing information. The reader is referred to the more extensive reviews of Luykx (1970), Nicklas (1971), Bajer and Molè-Bajer (1972), Forer (1969, 1974, 1978), Inoué and Ritter (1975), Fuge (1977, 1978), and McIntosh (1979) for a more comprehensive account of various aspects of this chapter. Because this chapter is primarily concerned with the force-generating mechanisms of mitosis, it also includes data from meiotic spindles in which these mechanisms do not appear to differ from those operating during mitosis.

II. TECHNICAL CONSIDERATIONS

Because many of the results obtained with the electron microscope require consideration of a similar set of technical limitations, it is most logical to discuss these first and thus establish the foundations upon which the rest of the work are built. There are four general types of problems involved in the interpretation of electron microscopic data on mitosis: loss of material, relocation of material, loss of three-dimensional data, and lack of qualitative (molecular composition) information.

A. Loss of Material

Loss of cellular material during processing for electron microscopy is a multifaceted problem. At a gross level there is an extensive literature describing the various quantities of material, largely lipids and proteins, which are extracted during fixation, dehydration, and embedding (reviewed in Hayat, 1970). In one respect this type of loss is probably essential in obtaining any information since complete retention of material, especially in low-water-content cells such as yeast cells, would yield such an opaque image (even in thin sections) that it would be impossible to differentiate cell "structure" from "background." Indeed, in order to obtain useful images of some cells (e.g., yeasts), it is necessary to rupture or remove the cell wall to permit ingress of fixative and egress of background material. While many of the lost materials are probably low-molecular-weight compounds and enzymes of relatively little significance to the structural components of the cell (and whose loss is therefore of little consequence in a structural investigation), this is not always so. A specific type of loss of potential significance in mitosis is the demonstrated instability of filamentous actin following osmium fixation (Maupin-Szamier and Pollard, 1978). Clearly these considerations indicate that a negative observation is of very dubious value.

It may be argued that the preservation of a specific structure in one part of a cell lends credibility to the observed absence of that structure elsewhere in the same cell (e.g., LaFountain, 1975). Unfortunately this is not necessarily so. For example, variation in associated proteins may produce differential stability (Maupin-Szamier and Pollard, 1978), or different cellular micro-environments may respond differently to fixation (Heath and Heath, 1978). For whatever reason, it is clear that differently located microtubules in a single cell can have differing stabilities (e.g., Behnke and Forer, 1967) and similarly located microtubules in different cells can have different properties (e.g., Heath, 1975a). Again, it is clear that negative data are of limited value.

While the complete loss of an entire population, or subpopulation, of cellular constituents is a serious problem, the partial loss of structure can have equally important consequences. One of the main contributions of electron microscopy has been a detailed description of the microtubular architecture of the mitotic spindle. Clearly, if an entire category of microtubules is lost during processing, the results will be invalid, but, equally, if there are artifactual length changes in the residual population, one would arrive at an erroneous conclusion. The primary approach in investigating this potential problem has been to examine changes in spindle birefringence during fixation on the assumption that there is some correlation between birefringence and microtubule content. The validity of this assay is doubtful due to conflicting data and opinions concerning the nature of the correlation. Some authors (e.g., Sato et al., 1975; Sato, 1975; Inoué et al., 1975; Salmon, 1975a,b) consider that spindle birefringence is entirely due to spindle microtubules, whereas other workers present data to show that other unidentified components are responsible for at least part of the birefringence image (e.g., McIntosh et al., 1975a; Forer, 1976; Forer and Zimmerman, 1976a,b; Forer et al., 1976; Forer and Brinkley, 1977). Further uncertainty lies in the fact that one is measuring form birefringence, so that changes in orientation as well as in quantity would be detected. Unfortuantely, at present we lack a better indicator of the quality of preservation of microtubules, so that one must accept these limitations and use the data as best one may.

The reported changes in birefringence during fixation lead one to conclude that the effects of fixation alone vary between different cells of one type, between different cell types, and between different stages of mitosis in a single cell type. Thus McIntosh et al. (1975a) report between 0 and 50% loss of birefringence during glutaraldehyde fixation of mammalian cells, Nicklas et al. (1979a) found losses of 10 to 30% in insect cells, and Inoué and Sato (1967) noted up to 50% loss in urchin eggs. LaFountain (1974, 1976) found no loss at anaphase but 50% loss at metaphase in craneflies. All of these reports are for total spindle (or half-spindle) birefringence; it remains to be seen if changes occur differentially in different parts of the spindle (e.g., kinetochore fibers versus continuous fibers). Interestingly, Nicklas *et al.* (1979a) report that the common glutaraldehyde–phosphate buffer fixative gave equivalent or superior preservation of birefringence relative to more exotic media designed to enhance microtubule stability. It should be noted that the above data apply only to the initial fixation. There is evidence (Forer and Blecher, 1975) that microtubule structure can also be affected by dehydration, so that the correlation between the living cell and the embedded one may be less than the above data indicate.

Apart from the ambiguities and variations noted above, the birefringence assay for quality of fixation is not adequately sensitive. Minor length changes in a subpopulation of microtubules would go undetected but would markedly influence the final data. Furthermore, since birefringence is typically measured in only one part of the spindle, one may also have undetected changes in different parts of the spindle. At present, only serial section-based reconstructions of microtubule populations have the necessary resolution to assay for length changes, but this requires fixed material, the quality of which is being assayed. In principle, one solution to this problem is that if one obtains the same data after several different fixation procedures, one has some confidence in the results. This approach has not yet been used on spindles, but a number of studies of cytoplasmic microtubules are relevant. Hardham and Gunning (1978), Chalfie and Thomson (1979), and Goldstein and Entman (1979) report comparable microtubule lengths in diverse cell types following fixation in either a simple glutaraldehyde–buffer mixture or a complex glutaraldehyde–microtubule polymerization-supporting medium. In contrast, Luftig *et al.* (1977) and Seagull and Heath (1980) found more or longer microtubules following the use of microtubule polymerization-supporting medium. Apart from the obvious possibility of intercell variability, the conclusions from this data are not simple because in both cases of increased microtubule content it was necessary to add GTP to obtain the effect. As argued by Seagull and Heath (1980), there is a distinct chance of artifactual polymerization due to elevated pools of GTP-charged tubulin or to transient effluxes of Ca^{2+} from permeabilized cell membranes generating polymerization stimulating low Ca^{2+} levels in the cytoplasm (Weisenberg, 1972).

Clearly, the currently available data are totally inadequate to validate or negate the reliability of microtubule preservation in mitotic spindles. An optimistic view is that one can achieve perfect preservation with a simple glutaraldehyde–phosphate buffer–osmium tetroxide regimen, but there is ample evidence to doubt that this occurs in many cells [see, for example, the much lower microtubule counts obtained by Forer and Brinkley (1977) than obtained by LaFountain (1976) for the same species]. An important direction for future research must be to analyze the detailed structure of spindles prepared in diverse ways so that one can identify the degree of artifact present in conventional preparations.

B. Relocation of Material

The problem of artifactual relocation of material in cells during preparation for electron microscopy occurs at two levels of resolution. On the gross level, one is concerned about relocation of structures from cytoplasm to spindle and vice versa. A currently important example of this type of relocation is the actin found in spindles (Forer, 1978). In order to demonstrate the presence and orientation of actin filaments using subfragments of the myosin molecule (e.g., HMM or S1), one must use highly disruptive procedures during which there is an excellent possibility that cytoplasmic actin can be moved into the spindle. Forer *et al.* (1979) have attempted to answer this criticism by demonstrating a more consistent pattern of polarity and location that would be expected of artifactually relocated actin. Future approaches to resolving this type of problem might include observation of fluorescently labeled molecules during fixation (e.g., Wang and Taylor, 1979) and the development of less disruptive fixation procedures which will still permit ingress of HMM or S1. It should be remembered that this level of relocation is not only of concern to electron microscopists; the fluorescent antibody studies face similar problems (e.g., Chapter 10 by Aubin, this volume). It is worth reiterating that cells in which mitosis is effected in the presence of an intact nuclear envelope (e.g., lists in Heath, 1980a) have a potential advantage with respect to this problem. Translocation of material across the nuclear envelope is less likely to occur than translocation in the absence of the envelope.

While gross translocations are an obvious problem which in principle can be relatively easily monitored, finer ones are no less serious in their effect but are harder to control. At least two major hypothetical mitotic motors make explicit predictions about intermicrotubule interactions (McIntosh *et al.*, 1969; Bajer, 1973). These predictions can be directly tested (e.g., Heath, 1974b; Oakley and Heath, 1978) but the tests are highly dependent on the accuracy of preservation of intermicrotubule distances. Relatively little spindle expansion or contraction could increase or decrease intermicrotubule distances beyond critical values and invalidate the results. A similar problem applies to the reported changes in microtubule orientation occurring during anaphase (see Section III,A,1,c). The fact that these changes are mitotic-stage specific may be no more than an indication of increased susceptibility to artifactual reorientation. Both of these types of relocation are difficult to control for, and again, one must rely on the old adage that similar results from dissimilar processing are most believable. Unfortuantely, the amount of work required for these types of analysis has so far precluded a comparison of different preparative techniques.

C. The Third Dimension

The most detailed and valuable data on spindle architecture have come from thin-sectioned material. Apart from the limitations discussed above, such techniques are essentially two-dimensional [although at high resolution a typical thin section has a substantial thickness (ca. 70 nm) which must always be considered]. In order to obtain data on the third dimension, one can go to higher-accelerating voltages or serial section reconstructions. In the high-voltage electron microscope, one can use sections up to about 1 μm in thickness and obtain three-dimensional images by use of stereo-pair micrographs (e.g., McIntosh et al., 1975b; Bajer et al., 1975; Peterson and Ris, 1976; McIntosh et al., 1979b). These images are excellent for obtaining an overview of the spindle (Fig. 1), but even the elegant pictures of the small yeast spindle (Peterson and Ris, 1976) lack adequate vertical spatial resolution to determine unequivocally microtubule organization. For this reason, there is increasing use of the higher-resolution technique of serial reconstructions from ultrathin sections (e.g., Manton et al., 1969a,b, 1970a,b; Fuge, 1971, 1973, 1974, 1980; Jensen and Bajer, 1973; Heath, 1974b, 1978; McIntosh et al., 1975a, 1979a; Moens, 1976; Dietrich, 1979; Tippit et al., 1980a; Heath and Heath, 1976; Oakley and Heath, 1978; numerous diatom papers listed in Pickett-Heaps and Tippit, 1978; McDonald et al., 1979). Some reports have used relatively short series to ask specific questions about structures such as single kinetochore fibers but, increasingly, quantitative data from whole spindles are becoming available (e.g., Heath, 1974b; Peterson and Ris, 1976; Oakley and Heath, 1978; McIntosh et al., 1979a; McDonald et al., 1979; Tippit et al., 1980a). The wealth of data obtainable from an entire spindle can become overwhelming, so that computer technology has been introduced in order to handle the data (Moens, 1978; McIntosh et al., 1979a; McDonald et al., 1979; Heath, 1981a). The computer has proven useful in two respects: (a) it can manipulate the raw data and generate more easily comprehensible visual displays (Fig. 2), and (b) it can be used as a bookkeeper in order to ask quantitative questions about intermicrotubule arrangements (McDonald et al., 1979).

Thus, one can obtain useful three-dimensional data in a number of ways. The essential point to remember is that all of these data are only as good as the original preparation, faults of which were discussed earlier.

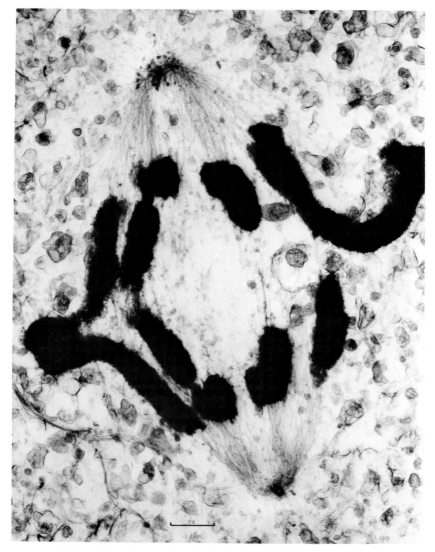

Fig. 1. High-voltage electron micrograph of a 0.75-μm-thick section of a lysed mammalian cell anaphase spindle. The general construction of the spindle is evident, but the difficulty of ascertaining the detailed arrangements and lengths of individual microtubules is equally apparent. Bar = 1 μm. From McIntosh *et al.* (1975b); courtesy of the authors and Raven Press.

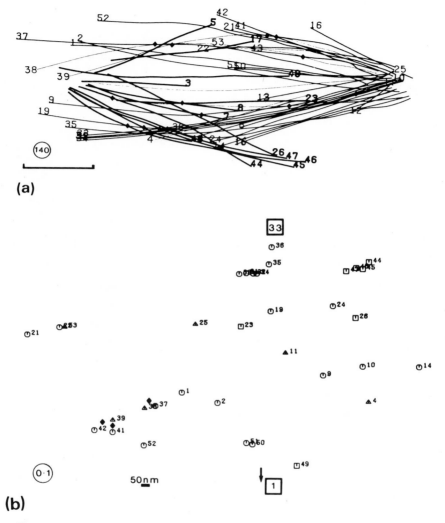

Fig. 2. Computer reconstructions of the non-kinetochore microtubules of an anaphase *Saprolegnia* spindle. The spindle was serially sectioned parallel to its long axis. In (a) the microtubules are displayed after the array was rotated 140° about its long axis. Points of close interaction between interdigitating microtubules from opposite poles or between inter-digitating microtubules and continuous or free microtubules are designated by diamonds. In (b) the same spindle is shown in a reconstructed cross section located close to the spindle equator. Interdigitating microtubules from opposite poles are designated by octagons and squares, while free and continuous microtubules are designated by triangles. Diamonds indicate points equivalent to those shown in (a). Bar in (a) = 1 μm. From Heath (1981a); courtesy of Cambridge University Press.

D. Lack of Compositional Information

The point of this section should be redundant, but since it is not always clearly remembered, it is worth making. Electron microscopy is essentially only a morphological tool. In the case of microtubules, one has sufficient morphological information to provide unambiguous identification (although not to describe adequately their complement of associated proteins), and likewise, HMM or S1 decoration is reasonably diagnostic for actin. However, undecorated filaments (e.g., Oakley and Heath, 1978; Schibler and Pickett-Heaps, 1980) do not possess a sufficiently distinctive morphology to identify their composition. Likewise, the commonly observed intermicrotubule cross-bridges observed in diverse cellular locations (McIntosh, 1974) may or may not be dynein. Thus one must always await indirect analysis via selective extraction, antibody localization, enzyme digestion, etc. before accepting the composition of a structure as demonstrated.

The rest of this chapter will consider some of the more interesting data obtained via electron microscopy of spindles. By and large, the data will be discussed at face value, but in most cases some of the above cautionary remarks apply. They will be mentioned specifically only when some form of artifact seems especially likely to have occurred.

III. MICROTUBULES

With the notable exception of one of the two nuclei in some dinoflagellates (Tippit and Pickett-Heaps, 1976), all eukaryotes adequately examined contain an array of microtubules associated with karyokinesis. These microtubules may be entirely extranuclear, but they are most commonly intermingled with the chromosomes, either inside a variously persistent nuclear envelope or free in the cell following envelope breakdown (reviewed in Heath, 1980a). While these microtubules may comprise as little as 1% of the spindle (Forer, 1969, 1974), they are undeniably the most studied part of the electron microscopist's spindle. As such, they will form the major part of this review and will be considered in terms of their arrangements, synthesis, and interactions.

A. Microtubule Arrangements

There are five currently described distinct types of microtubules which have been identified in most suitably examined spindles. Kinetochore or chromosomal microtubules have one end attached to a kinetochore and the opposite end extending for some undefined distance (see Section III,A,1,b) toward the spindle pole. Non-kinetochore microtubules are subdivided into four types. Continuous microtubules run continuously from one spindle pole to the other. Interdigitating microtubules have one end inserted at the spindle pole and the other end terminating at some distance past the equator of the spindle in the opposite half-spindle. Thus the two arrays of interdigitating microtubules form a pole-to-pole bundle which is overlapped at its center. Polar microtubules are like short interdigitating microtubules; they each extend from only one spindle pole, but they do not cross the equator, and each lies entirely in one half-spindle. Free microtubules lie free in the spindle, with neither end inserted into any recognizable component of the spindle. This terminology is a hybrid of those used by Bajer and Molè Bajer (1972), Heath (1974b), and McIntosh *et al.* (1975b).

1. Kinetochore Microtubules

Kinetochore microtubules are an almost universal component of spindles (for a discussion of possible exceptions, see Heath, 1980a). However, there is substantial variability in their numbers and arrangements which may be informative.

a. Numbers. In many fungi and protists there is only one kinetochore microtubule per kinetochore (listed in Heath, 1980a), whereas the maximum noted is $\sim$ 120 in *Haemanthus* (Jensen and Bajer, 1973). This range has led to the obvious suggestion that there may be a correlation between chromosome size (small in fungi and protists, large in *Haemanthus*) and number of kinetochore microtubules (Fuge, 1977, 1978; Heath, 1978). Moens (1979) has tested this suggestion in cells having a 10-fold range of chromosome volumes and found that there is no correlation between chromosome volume and number of chromosomal microtubules. He suggests that the number of kinetochore microtubule reflects the karyotypic history (e.g., chromosome fusions) of the cell rather than any force requirements. However, since there is over a 100-fold range in DNA per chromosome (e.g., Heath, 1980a), considering this variation and the undoubtedly variable quantities of chromosomal proteins, it may still be that over this broader "load range" there is a correlation with the number of kinetochore microtubules per kinetochore. When seeking such a correlation, it is worth recalling the potential artifact problem. Thus Forer and Brinkley (1977) report only 25% of the number of kinetochore microtubules found in the same cell type by LaFountain (1976). Clearly, one should observe caution in comparing the results from different studies.

When kinetochore fibers undergo the transition from metaphase to anaphase, it is quite possible that they show some instructive change in kinetochore microtubule numbers. Changes in kinetochore microtubule numbers between metaphase and anaphase have been reported in *Haemanthus* (Jensen and Bajer, 1973). However, in all lower eukaryotes with single kinetochore microtubules per kinetochore (e.g., Heath, 1980a) and in craneflies (LaFountain, 1976; Fuge, 1980), the numbers do not change. It seems most likely that the change reported by Jensen and Bajer (1973) is an indication of greater lability of the kinetochore microtubules during anaphase and their consequent loss during fixation. At present, it seems most probable that the onset of anaphase does not involve a change in kinetochore microtubule numbers. However, Moens' (1979) data do suggest a possible increase in the number of microtubules during metaphase, which reaches a maximum immediately before anaphase. Such a possibility should be investigated by further research.

   b. Lengths.   It is difficult to generalize on the lengths of kinetochore microtubules during mitosis because in many cases the required serial sections have not been obtained. However, it is clear that when there are only one or two kinetochore microtubules per kinetochore, those microtubules do indeed extend from the kinetochore entirely to the pole throughout mitosis (e.g., Heath, 1974b, 1978; Roos, 1975; Heath and Heath, 1976; Moens, 1976; Oakley and Heath, 1978) (Fig. 3). The data from other cells are more ambiguous. LaFountain (1976) suggests that in craneflies the kinetochore microtubules extend completley to the spindle poles during metaphase and anaphase; but this suggestion was based on single sections which could not describe the behavior of the entire population of kinetochore microtubules. The similarly ambiguous data of Jensen and Bajer (1973), Coss and Pickett-Heaps (1974), and Dietrich (1979) support this suggestion. Conversely, Fuge (1974, 1980) and McIntosh *et al.* (1975a) find that many of the kinetochore microtubules do not reach the poles, especially during anaphase. These data are in agreement with the observed decline in kinetochore fiber birefringence near the poles (e.g., Forer, 1976), but again, one must consider the real possibility of artifact in a labile structure such as the anaphase kinetochore fiber. All one can conclude at present is that for those kinetochore microtubules that extend to the poles, their deploymerization must be at a rate comparable to anaphase movement, with all the obvious implications for control of anaphase. Conversely, if there really are shorter kinetochore microtubules, it is evident that they may have some other function which must be considered in hypothetical models for mitosis.

   c. Arrangements.   Analysis of the arrangement of kinetochore microtubules may be expected to provide information on two important questions: (1) With what do kinetochore microtubules interact? (2) Are there any changes in arrangement associated with the onset of anaphase movement?

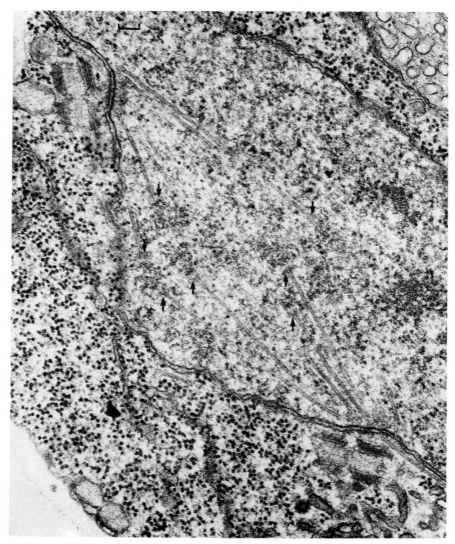

Fig. 3. Median longitudinal section of a metaphase *Saprolegnia* spindle showing kinetochore microtubules extending completely from the kinetochores (arrows) to the spindle poles. Note the convergence of the microtubules at the poles. Bar = 0.1 μm. From Heath (1978); courtesy of Academic Press.

Among the smaller spindles studied, there appear to be two types of
arrangement of the kinetochore microtubules. The kinetochore microtubules may lie around the periphery of a central group of non-kinetochore
microtubules (e.g., Roos, 1975; Moens, 1976; Heath and Heath, 1976) (Fig.
4), or they may be more completely interspersed with the non-kinetochore
microtubules (e.g., Heath, 1974b, 1978; Oakley and Heath, 1978) (Fig. 3).
However, there is little obvious effective difference between these arrangements because, in all genera except *Cryptomonas* (Oakley and Heath, 1978),

Fig. 4. Median longitudinal section of a metaphase *Uromyces* spindle showing the
central bundle of non-kinetochore microtubules with the peripherally arranged chromatin and
kinetochore microtubules (arrows). Note the multilayered polar nucleus-associated organelles. Bar = 1 μm. From Heath and Heath (1976); courtesy of Rockefeller University Press.

the convergent nature of each half-spindle is such that, at least at the spindle poles, the kinetochore microtubules come close to the non-kinetochore microtubules (e.g., Figs. 3 and 4), so that they could interact with one another, as postulated by numerous hypotheses. However, as noted by Heath (1974b) and Roos (1975), these spindles are constructed such that lateral interactions along the entire length of the kinetochore microtubules and adjacent non-kinetochore microtubules are highly unlikely to occur. In the exception. *Cryptomonas* (Oakley and Heath, 1978), the spindle has much broader poles and less than 12% of the kinetochore microtubules come within 50 nm of non-kinetochore microtubules, thus making it unlikely that there is any functionally significant interaction between the two types of microtubules. In all of these spindles, there are no reports of any change in arrangement during anaphase.

In larger spindles, the arrangement of the kinetochore microtubules is more complex but fundamentally similar. Thus the kinetochore microtubules typically form bundles emanating from each kinetochore (e.g., Jensen and Bajer, 1973; Coss and Pickett-Heaps, 1974; McIntosh *et al.*, 1975a,b; Dietrich, 1979). These bundles interact with bundles of non-kinetochore microtubules, and individual kinetochore microtubules leave the bundles and enter non-kinetochore microtubule bundles, and vice versa. This mixing may be so extensive that the kinetochore microtubule bundles are barely discernible as discrete populations (Fuge, 1971; LaFountain, 1976; Forer and Brinkley, 1977). Clearly, at least at a simplistic level, the intermixing of the two categories of microtubule is consistent with, but does not prove, their interaction in a force-generating system. However, these more complex kinetochore microtubule bundles have revealed some interesting apparent changes during anaphase which *may* prove to be instructive in terms of mitotic mechanisms. In *Haemanthus* there is an increased splaying of the kinetochore microtubules during anaphase (Jensen and Bajer, 1973; Bajer *et al.*, 1975). A comparable divergence of kinetochore fibers is seen *in vivo* by Hard and Allen (1977), and a marked reversal to a close-packed array of kinetochore mocrotubules occurs at high and low temperatures when chromosome movements cease (Lambert and Bajer, 1977; Rieder and Bajer, 1978). These results support the real, as opposed to artifactual, nature of the splaying. A comparable, but smaller, increase in divergence of the most divergent kinetochore microtubules occurs in crane-flies (LaFountain, 1976) too, and Fuge (1980) reports an increase in skewed kinetochore microtubules, but without significant bundle divergence in the same cells, thus supporting the generality of the splaying phenomenon. In contrast, however, no sign of splaying has been observed in *Paeonia* (Dietrich, 1979), *Oedogonium* (Coss and Pickett-Heaps, 1974), or mammalian cells (McIntosh *et al.*, 1975a,b). Interestingly, Nicklas *et al.* (1979b) reported differential splaying in opposing half-spindles of grasshopper cells which were exposed to a temperature gradient. The kinetochore microtubules were more splayed in the cooler half-spindle, but the chromosomes on *both* half-spindles moved at the *same* rate. Clearly, further work is needed to establish the range of occurrence and significance of the splaying phenomenon.

2. Non-Kinetochore Microtubules

One of the major achievements of electron microscopy of mitosis has been the demonstration that the continuous spindle fibers are not usually composed of continuous microtubules but rather of populations of interdigitating microtubules In a minority of reported species [most clearly in *Uromyces* (Heath and Heath, 1976), and probably *Barbulanympha* (Inoué and Ritter, 1978) and possibly a few other species listed in Heath (1980a)] the nonkinetochore microtubules appear to be predominantly continuous microtubules throughout mitosis. However, in the majority of suitably investigated species, continuous microtubules are either absent or form a very small percentage of the population. The preponderance of interdigitating microtubules is most clearly seen in the diatoms (Manton *et al.*, 1969a,b, 1970a,b and the numerous papers reviewed in Pickett-Heaps and Tippit, 1978), where they form two very clearly defined interdigitating half-spindles which slide apart (with or without concomitant additional microtubule polymerization) during anaphase. This arrangement can also be detected *in vivo* by polarization microscopy (Pickett-Heaps and Tippit, 1978), thus demonstrating its reality. Within the zone of overlap of the interdigitating microtubules there is "evidence for a specific interaction between antiparallel microtubules" (McDonald *et al.*, 1979, p. 443). An essentially comparable system seems to occur in numerous other cell types (e.g., Brinkley and Cartwright, 1971; McIntosh and Landis, 1971; Oakley and Heath, 1978; Tippit *et al.*, 1980a). In each of these other cell types the regions of overlap are not as regular as in the diatoms, and at least in the cells studied by Tippit *et al.* (1980a) the interdigitating microtubules seem to form a number of separate "mini-spindles." However, the important point is that in each case the interdigitating microtubules come close enough together to interact by intermicrotubule cross-bridges. In contrast to this situation, in oomycete fungi interdigitating microtubules lie farther apart, so that any force generation via intermicrotubule cross-bridges is unlikely at anaphase (Heath, 1974b, 1978, 1981a) (Fig. 2). By telophase, the median constriction of the nucleus brings the antiparallel interdigitating microtubules into close proximity with one another, so that then an interaction becomes formally possible again.

Apart from the oomycetes, it is clear that the arrangement of most nonkinetochore microtubules favors an intermicrotubule sliding hypothesis for force generation during anaphase. It should also be remembered that in a number of spindles built on the interdigitating microtubule pattern, apparent sliding is also accompanied by microtubule polymerization. Thus, when continuous microtubules dominate the non-kinetochore microtubule population, sliding with concomitant polymerization may still be the force-generating mechanism (Heath, 1978).

The role of the polar microtubules remains obscure. They exist in most reported spindles and could interact with the nuclear matrix (Heath, 1975b), or they could represent partially artifactually depolymerized interdigitating microtubules. Similar comments can be made for the free microtubules, but they at least have been reported to participate with interdigitating microtubules to form interpolar bundles of microtubules (Oakley and Heath, 1978). Clearly more work is required to elucidate the true role of these types of non-kinetochore microtubules.

B. Microtubule Synthesis

From the outset of ultrastructural analysis of mitosis, it was evident that centrioles (or something associated with centrioles) and kinetochores were foci for various types of spindle microtubules. Such structures are evidently microtubule-organizing centers, as envisaged by Pickett-Heaps (1969). More detailed electron microscopy has played an essential role in analyzing these microtubule-organizing centers, so that we now have a good idea of their structure, function, and composition even if we lack detailed analysis of their moelcular architecture. These data will be discussed in terms of the structures which lie at the spindle poles and the kinetochores.

1. Polar Structures

One of the best-known features of mitosis is that in many animal cells there are a pair of orthogonally arranged centrioles lying at each spindle pole. What is less well known is that centrioles are merely a subtype of nucleus-associated organelle (Girbardt and Hädrich, 1975), other forms of which are a universal feature of mitosis in all cell types. Indeed, since centrioles are absent from the spindles of most fungi, protists, and vascular plants and from the spindles of many algae, it is likely that ⅓ to ½ of known eukaryotic species have some type of nucleus-associated organelle *other* than centrioles. Detailed reviews of the range of nucleus-associated organelle types and their functions have been published (Heath, 1980a, 1981b; Peterson and Berns, 1980), so that an overview will suffice here. The one common feature of all nucleus-associated organelles is that they have the form of variously arranged osmiophilic material (although it must be admitted that such material is not evident in most of the higher plant cells examined).

Centriole-bearing cells are no exception to this rule, since undoubtedly it is the osmiophilic material around the centrioles which has the microtubule-organizing center properties (e.g., Gould and Borisy, 1977). While admittedly extrapolating from very few data, it does seem likely that the following points summarize our knowledge of the behavior of probably all polar nucleus-associated organelles. (1) They act as initiators of microtubule polymerization, both *in vivo* and *in vitro*. (2) They can specify the number and orientation, but not length, of the initiated microtubules (Gould and Borisy, 1977; Hyams and Borisy, 1978). (3) The material active in initiation is proteinaceous and may also contain both RNA and DNA (for a summary of these data, see Peterson and Berns, 1980). (4) The capacity of the nucleus-associated organelles to nucleate microtubule polymerization varies during the cell cycle (Snyder and McIntosh, 1975; Gould and Borisy, 1977; Hyams and Borisy, 1978; Telzer and Rosenbaum, 1979), which suggests that control of spindle formation is at least partly mediated, directly or indirectly, via the nucleus-associated organelle. What clearly remains to be determined is the way in which these often seemingly simply organized structures are capable of exerting such precise control over the form of the typically complex and reproducible spindle.

2. Kinetochores

The preceding discussion on the nucleus-associated organelles was deliberately vague about the types of spindle microtubules which are nucleated at the spindle poles. Undoubtedly all non-kinetochore microtubules originate there, with the possible exception of the free microtubules. Even these could be nucleated at the poles and "exported" as free fragments, as suggested by Gunning *et al.* (1978). What remains unclear is the origin of the kinetochore microtubules. Undoubtedly, kinetochores can act as microtubule-organizing centers *in vitro* (McGill and Brinkley, 1975; Telzer *et al.*, 1975; Snyder and McIntosh, 1975; Gould and Borisy, 1978; Bergen *et al.*, 1980). Similarly, in a number of cells there is evidence that at least part of the kinetochore microtubule population forms from the kinetochores *in vivo* too (Bajer and Mole-Bajer, 1969; Pickett-Heaps and Fowke, 1970; Lambert and Bajer, 1975; Witt *et al.*, 1979). However, Pickett-Heaps and Tippit (1978) have resurrected the idea the kinetochores merely "trap" microtubules which are formed from the spindle poles. Since the majority of the microtubules which they describe as kinetochore microtubules do not in fact terminate in the chromatin (Pickett-Heaps *et al.*, 1980; Tippit *et al.*, 1980b), it is doubtful whether they are strictly comparable to the more common kinetochore microtubules. Similarly, the first microtubules associated with disconnected chromosomes in grasshopper cells could well be trapped (Nicklas *et al.*, 1979a), but again, it is not clear that these are comparable to normal

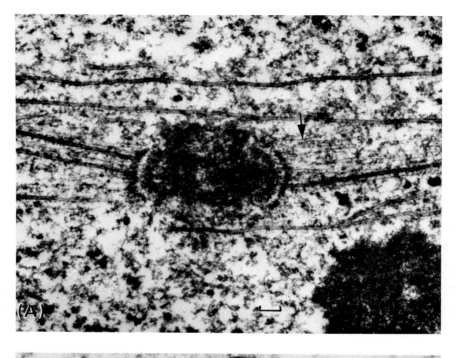

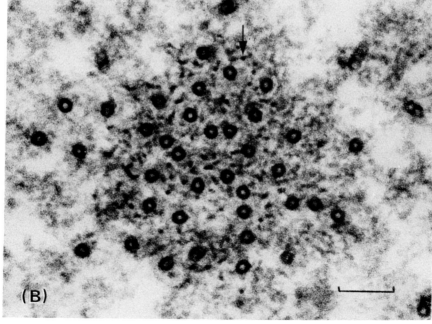

kinetochore microtubules. Thus, while it is possible that some kinetochore microtubules are trapped, it seems most likely that the bulk of normal kinetochore microtubules are indeed initiated from the kinetochores.

As emphasized by Borisy (1978) and Bergen et al. (1980), the site of microtuble initiation is likely to be important since it will control the polarity of the formed microtubules. This polarity, in turn, is critical to several hypothetical models for mitosis. While Bergen et al. (1980) have provided indirect evidence for the polarities of the various types of spindle microtubules, the question will be most effectively and unambiguously answered when polarities are directly analyzed by techniques such as the dynein decoration procedure employed by Haimo et al. (1979).

While there are many more aspects to the behavior of microtubules during mitosis than discussed here, the above points highlight some of the most important features. The one remaining aspect of their behavior, that of how and with what microtubules interact, is perhaps best delayed until other components of the spindle are described.

IV. ACTIN FILAMENTS

As discussed previously, while there may be uncertainties regarding the exact numbers and arrangements of microtubules, it is at least certain that they are part of the spindle. There remains much less certainty about actin. The total evidence for the role of actin in mitosis has been critically reviewed (Forer, 1978), so that only a summary of the ultrastructural evidence need be given here. There are now a number of reports of filaments 6–8-nm in diameter in the spindles of various regularly fixed cells (e.g., Bajer and Molè-Bajer, 1969; McIntosh and Landis, 1971; McIntosh et al., 1975b; Euteneuer et al., 1977; Forer and Brinkley, 1977; Oakley and Heath, 1978; and, most spectacularly, Schibler and Pickett-Heaps, 1980) (Fig. 5). While such filaments are morphologically similar to actin filaments, they are not proven to be such. However, there are also numerous reports of HMM or S1 binding filaments in spindles (e.g., Gawadi, 1971, 1974; Forer and Behnke, 1972; Hinkley and Telser, 1974; Forer and Jackson, 1976; Schloss et al., 1977; Forer et al., 1979). If one considers (a) the location of the filaments in well-fixed cells, (b) the positive identification from the rather disrupted HMM-treated cells, and (c) the specific, non-random location and orienta-

---

Fig. 5. Longitudinal (a) and transverse (b) sections of kinetochore fibers of *Oedogonium*. Note the fibers (arrows) intermixed with the microtubules and the osmiophilic material permeating the entire fiber in (b). Bars = 0.1 μm. From Schibler and Pickett-Heaps (1980); courtesy of Wissenschaftliche Verlagsgesellschaft.

tion reported by Gawadi (1974) and Forer *et al.* (1979), it seems likely that actin is indeed a real component of the spindle. This conclusion is supported by evidence from various light microscopic techniques reviewed in Forer (1978) and Forer *et al.* (1979). The interesting additional information obtained by electron microscopy is that the preserved filaments are most abundant in the kinetochore fibers (Fig. 5), where they seem to be attached to the chromosomes. In this location they are dominantly oriented with the HMM-produced arrowheads facing the chromosome, which is consistent with their interaction with kinetochore microtubules to effect chromosome-to-pole movement (Forer *et al.*, 1979). However, actin is also typically located in the interzone at anaphase. Its role in this location is still uncertain. Unfortunately at present, reliable data on location, changes in arrangement associated with movement, and sites of polymerization comparable to the microtubule data are still lacking for actin. As a final point concerning the question of spindle actin, one should recall the important warning that actin filaments are demonstrably fixation labile (Maupin-Szamier and Pollard, 1978; Gicquaud *et al.*, 1980). Thus the relatively few reports of actin in non-HMM-treated spindles compared with the abundance of good general ultrastructural studies of mitosis do not provide a significant argument against the universality of actin as a spindle component.

While microtubules and actin are the primary cytoskeletal structures which have been identified in spindles, it is worth reemphasizing the point (Forer, 1969; McIntosh *et al.*, 1975b) that isolated mitotic apparatuses, which have already lost $\sim$ 95% of their *in vivo* mass, are quite capable of retaining their general morphology when microtubules have been removed by various techniques (e.g., Kane and Forer, 1965; Bibring and Baxandall, 1968, 1971; Forer *et al.*, 1976). These observations draw attention to the fact that there is much more to a spindle than just the microtubules and actin (and their known associated proteins; see Section V). An especially interesting example of this other material in the spindle is the darkly stained material surrounding the microtubules in the equatorial overlap region of diverse telophase spindles (e.g., McIntosh *et al.*, 1975b) and in some kinetochore fibres (e.g., McIntosh *et al.*, 1975b, and especially Schibler and Pickett-Heaps, 1980) (Fig. 5). Clearly, an important avenue for future research is the characterization of the rest of the structural material of the spindle.

V. INTERACTIONS OF CYTOSKELETAL COMPONENTS

As outlined above, spindles contain two of the major cytoskeletal components, microtubules and actin. The arrangements of these components have already been described. The purpose of this section is to focus more specifically on the interactions between (a) microtubules themselves, (b) microtubules and microfilaments, and (c) microtubules and chromatin.

A. Intermicrotubule Interactions

The best-known type of microtubule interaction has the morphological form of an intermicrotubule cross-bridge which is typically 3–7 nm wide and 10–40 nm long (McIntosh, 1974). Such cross bridges have been reported in mitotic spindles (e.g., Wilson, 1969; Hepler *et al.*, 1970; McIntosh and Landis, 1971; Brinkley and Cartwright, 1971; McIntosh, 1974; Ritter *et al.*, 1978), but to date there has been no systematic study of their distribution throughout the spindle or of any possible changes in their arrangement or frequency at different stages of mitosis. Undoubtedly, part of the reason for the lack of such data is that these bridges are frequently poorly resolved (e.g., McIntosh, 1974), so that a description of their distribution would require a large amount of subjective decision making. In addition, these bridges are probably fixation labile. For example, in one of the most regular arrays of precisely spaced spindle microtubules known, in the overlap region of the equator of diatom spindles, cross bridges are seen only in "unusually well-preserved" spindles (McDonald *et al.*, 1979, p. 456). However, because reported cross bridges do not exceed 40 nm in length, one can analyze spindles for the *potential* to form cross-bridges by measuring the distribution of intermicrotubule distances in the range which would permit the formation of cross-bridges, i.e., less than 50 nm. Only one such analysis is known for entire spindles (Oakley and Heath, 1978), and there it was found that the non-kinetochore microtubules could be extensively cross-bridged to one another effectively to form a continuous array from pole to pole, but the kinetochore microtubules rarely (~ 12%) came close enough to cross-bridge with non-kinetochore microtubules. A similar approach to the non-kinetochore microtubule population in *Saprolegnia* can be used to argue *against* a role for cross-bridges in spindle elongation (Heath, 1981a) (Fig. 2). Unfortunately, using distances rather than preserved cross-bridges is valuable only in a negative sense because the potential for cross-bridge formation does not prove that it occurs.

Further discussion of cross-bridges in spindles requires a brief consideration of their nature in other systems, a detailed review of which is beyond the scope of this chapter. However, the essential point which can be derived from the review of Dustin (1978) and the papers in Roberts and Hyams (1979) is that cross-bridges come in two general types: active mechanochemical ones which generate intermicrotubule sliding, the best known of which is dynein, and more rigid ones which form static, long-term interconnections to generate a strong cross-linked array of microtubules for some mechanical

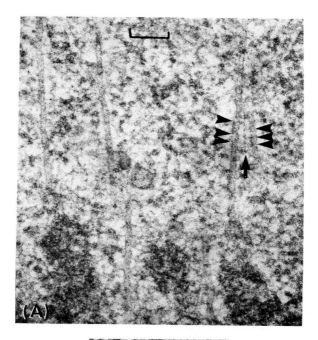

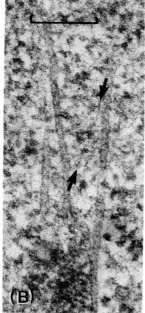

role. There is some ambiguous evidence for the existence of both types in spindles. During spindle elongation in diatoms, it seems fairly certain that non-kinetochore microtubules slide past one another at the spindle equator (e.g., McDonald *et al.*, 1977; Pickett-Heaps and Tippit, 1978), where cross-bridges probably exist (McDonald *et al.*, 1979). This type of sliding is probably widespread (e.g., McIntosh and Landis, 1971; Oakley and Heath, 1978; Tippit *et al.*, 1980a), but the data are most convincing and the system is most ordered in the diatoms. However, in contrast, it seems entirely likely that in those spindles composed primarily of true continuous microtubules (e.g., Heath and Heath, 1976; possibly also Ritter *et al.*, 1978 and Inoué and Ritter, 1978), these microtubules are statically linked together for extra rigidity, as suggested elsewhere (Heath, 1980a). Unfortunately, the data do not exclude the possibility of sliding with coordinated polymerization to maintain the continuity of the continuous microtubules. Clearly, further data on the existence of either type of cross-bridge are needed from diverse spindle types.

B. Microtubule–Microfilament Interactions

Since the data for the existence of microfilaments in spindles are sparse (Section IV), it is not surprising that the data for an interaction between filaments and tubules are even more sparse. However, there is enough information to make the possibility worth analyzing, expecially as such an interaction is explicitly predicted by the mitotic models of Forer (1974) and Oakley and Heath (1978). The evidence for this interaction is essentially as follows: (a) In all reported spindles (see Section IV), many of the observed filaments come very close (within ∼ 10 nm) to the surface of microtubules. However, there are no statistical data to show that they come closer than would be expected by chance, although subjectively this does appear to be the case. (b) The filaments may be linked to the microtubules by regularly spaced cross-bridges (Oakley and Heath, 1978) (Fig. 6). (c) When microtubules are variously bent, the associated parallel microfilaments closely follow the contours of the microtubules (Forer and Jackson, 1979). (d) The stabilization of actin filaments seems to effect a concomitant stabilization of microtubules (Forer and Behnke, 1972; Forer, 1974). (e) Actin and microtubules can be cross-bridged to each other *in vitro* (Griffith and Pollard, 1978). These data all support the idea that microfilaments do interact with

Fig. 6. (a,b)Possible cross-bridge-mediated interactions between microfilaments (arrows) and spindle microtubules in a metaphase spindle of *Cryptomonas*. Cross-bridges are indicated by arrowheads. Bars = 0.1 μm. From Oakley and Heath (1978); courtesy of the Company of Biologists.

spindle microtubules but, apart from an unmeasured subjective impression that they interact more frequently with kinetochore microtubules than with non-kinetochore microtubules, there is no information on the nature of the interaction or on the distribution of interactions through the spindle and through the stages of mitosis. Such information will be usefully available only when we have much more confidence in the technical ability to preserve both filaments and their sites of interaction.

C. Microtubule–Chromatin Interactions

The most obvious and probably dominant way in which chromatin interacts with microtubules is via the knietochore. Kinetochores vary substantially in size and detailed structure, but their one main common feature is that they represent a differentiated part of the chromatin into which the ends of kinetochore microtubules insert. Their role in kinetochore microtubule formation was discussed in Section III,B,2 and, since there are few data on the way in which the kinetochore microtubules are physically connected to them, this topic will not be pursued further. What should be emphasized are the possible alternative interactions between chromatin and microtubules. There is now considerable evidence of an interaction in which chromatin appears to slide along the surface of microtubules. While the morphology is consistent with sliding, the demonstration of apparent steady-state, opposite-end assembly–disassembly of microtubules (Margolis and Wilson, 1978; but see also Bergen and Borisy, 1980) means that in fact one may not be observing sliding but rather movement of a fixed lateral attachment site powered by turnover of microtubule subunits. Whichever system is operating, it is quite evident that in craneflies (Fuge, 1972, 1973, 1974, 1975) there is a force producing, or transmitting, *lateral* interaction between non-kinetochore microtubules and chromatin. As noted by Heath (1974a), a similar situation prevails in the diatom *Lithodesmium* (Manton *et al.*, 1969a,b, 1970a,b) and possibly in *Prymnesium* (Manton, 1964). The more extensive studies of a range of diatoms (summarized by Pickett-Heaps and Tippit, 1978) reveal that the phenomenon is common among these organisms. The detailed mechanisms involved and the full range of organisms in which it occurs are unclear at present, but current evidence suggests that it is a phenomenon which deserves more attention in future studies.

VI. CONCLUSIONS AND FUTURE QUESTIONS

The preceding account focused on aspects of the spindle which may be involved in production and control of the forces needed to effect chromosome movement. By mandate and choice, it has dealt with the data which have been primarily or exclusively obtained via the electron microscope. There are, of course, many more data on spindle formation and composition (e.g., the vesicular system reviewed by Harris, Chapter 2, this volume) which have been obtained by electron microscopy and which must be sought through the broader reviews mentioned earlier. Again, as emphasized at the outset, the potential ambiguities due to technical limitations must be remembered in all ultrastructural studies. However, from the above summary, one can recognize important areas of ultrastructural research which require more work before a complete understanding of mitosis is possible. These include: (a) development of improved techniques to ensure accurate preservation of all spindle components; (b) more detailed quantitative work to describe how the components relate to one another and how the relationships alter during spindle action; (c) more thorough analysis of the "matrix" of the spindle, more than 95% which is undescribed; (d) a complete description of the polarities of the microtubules and microfilaments throughout the spindle; (3) extension of spindle analysis to a greater phylogenetic range of organisms to seek patterns of evolution and common denominators in spindle structure.

REFERENCES

Bajer, A. (1973). Interaction of microtubules and the mechanism of chromosome movement (zipper hypothesis) 1. General principle. *Cytobios* **8**, 139–160.

Bajer, A., and Molè-Bajer, J. (1969). Formation of spindle fibers, kinetochore orientation, and behaviour of the nuclear envelope during mitosis in endosperm. Fine structural and *in vitro* studies. *Chromosoma* **27**, 448–484.

Bajer, A., and Molè-Bajer, J. (1972). Spindle dynamics and chromosome movements. *Int. Rev. Cytol., Suppl.* **3**, 1–271.

Bajer, A. S., Molè-Bajer, J., and Lambert, A. M. (1975). Lateral interaction of microtubules and chromosome movements. *In* "Microtubules and Microtubule Inhibitors" (M. Borgers and M. de Brabander, eds.), pp. 393–423. North Holland Publ., Amsterdam.

Behnke, O., and Forer, A. (1967). Evidence for four classes of microtubules in individual cells. *J. Cell Sci.* **2**, 169–192.

Bergen, L. G., and Borisy, G. G. (1980). Head-to-tail polymerization of microtubules *in vitro*. Electron microscope analysis of seeded assembly. *J. Cell Biol.* **84**, 141–150.

Bergen, L. G., Kuriyama, R., and Borisy, G. G. (1980). Polarity of microtubules nucleated by centrosomes and chromosomes of Chinese hamster ovary cells *in vitro*. *J. Cell Biol.* **84**, 151–159.

Bibring, T., and Baxandall, J. (1968). Mitotic apparatus: the selective extraction of protein with mild acid. *Science* **161**, 377–379.

Bibring, T., and Baxandall, J. (1971). Selective extraction of isolated mitotic apparatus. Evidence that typical microtubule protein is extracted by organic mercurial. *J. Cell Biol.* **48**, 324–339.

Borisy, G. G. (1978). Polarity of microtubules of the mitotic spindle. *J. Mol. Biol.* **124**, 565–570.

Brinkley, B. R., and Cartwright, J. (1971). Ultrastructural analysis of mitotic spindle elongation in mammalian cells *in vitro*. Direct microtubule counts. *J. Cell Biol.* **50**, 416–431.

Chalfie, M., and Thomson, J. N. (1979). Organization of neuronal microtubules in the nematode *Caenorhabditis elegans*. *J. Cell Biol.* **82**, 278–288.

Coss, R. A., and Pickett-Heaps, J. D. (1974). The effects of isopropyl N-phenyl carbamate on the green alga *Oedogonium cardiacum* 1. Cell division. *J. Cell Biol.* **63**, 84–98.

Dietrich, J. (1979). Reconstructions tridimensionnelles de l'appareil mitotique à portir de coupes sériées longitudinales de méiocytes polliniques. *Biol. Cellulaire* **34**, 77–82.

Dustin, P. (1978). "Microtubules." Springer-Verlag, Berlin and New York.

Euteneuer, U., Bereiter-Hahn, J., and Schliwa, M. (1977). Microfilaments in the spindle of *Xenopus laevis* tadpole heart cells. *Cytobiologie* **15**, 169–173.

Forer, A. (1969). Chromosome movements during cell division. *In* "Handbook of Molecular Cytology" (A. Lima-de-Faria, ed.), pp. 553–601. North Holland Publ., London.

Forer, A. (1974). Possible roles of microtubules and actin-like filaments during cell division. *In* "Cell Cycle Controls" (G. M. Padilla, I. L. Cameron, and A. M. Zimmerman, eds.), pp. 319–335. Academic Press, New York.

Forer, A. (1976). Actin filaments and birefringent spindle fibers during chromosome movements. *In* "Cell Motility" (R. Goldman, T. Pollard, and J. Rosenbaum, eds.), pp. 1273–1293. Cold Spring Harbor Lab., Cold Spring Harbor, New York.

Forer, A. (1978). Chromosome movements during cell division: possible involvement of actin filaments. *In* "Nuclear Division in the Fungi" (I. B. Heath, ed.), pp. 21–88. Academic Press, New York.

Forer, A., and Behnke, O. (1972). An actin-like component in spermatocytes of a crane fly (*Nephrotoma suturalis* Loew). 1. The spindle. *Chromosoma* **39**, 145–173.

Forer, A., and Blecher, S. R. (1975). Appearances of microtubules after various fixative procedures, and comparison with the appearances of tobacco mosaic virus. *J. Cell Sci.* **19**, 579–605.

Forer, A., and Brinkley, B. R. (1977). Microtubule distribution in the anaphase spindle of primary spermatocytes of a crane fly (*Nephrotoma suturalis*). *Can. J. Genet. Cytol.* **19**, 503–519.

Forer, A., and Jackson, W. T. (1976). Actin filaments in the endosperm mitotic spindles in a higher plant, *Haemanthus katherinae* Baker. *Cytobiologie* **12**, 199–214.

Forer, A., and Jackson, W. T. (1979). Actin in spindles of *Haemanthus katherinae* endosperm. 1. General results using various glycerination methods. *J. Cell Sci.* **37**, 323–347.

Forer, A., and Zimmerman, A. M. (1976a). Spindle birefringence of isolated mitotic apparatus analysed by treatments with cold, pressure, and diluted isolation medium. *J. Cell Sci.* **20**, 329–339.

Forer, A., and Zimmerman, A. M. (1976b). Spindle birefringence of isolated mitotic apparatus analysed by pressure treatment. *J. Cell Sci.* **20**, 309–327.

Forer, A., Jackson, W. T., and Engberg, A. (1979). Actin in spindles of *Haemanthus katherinae* endosperm. II. Distribution of actin in chromosomal spindle fibres, determined by analysis of serial sections. *J. Cell Sci.* **37**, 349–371.

Forer, A. Kalnins, V. I., and Zimmerman, A. M. (1976). Spindle birefringence of isolated mitotic apparatus: Further evidence for two birefringent spindle components. *J. Cell Sci.* **22**, 115–131.

Fuge, H. (1971). Spindelbau, Mikrotubuli-verteilung und Chromosomen struktur während der I. meiotischen Teilung der Spermatocyten von *Pales ferruginea*. Eine electronenmikroskopische Analyse. *Z. Zellforsch. Mikrosk, Anat.* **120**, 579–599.

Fuge, H. (1972). Morphological studies on the structure of univalent sex chromosomes during anaphase movement in spermatocytes of the crane fly *Pales ferruginea*. *Chromosoma* **39**, 403–417.

Fuge, H. (1973). Verteilung der Mikrotubuli in Metaphase-und Anaphase-Spindeln der Spermatocyten von *Pales ferruginea*. Eine quantitative Analyse von Serienquerschnitten. *Chromosoma* **43**, 109–143.

Fuge, H. (1974). The arrangement of microtubules and the attachment of chromosomes to the spindle during anaphase in tipulid spermatocytes. *Chromosoma* **45**, 245–260.

Fuge, H. (1975). Anaphase transport of akinetochoric fragments in tipulid spermatocytes. Electron microscopic observations on fragment-microtubule interactions. *Chromosoma* **52**, 149–158.

Fuge, H. (1977). Ultrastructure of mitotic cells. *In* "Mitosis Facts and Questions" (M. Little, N. Paweletz, C. Petzelt, H. Ponstingl, D. Schroeter, H.-P. Zimmerman, eds.), pp. 51–77. Springer-Verlag, Berlin and New York.

Fuge, H. (1978). Ultrastructure of the mitotic spindle. *Int. Rev. Cytol., Suppl.* **6**, 1–58.

Fuge, H. (1980). Microtubule disorientation in anaphase half-spindles during autosome segregation in crane fly spermatocytes. *Chromosoma* **76**, 309–328.

Gawadi, N. (1971). Actin in the mitotic spindle. *Nature (London)* **234**, 410.

Gawadi, N. (1974). Characterization and distribution of microfilaments in dividing locust testis cells. *Cytobios* **10**, 17–35.

Gicquaud, C., Gruda, J., and Pollender, J.-M. (1980). La phalloidine protège la F-actine contre less effets destructeurs de l'acide osmique et du permanganate. *Eur. J. Cell Biol.* **20**, 234–239.

Girbardt, M., and Hädrich, H. (1975). Ultrastruktur des Pilzkernes III. Genese des Kernassoziierten Organells (NAO = "KCE"). *Z. Allg. Mikrobiol.* **15**, 157–167.

Goldstein, M. A., and Entman, M. L. (1979). Microtubules in mammalian heart muscle. *J. Cell Biol.* **80**, 183–195.

Gould, R. R., and Borisy, G. G. (1977). The pericentriolar material in Chinese hamster ovary cells nucleates microtubule activity. *J. Cell Biol.* **73**, 601–615.

Gould, R. R., and Borisy, G. G. (1978). Quantitative initiation of microtubule assembly by chromosomes from Chinese hamster ovary cells. *Exp. Cell Res.* **113**, 369–371.

Griffith, L. M., and Pollard, T. D. (1978). Evidence for actin filament-microtubule interaction mediated by microtubule associated proteins. *J. Cell Biol.* **78**, 958–965.

Gunning, B. E. S., Hardham, A. R., and Hughes, J. E. (1978). Evidence for initiation of microtubules in discrete regions of the cell cortex in *Azolla* root-tip cells, and an hypothesis on the development of cortical arrays of microtubules. *Planta* **143**, 161–179.

Haimo, L. T., Telzer, B. R., and Rosenbaum, J. L. (1979). Dynein binds to and crossbridges cytoplasmic microtubules. *Proc. Natl. Acad. Sci.* **76**, 5759–5763.

Hard, R., and Allen, R. D. (1977). Behaviour of kinetochore fibres in *Haemanthus katherinae* during anaphase movements of chromosomes. *J. Cell Sci.* **27**, 47–56.

Hardham, A. R., and Gunning, B. E. S. (1978). Structure of cortical microtubule arrays in plant cells. *J. Cell Biol.* **77**, 14–34.

Hayat, M. A. (1970). "Principles and Techniques of Electron Microscopy: Biological Applications," Vol. 1. Van Nostrand Reinhold, New York.

Heath, I. B. (1974a). Genome separation mechanisms in prokaryotes, algae, and fungi. *In* "The Cell Nucleus" (H. Busch, ed.), Vol. 2, pp. 487–515. Academic Press, New York.

Heath, I. B. (1974b). Mitosis in the fungus *Thraustotheca clavata*. *J. Cell Biol.* **60**, 204–220.

Heath, I. B. (1975a). The effect of antimicrotubule agents on the growth and ultrastructure of the fungus *Saprolegnia ferax* and their ineffectiveness in disrupting hyphal microtubules. *Protoplasma* **85**, 147–176.

Heath, I. B. (1975b). The possible significance of variations in the mitotic systems of the aquatic fungi (Phycomycetes). *BioSystems* **7**, 351–359.

Heath, I. B. (1978). Experimental studies of mitosis in the fungi. *In* "Nuclear Division in the Fungi" (I. B. Heath, ed.), pp. 89–176. Academic Press, New York.

Heath, I. B. (1980a). Variant mitoses in lower eukaryotes: indicators of the evolution of mitosis? *Int. Rev. Cytol.* **64**, 1–80.

Heath, I. B. (1980b). The behaviour of kinetochores during mitosis in the fungus *Saprolegnia ferax. J. Cell Biol.* **84**, 531–546.

Heath, I. B. (1981a). Mechanisms of nuclear division in fungi. *In* "The Fungal Nucleus" (K. Gull and S. Oliver, eds.), Cambridge Univ. Press, Cambridge, pp. 85–112.

Heath, I. B. (1981b). Nucleus associated organelles in fungi. *Int. Rev. Cytol.* **69**, 191–221.

Heath, I. B., and Heath, M. C. (1976). Ultrastructure of mitosis in the cowpea rust fungus *Uromyces phaseoli* var. *vignae. J. Cell Biol.* **70**, 592–607.

Heath, I. B., and Heath, M. C. (1978). Microtubules and organelle motility in the rust fungus *Uromyces phaseoli* var *vignae. Cytobiologie* **16**, 393–411.

Hepler, P. K., McIntosh, J. R., and Cleland, S. (1970). Intermicrotubule bridges in mitotic spindle apparatus. *J. Cell Biol.* **45**, 438–444.

Hinkley, R., and Telser, A. (1974). Heavy meromyosin-binding filaments in the mitotic apparatus of mammalian cells. *Exp. Cell Res.* **86**, 161–164.

Hyams, J. S., and Borisy, G. G. (1978). Nucleation of microtubules *in vitro* by isolated spindle pole bodies of the yeast *Saccharomyces cerevisiae. J. Cell Biol.* **78**, 401–414.

Ionué, S., and Ritter, Jr., H. (1975). Dynamics of mitotic spindle organization and function. *In* "Molecules and Cell Movement" (S. Inoue and R. E. Stephens, eds.), pp. 3–30. Raven, New York.

Inoué, S., and Ritter, H. (1978). Mitosis in *Barbulanympha*. II. Dynamics of a two-stage anaphase, nuclear morphogenesis, and cytokinesis. *J. Cell Biol.* **77**, 655–684.

Inoué, S., and Sato, H. (1967). Cell motility by labile association of molecules. *J. Gen. Physiol. Suppl.*, Pt. 2. **50**, 259–292.

Inoué, S., Fuseler, J., Salmon, E., and Ellis, G. W. (1975). Functional organization of mitotic microtubules: physical chemistry of the *in vivo* equilibrium system. *Biophys. J.* **15**, 725–744.

Jensen, C., and Bajer, A. (1973). Spindle dynamics and arrangement of microtubules. *Chromosoma* **44**, 73–89.

Kane, R. E., and Forer, A. (1965). The mitotic apparatus. Structural changes after isolation. *J. Cell Biol.* **25**, 31–39.

LaFountain, J. R. (1974). Birefringence and fine structure of spindles in spermatocytes of *Nephrotoma suturalis* at metaphase of the first meiotic division. *J. Ultrastruct. Res.* **46**, 268–278.

LaFountain, J. R. (1975). What moves chromosomes, microtubules or microfilaments? *BioSystems* **7**, 363–369.

LaFountain, J. R. (1976). Analysis of birefringence and ultrastructure of spindles in primary spermatocytes of *Nephrotoma suturalis* during anaphase. *J. Ultrastruct. Res.* **54**, 333–346.

Lambert, A. M., and Bajer, A. (1975). Fine structure dynamics of the prometaphase spindle. *J. Microscopie* **23**, 181–194.

Lambert, A. M., and Bajer, A. S. (1977). Microtubule distribution and reversible arrest of chromosome movements induced by low temperature. *Cytobiologie* **15**, 1–23.

Luftig, R. B., McMillan, P. N., Weatherbee, J. A., and Weihing, R. R. (1977). Increased visualization of microtubules by an improved fixation procedure. *J. Histochem. Cytochem.* **25**, 175–187.

Luykx, P. (1970). Cellular mechanisms of chromosome distribution. *Int. Rev. Cytol., Suppl.* **2**, 1–173.

Manton, I. (1964). Observations with the electron microscope on the division cycle in the flagellate *Prymnesium parvum* Carter. *J. R. Microsc. Soc.* **83**, 317–325.

Manton, I., Kowallik, K., and von Stosch, H. A. (1969a). Observations of the fine structure and development of the spindle at mitosis and meiosis in a marine diatom (*Lithodesmium undulatum*). I. Preliminary survey of mitosis in spermiogenesis. *J. Microsc. (Oxford)* **89**, 295–320.

Manton, I., Kowallik, K., and von Stosch, H. A. (1969b). Observations of the fine structure and development of the spindle at mitosis and meiosis in a marine diatom (*Lithodesmium undulatum*). II. The early meiotic stages in male gametogenesis. *J. Cell Sci.* **5**, 271–298.

Manton, I., Kowallik, K., and von Stosch, H. A. (1970a). Observations of the fine structure and development of the spindle at mitosis and meiosis in a marine diatom (*Lithodesmium undulatum*). III. The later stages of meiosis I in male gametogenesis. *J. Cell Sci.* **6**, 131–157.

Manton, I., Kowallik, K., and von Stosch, H. A. (1970b). Observations of the fine structure and development of the spindle at mitosis and meiosis in a marine diatom (*Lithodesmium undulatum*). IV. The second meiotic division and conclusion. *J. Cell Sci.* **7**, 407–444.

Margolis, R. L., and Wilson, L. (1978). Opposite end assembly and disassembly of microtubules at steady state *in vitro*. *Cell* **13**, 1–8.

Maupin-Szamier, I. P., and Pollard, T. D. (1978). Actin filament destruction by osmium tetroxide. *J. Cell Biol.* **77**, 837–852.

McDonald, K., Pickett-Heaps, J. D., McIntosh, J. R., and Tippit, D. H. (1977). On the mechanism of anaphase spindle elongation in *Diatoma vulgare*. *J. Cell Biol.* **74**, 377–388.

McDonald, K. L., Edwards, M. K., and McIntosh, J. R. (1979). Cross-sectional structure of the central mitotic spindle of *Diatoma vulgare*. Evidence for specific interactions between antiparallel microtubules. *J. Cell Biol.* **83**, 443–461.

McGill, M., and Brinkley, B. R. (1975). Human chromosomes and centrioles as nucleating sites for the *in vitro* assembly of microtubules from bovine brain tubulin. *J. Cell Biol.* **67**, 189–199.

McIntosh, J. R. (1974). Bridges between microtubules. *J. Cell Biol.* **61**, 166–187.

McIntosh, J. R. (1979). Cell division. *In* "Microtubules" (K. Roberts and J. S. Hyams, eds.), pp. 428–441. Academic Press, New York.

McIntosh, J. R., and Landis, S. C. (1971). The distribution of spindle microtubules during mitosis in cultured human cells. *J. Cell Biol.* **49**, 468–497.

McIntosh, J. R., Hepler, P. K., and Van Wie, D. G. (1969). Model for mitosis. *Nature (London)* **224**, 659–663.

McIntosh, J. R., Cande, W. Z., Snyder, J. A., and Vanderslice, K. (1975a). Studies on the mechanisms of mitosis. *Ann. N. Y. Acad. Sci.* **253**, 407–427.

McIntosh, J. R., Cande, W. Z., and Snyder, J. A. (1975b). Structure and physiology of the mammalian mitotic spindle. *In* "Molecules and Movement" (S. Inoué and R. E. Stephens, eds.), pp. 31–76. Raven, New York.

McIntosh, J. R., McDonald, K. L., Edwards, M. K., and Ross, B. M. (1979a). Three-dimensional structure of the central mitotic spindle of *Diatoma vulgare*. *J. Cell Biol.* **83**, 429–442.

McIntosh, J. R., Sisken, J. E., and Chu, L. K. (1979b). Structural studies on mitotic spindles isolated from cultured human cells. *J. Ultrastruct. Res.* **66**, 40–52.

Moens, P. B. (1976). Spindle and kinetochore morphology of *Dictyostelium discoideum*. *J. Cell Biol.* **68**, 113–122.

Moens, P. B. (1978). Computer-assisted analysis and display of nuclear structures. *In* "Proc. 9th Int. E.M. Congress," (J. Sturgess, ed.), Vol. 3, pp. 557–563. Microscopical Society of Canada, Toronto.

Moens, P. B. (1979). Kinetochore microtubule numbers of different sized chromosomes. *J. Cell Biol.* **83**, 556–561.

Nicklas, R. B. (1971). Mitosis. *Adv. Cell Biol.* **2**, 225–297.

Nicklas, R. B., Brinkley, B. R., Pepper, D. A., Kubai, D. F., and Rickards, G. K. (1979a). Electron microscopy of spermatocytes previously studied in life: Methods and some observations on micromanipulated chromosomes. *J. Cell Sci.* **35**, 87–104.

Nicklas, R. B., Kubai, D. F., and Ris, H. (1979b). Electron microscopy of the spindle in locally heated cells. *Chromosoma* **74**, 39–50.

Oakley, B. R., and Heath, I. B. (1978). The arrangement of microtubules in serially sectioned spindles of the alga *Cryptomonas. J. Cell Sci.* **31**, 53–70.

Peterson, S. P., and Berns, M. W. (1980). The centriolar complex. *Int. Rev. Cytol.* **64**, 81–106.

Peterson, J. B., and Ris, H. (1976). Electron-microscopic study of the spindle and chromosome movement in the yeast *Saccharomyces cerevisiae. J. Cell Sci.* **22**, 219–242.

Pickett-Heaps, J. D. (1969). The evolution of the mitotic apparatus: an attempt at comparative ultrastructural cytology in dividing plant cells. *Cytobios.* **1**, 257–280.

Pickett-Heaps, J. D., and Fowke, L. C. (1970). Cell division in *Oedogonium.* II. Nuclear division in *O. cardiacum. Aust. J. Biol. Sci.* **23**, 71–92.

Pickett-Heaps, J. D., and Tippit, D. H. (1978). The diatom spindle in perspective. *Cell* **14**, 455–467.

Pickett-Heaps, J. D., Tippit, D. H., and Leslie, R. (1980). Light and electron microscopic observations on cell division in two large pennate diatoms, *Hantzschia* and *Nitschia.* II. Ultrastructure. *Eur. J. Cell Biol.* **21**, 12–27.

Rieder, C. L., and Bajer, A. S. (1978). Effect of elevated temperatures on spindle microtubules and chromosome movements in cultured newt lung cells. *Cytobios* **18**, 201–234.

Ritter, H., Inoué, S., and Kubai, D. F. (1978). Mitosis in *Barbulanympha.* I. Spindle structure, formation, and kinetochore engagement. *J. Cell Biol.* **77**, 638–654.

Roberts, K., and Hyams, J. S. (1979). "Microtubules." Academic Press, New York.

Roos, U.-P. (1975). Mitosis in the cellular slime mould *Polysphondylium violaceum. J. Cell Biol.* **64**, 480–491.

Salmon, E. D. (1975a). Pressure induced depolymerization of spindle microtubules. I. Changes in birefringence and spindle length. *J. Cell Biol.* **65**, 603–614.

Salmon, E. D. (1975b). Pressure induced depolymerization of spindle microtubules. II. Thermodynamics of *in vivo* spindle assembly. *J. Cell Biol.* **66**, 114–127.

Sato, H. (1975). The mitotic spindle. *In* "Aging Gametes" (R. J. Blandau, ed.), pp. 19–49. Karger, Basel.

Sato, H., Ellis, G. W., and Inoué, S. (1975). Microtubular origin of mitotic spindle form birefringence. Demonstration of the applicability of Wiener's equation. *J. Cell Biol.* **67**, 501–517.

Schibler, M. J., and Pickett-Heaps, J. D. (1980). Mitosis in *Oedogonium:* spindle microfilaments and the origin of the kinetochore fiber. *Eur. J. Cell Biol.* **22**, 687–698.

Schloss, J. A., Milsted, A., and Goldman, R. D. (1977). Myosin subfragment binding for the localization of actin-like microfilaments in cultured cells: A light and electron optical study. *J. Cell Biol.* **74**, 794–815.

Seagull, R. W., and Heath, I. B. (1980). The organization of cortical microtubule arrays in the radish root hair. *Protoplasma* **103**, 205–229.

Snyder, J. A., and McIntosh, J. R. (1975). Initiation and growth of microtubules from mitotic centers in lysed mammalian cells. *J. Cell Biol.* **67**, 744–760.

Telzer, B. R., and Rosenbaum, J. L. (1979). Cell cycle-dependent, *in vitro* assembly of microtubules onto the pericentriolar material of HeLa cells. *J. Cell Biol.* **81**, 484–497.

Telzer, B. R., Moses, M. J., and Rosenbaum, J. L. (1975). Assembly of microtubules onto kinetochores of isolated mitotic chromosomes of HeLa cells. *Proc. Natl. Acad. Sci. U.S.A.* **72**, 4023–4027.

Tippit, D. H., and Pickett-Heaps, J. D. (1976). Apparent amitosis in the binucleate dinoflagellate *Peridinium balticum*. *J. Cell Sci.* **21**, 273–289.

Tippit, D. H., Pillus, L., and Pickett-Heaps, J. D. (1980a). The organization of spindle microtubules in *Ochromonas danica*. *J. Cell Biol.* **87**, 531–545.

Tippit, D. H., Pickett-Heaps, J. D., and Leslie, R. (1980b). Cell division in two large pennate diatoms *Hantzschia* and *Nitzschia*. III. A new proposal for kinetochore function during prometaphase. *J. Cell Biol.* **86**, 402–416.

Wang, Y-L., and Taylor, D. L. (1979). Distribution of fluorescently labeled actin in living sea urchin eggs during early development. *J. Cell Biol.* **81**, 672–679.

Weisenberg, R. C. (1972). Microtubule formation *in vitro* in solutions containing low calcium concentrations. *Science N.Y.* **177**, 1104–1105.

Wilson, H. J. (1969). Arms and bridges on microtubules in the mitotic apparatus. *J. Cell Biol.* **40**, 854–859.

Witt, P. L., Ris, H., and Borisy, G. G. (1979). Nucleation of microtubules by kinetochores in Chinese hamster ovary cells. *J. Cell Biol.* **83**, 379a.

12

Mitosis: Studies of Living Cells—A Revision of Basic Concepts

A. S. BAJER and J. MOLÈ BAJER

I. INTRODUCTION

The purpose of this chapter is to draw attention (1) to certain features of a monopolar spindle, especially its ability for chromosome transport, and (2) to new conceptual interpretations of the course of mitosis derived from these studies. Both of these points are discussed in Section III. In addition, Section II draws attention to the subtle relationship between the spatial arrangement of microtubules and chromosome movement. The observations presented here are mostly unpublished data on monopolar spindles in the newt. We have found that in newt lung epithelial cells the monopolar spindle, i.e., a half-spindle alone, has a practically unlimited capability for migration comparable to the separation of spindle poles (elongation of the spindle)

MITOSIS/CYTOKINESIS
Copyright © 1981 by Academic Press, Inc.
All rights of reproduction in any form reserved.
ISBN 0-12-781240-7

during standard anaphase. Kinetochore fibers of such a spindle are directed to the single pole, and the kinetochore fibers are capable of chromosome transport toward and away from this pole. These movements are comparable to normal kinetochore-to-pole movements. Thus, a bipolar spindle is required only for the alignment of chromosomes at the equatorial region, i.e., a metaphase plate is needed for regular segregation of genetic material but not for chromosome movement itself. Furthermore, we believe that anaphase-like prometaphase demonstrates that the role of interzonal microtubules is to slow down chromosome movement and not to push two half-spindles apart. Comparison of the monopolar spindle movement with standard anaphase movement suggests that chromosomes themselves determine the life span of the kinetochore fiber. Thus, the development of the spindle after breaking of the nuclear envelope is regulated by the chromosomes (kinetochores) themselves. We maintain that these interpretations, based on observations of living cells, apply to all standard astral spindles.

II. THEORETICAL FRAMEWORK

A. Development of the Mitotic Spindle: Order versus Disorder

The purpose of these general remarks is to draw attention to the basic relationship between chromosome movement and microtubular spindle organization. The conclusions are based largely on unpublished data now being gathered on the minimum microtubular structure of the functional *standard* anastral spindle of *Haemanthus* endosperm and the astral spindle of the Oregon newt *Taricha granulosa granulosa*. What emerges from these studies is the ability to define two extreme structural conditions in which the spindle does not function.

Mitosis can be viewed descriptively as sequential equilibrium states* that are followed by a rapid disturbance, e.g., breaking of the nuclear envelope or separation of kinetochores at the start of anaphase. Each equilibrium state may last quite a long time. Disturbance of an equilibrium or formation of a new equilibrium results either in formation of a new structure (the spindle in prophase, the phragmoplast or mid-body in telophase), in movements (the

---

*A true equilibrium is probably never reached in the spindle. The spindle is in fact not in a steady state but in quasi-static process, i.e., in a process proceeding slowly enough that the system (the spindle) passes through a continuous sequence of equilibrium states. A quasi-static process may or may not be reversible.

movement of chromosomes in prometaphase or anaphase), or both (e.g., the movement of centrioles). During cell division there are three equilibrium states characterized by the presence of a comparatively regular structure.

1. Clear zone formation in anastral mitosis (this stage can last for weeks at low temperatures in natural conditions; Lambert, 1970, 1980). The clear zone is a cytoplasmic component of the mitotic spindle which appears before breaking of the nuclear envelope in anastral spindles; the clear zone can be compared to very early stages of spindle formation in animal cells.

2. Metaphase spindle. The metaphase is characterized by kinetochore microtubules that tend to be, but never are, arranged in parallel arrays. Metaphase can last a very long time, e.g., in some marine animals.

3. Formation of a phragmoplast or mid-body during cytokinesis.

We believe that changes of microtubule arrangement result from two superimposed and functionally related processes: assembly–disassembly and lateral association/dissociation. Regulatory factors for these processes in living cells are largely unknown.

Evidence for microtubule assembly/disassembly is well established (Inoué et al., 1975). The evidence for lateral association/dissociation of microtubules is not as straightforward and cannot yet be presented as numerical data (Bajer, 1977). The major evidence for lateral association/dissociation is based on studies of chromosome behavior in living cells, combined with electron microscopy (EM) studies. Thus, e.g., the observation of large numbers of microtubules arranged in doublets (*Haemanthus* endosperm, Bajer, 1977, Figs. 9-5 and 9-8; Bajer and Molè-Bajer, 1975, Fig. 2) among irregular arrays is improbable unless some force brings and/or holds them together. Additional evidence for the lateral association of microtubules has been obtained from studies *in vitro* (Haimo et al., 1979) which shows that dynein promotes the lateral association of brain microtubules. The arrangements of microtubules of the functional spindle depend most likely on the subtle threshold of the numerical ratio between kinetochore and nonkinetochore microtubules and the rate of their assembly/disassembly. There are, however, two extreme arrangements of kinetochore fibers in the anastral spindle of *Haemanthus* when chromosome movements cease: very regular (parallel) and disorganized (Bajer et al., 1975). Numerical or spatial definition of these two extreme conditions is not an easy task because of variability between cells and because the large numbers of microtubules do not permit reliable three-dimensional reconstruction. It should be stressed that we are not able to disassemble microtubules of the spindle with any treatment without initial, often short-lasting, rearrangement of microtubules.

A useful aspect of these suggestions is that they allow an assessment of spindle function (understood as the ability to move chromosomes) in terms of the arrangement of microtubules. Disturbance of the equilibrium is always functionally connected with a change in the arrays of microtubules from approximately parallel to divergent/irregular (or irregular to parallel) when a new equilibrium is reached. This relation is rather straight-forward during mitosis. Chromosome movements gradually cease during pro-metaphase, when the microtubules of the kinetochore fibers become approximately parallel (metaphase equilibrium). Chromosome movements start again at the onset of anaphase, when the equilibrium is abruptly disturbed by the splitting of chromosomes. During anaphase, disassembly and rearrangement of microtubules occur simultaneously. Under standard conditions, there are more microtubules in the astral and anastral spindles than are needed to move the chromosomes; chromosome velocity in such conditions is determined not by the number of microtubules but by the rate of reorganization of the spindle structure, as suggested by Taylor (1959, 1963). If the microtubule reorganizational changes proceed quickly and the microtubular structure is rapidly shifted toward a new equilibrium, e.g., by temperature shock or drugs (Lambert and Bajer, 1977; Bajer *et al.*, 1975), and if non-kinetochore microtubules, which are probably non-anchored and therefore less stable disassemble first, then kinetochore microtubules become parallel and chromosome movements are arrested. Chromosome movements start again when this equilibrium is disturbed. Microtubule arrays can be followed with the polarizing microscope, but only if the thickness of the structure is known and if changes in orientation do not occur at the same time; if so, one can estimate the rate of assembly/disassembly of microtubules. Assembly/disassembly of microtubules in the anastral spindle is most pronounced at the polar regions, where microtubules terminate and are not arranged in an orderly fashion. Studies using the polarizing microscope, which detects only the parallel component of the spindle, give the impression that the spindle is a highly regular structure. In standard metaphase, such organization exists only in a small region of the spindle (the kinetochore fiber close to the kinetochore), which very likely transmits but does not produce the force for chromosome movement.

B. Conceptual Conclusions

Pronounced spatial rearrangements of microtubules may occur rapidly in experimental conditions. Chromosomes move only when a kinetochore fiber changes its arrangement from approximately parallel to non-parallel. Movements are arrested in these two extreme conditions (parallel and disorganized). For different types of spindles (astral versus anastral) the changes occur predominantly in different areas of the kinetochore fibers. Microtubule rearrangement may occur rapidly under a variety of experimental conditions, and unless monitored, a distorted image of spindle structure, especially in EM, may be obtained. Parallel arrangements of microtubules and their assembly/disassembly can be assessed in living cells using the polarizing microscope, but these techniques cannot be used to assess microtubule reorientation (change of arrangement) which may occur simultaneously with their assembly/disassembly.

III. UNPUBLISHED OBSERVATIONS

A. The Mitotic Spindle in the Newt: Kinetochore Fibers Form the Central Spindle

The existence of structural variations of the mitotic apparatus in different tissues of the same organism is conceptually uncomfortable and technically troublesome. Variations within the same tissue, which also occur, are greater in some tissues than in others. The spindles of epithelial cells of the newt *T. granulosa granulosa*, studied in this chapter, are especially variable. These flat cells with large chromosomes may pose mechanical problems for chromosome distribution, and deviations are probably more common than in other tissues. This permits one to follow chromosome movements in unique conditions comparable to intricate microsurgery performed by the cell itself. Furthermore, these cells can be observed directly in a Rose chamber (Rose *et al.*, 1958) and therefore are not disturbed by any experimental manipulation. Survival of such cells is 100%, which shows that the movements are not the result of irreversible processes which often occur in dying or experimentally treated cells.

1. Observations and Interpretations

In newt epithelium and coracoid bone fibroblasts (Molè-Bajer, 1975), centriolar pairs usually start migrating before the breaking of the nuclear envelope. Migration continues in prometaphase and results in the elongation of the prometaphase spindle (Fig. 1). The movement of asters, especially in early stages, is often unpredictable. Since a majority of chromosomes in newt epithelium are initially monopolar oriented (Molè-Bajer *et al.*, 1975), as also observed in PtK$_2$ cells (Roos, 1976), the asters seem to form two separate prometaphase half-spindles that may be aligned with each other at different angles. Their long axes are determined by the majority of monopolar-oriented chromosomes attached to that particular pole.

It is seldom possible to establish in living cells, with an accuracy of 1–5 minutes, the moment when the kinetochore fiber of a particular chromosome is formed.

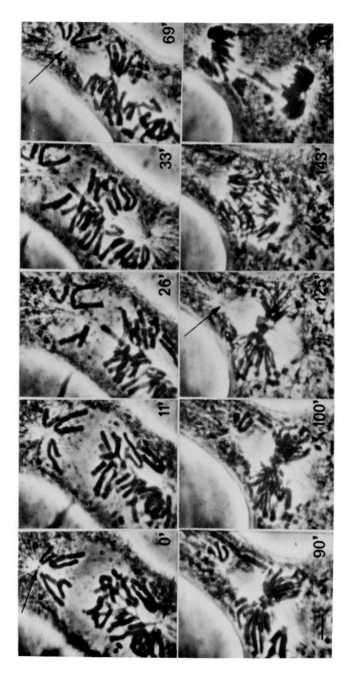

Fig. 1. Shortening of the spindle during prometaphase-metaphase. Time in minutes on micrographs. At the beginning of prometaphase, numerous monopolar-oriented chromosomes are located close to the spindle poles with visible centrosomes (arrows at 0', 69', 125'). The axis of the spindle is established by bipolar-oriented chromosomes, and a regular metaphase plate forms. This is followed by anaphase with chromosome bridges. This cell shows considerable spindle shortening (about 1/3) in prometaphase. The spindle elongates only slightly during anaphase. Micrographs from 16-mm film. Bar = 20 μm.

During standard prometaphase, usually only a few chromosomes establish bipolar kinetochore fibers within minutes after the breaking of the nuclear envelope, and only when chromosomes become bipolar does the spindle behave as a single structural unit. The number of bipolar-oriented chromosomes increases rapidly during prometaphase.

Immunocytochemical visualization of microtubules combined with EM studies (Bajer *et al.*, 1980) in the newt demonstrates that the asters separate without any interconnecting microtubules (central spindle). As mitosis progresses, in early stages of aster separation, the astral microtubules may be bent and seem to trail the centrioles like the tail of a comet. This rare configuration can be easily interpreted in EM and polarizing microscopes as the central spindle. However, our combined observations with the EM and light microscopes failed to demonstrate the existence of a central spindle in the newt, i.e., in the object considered classical for demonstration of this structure. The few bipolar kinetochore fibers which form first are the initial bipolar spindle in prometaphase. This initial thin bipolar spindle, composed of bipolar kinetochore fibers and non-kinetochore microtubules of unclear origin, can easily be interpreted as the central spindle, especially in the polarizing microscope.

2. Conclusions

In the newt the early prometaphase spindle is composed of two half-spindles with asters connected (held together) by kinetochore fibers. The movements of asters are independent and autonomous until they become connected by bipolar kinetochore fibers. This conclusion is based on observations of the initial angle between two groups of monopolar chromosomes in early prometaphase. Observations on migration of monopolar spindles (Section III,B) support this conclusion. Asters anchor and focus the spindle microtubules toward one center. The mechanism of aster mobility is not clear (see literature in Harris, 1978; Wolf, 1975, 1978).

B. Monopolar Spindle and Anaphase-like Prometaphase

Two extremes in the malfunction of asters are (1) their total failure to separate (migrate) and (2) their precocious and too distant separation. These malfunctions result in (1) monopolar division followed by the restitution of the nucleus (Fig. 4B, 6A–C, D–E), or (2) anaphase-like prometaphase (a new term, see Figs. 5B, C, 7).

1. Observations

a. **Monopolar Division.**   One of the asters may dissociate from the spindle in prophase, i.e., before kinetochore fibers are formed (Bajer *et al.*,

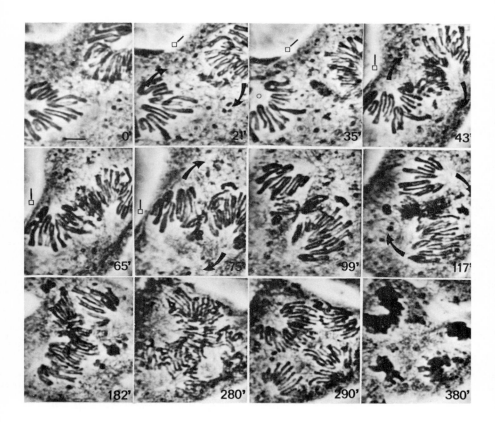

Fig. 2. Anaphase-like prometaphase. Time in minutes on micrographs. Two prometa-phase monopolar half-spindles containing 8 and 14 chromosomes migrate away from each other. They rotate clockwise (thick curved arrows) and establish a thin bipolar spindle (be-tween the arrows at 75′). This process is very slow and takes more than 20 minutes (43′–75′). Monopolar-oriented chromosomes are attached to opposite sides of the bipolar spindle (seen clearly between 75′ and 117′ and the whole spindle continues rotation; see text). Finally, an irregular metaphase plate is formed and multipolar division follows (280′–380′). Note: The cell was rotated between 35′ and 43′; the cell axis is marked by the line with the rectangle. The spindle pole of one monopolar half-spindle is marked by the circle at 35′. Micrographs from 16-mm film. Thick bar = 20 μm.

1980). The remaining aster forms a monopolar spindle (Fig. 3A). A "monopolar figure" (Paweletz and Mazia, 1979; Mazia *et al.*, 1981) is formed when the migration of centrosomes fails. A characteristic feature of chromo-somes in all monopolar spindles is their continuous oscillation toward aad away from the single pole. The kinetochore fibers of these chromosomes are

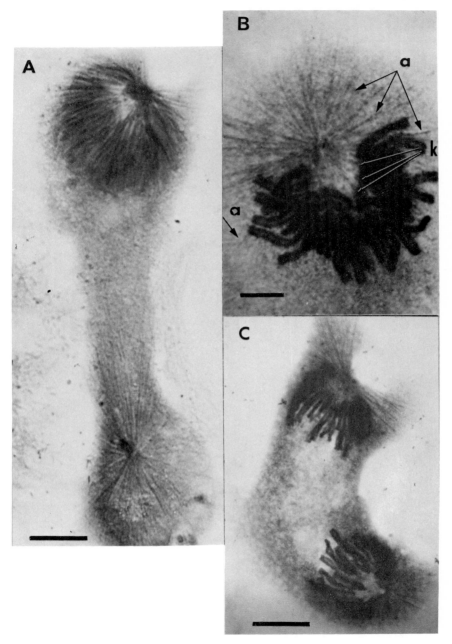

Fig. 3. Behavior of asters in the newt mitosis (A–B). Monopolar spindles that were formed when the aster dissociated from the spindle (A) and asters did not separate (B). PAP immunocytochemical (De Brabander *et al.*, 1977) demonstration of microtubules. Chromo-

connected to this single pole until reorientation. Immunochemical studies show (Fig. 3B) that astral microtubules are found behind kinetochores. Functional bipolar fibers do not form until reorientation, which usually occurs during transformation into the bipolar spindle. This takes place in about 85% of monopolar prometaphase spindles, which finally form normal bipolar spindles, followed by a standard anaphase. In such cases, the transitory monopolar spindle may be considered a monopolar phase of standard prometaphase.

b. Anaphase-like Prometaphase. In rare instances in early prometaphase, there is a delay in the formation of bipolar kinetochore fibers, and all chromosomes remain monopolar but orient toward two separated centrosomes. In such cells, two prometaphase half-spindles containing a random number of monopolar-oriented double-chromatid chromosomes migrate away from each other, often for a distance two or three times greater than in standard anaphase separation (Figs. 2–35 and 4A–D). A bipolar spindle does not form, and the cell may even be torn apart. This process resembles cleavage, although it is much slower and is not accompanied by the formation of a mid-body.

The movement of poles apart from each other has several features of standard anaphase. The life span of a kinetochore fiber in anaphase-like prometaphase (double-chromatid chromosome) may be 10 to 15 times greater than in standard anaphase (single-chromatid chromosomes). It may be therefore related to kinetochore structure (as suggested by Lyubskii, 1974a, b, and Alov and Lyubskii,1974).The splitting of kinetochores at the onset of standard anaphase is followed by final and fast disassembly of kinetochore fibers. Such splitting does not occur during anaphase-like prometaphase, and oscillatory movement of kinetochores toward and away from the centrioles continues during migration of prometaphase half-spindles; during this process, kinetochores can come closer to the pole than during standard anaphase. This finding can be generalized: the shortest distance between a kinetochore and the pole is achieved in some standard pro-

---

somes counterstained with toluidine blue. (A) Monopolar spindle with well-developed aster. The second chromosome-free aster dissociated from the spindle and migrated away. (B) Monopolar spindle. Pronounced aster with all chromosomes connected to one pole. Some kinetochore fibers (arrow k) are seen. Astral microtubules (arrow a) stretch behind chromosome arms, and therefore distal kinetochores are immersed in the tubulin pool. There are few or no microtubules connected to the distal kinetochores from the centrioles until reorientation, which occurs in about 85% of cells. (C) Anaphase-like prometaphase. Two monopolar spindles move away. Well-developed asters and no interzonal microtubules are seen. False interzone between chromosome group may contain some non-polymerized fibrils. Distribution of chromosomes is 12:10 in C. Bar on A–C = 10 μm.

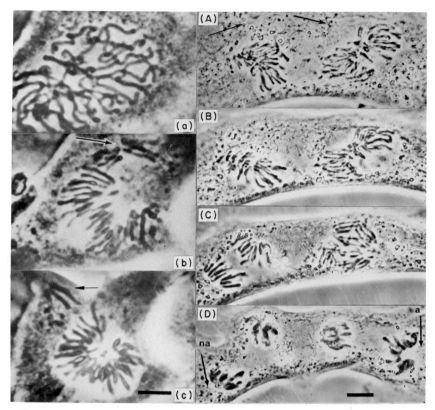

Fig. 4.   Monopolar division with migration of two chromosomes outside the spindle, and spindles with and without asters. (a–c): (a) Early prometaphase. (b) Monopolar spindle with two chromosomes eliminated (arrow). These two chromosomes show bipolar orientation. They are in a slightly delayed chromosomal cycle as seen especially in (c), where the main group of chromosomes is in early telophase, while the length of the eliminated chromosomes corresponds to early anaphase. It is not possible to determine whether there is an additional aster toward which these two chromosomes migrate. The shape of the cell suggests that this is not the case. Monopolar-oriented chromosomes oscillate but do not approach the single pole before formation of the restitution nucleus as closely as centrophilic chromosomes. Time after a (0), b (2 hr), c (2 hr 15 min). Bar = 10 μm. (A–D) Anaphase-like prometaphase followed by formation of two bipolar spindles. During anaphase-like prometaphase two monopolar prometaphase half-spindles, one of them containing 8 chromosomes [they can be counted in (C)], moved away from each other. They established two bipolar spindles, and each of them entered normal anaphase. It is unlikely that all poles contain centrioles and asters, but this cannot be determined on the basis of phase contrast micrographs. There are arrangements of granules in somewhat astral fashion; arrows in A may indicate that two asters are located between two bipolar spindles. It is unlikely, however, that the pole marked (arrow na—no aster, arrow a marks aster in D) contains an aster and a centriole. This is indicated by the rather diffuse chromosome movements. Cell observed in Rose chamber. Time after A (0 min), B (25 min), C (40 min), D (49 min). Bar = 10 μm.

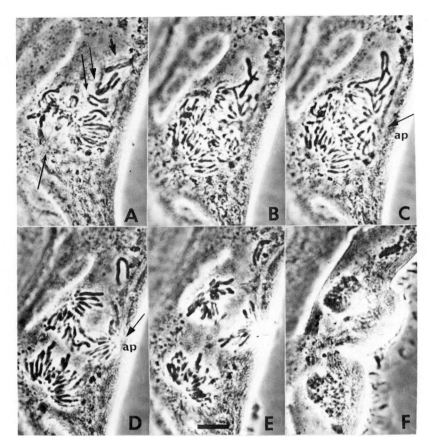

Fig. 5. Chromosome movements outside the spindle and formation of additional centers. Bipolar spindle with asters (arrows in A indicate the most probable position of asters with centrioles). A few (three or four) chromosomes do not return to the plate before the start of anaphase. These chromosomes (arrows in A) are bipolar oriented, with one polar region (close to the cell periphery) almost certainly without a centriole and an aster. During anaphase tripolar anaphase takes place, i.e., one sister chromosome group is split into two groups and only one migrates toward the main pole (arrow in A). The two centers are aligned perpendicular to the long axis of the main spindle. During anaphase (B–E) all (except one centrophilic) sister chromosomes migrate toward the aster and the additional pole (arrow ap in C–D). The particular single chromosome (short arrow in A), with its kinetochore leading and therefore probably most well-developed kinetochore fiber, moves against the astral microtubules of the main spindle. The single chromosome migrates over a considerable distance. The arrangement of granules does not indicate the existence of an established aster toward which the third kinetochore is moving. This possibility, however, cannot be excluded on the basis of light microscopic observation; cf. Fig. 9. Time after A (0 min), B (13 min), C (15 min), C (19 min), D (25 min), F (48 min). Bar = 10 μm.

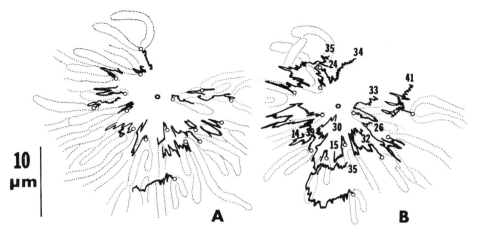

10 µm

A   **B**

Fig. 6. The paths of kinetochores in a monopolar spindle which later transform into a bipolar one. Normal metaphase is followed by regular chromosome distribution. (A) shows the paths during 10 minutes and (B) at various lengths of time (minutes marked). Variation in behavior between different chromosomes is clearly shown; chromosomes at the edge oscillate less than those surrounded by others on both sides. Some chromosomes also move sideways. The direction of oscillation is toward and away from the centrosome (circle). Also, see Fig. 7.

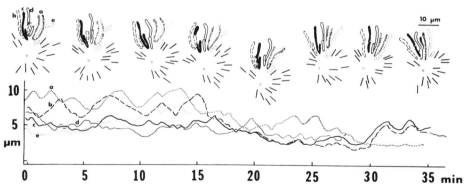

Fig. 7. Chromosome oscillation in monopolar spindle. The distance from the centrosome (0) is plotted against the time. The drawing shows the position of chromosomes at the same time. The position of some other chromosomes is marked by lines. The paths of some chromosomes in the same cell are shown in Fig. 6. There is a transitory correlation in the movement between some chromosomes which are close together, e.g., chromosomes b and c move synchronously in time (5–10 and 17–34 minutes). These two chromosomes are superimposed. Chromosome a is influenced by its neighbor (e) in 12 minutes. Occasionally more than two chromosomes at the time move synchronously (c, d, e at about 22 minutes). Synchrony of movement is interpreted by temporary lateral association of kinetochore fibers. These graphs were generated by projection of the film with a speed of one frame/sec on chart recorder advanced 12.5 cm/min.

metaphases by single monopolar-oriented chromosomes termed "cen-trophilics" (Zirkle, 1957) that are common features of prometaphase in a variety of tissues.

During anaphase-like prometaphase in the newt, one or both separating half-spindles rotate (usually at the cell periphery) and migrate toward the center of the cell, where the two prometaphase half-spindles fuse and a bipolar spindle may be established. Normal anaphase often follows. Several variations of this process (see, e.g., Brenner *et al.*, 1977) have been observed by us in over 30 cases (Fig. 1 illustrates such variations). A more detailed analysis of this process will be given elsewhere (Bajer, *J. Cell Biol.*, in press).

2. Summary

Asters migrate independently of each other over a long distance, and one of them may dissociate from the spindle and migrate to the cell periphery. The remaining aster becomes part of a monopolar spindle after breaking of the nuclear envelope. Monopolar chromosome orientation is established as a rule first at the onset of prometaphase, and if bipolar kinetochore fibers do not form, the asters may separate precociously and the two monopolar half-spindles may move apart in an anaphase-like fashion. During the astral movements, monopolar-oriented chromosomes show an oscillatory motion similar to that in standard prometaphase and anaphase, i.e., they have the ability to move toward and away from the centrosome. Thus chromosome movement is superimposed on the migration of the aster, as in standard anaphase. Thus, the two most characteristic features of a monopolar spindle are (1) oscillatory movements of the monopolar-oriented chromosomes and (2) migration of the asters.

3. Interpretations

Anaphase-like-prometaphase is a rare event usually seen only in late stages before the chromosomes return to the metaphase plate (Rieder, 1977). The process was observed by Galeotti in 1893 (see Wilson, 1928, Fig. 73) and in PtK cells in numerous laboratories (Brenner *et al.*, 1977; M. De Braban-der, Belgium; W. Heneen, Sweden; C. Rieder, United States, personal communications). If anaphase-like prometaphase is followed by cytokinesis, chromosomes are distributed randomly.

Anaphase-like prometaphase begins shortly after breaking of the nuclear envelope in cells with persistent monopolar orientation of all chromosomes. Therefore, it is argued that the mitotic distribution achieved by rapid movement may not be a preferable mechanism. A more sluggish movement is certainly more safe, as it permits reorientation and/or formation of bipolar-oriented fibers which then align kinetochores at the equatorial re-gion, a necessary condition for standard chromosome distribution.

Chromosome oscillation during the migration of whole half-spindles in monopolar division and anaphase-like prometaphase demonstrate two points. (1) A kinetochore with a single spindle fiber pointing to one pole can move toward and away from that spindle pole. (2) Spindle poles can move apart in the absence of any microtubular connections in the interzone. Thus, both in monopolar division and in anaphase-like prometaphase, the motive force for chromosome movement and polar separation is located in front of and not behind kinetochores. The data exclude the model of chromosome movement and spindle elongation proposed by Margolis *et al.* (1978).

The question arises, however, of whether the mechanism of standard anaphase is the same as that of anaphase-like prometaphase. During both processes, chromosomes oscillate back and forth and chromosome velocities are very similar. Oscillatory movement superimposed on the migration toward the poles during standard anaphase is seen in numerous films on mitosis (e.g., Bloom and Zirkle, ca. 1954) but has not been stressed in the literature (cf. Rieder, 1977). Anaphase-like prometaphase is, however, a prolonged, non-standard "double monopolar" prometaphase that is different from the usual spindle in some chromosome structure (double in anaphase-like prometaphase versus single in normal anaphase) and life span of the kinetochore fiber. Kinetochore fibers shorten and disassemble finally and irreversibly at the end of standard anaphase, in contrast to anaphase-like prometaphase, in which they both shorten and elongate as the chromosomes oscillate. However, during typical prometaphase in the newt, kinetochore fibers of centrophilic chromosomes (Zirkle, 1957) can shorten even more than during standard anaphase. Therefore, the degree of shortening of a kinetochore fiber is not an essential feature of anaphase. We believe, therefore, that the molecular mechanisms of kinetochore movement in anaphase-like prometaphase, prometaphase, and standard anaphase are the same (Östergren, 1951). We feel justified therefore in comparing the separation of poles (spindle elongation) and kinetochore-to-pole movement in anaphase-like prometaphase and standard anaphase.

C. Anaphase of Chromosomes Outside the Spindle

At the start of anaphase (in about 5% of the cells) one or a few chromosomes may be located close to the centrioles in the astral region, either at the edge of the spindle or between the spindle pole and the cell periphery (Figs. 4A–C, 5). If it is assumed that all microtubules attached to a particular kinetochore and/or aster are not of mixed polarity, then such a system offers an opportunity to study the relation of microtubule polarity to chromosome movement.

1. Observations

a. **Multiple Chromosome Configuration.**   Two variations exist in the arrangement of chromosomes which start anaphase outside the metaphase plate.

1. An additional center (pole) exists, and chromosomes are within (inside) minute accessory spindles. Chromosomes undergo characteristic movements toward the main and accessory poles, which leaves no doubt that bipolar fibers are well developed. The behavior of such chromosomes is comparable to that of a standard metaphase arrangement in less crowded conditions.

2. No additional centers resembling asters are detected in light microscopy, although this is not rigorous evidence of their absence. Chromosomes which are left in the astral region during metaphase are almost motionless, and their behavior is often difficult to define clearly; under the light microscope, such chromosomes often cannot be classified as having either a typical bipolar or monopolar orientation. During anaphase, one sister chromosome group moves toward the astral center (centrosome), while the other migrates in the opposite direction toward the cell periphery, sometimes for a long (20 μm or so) distance. During a long migration, chromosomes may oscillate back and forth, as during normal anaphase. EM data detect very few non-kinetochore microtubules. Thus, at least in some cases, such a half-spindle is composed primarily of kinetochore microtubules.

b. **Single Chromosomes.**   In *Taricha* lung epithelium cells, in contrast to heart fibroblasts (Zirkle, 1970), the return of monopolar-oriented chromosomes to the metaphase plate is not a prerequisite for the start of anaphase. However, as pointed out by Zirkle (1970), there are unknown mechanisms which delay somewhat the start of anaphase if all chromosomes are not at the equatorial region.

Arms of single chromosomes left in the astral region at the start of anaphase point away from centrioles and either oscillate or are motionless. Such chromosomes often start migration when the main chromosome groups are in mid- or late anaphase at a time that coincides with growth of the aster. One chromosome always moves toward the centriole while its sister migrates in the opposite direction, often along poorly defined astral rays which seem to serve as "guiding tracks" determining the direction of migration. The condensation cycle of chromosomes that are outside the metaphase plate at the start of anaphase is slightly delayed. The migration of these chromosomes is of various lengths and is often not synchronous with that of the main group. These chromosomes show either clear bipolar orientation, i.e., are in minute accessory spindles formed by additional poles, or show undefined orientation in the light microscope.

   c. Anchorage of the Kinetochore Fiber.   To exert a pulling action, a kinetochore fiber must be anchored. Although the molecular mechanism of anchorage is not known, there seem to be a variety of ways by which anchorage may be achieved.

   OBSERVATIONS.   Two basic types of spindle pole organization exist in the newt, namely, with and without asters (Fig. 4). Newt asters are very well developed (Fig. 3), but their size is underestimated with any technique except immunocytochemical reaction (Bajer *et al.*, 1980). When the asters migrate or when, as is often the case, they execute slight rocking movements, kinetochores follow their movements and seem to be anchored or mechanically connected to the polar region of the aster, where microtubules are very irregularly arranged. In rare cases when asters are not present, the distance of maximum separation of chromosomes and spindle elongation is often comparable to anaphase in spindles with asters, as in crane fly spermatocytes (Dietz, 1959), except that often in the newt during anaphase, there are no backward oscillatory movements. In the newt there are a variety of filaments on the surface of the spindle. In the absence of asters, microtubules and filaments intermingle in the polar region (see also Euteneuer *et al.*, 1977). EM studies do not permit us to state whether they are microfilaments or intermediate filaments.

2. Conclusions and Interpretations

   Studies *in vitro* demonstrate the existence of head-to-tail assembly/disassembly of microtubules (Margolis *et al.*, 1978; Bergen and Borisy, 1980, Kirschner 1980). Thus, the polarity of microtubules may be important in regulating the rate of assembly/disassembly at different ends. The movement of chromosomes that start anaphase show that chromosomes may move in both directions along astral microtubules, i.e., toward and away from centrioles. If all astral and all kinetochore microtubules are of the same polarity, then chromosomes may move both with and against the specific polarity of astral microtubules. Such conclusions can be also drawn from the elegant work of Warner and Mitchell (1981) on polarity of microtubules, or determined by arrangements of dynein arms in flagella. Furthermore, the pattern of movements suggests that one kinetochore fiber is used successively as a structural support for movements of neighboring chromosomes. Such conclusions can also be derived from studies of chromosome movements in experimental conditions (chloral hydrate, Molè-Bajer, 1967, 1969) in which only kinetochore fibers without any additional microtubules are present. The movement is irregular, but time-lapse analysis suggests that, as in the newt, one kinetochore fiber serves as a mechanical support for its neighbor. The jerky movements of such chromosomes is predicted by the zipper hypothesis (Bajer, 1973, 1977; Novitski and Bajer, 1978), which suggests that, basically, one microtubular system (kinetochore fiber) can use another microtubular system (another kinetochore fiber and/or non-kinetochore microtubules) of the same or opposite polarity as a support (framework) for the movement. This assumption would explain oscillatory motion during anaphase and the capability for long migration of multiple chromosomes in the astral region. It should be stressed again that light microscopic observations cannot rigorously exclude the existence of additional centers in the newt, and EM data show that a depleted cytoplasmic microtubular cytoskeleton is still present. It is clear, however, that a regularly organized additional microtubular system is not needed for short or long migration of chromosomes. It is possible to state numerous models based on microtubules of the same polarity. It is difficult to discriminate between the various models on the basis of present data. Moreover, it is questionable whether novel models are useful before more data are obtained.

Kinetochore fibers "pull" toward a variety of structures. Therefore, either the anchorage of kinetochore fibers is variable or they are anchored along their entire length; at present, we consider the latter to be less likely than the former.

IV. OVERALL CONCLUSIONS

It is not obvious to us from the microtubular structure of the mitotic apparatus how the system works. Thus, parallel arrays of microtubules in the interzone are not evidence of active (force-producing) or passive (slipping out of register) sliding. The standard spindle is not a regularly organized system and becomes disorderly at the end of anaphase. The system operates on the basis of difficult-to-define thresholds (e.g., the number of non-kinetochore microtubules which may be needed for movement and/or anchorage) and tolerates a wide range of structural changes. On the level of EM morphology, three processes occur simultaneously: assembly/disassembly (Inoué et al., 1975), lateral association/dissociation (zipping, Bajer, 1973; Novitski and Bajer, 1977), and translation along the long axis of the spindle (sliding, McIntosh et al., 1975).

The data presented here demonstrate the functional autonomy of each half-spindle and the variability of anchorage of kinetochore fibers. A variety of more or less complicated hypotheses are possible (e.g., "piles in the mud":

Bajer and Molè-Bajer, 1979) that can explain the functional autonomy of each half-spindle; mechanisms may even be different in astral and anastral spindles. Furthermore, we find no evidence for the presence of a central spindle in the newt. The central spindle is generally thought to transform into interzonal structures, such as the mid-body, often called the "stem body." The origin of the latter is not well understood in terms of microtubule assembly and rearrangement. However, "stem body" (Stemkörper) is a conceptual term for the structure in the interzone which pushes chromosome groups apart. This interpretation was developed by Belar (1929) and subsequently abandoned by him (Belar and Huth, 1933) on the basis of studies of monopolar division. The assumption that anaphase-like prometaphase is comparable to standard anaphase leads to a new conceptual role for the interzone in the astral spindle. We believe that if microtubular struts are present in the interzone of standard anaphase, they do not support kinetochore fibers mechanically or structurally and are not needed for chromosome movements. Their role is to slow down kinetochore migration which might be required in some cell types to prevent precocious cytokinesis.

The maximum distance of chromosome migration is related to the length of the metaphase spindle, and this, too, argues against an active role for the interzone in polar separation and kinetochore-to-pole movement. Often chromosomes may move two to three times the length of the metaphase spindle (Roth et al., 1966; Allenspach and Roth, 1967), and they may even move five times the length of the spindle in non-standard divisions. This can be explained by the autonomous movement of each half-spindle, and therefore telescopic sliding (Inoué and Ritter, 1978) may be a passive slipping out of register. It is doubtful that the structural principle of chromosome movement would be different in different cells of the same organism or would change between mid- and late anaphase. Thus, an increase of velocity may occur in mid-anaphase when the slowing mechanism does not operate (i.e., when interzonal microtubules slip out of register). The functional consequence of the autonomous movement of each half-spindle (front wheel drive) is that the stronger the interzonal connections, the longer (timewise) and further (lengthwise) half-spindles migrate. Any experimental manipulation of the interzone (i.e., heating) should influence chromosome movements and is consistent with this assumption, and not against it as argued by Nicklas (1979).

The main conceptual interpretations concerning the astral spindle of the newt emerging from present data are given below.

1. Each half-spindle has an independent, autonomous ability for migration, both in prometaphase and in anaphase. The maximum distance of migration is determined primarily by the life span of kinetochore fibers and

by mechanical conditions (the size and shape of the cell, etc.). The basic mechanism of movement is the same during prometaphase and anaphase, although the kinetochore fibers of prometaphase–metaphase (double-chromatid) chromosomes can be functional at least 10 times longer than kinetochore fibers of anaphase (single-chromatid) chromosomes. Chromosomes (kinetochores) regulate and determine the final assembly/disassembly of kinetochore fibers through a still unknown mechanism.

2. Interzonal struts (microtubules) are not needed for chromosome-to-pole movement and separation of the poles. The presence of interzonal microtubules slows down chromosome movements, preventing premature cytokinesis, and delays disassembly of kinetochore microtubules. The final result is that the migration is longer in the presence of abundant interzonal microtubules.

3. The polarity of microtubules may play an essential role in the assembly/disassembly of microtubules and thus regulate the rate of chromosome movement. Kinetochore fibers can, however, transport chromosomes with and against astral microtubules, which suggests that the polarity of microtubules is not a *sine qua non* condition for chromosome movement.

4. Microfilaments and/or intermediate filaments are abundant on the surface of the spindle (newt) and at the cell periphery. They have not been found inside the spindle. Microtubules intermingle with filaments, and this arrangement may represent one way of anchoring asters and microtubules within the cell.

5. The force for chromosome movement is located in front of the kinetochores. This is consistent with several interpretations, some of them not mutually exclusive but difficult to distinguish on the basis of existing data.

ACKNOWLEDGMENTS

We would like to thank Dr. M. De Brabander (Beerse, Belgium), Dr. W. Heneen (Lund, Sweden), and Dr. C. Rieder (Albany, N.Y.) for permission to cite their unpublished results. Our thanks are also extended to C. Cypher for constructive criticism, and to S. Paulaitis and M. Stone for stimulating help during experiments, all from Eugene, Ore. Our work was supported by National Institute of Health Grants GM 21741 and GM 26121.

REFERENCES

Allenspach, A. L., and Roth, L. E. (1967). Structural variations during mitosis in the chick embryo. *J. Cell Biol.* 33, 179–196.
Alov, I. A., and Lyubskii, S. L. (1974). Experimental study of the functional morphology of the kinetochore in mitosis. *Byull. Eksp. Biol. Med.* **78**, 91–94.

Bajer, A. (1973). Interaction of microtubules and the mechanisms of chromosome movement (zipper hypothesis). *Cytobios* **8**, 139–160.

Bajer, A. (1977). Interaction of microtubules and the mechanism of chromosome movements (zipper hypothesis): II. Dynamic architecture of the spindle. *In* "Mechanisms and Control of Cell Division" (T. L. Rost and E. M. Gifford, Jr., eds.), pp. 233–258. Dowden, Hutchinson and Ross, Stroudsburg, Pa.

Bajer, A., and Molè-Bajer, J. (1975). Lateral movements in the spindle and the mechanism of mitosis. *In* "Molecules and Cell Movement" (S. Inoué and R. E. Stephens, eds), pp. 77–96. Rowen Press, New York.

Bajer, A., and Molè-Bajer, J. (1979). Anaphase-stage unknown. *In* "Cell Motility: Molecules and Organization" (S. Hatano, H. Ishikawa, and H. Sato, eds.), pp. 569–591. Univ. of Tokyo Press, Tokyo.

Bajer, A., Molè-Bajer, J., and Lambert, A. M. (1975). Lateral interaction of microtubules and chromosome movements. *In* "Microtubules and Microtubule Inhibitors" (M. Borgers and M. De Brabander, eds.), pp. 393–423. North Holland Publ., Amsterdam.

Bajer, A. S., De Brabander, M., Molè-Bajer, J., De Mey, J., Geuens, G., and Paulaitis, S. (1980). Aster migration and functional autonomy of monopolar spindle. *In* "Microtubules and Microtubule Inhibitors" (M. De Brabander and J. De Mey, eds.), Elsevier, Amsterdam, 399–425.

Belar, K. (1929). Beiträge zur Kausalanalyse der Mitose. II. *Arch. Entwicklungsmech. Org.* **118**, 359–480.

Belar, K., and Huth, W. (1933). Zur Teilungsautonomie der Chromosomen. *Z. Zellforsch. Mikrosk. Ana.* **17**, 51–66.

Bergen, L. G., and Borisy, G. G. (1980). Head-to-tail polymerization of microtubules *in vitro*. *J. Cell Biol.* **84**, 141–150.

Bloom, W., and Zirkle, R. E. (ca. 1954) (film not dated). Mitosis in the newt cels cells. 16 mm film. Univ. Chicago Bookstore, Chicago, Ill.

Brenner, S., Branch, A., Meredith, S., and Berns, M. W. (1977). The absence of centrioles from spindle poles of rat kangaroo (PtK$_2$) cells undergoing meiotic-like reduction division *in vitro. J. Cell Biol.* **72**, 368–379.

De Brabander, M., De Mey, J., Joniau, M., and Geuens, S. (1977). Immunocytochemical visualization of microtubules and tubulin and the light- and electron-microscopy level. *J. Cell Sci.* **28**, 283–301.

Dietz, R. (1959). Chromosomesfreie Spindelpole in Tipuliden-spermatozyten. *Z. Naturforsch.* **14b**, 7–49–572, 749–752.

Euteneuer, U., Bereiter-Hahn, J., and Schliwa, M. (1977). Microfilaments in the spindle of *Xenopus laevis* tadpole heart cells. *Cytobiologie. (Eur. J. Cell. Biol.)* **15**, 169–173.

Haimo, L. T., Telzer, B. R. and Rosenbaum, J. L. (1979). Dynein binds to and crossbridges cytoplasmic microtubules.*Proc. Natl. Acad. Sci. U.S.A.* **76**, No. 11, 5759–5763.

Harris, P. (1978). Triggers, trigger waves, and mitosis: A new model. *In* "Cell Cycle Regulation" (J. R. Jeter, I. L. Cameron, G. M. Pallida, and A. M. Zimmerman, eds.), pp. 75–104. Academic Press, New York.

Inoué, S., and Ritter, Jr., H. (1978). Mitosis in barbulanympha. II. Dynamics of a two-stage anaphase, nuclear morphogenesis and cytokinesis. *J. Cell Biol.* **77**, 655–684.

Inoué, S., Fuseler, J., Salmon, E. D., and Ellis, G. W. (1975). Functional organization of mitotic microtubules. *Biophys. J.* **15**, 725–744.

Kirschner, M. W. (1980). Implications of treadmilling for the stability and polarity of actin and tubulin polymers *in vivo. J. Cell. Biol.* **86**, 330–334.

Lambert, A. M. (1980). The role of chromosomes in anaphase- trigger and nuclear envelope activity in spindle formation. *Chromosoma* **76**, 295–308.

Lambert, A. M. (1970). Etude de Structure Cinetiques En Rapport anec la Rupture de la Membrane Nucleaire en Debut de Nerose chez mnium Hornum. *In C. R. Acad. Sc. Paris*, **270D**, 481–484.

Lambert, A. M., and Bajer, A. S. (1977). Microtubules distribution and reversible arrest of chromosome movements induced at low temperature. *Cytobiologie* **15**, 1–23.

Lyubskii, S. L. (1974a). Changes in kinetochore ultrastructure produced by the stathmokinetic action of cooling. *Byull. Eksp. Biol. Med.* **78**, 80–83.

Lyubskii, S. L. (1974b). Changes in the ultrastructure of the kinetochore during mitosis. *Byull Eksp. Biol. Med.* **77**, 113–116.

Margolis, R. L., Wilson, L., and Kiefer, B. I. (1978). Mitotic mechanism based on intrinsic microtubule behavior. *Nature (London)* **272**, 450–452.

McIntosh, J. R., Cande, W. Z., and Snyder, J. A. (1975). Structure and physiology of the mammalian mitotic spindle. *In* "Molecules and Cell Movement" (S. Inoué and R. E. Stephens, eds.), p. 31–76. Raven, New York.

Mazia, D., Paeletz, N., Sluder, G., Finze, E. M. (1981). Cooperation of kinetochores and ole in the establishment of mono-polar mitotic apparatus. *Proc. Natl. Acad. Sci. U.S.A.* **78**, 377–381.

Molè-Bajer, J. (1967). Chromosome movements in chloral hydrate treated endosperm cells *in vitro*. *Chromosoma* **22**, 465–480.

Molè-Bajer, J. (1969). Fine structural studies of apolar mitosis. *Chromosoma*. **26**, 427–448.

Molè-Bajer, J. (1975). The role of centrioles in the development of the astral spindle (newt). *Cytobios* **13**, 117–140.

Molè-Bajer, J., Bajer, A., and Owczarzak, A. (1975). Chromosome movements in prometaphase and aster transport in the newt. *Cytobios* **13**, 45–65.

Nicklas, R. B. (1979). Chromosome movement and spindle birefringence in locally heated cells: Interaction versus local control. *Chromosoma* (Berl) **74**, 1–37.

Novitski, C. E., and Bajer, A. S. (1977). Interaction of microtubules and the mechanism of chromosome movement (zipper hypothesis). III. Theoretical analysis of energy requirements and computer simulation of chromosome movement. *Cytobios* **18**, 173–182.

Östergren, G. (1951). The mechanism of co-orientation in bivalents and multivalents. The theory of orientation by pulling. *Hereditas* **37**, 85–156.

Paweletz, N., and Mazia, D. (1979). Fine structure of the mitotic cycle of the unfertilized sea urchin eggs activated by ammoniacal sea water. *Eur. J. Cell Biol.* **20**, 37–44.

Rieder, C. L. (1977). An *in-vitro* light and electron microscopic study of anaphase chromosome movement in normal and temperature elevated earicha lung cells. Ph.D. Thesis, Univ. of Oregon, Eugene, Oregon.

Roos, U. (1976). Light and electron microscopy of rat kangaroo cells in mitosis. *Chromosoma* **54**, 363–385.

Rose, G. G., Pomerat, C. M., Shindler, T. O., and Trunnell, J. B. (1958). A cellophane-strip technique for culturing tissue in multipurpose chambers. *J. Biophys. Biochem. Cytol.* **4**, 761–764.

Roth, L. E., Wilson, H. T., and Chakhaborty, T. (1966). Anaphase structure in mitotic cells typified by spindle elongation. *J. Ultrastruct. Res.* **14**, 460–483.

Taylor, E. W. (1959). Dynamics of spindle formation and its inhibition by chemicals. *J. Biophys. Biochem. Cytol.* **6**, 193–196.

Taylor, E. W. (1963). Brownian and saltatory movements of cytoplasmic granules and the movement of anaphase chromosomes. *Proc. Int. Cong. Rheol., Rhode Island*, pp. 175–191.

Warner, F., and Mitchell, D. R. (1981). Polarity of dynein–microtubule interactions *in vitro* cross-bridging between parallel and antiparallel microtubules. *J. Cell Biol.* **89**, 35–44.

Wilson, E. B. (1928). "The Cell in Development and Heredity," 3rd Ed., pp. 1232. Macmillan, New York.

Wolf, R. (1975). A new migration mechanism of cleavage nuclei based on the cytaster. *Verh. Dtsch. Zool. Ges., 1974,* 174–178.

Wolf, R. (1978). The cytaster, a colchicine-sensitive migration organelle of cleavage nuclei in an insect egg. *Dev. Biol.* **62,** 464–472.

Zirkle, R. E. (1957). Partial cell irridation. *Adv. Biol. Med. Phys.* **5,** 103–146.

Zirkle, R. E. (1970). Ultraviolet-microbeam irradiation of newt cell cytoplasm: Spindle destruction, false anaphase, and delay of true anaphase. *Radiat. Res.* **41,** 516—537.

13

Studies of Mitotic Events Using Lysed Cell Models

JUDITH A. SNYDER

I. INTRODUCTION

The mitotic apparatus or spindle is a biological machine designed to segregate duplicated genetic material to daughter cells during the process of cell division. It appears to be uniquely designed for this task since it rapidly forms within minutes at the onset of division and dissolves with completion of chromosome separation. Because of the transient nature of the mitotic apparatus, appearing only briefly in the cell cycle, it is not surprising that the mitotic apparatus is a very labile structure. These labile properties have

MITOSIS/CYTOKINESIS
Copyright © 1981 by Academic Press, Inc.
All rights of reproduction in any form reserved.
ISBN 0-12-781240-7

made it a very difficult entity to study. Its structure and function are sensitive to changes in temperature, ion concentration, pH, and certain drugs called "spindle poisons." Partly because of these properties, it has not yielded easily to attempts to solve two of the major mysteries it still holds: (1) how does the spindle form so quickly at the onset of mitosis? and (2) what is the biochemical motive force that causes chromosomes to congress to a central region of the spindle at metaphase and then separate during anaphase, thus creating two daughter cells with the same genome? The purpose of this chapter is to summarize what is known about spindle assembly, chromosome motion, and cytokinesis using lysed cell models.

Investigators in the field of mitosis have developed a variety of terms to describe the cell's mitotic apparatus. The terms are defined here as they relate to the context of this review.

1. Mitotic apparatus—the spindle including asters, chromosomes, centrioles, and the matrix associated with the spindle.
2. Spindle—the fibrous elements, mainly composed of two classes of microtubules: those that attach directly to chromosomes (kinetochore microtubules) and those that do not (non-kinetochore microtubules).
3. Spindle poles—the region at each end of the spindle on which microtubules focus. In mammalian cells this region usually contains a pair of centrioles.
4. Mitotic centers—regions in the cell capable of initiating the nucleation of microtubules to form an aster. In animal cells they ultimately form spindle poles.
5. Microtubule-organizing center—any region of the cell capable of nucleating microtubules; In this chapter, it is restricted to the centrosome and possibly the kinetochores.
6. Centrosome—the centriolar duplex and surrounding osmiophilic material that acts as a microtubule-organizing center.
7. Centriolar region—the osmiophilic region immediately surrounding the centrioles in mammalian cells.
8. Kinetochore—the specialized region of a chromosome that attaches to the spindle fibers.
9. Chromosomes—the organelles that form at the onset of mitosis, containing the duplicated genetic material in the form of sister chromatids.
10. Kinetochore microtubules—those spindle microtubules that end on the kinetochores of chromosomes.
11. Non-kinetochore microtubules—all spindle microtubules that do not end on kinetochores. These include a small percentage of microtubules that extend from pole to pole and a larger number that extend from a pole and end in the spindle domain.

Prior to the discovery by Weisenberg (1972) of *in vitro* microtubule polymerization, attempts to isolate mitotic cells that retained spindle structure and function were largely unsuccessful. The initial strategy was to isolate the mitotic apparatus in solutions that were poor solvents for proteins. Mazia and Dan (1952) were the first to isolate the mitotic apparatus from sea urchin zygotes by fixing mitotic cells in cold ethanol and removing the cytoplasm with detergent. Following successes of muscle physiologists using glycerinated models for muscle contraction, Hoffman-Berling (1954) was able to show both chromosome motion and spindle elongation in glycerinated mitotic cells. Numerous investigators improved on spindle isolation procedures (Kane, 1965; Forer and Zimmerman, 1976; for reviews, see McIntosh, 1977; Sakai, 1978b), but none of these spindle isolates retained the normal physiological properties of being sensitive to cold and calcium when kept in the original isolation medium.

Goode and Roth (1969) defined conditions of pH, ionic strength, and composition that stabilized the mitotic apparatus of a giant amoeba; this work came the closest to studying the spindle under physiological conditions. They found they could get limited chromosome motion if ATP was added, suggesting that this mitotic apparatus was more *in vivo*-like than others.

II. USE OF *IN VITRO* MICROTUBULE POLYMERIZATION TO STUDY THE MITOTIC APPARATUS

Weisenberg's discovery in 1972 of *in vitro* assembly of microtubules from tubulin subunits opened a new approach in the study of the role of microtubules in cellular events. He assembled neuronal microtubules from their subunits by choosing buffer conditions that chelated divalent cations. This method has since been used by many investigators to learn more about microtubule biochemistry, particularly its assembly characteristics (for reviews, see Kirschner, 1978; Scheele and Borisy, 1979; Timasheff and Griffith, 1980). Assembly *in vitro* occurs by a condensation polymerization mechanism (Johnson and Borisy, 1977), and it is not known whether a comparable mechanism occurs *in vivo*. It is likely that understanding the intrinsic properties of microtubules elucidated by *in vitro* assembly studies will help greatly in understanding how microtubules function as cytoskeletal elements, their involvement in cell motility, and the role they may play in chromosome motion.

Both the assembly characteristics and the stability of microtubules formed *in vitro* are affected by the presence of other proteins. Analysis of the microtubules that polymerize *in vitro* show that tubulin comprises only about 80% of the microtubules formed by the purification scheme of temperature-dependent cycles of assembly–disassembly (for reviews, see Kirschner 1978; Scheele and Borisy, 1979; Timasheff and Griffith, 1980). Polyacrylamide gel electrophoresis indicates that the remaining 20% of the protein is represented predominantly by high-molecular-weight species and to a lesser extent by numerous other proteins. These proteins are called "microtubule-associated proteins" if they retain constant stoichometry with tubulin during successive cycles of assembly–disassembly. Two of these proteins, a pair of high-molecular-weight microtubule-associated proteins called "HMW" (Murphy and Borisy, 1975) or MAP_1 and MAP_2 (Sloboda *et al.*, 1976) and another called "tau" (Weingarten *et al.*, 1975) have been implicated in enhancing the stability of microtubules and the rate and extent of microtubule polymerization *in vitro*. Farrell *et al.* (1979) have shown that pure tubulin isolated from highly stable flagellar microtubules and polymerized *in vitro* gives microtubules that are more labile than the flagellar tubules from which they were isolated. Thus the microtubule-associated proteins may have a strong influence on the properties and hence the function of microtubules. The results of the lysed cell model experiments presented here should be interpreted with the understanding that a wide range of purity of the microtubule protein used by various investigators exists.

III. USE OF THE LYSED CELL MODEL TO STUDY SPINDLE ASSEMBLY

The lysed cell model was developed to investigate the properties of the mitotic apparatus under conditions that more closely approximate those found *in vivo*. Due to the nature of the equilibrium between tubulin subunits and microtubules, it was quickly discovered that addition of exogenous microtubule protein to lysed mitotic cells would stabilize the spindle microtubules by preventing the equilibrium from favoring disassembly (Cande *et al.*, 1974; Inoué *et al.*, 1974; Rebhun *et al.*, 1974).

The lysed cell system using mitotic mammalian tissue culture cells has proven to be a useful tool in studying the effects of exogenous microtubule protein on spindle stability and function (Cande *et al.*, 1974; Rebhun *et al.*, 1974; Snyder and McIntosh, 1975). Using microtubule protein dissolved in cold assembly buffer and 0.1% Triton X-100, exogenous microtubule protein can be rapidly introduced into mitotic cells. This reasonably high detergent concentration allows macromolecules to enter the cell easily with a concomitant loss of cytoplasm. Some of the lysed cell models used to investigate microtubule initiation showed so little remaining cytoplasm (McGill and Brinkley, 1975; Snyder and McIntosh, 1975) that lysis conditions and experimental results were similar to those reported in isolates of microtubule-organizing centers (Telzer *et al.*, 1975; Gould and Borisy, 1977).

Both the birefringence and the structure of the spindle can be altered when cells are lysed in exogenous tubulin (Fig. 1). The amount of birefringence exhibited by the stabilized spindle is dependent on the concentration of polymerizable subunits used in the buffered lysis medium (Snyder and McIntosh, 1975). Introduction of a high concentration of microtubule protein subunits (5–10 mg/ml) resulted in a spindle with increased birefringence. Metaphase cells lysed in buffers containing microtubule protein that had been allowed to polymerize and come to equilibrium, maintained spindle birefringence or showed a slight loss of spindle birefringence, irrespective of total protein in the lysis buffer. After mitotic cells were lysed in buffers containing subunit concentrations below the critical concentration for *in vitro* assembly (0.2 mg/ml), there was rapid loss of birefringence. Lysis of mitotic cells into assembly buffer alone resulted in loss of all birefringence, with the poles moving inward until they stopped at the metaphase plate (Cande *et al.*, 1974).

Changes in birefringence can be correlated with changes in microtubule

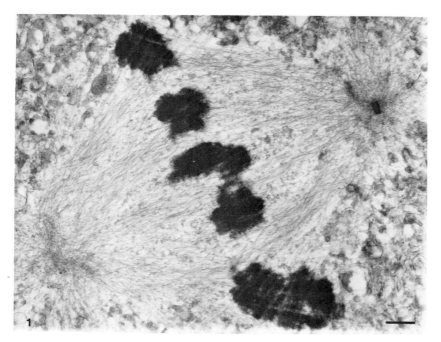

Fig. 1. High-voltage electron micrograph of a metaphase PtK$_1$ cell lysed in 8 mg/ml porcine brain microtubule protein. These is an approximate tripling of microtubule number which is accompanied by an increase in average tubule length. Some of the microtubules extend beyond the normal domain of the spindle. Bar = 1 μm. X10,100.

number within the spindle. Sato *et al.* (1975) showed that the amount of birefringence and microtubule number are directly related. A profile of microtubule distribution within the spindle can be made by counting microtubules seen in serial cross sections of electron micrographs. The number of microtubules seen in each cross section is plotted against their location in the spindle, giving a relative distribution of spindle microtubules (McIntosh and Landis, 1971). Analyses of this kind have shown that an increase in birefringence showed a concomitant increase in microtubule number. Those spindles that increased in microtubule number had a relatively high proportion of microtubules near the poles. Spindles that lost birefringence but did not dissolve completely, showed only kinetochore microtubules. This is consistent with the finding of others that kinetochore microtubules are more stable to other perturbations, such as cold, spindle poisons, and pressure (Brinkley and Cartwright, 1975). Studies of this kind have documented the importance of the equilibrium between spindle fibers and tubulin subunits and how these dynamics affect spindle structure.

Increases in spindle birefringence above normal *in vivo* levels by the addition of exogenous heterologous microtubule protein have also been documented in the mitotic apparatus isolated from marine eggs. Mitotic spindles isolated in buffers containing 15–20 mg/ml microtubule protein became wider, the length of astral fibers increased, and the length of the spindle increased by up to 40% (Rebhun *et al.*, 1974). Inoué *et al.* (1974) showed similar increases in birefringence in the *Chaetopterus* mitotic apparatus and concluded that they were due to the incorporation of heterologous tubulin into the spindle and asters. Increases in aster size in both lysed cell spindles and isolated spindles led investigators to study the possibility that exogenous tubulin could be nucleated at the centrosome in much the same fashion asters are formed *in vivo*.

Appropriately, Weisenberg was the first to use the technique of *in vitro* microtubule polymerization to probe the role of the centriolar region in assembling microtubules to form asters (Weisenberg, 1973). Supernatants of homogenized marine eggs were added to crude homogenates of similar meiotic oocytes *in vitro*. The microtubule protein from the supernatant of one homogenate was capable of forming a normal aster from a microtubule-organizing center found in the crude homogenate. Interestingly, only crude homogenates from activated eggs (those eggs in which the germinal vesicle had broken down) could form asters; those from inactivated eggs could not. In activated eggs the size of the aster produced depended on the time after the egg was activated (Weisenberg and Rosenfeld, 1975). At progressively longer times, up to 15 minutes, the size of the aster gradually increased, suggesting a ripening effect for microtubule initiation is present. There was a change in an organizing region, not a change in the tubulin pool, that triggered aster assembly in activated marine eggs.

Increases in birefringence and hence microtubule number in mammalian spindles lysed in high concentrations of exogenous microtubule protein may be due to the addition of subunits to microtubule fragments rather than to initiation at microtubule-organizing centers. To distinguish between these two possibilities, mitotic cells were treated with Colcemid, a potent microtubule assembly inhibitor (Brinkley *et al.*, 1967). Lysis of these Colcemid-arrested mitotic cells in the presence of exogenous microtubule protein resulted in the formation of two bright birefringent regions, as observed with polarization light optics (Fig. 2). Examination of these birefringent regions with electron microscopy revealed two asters (Fig. 2e) (Snyder and McIntosh, 1975). As two adjacent asters form from exogenous microtubule protein in this lysed cell system, there is no change in their position relative to one another. McGill and Brinkley (1975), performing similar experiments on Colcemid-blocked mitotic HeLa cells, also found that the centriolar region and, to some degree, kinetochores were the sites at which microtubule assembly occurred.

In PtK$_1$ cells the number and length of microtubules seen at kinetochores in cells lysed with exogenous microtubule protein are less than those seen *in vivo*. This may be explained by something in the lysis conditions that interferes with assembly, such as free calcium ions or loss of template structure due to Colcemid. In other studies using chromosomes isolated from HeLa cells, the kinetochore has been shown to act as a microtubule-organizing center when exogenous tubulin is supplied (Telzer *et al.*, 1975). Even the number of microtubules initiated from chromosomes isolated from CHO cells is comparable to that seen *in vivo* (Gould and Borisy, 1978). However, kinetochores may act as traps for centrosomal microtubules and may not be true nucleating structures (for a recent discussion see Tippet *et al.*, 1980). This would explain the differences in initiation capacity between the centrosome and kinetochore in lysed cells.

The studies described above all have in common the use of microtubule protein which has the liability of spontaneous nucleation *in vitro*. Tubulin purified to some degree by high-speed centrifugation or ion exchange chromatography is incapable of spontaneous nucleation over reasonable periods of time but will form microtubules at microtubule-organizing centers in lysed mitotic PtK cells (Snyder and McIntosh, 1975). The centriolar region and kinetochores must serve a microtubule initiator function in lysed mitotic cell preparations. The centriolar region at the onset of mitosis may contain microtubule-associated proteins (Mascardo and Sherline, 1980) which may help to stimulate the rapid assembly and elongation of mi-

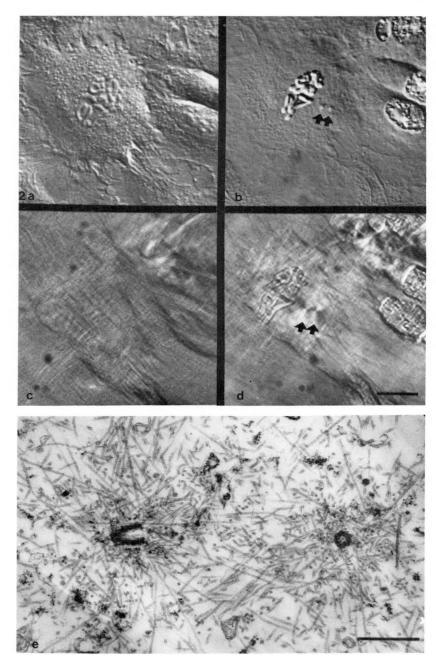

crotubules. Microtubule-associated proteins are not a stringent requirement for assembly since pure tubulin can assemble *in vitro* in the presence of high magnesium or the addition of polyethylene glycol or dimethyl sulfoxide (for a review, see Scheele and Borisy, 1979).

The ability of the centriole to initiate microtubule assembly is a time-dependent process in mammalian cells. Ripening or maturation takes place quickly between prophase and prometaphase. Colcemid-blocked cells lysed at various times in the mitotic cycle with the same concentration of microtubule protein show dramatic differences in the number and length of microtubules initiated at the mitotic centers (Snyder and McIntosh, 1975; Telzer and Rosenbaum, 1979). Interphase and prophase cells do not differ in the number of microtubules initiated at the centrosome even though mitotic events such as chromosome condensation and rounding of the cell have been triggered in prophase cells. Increased microtubule initiation capacity occurs at the time of nuclear envelope breakdown (Snyder and McIntosh, 1975). This parallels the finding of Weisenberg who found that asters can be formed in activated marine eggs only after the germinal vesicle has broken down (Weisenberg, 1973). These observations have led to speculation that some factor(s) from the nucleus confers initiation capacity on the mitotic center. If the kinetochore is a true organizing center, it may differ in this regard since in PtK cells kinetochore development is not complete at the time of nuclear envelope breakdown (Roos, 1973), and the ability to act as a microtubule-organizing center may be regulated by the further development of the kinetochore.

A. Nature of Osmiophilic Material Surrounding the Centrosome

In mammalian cells, centrioles themselves do not apparently play an active role in microtubule initiation (McGill and Brinkley, 1975; Gould and Borisy, 1977). Rather, microtubules are nucleated from an amorphous substance surrounding the centrioles that apparently serves as a template for microtubule assembly (Gould and Borisy, 1977). The amount of this substance increases during the cell cycle until it reaches a maximum concentra-

---

Fig. 2. (a, c) Differential interference contrast and polarization light micrographs of Colcemid-blocked mitotic PtK$_1$ cell prior to lysis. (b, d) Differential interference contrast and polarization images of the same cell after lysis in 8 mg/ml microtubule protein and warmed to 37°C for 15 minutes. Arrows mark the position of the mitotic centers. Bar = 10 μm. X1300. (e) Electron micrograph of the mitotic centers from the same cell shown in (a–d). Numerous microtubules emanate from the centriolar regions. The cell contents are well extracted under these lysis conditions. Bar = 1. μm. From Snyder and McIntosh (1975).

tion at prometaphase (Robbins *et al.*, 1968). Only a portion of this material is retained in mitotic cells lysed in the microtubule assembly buffer described above. Examination of this region in lysed Colcemid-blocked mitotic CHO cells (Gould and Borisy, 1977) reveals a large number of virus- or clathrin-like particles embedded in amorphous material. These particles, first described by Wheatley (1974) in CHO cells, have been studied in detail by Gould and Borisy (1977): They apparently serve as markers for microtubule initiation capacity. These virus-like particles tend to aggregate at the mitotic centers and the kinetochores as the mitotic centers mature in their microtubule initiation capacity. It is not clear what role, if any, these particles serve in cells. Since they are not ubiquitous in all cell lines, it is unlikely that they play a direct role in microtubule initiation.

Even though the region surrounding the centrioles seems diffuse and amorphous, it is sufficiently well anchored to remain *in situ* after extensive lysis. Perfusion of the lysis mixture under a glass cover slip on which cells are subcultured shows maximal extraction of cytoplasmic contents in approximately 2 minutes. At this time the plasmalemma is in remnants, membrane-bound organelles are destroyed, and only fragments of rough endoplasmic reticulum remain (Snyder, 1980). The position of the centrioles and the chromosomes remains unchanged.

Lysis of Colcemid-blocked mitotic PtK$_1$ cells for periods of 2 to 8 minutes prior to addition of microtubule protein does not affect the number of microtubules that can be initiated. There is little loss of amorphous material and associated particles in the centriolar region under these conditions (Snyder and McIntosh, 1976; Pepper and Brinkley, 1980; Snyder, 1980). Addition of microtubule protein to these Colcemid-blocked and detergent-extracted cells up to 8 minutes after lysis results in aster formation that is comparable in microtubule number and length to asters formed when microtubule protein and detergent are added simultaneously. The addition of microtubule protein after approximately 15 minutes of lysis shows that nucleation capacity is abolished, presumably due to the diffusion of the initiation substance away from the centrosome (Snyder, 1980). Close examination of the centriolar region of these cells shows a loss of the particles and amorphous material.

1. Introduction of Molecules that Block Microtubule Initiation

Because initiation capacity is not altered by lysis for at least 8 minutes in Colcemid-blocked mitotic cells, there is a reasonable amount of time in which to add other substances which might affect microtubule initiation capacity. Various enzymes have been added to the lysis mixture for 5 to 10 min prior to addition of microtubule protein (Pepper and Brinkley, 1980; Snyder, 1980). Enzymes specific for digestion of DNA and polysaccharides have no effect on centrosomal initiation capacity; however DNase I activity affects kinetochore structure (Pepper and Brinkley, 1980). Enzymes specific for RNA and protein greatly diminished microtubule initiation capacity in cultured mammalian cells (Snyder and McIntosh, 1976; Pepper and Brinkley, 1980; Snyder, 1980). Incubation of cells for 5 minutes in RNase A or T_2 at concentrations ranging from 250 to 500 μg/ml prior to the addition of microtubule protein showed greatly reduced aster formation (see Fig. 3). Most of the amorphous material disappears from the centriolar region, and the

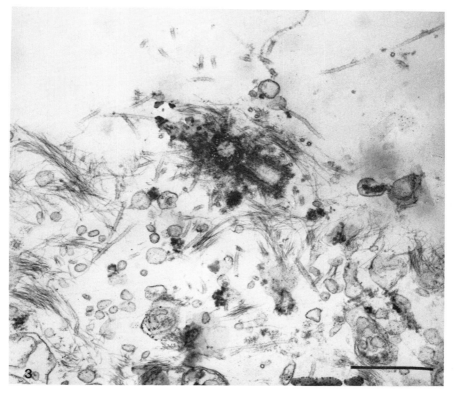

Fig. 3. Colcemid-blocked mitotic PtK$_1$ cell lysed in PIPES buffer containing 500 μg/ml RNase A for 5 minutes prior to the addition of 5 mg/ml microtubule protein and warmed for 10 minutes at 37°C. There is a substantial loss of microtubule initiation capacity and pericentriolar material. The position of the centrioles is outside the center of the chromosome rosette at the margin of the cell. Bar = 0.5 μm. X48,500. From Snyder, 1980.

312 Judith A. Snyder

centrioles themselves are no longer anchored near the nuclear envelope. Instead the centrioles are usually found near the cell margin.

These results are consistent with the investigations of others. Zacroff *et al.* (1976) showed that the structure of marine egg asters formed *in vitro* were sensitive to various RNAs and RNases. Some of the strongest evidence showing a functional role for RNA in the production of asters has come from the microinjection work of Heidemann *et al.* (1977). Injection of basal bodies isolated from *Chlamydamonas* into mature frog oocytes resulted in aster formation associated with the centrioles. If, however, the basal bodies were digested with RNase or trypsin prior to injection, asters did not form.

The effect of tubulin antibody on the nucleation capacity of kinetochores and centrosomes has also been investigated. Brief treatment of Colcemid-blocked lysed mitotic cells with tubulin antibody before addition of pure tubulin resulted in almost complete blockage of microtubule nucleation with the centrosomal region being more sensitive than kinetochores (Pepper and Brinkley, 1979).

Lysis of cells in acidic buffers for 5 minutes prior to addition of microtubule protein (near neutral pH) also blocks microtubule assembly. Addition of citrate buffer at pH 5.5 for 5 minutes will block nucleation without the loss of pericentriolar material. Since the pKs of RNAs are much lower than 5.5, only the protein component is affected by this pH change (Snyder, 1980). This gives further evidence that the microtubule-organizing center contains a ribonucleo-protein complex.

IV. USE OF THE LYSED CELL SYSTEM FOR INVESTIGATION OF CHROMOSOME MOTION

Separation of chromosomes during anaphase can be due to at least two separate events. One event called "anaphase A," by Inoué and Ritter (1975), involves the shortening of the chromosome-to-pole distance. The other, "anaphase B," is characterized by elongation of the spindle, causing a further increase in the distance between the chromosomes. In some cases (Inoué and Ritter, 1975), this lengthening approaches five times the original interpolar distance.

The connection of microtubules directly to chromosomes and the constant rearrangement of microtubules during mitosis suggest that microtubules play a role, whether direct or indirect, in chromosome motion. Using intrinsic properties of microtubules, several models have been proposed to account for the movement of chromosomes, particularly during anaphase.

A. Models for Chromosome Motion

Due to the nature of the equilibrium of the microtubule with its subunits, several investigators have postulated models based on assembly and disassembly characteristics of microtubules in the spindle to account for chromosome motion. For example, Inoué and Sato (1967) proposed that the force needed to move chromosomes comes from the depolymerization of kinetochore microtubules near the spindle poles, and the rate at which chromosomes move is governed by the rate of depolymerization (Inoué and Ritter, 1975). A similar mechanism of anaphase motion has been proposed by Dietz (1972), but depolymerization is thought to occur along the length of the microtubule, not at the ends of microtubules. Margolis and Wilson (1978a) have proposed that chromosome motion is dependent on the movement of subunits along the length of a microtubule. Subunits shuttle through a microtubule, with the dimer preferentially adding to one end and coming off at the opposite end (Margolis and Wilson, 1978b). At the onset of anaphase, elongation of kinetochore microtubules ceases and net disassembly occurs to allow chromosomes to migrate poleward. An ATP-dependent sliding of anti-parallel non-kinetochore microtubules (anaphase B) is proposed to drive the poles apart (Margolis and Wilson, 1978a).

An alternate hypothesis for the role of microtubules in chromosome motion based on structural properties of microtubules has been proposed by Bergen and Borisy (1980). By measuring the rates of microtubule growth from isolated microtubule organizing centers they have predicted that one end of a microtubule is both the preferred end for subunit addition and subunit loss. These results, though contradictory to others, also suggest that kinetochore and nonkinetochore microtubules are arranged anti-parallel in the half spindle (Bergin et al., 1980). Knowledge of the polarity of the classes of microtubules in a half spindle is essential in understanding anaphase motions.

McIntosh et al. (1969) have predicted that the force necessary to move chromosomes is derived from a sliding of antiparallel microtubules against each other. These antiparallel microtubules would be linked with bridge molecules, analogous to the flagellar ATPase dynein, and would serve to transduce chemical energy into motion. This sliding of antiparallel microtubules over one another would account for early anaphase motions. Though this theory has gained popularity the author himself has suggested it be discarded. Euteneuer and McIntosh (1980) have convincingly demonstrated that both kinetochore and nonkinetochore microtubules display the same polarity in a half spindle. Though similar polarity of microtubules exists in the half spindle specific interactions between microtubules and movement of subunits within a microtubule are likely to be important in the mechanism of mitosis.

The lysed cell system is being used to try to determine how addition of exogenous microtubule protein and micromolecules affect the rate of chromosome motion. Cande *et al.* (1974) have developed a permeabilized cell model to determine the molecular requirements of chromosome motion. In PtK $_1$ cells lysed in anaphase in low (0.05%) concentrations of Triton X-100 and polymerizable microtubule protein, chromosomes were shown to continue moving, though not at *in vivo* rates.

It should be pointed out that electron microscopy of lysed anaphase cells that have exhibited chromosome movement shows less extraction of the cytoplasm than for cells used in studies on the initiation of microtubules at microtubule-organizing centers. Use of different lysis conditions has led to an improved ability to move chromosomes in lysed anaphase cells (Cande, 1978). Using a buffered lysis mixture of non-ionic detergent, Brij 58, calcium chelators, and Carbowax 20M, anaphase motions and spindle elongation occur for up to 10 minutes after lysis. These lysis conditions allow cells to be made permeable to small ions, nucleotides, and small proteins. Brij 58, unlike Triton X-100, tends to destroy the plasmalemma completely, with little effect on the cytoplasm. Membranous organelles in these permeabilized cells do not appear to be noticeably disrupted, though in some cases the mitochondria are swollen. Chromosomes move at 70% of the *in vivo* rate, while spindle elongation occurs at 40% of the *in vivo* rate. The slower *in vivo*-like spindle elongation in this system may be due to an alteration in the assembly–disassembly characteristics of spindle microtubules. Cande has evidence that addition of taxol, a drug shown to increase microtubule polymerization (Schiff *et al.*, 1979) will also block spindle elongation in anaphase (Cande *et al.*, 1981).

Lysis of anaphase cells in 4 to 5 mg/ml microtubule protein inhibits the rates of chromosome-to-pole movement and spindle elongation by about 45%. It is not clear if these results are an artifact of non-physiological microtubule growth in the spindle or perhaps the absence of a dynein-like component in the lysis medium.

The movement of chromosomes in this lysed cell system is not strictly dependent on the addition of ATP, although addition of this nucleotide increases the rates of movement. Addition of inhibitors of oxidative phosphorylation can retard chromosome motion by as much as 75%. For chromosome motion to occur in this model system, the birefringence of the spindle must be maintained at normal levels. Increase in the detergent concentration results in loss of spindle structure and hence chromosome movements (Cande, 1978). Similar results in moving chromosomes in the isolated mitotic apparatus have been achieved by Sakai (1978a); for a review, see Sakai (1978b).

B. Evidence for Dynein in the Spindle

Having established that the rate of *in vitro* chromosome motion depends on addition of ATP, investigators have looked for a dynein-like molecule in the spindle. An antibody to fragment A dynein obtained from a flagellar source has been applied to spindle isolates *in vitro*. The antibody caused chromosome-to-pole motion to stop, and in some instances the chromosomes moved back to the metaphase plate (Sakai *et al.*, 1976). If this antibody is applied to sea urchin zygotes in mitosis or to an isolated mitotic apparatus, its localization by indirect immunofluorescence shows strong staining in the spindle (Mohri *et al.*, 1976). However, Sakai (1978b) has pointed out that dynein may be present in large amounts in the cell for later use in cilia production.

Proteins that comigrate on PAGE gels with ciliary dynein have been shown in pellets of isolated spindles from marine eggs (Salmon and Jenkins, 1977; Murphy, 1980). It seems reasonable that some dynein-like species will be found since cross-bridges seen in spindles (McIntosh, 1974) may be analogous to the HMW arms seen on the surface of microtubules polymerized *in vitro* (Murphy and Borisy, 1975).

Cande (1978) used the lysed cell model described above to see if addition of inhibitors of dynein activity have any effect on chromosome motion. The motion of chromosomes during anaphase was stopped by the addition of vanadate. Vanadate in the V^+ oxidation state is a potent inhibitor of the oubain-sensitive Na^+, K^+-ATPase (Cantley *et al.*, 1977). Gibbons *et al.* (1978) have shown that the addition of vanadate to demembranated flagellar axonemes blocks flagellar motility. Addition of 10–100 μM of vanadate to lysed mitotic cells in the process of anaphase will reversibly inhibit both chromosome motion and spindle elongation (Cande and Wolniak, 1978). In the presence of vanadate, spindles change to a barrel configuration, suggesting the loss of cross-bridges required for normal spindle shape. This is the best evidence to date for the role of a dynein-like molecule involved in spindle shape, the movement of chromosomes, and the sliding of microtubules during spindle elongation (Fig. 4).

C. Does Actin Have a Functional Role in the Spindle?

In the search for mechanisms controlling chromosome movement, an actin-mediated model has been suggested. In the last decade the increased

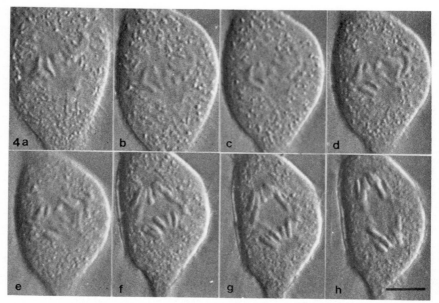

Fig. 4. Micrographs of anaphase PtK₁ cells demonstrating the reversible inhibition of chromosome separation and spindle elongation by 100-μM vanadate. After 4 minutes in vanadate, a solution containing 2.5 mM norepinephrine is added at (a) 0.1 minute before lysis, (b) 1 minute after lysis, (c) 2 minutes (d) 3.8 minutes, (e) 4.2 minutes, (f) 6 minutes, (g) 8 minutes, and (h) 12.5 minutes after lysis. Bar = 10 μm. X2000. From Cande and Wolniak (1978).

interest in the involvement of muscle-like proteins in non-muscle cell motility has promoted studies demonstrating the localization of actin in the mitotic spindle.

As early as 1965, Aronson showed that spindles isolated from sea urchin zygotes bound fluorescent heavy meromyosin. In using heavy meromyosin as a specific marker for actin, as well as its polarity the cell must be lysed, usually by glycerination. Using this technique, numerous workers have confirmed the existence of actin filaments in the spindle (for a review, see Sakai, 1978b).

Sanger (1975) introduced fluorescent heavy meromyosin into mitotic tissue culture cells to study the distribution of actin in the spindle. The fluorescent image showed staining between chromosomes and poles, at kinetochores, and the cleavage furrow at cytokinesis. These results were confirmed (Cande *et al.*, 1977) using the indirect immunofluorescent antibody technique with antiactin antibody introduced into cells that had been mildly treated with detergent and fixed with formaldehyde. The fluorescent staining was found predominantly between chromosomes and poles, with little fluorescence in the interzonal region at anaphase. Addition of the antibody to cells not lysed prior to fixation showed a diffuse fluorescence throughout the cell. The results of Herman and Pollard (1978) differed from those of Cande and Sanger in that the localization of fluorescent heavy meromyosin in mitotic cells was concentrated in the interzone during anaphase but, surprisingly, no staining at the cleavage furrow at cytokinesis was seen.

Recently, a functional test for the role of actin in the spindle has been established (Meeusen and Cande, 1979a; Cande *et al.*, 1981). Preliminary evidence does not support a functional role of actomyosin in chromosome movement. A type of biochemical probe has been prepared which can test for any actomyosin-mediated cellular process. Treatment of rabbit skeletal heavy meromyosin or subfragment 1 (S_1) with N-ethylmaleimide produces a type of myosin (designated N-ethylmaleimide-heavy meromyosin or N-ethylmaleimide-S_1, respectively) that binds actin and inhibits the actin activation of native heavy meromyosin-ATPase activity. It inhibits the contraction of glycerinated muscle myofibrils but has no effect on the beating of demembranated cilia (Meeusen and Cande, 1979a). Further, microinjection of N-ethylmaleimide-heavy meromyosin into amphibian eggs blocks cleavage in the zone of injection, yet the cytoplasm contains functional spindles (Meeusen and Cande, 1979b).

Introduction of N-ethylmaleimide-S_1 by gentle detergent lysis to cells undergoing anaphase has no effect on chromosome movement. Similarly, addition of high concentrations of cytochalasin B, a drug that depolymerizes actin (Brown and Spudich, 1979), and phalloidin, a drug that stabilizes actin (Wieland and Govindan, 1974), to the lysis medium has no effect on chromosome-to-pole movement or spindle elongation (Cande *et al.*, 1981).

Another biochemical probe developed to test the possible role of myosin in chromosome motion is the introduction of myosin antibodies into dividing cells. Antimyosin microinjected into dividing marine eggs blocks cleavage but not mitosis (Mabuchi and Okuno, 1977; Kiehart *et al.*, 1977).

It seems unlikely that actin and other non-muscle contractile proteins are a functional part of the spindle. Their occurrence in the spindle may be due to normal relocation during prophase or may be the result of cell perturbation by glycerination or detergent treatment. It can be argued that many fluorescently labeled proteins introduced into mitotic cells seem to end up in the spindle because the spindle can act as a trap for proteins. Further doubt has been cast on the interpretation of immunofluorescent staining patterns found using antibodies. Aubin *et al.* (1979) have shown that the apparent localization of contractile proteins using immunocytochemical techniques in mitotic cells can be varied by the method of cell preparation and the purity of the antibody used.

V. MODELS FOR CYTOKINESIS USING LYSED CELLS

Ptk$_1$ cells lysed late in cell division in a medium containing polyethylene glycol and the non-ionic detergent Brij 58 can continue to undergo cleavage after lysis. This lysed model has been used to determine the effects of calcium and nucleotides on the furrowing process (Cande, 1980).

The optimal calcium concentration for maintenance of cleavage after lysis was found to range from 0.1 to 0.6 μM. Concentrations of calcium greater than 1 μM blocked the furrowing process by enhancing relaxation rather than constriction. Cleavage continued in 10^{-8} M calcium, but the extent of constriction was approximately 50% less than in control cells. Calcium levels required for furrowing are lower than those needed to generate tension in muscle but are comparable to levels that stimulate solation of actin gels (Hellewell and Taylor, 1979).

Nucleotides in the lysis buffer, particularly Mg–ATP, stimulated furrowing with maximal constriction at 5 mM Mg–ATP; but higher concentrations blocked furrowing. The cleavage that occurred in the absence of added nucleotide is about half that when 5 mM Mg–ATP is added.

Cleavage is blocked by phalloidin, cytochalasin B, and N-ethylmaleimide-S$_1$ (Cande, 1980). It is not clear how each of these drugs affects the contractile ring, either by stabilizing, disrupting, or solubilizing it, but all show the same result of blocking cytokinesis, though some act more rapidly than others.

VI. SUMMARY AND THE FUTURE OF THE LYSED CELL MODEL IN STUDYING MITOTIC EVENTS

A. The Lysed Cell Model and Spindle Assembly

The lysed cell system has proven to be a valuable tool in studying mechanisms of spindle assembly, chromosome motion, and cytokinesis. Both centrosomes and kinetochores in mammalian tissue culture cells play a role in spindle assembly and exhibit both spatial and temporal control of microtubule initiation. How these regions are actually structured to form a template for microtubule assembly is not yet clear, but it appears that a nucleoprotein complex is necessary and tubulin may form part of this template. Temporal control is exhibited by the ripening or maturation of the centrosome's microtubule initiation capacity at the onset of mitosis. In both marine eggs and mammalian cells, the ripening process is triggered by nuclear envelope breakdown.

Continued use of the lysed cell model in studying spindle assembly should remain a productive avenue of research. Since the function of a microtubule-organizing center is not easily perturbed by cell lysis, many experiments are left to be done. How does a microtubule-organizing center initiate microtubules from pure tubulin that will not assemble *in vitro?* Is it a region of high microtubule-associated protein concentration, and if so, what role do the microtubule-associated proteins play in temporal control of assembly? Experiments similar to those of Pepper and Brinkley (1979), using antibodies to microtubule-associated proteins instead of tubulin, may add considerable information as to the participation of other proteins in microtubule assembly *in vivo*.

The site of subunit addition to microtubules growing from an aster or kinetochores needs to be identified in order to understand how the spindle assembles and elongates during mitosis. The work of Summers and Kirschner (1979) and Bergen *et al.* (1980) using isolated microtubule-organizing centers predicts the rates and ends of microtubules that subunits add to and depolymerize from *in vitro*. Heidemann *et al.* (1980) have demonstrated in lysed mitotic cells that exogenous tubulin adds to the end of the microtubule distal to the mitotic center. The efforts of these workers are important contributions towards the understanding of the role of the microtubule in mitosis. This same question should be asked in an *in vivo* system. Experiments might be done with fluorescein-labeled tubulin (Travis *et al.*, 1980; Keith *et al.*, 1981) to determine how microtubules form and elongate during mitosis.

B. The Lysed Cell Model, Chromosome Motion, and Cytokinesis

The use of lysed cell models in studying anaphase motions and spindle elongation has directed us to look for a dynein-like molecule as a bridge necessary for the sliding of microtubules past one another. We have strong evidence that ATP is a requirement for chromosome motion and that an ATPase similar to dynein plays a functional role in this process. A functional role of actomyosin mediation of chromosome motion seems implausible.

Spindle elongation in lysed anaphase cells requires addition of exogenous tubulin. Sliding of overlapping microtubules from opposite poles past one another during anaphase may not completely account for elongation of the spindle during anaphase B. It seems necessary to invoke an elongation of non-kinetochore microtubules by polymerization to push the poles apart completely.

The role of the contractile ring and how it may function has also been investigated using lysed cells. Cleavage in lysed cells is promoted by millimolar concentrations of Mg–ATP, submicromolar levels of Ca^{2+}, and neutral pH. Furrow relaxation can be enhanced rapidly by addition of high Ca^{2+}, alkaline pH, and cytochalasin B, but not as rapidly or to the same extent in the presence of ATP or N-ethylmaleimide-heavy meromyosin. These differences signal two modes of action and will probably be useful in separating out the molecular events involved in cytokinesis.

The future use of the gently lysed anaphase models will depend on achieving reliable lysis conditions. The major problem is the lack of confidence in the degree of cell lysis achieved during any particular experiment. There is not the precision one would like to have in knowing what is entering and exiting the cell, at what rate, and to what extent. For instance, Brij 58 and Triton X-100 detergents have very different effects on cell ultrastructure when used in similar concentrations. It is difficult to compare and evaluate the results in lysed cells models using these two detergents. Apart from different detergents acting on cells in dissimilar ways, the metabolic state of the cell itself can affect its susceptibility to detergents. Cells that are actively metabolizing are highly affected by detergents, compared to non-metabolizing cells (Komor et al., 1979). All these factors point to a large variability in the effects of detergent on cells and the need to find other methods to introduce molecules into cells with minimal perturbation.

A new approach may be to permeabilize cells with molecules, which are not as harsh on membranes as detergents. For instance low concentrations of lysolecithin will make cells leaky to small molecules, and the cells will eventually reseal themselves (Miller et al., 1979). A novel molecule that can insert itself into membranes and cause an increase in permeability is melletin. It is a soluble protein derived from bee venom in which 19 of the 20 N-terminal amino acid residues are hydrophobic and the other C-terminal residues are polar. This peptide most likely inserts its hydrophobic N-terminals into the lipid bilayer and may prove to be a reliable method for permeabilizing cells (for a review, see Wickner, 1979). Other methods of rapidly permeabilizing mammalian cells have been reviewed by Heppel and Makan (1977).

Efforts toward introducing macromolecules into cells with minimal extraction of cytoplasm may be pursued using various forms of the microinjection technique. Kiehart et al. (1977) have demonstrated the usefulness of microinjecting antibodies into sea urchin eggs to determine what classes of molecules are involved in various mitotic events.

Efforts are currently underway to "microinject" antibodies into tissue culture cells using cell fusion techniques (Schlegel and Rechsteiner, 1978). Red blood cell ghosts can be loaded with antibodies and fused to monolayer cultures (Eckert, 1980). This method has the advantage of using a small volume of material which to some extent can be controlled. One drawback may be the length of time it takes (about 15 minutes) for the antibody to get in and recognize its antigenic determinant. This method can be used with any protein, such as a fluorescein-labeled tubulin, and may turn out to be a valuable tool to help elucidate the roles of macromolecules in the process of mitosis and cytokinesis.

REFERENCES

Aronson, J. F. (1965). The use of a fluorescein-labeled heavy meromyosin for the cytological demonstration of actin. *J. Cell Biol.* **26**, 293–298.

Aubin, J. E., Weber, K., and Osborn, M. (1979). Analysis of actin and microfilament-associated proteins in the mitotic spindle and cleavage furrow by Ptk₂ cells by immunofluorescence microscopy. *Exp. Cell Res.* **124**, 93–110.

Bergen, L. G., and Borisy, G. G. (1980). Head to tail polymerization of microtubules *in vitro.* Electron microscope analysis of seeded assembly. *J. Cell Biol.* **84**, 141–150.

Bergen, L. G., Kuriyama, R., and Borisy, G. G. (1980). Polarity of microtubules nucleated by centrosomes and chromosomes of Chinese hamster ovary cells *in vitro. J. Cell Biol.* **84**, 151–189.

Brinkley, B. R., and Cartwright, J., Jr. (1975). Cold labile and cold stable microtubules in the mitotic spindle of mammalian cells. *Ann. New York Acad. Sci.* **253**, 428–439.

Brinkley, B. R., Stubblefield, E., and Hsu, T. C. (1967). The effects of colcemid inhibition and reversal on the fine structure of the mitotic apparatus of Chinese hamster cells *in vitro. J. Ultrastruct. Res.* **19**, 1–18.

Brown, S. S., and Spudich, J. A. (1979). Cytochalasin inhibits the rate of elongation of actin filament fragments. *J. Cell Biol.* **83**, 657–662.

Cande, W. Z. (1978). Chromosome movement in lysed cells. *In* "Cell Reproduction" (E. Dirksen and D. Prescott, eds), pp. 457–464. Academic Press, New York.

Cande, W. Z. (1980). A permeabilized cell model for studying cytokinesis using mammalian tissue culture cells. *J. Cell Biol.,* **87**, 326–335.

Cande, W. Z., and Wolniack, S. M. (1978). Chromosome movement in lysed mitotic cells is inhibited by vanadate. *J. Cell Biol.* **79**, 573–580.

Cande, W. Z., Snyder, J. A., Smith, D., Summers, K., and McIntosh, J. R. (1974). A functional mitotic spindle prepared from mammalian cells in culture. *Proc. Natl. Acad. Sci. U.S.A.* **71**, 1559–1563.

Cande, W. Z., Lazarides, E., and McIntosh, J. R. (1977). A comparison of the distribution of actin and tubulin in the mammalian mitotic spindle as seen by immunofluorescence. *J. Cell Biol.* **72**, 552–567.

Cande, W. Z., McDonald, K., and Meeusen, R. L. (1981). A permeabilized cell model for studying cell division: a comparison of anaphase chromosome movement and cleavage furrow constriction in lysed Ptk cells. *J. Cell Biol.* **88**, 618–629.

Cantley, L. C., Jr., Josephson, L., Warner, R. Yanagisawa, Lechene, C., and Guidotti, G. (1977). Vanadate is a potent (Na⁺,K⁺)-ATPase inhibitor found in ATP derived from muscle. *J. Biol. Chem.* **252**, 7421–7423.

Dietz, R. (1972). Die Assembly—Hypotheses der Chromosome bewegung und die vëranderunger der spindellänge während der anaphase I in spermatocyten von *Pales ferruginea*. *Chromosoma* **38**, 11–76.

Eckert, B. S. (1980), Disassembly of cytokeratin filaments in living in Ptk_1 cells by microinjection of antibody. *J. Cell Biol.* **87**, 218a.

Farrell, K. W., Morse, A., and Wilson, L. (1979). Characterization of the *in vitro* reassembly of tubulin derived from stable *Strongylocentratus purpuratus* outer doublet microtubules. *Biochemistry* **8**, 905–910.

Euteneuer, U., and McIntosh, J. R. (1980). Polarity of midbody and phragmoplast microtubules. *J. Cell Biol.* **87**, 509–515.

Euteneuer, U., and McIntosh, J. R. (1981). Structural polarity of kinetochore microtubules in Ptk_1 cells. *J. Cell Biol.* **89**, in press.

Forer, A., and Zimmerman, A. M. (1976). Spindle binefringence of isolated mitotic apparatus analyzed by treatments with cold, pressure, and diluted isolation medium. *J. Cell Sci.* **20**, 329–339.

Gibbons, I. R., Cosson, M. P., Evans, J. A., Gibbons, B. H., Houck, B., Martinson, K. H., Sale, W. S., and Tang, W. Y. (1978). Potent inhibition of dynein adenosine triphosphatase and of the motility of cilia and sperm flagella by vanadate. *Proc. Natl. Acad. Sci. U.S.A.* **75**, 2220–2224.

Goode, D., and Roth, L. E. (1969). The mitotic apparatus of a giant amoeba: Solubility properties and induction of elongation *Exp. Cell Res.* **58**, 343–352.

Gould, R. R., and Borisy, G. G. (1977). The pericentriolar material in Chinese hamster ovary cells nucleates microtubule formation. *J. Cell Biol.* **73**, 601–615.

Gould, R. R., and Borisy, G. G. (1978). Quantitative initiation of microtubule assembly by chromosomes from Chinese hamster ovary cells. *Exp. Cell Res.* **113**, 369–374.

Heidemann, S. R., Sander, G., and Kirschner, M. W. (1977). Evidence for a functional role of RNA in centrioles. *Cell* **10**, 337–350.

Heidemann, S. R., Zeive, G. W. McIntosh, J. R. (1980). Evidence for microtubule addition to the distal end of mitotic structures *in vitro*. *J. Cell Biol.* **87**, 152–159.

Hellewell, S. B., and Taylor, L. D. (1979). The contractile basis of ameboid movement. VI. The solation–contraction coupling hypothesis. *J. Cell Biol.* **83**, 633–648.

Heppel, L. A., and Makan, N. (1977). Methods of rapidly altering the permeability of mammalian cells. *J. Supramol. Struct.* **6**, 399–409.

Herman, I. M., and Pollard, T. D. (1978). Actin localization in fixed dividing cells stained with fluorescent heavy meromyosin. *Exp. Cell Res.* **114**, 15–25.

Hoffman-Berling, H. (1954). Die bedevtung des adenosintriphosphat für die zellund Kernteilungsbewegungen in der anaphase. *Biochim. Biophys. Acta* **15**, 226–236.

Inoué, S., and Ritter, H., Jr. (1975). Dynamics of mitotic spindle organization and function. *In* "Molecules and Cell Movement" (S. Inoué and R. E. Stephens, eds.), pp. 3–31. Raven, New York.

Inoué, S., and Sato, H. (1967). Cell motility by labile association of molecules. *J. Gen. Physiol.* **50**, 259–285.

Inoué, S., Borisy, G. G., and D. P. Kiehart, (1974). Growth and lability of *Chaetopterus* oocyte mitotic spindles isolated in the presence of porcine brain tubulin. *J. Cell Biol.* **62**, 175–184.

Johnson, K. A., and Borisy, G. G. (1977). Kinetic analysis of microtubule self-assembly *in vitro*. *J. Mol. Biol.* **117**, 1–31.

Kane, R. E., (1965). The mitotic apparatus. Physical-chemical factors controlling stability. *J. Cell Biol.* **25**, 137–144.

Keith, C. H., Feramisco, J. R., and Shelanski, M. (1981). Direct visualization of fluorescein-labeled microtubules *in vitro* and in microinjected fibroblasts. *J. Cell Biol.* **98**, 234–240.

Kiehart, D. P., Inoué, S., and Mabuchi, I. (1977). Evidence that force production in chromosome movement does not involve myosin. *J. Cell Biol.* **75**, 258a.

Kirschner, M. W. (1978). Microtubule assembly and nucleation. *Int. Rev. Cytol.* **54**, 1–71.

Komor, E., Weber, H., and Tanner, W. (1979). Greatly decreased susceptibility of non-metabolizing cells towards detergents. *Proc. Natl. Acad Sci. U.S.A.* **76**, 1814–1818.

Mabuchi, I., and Okuno, M. (1977). The effect of myosin antibody on the division of starfish blastomeres. *J. Cell Biol.* **74**, 251–263.

Margolis, R., and Wilson, L. (1978a). Mitotic mechanism based on intrinsic microtubule behavior. *Nature (London)* **272**, 450–452.

Margolis, R. L., and Wilson, L. (1978b). Opposite end assembly and disassembly of microtubules at steady state *in vitro*. *Cell* **13**, 1–8.

Mascardo, R. N., and Sherline, P. (1980). High molecular weight microtubule associated proteins (HMW's) are associated with the cytoplasmic microtubule organizing center. *J. Cell Biol.* **87**, 246a.

Mazia, D., and Dan, K. (1952). The isolation and biochemical characterization of the mitotic apparatus of dividing cells. *Proc. Natl. Acad. Sci. U.S.A.* **38**, 826–838.

McGill, M., and Brinkley, B. R. (1975). Human chromosomes and centrioles as nucleation sites for the *in vitro* assembly of microtubules from bovine brain tubulin. *J. Cell Biol.* **67**, 189–199.

McIntosh, J. R. (1974). Bridges between microtubules. *J. Cell Biol.* **61**, 166–187.

McIntosh, J. R. (1977). Mitosis *in vitro:* Isolates and models of the mitotic apparatus. *In* "Mitosis Facts and Questions" (M. Little *et al.*, eds), pp. 167–184. Springer-Verlag, Berlin and New York.

McIntosh, J. R., Helper, P. K., and Van Wie, D. G. (1969). Model for mitosis. *Nature (London)* **244**, 659–663.

McIntosh, J. R. and Landis, S. C. (1971). The distribution of spindle microtubules during mitosis in cultured human cells. *J. Cell Biol.* **49**, 468–497.

Meeusen, R. L., and Cande, W. Z. (1979a). N-ethylmaleimide-modified heavy meromyosin. A probe for actomyosin interactions. *J. Cell Biol.* **82**, 57–65.

Meeusen, R. L., and Cande, W. Z. (1979b). Effect of a specific actomyosin inhibitor on cleavage in amphibian eggs. *J. Cell Biol.* **83**, 317a.

Miller, M. R., Castellot, J. J., and Pardee, A. B. (1979). A general method for permeabilizing monolayer and suspension cultured animal cells. *Exp. Cell Res.* **120**, 421–425.

Mohri, H., Mohri, T., Mabuchi, I., Yazcki, I, Sakai, H., and Ogawa, K. (1976). Localization of dynein in sea urchin eggs during cleavage. *Dev. Growth Differ.* **18**, 391–398.

Murphy, D. B. (1980). Identification of microtubule associated proteins in the meiotic spindle of surf clam oocytes. *J. Cell Biol.* **84**, 235–245.

Murphy, D. B., and Borisy, G. G. (1975). Association of high molecular weight proteins with microtubules and their role in microtubule assembly *in vitro*. *Proc. Natl. Acad. Sci. U.S.A.* **72**, 2696–2700.

Pepper, D. A., and Brinkley, B. R. (1979). Microtubule initiation at kinetochores and centrosomes in lysed mitotic cells. *J. Cell Biol.* **82**, 585–591.

Pepper, D. A., and Brinkley, B. R. (1980). Tubulin nucleation and assembly in mitotic cells: evidence for nucleic acids in kinetochores and centrosomes. *Cell Motility* **1**, 1–15.

Pickett-Heaps, J. D. (1969). The evolution of the mitotic apparatus: An attempt at comparative ultrastructural cytology in dividing plant cells. *Cytobios* **3**, 257–280.

Rebhun, L. I., Rosenbaum, J., Lefebvre, P., and Smith G. (1974). Reversible restoration of the birefringence of cold-treated, isolated mitotic apparatus of surf clam eggs with chick brain tubulin. *Nature (London)* **249**, 113–115.

Robbins, E., Jentzsch, G., and Micali, A. (1968). The centriole cycle in synchronized Hela cells. *J. Cell Biol.* **36**, 329–339.

Roos, U. P. (1973). Light and electron microscopy of rat kangaroo cells in mitosis. II. Kinetochore structure and function. *Chromosoma* **41**, 195–220.

Sakai, H. (1978a). Induction of chromosome motion in the isolated mitotic apparatus as a function of microtubules. *In* "Cell Reproduction" (E. R. Dirksen and D. M. Prescott, eds.), pp. 425–432. Academic Press, New York.

Sakai, H. (1978b). The isolated mitotic apparatus and chromosome motion. *Int. Rev. Cytol.* **55**, 22–48.

Sakai, H., Mabuchi, I., Shimoda, S., Kiriyama, R., Ogawa, K., and Mohri, H. (1976). Induction of chromosome motion in the glycerol-isolated mitotic apparatus: Nucleotide specificity and the effects of anti-dynein and myosin sera on the motion. *Dev. Growth Differ.* **18**, 211–219.

Salmon, E., and Jenkins, R. (1977). Isolated mitotic spindles are depolymerized by micromolar calcium and show evidence of dynein. *J. Cell Biol.* **75**, 295a.

Sanger, J. W. (1975). Presence of actin during chromosomal movement. *Proc. Natl. Acad. Sci. U.S.A.* **72**, 2451–2455.

Sato, H., Ellis, G. W., and Inoué, S. (1975). Micortubular origin of mitotic spindle form birefringence. Demonstration of the applicability of Wiener's equation. *J. Cell Biol.* **67**, 501–517.

Scheele, R. B., and Borisy, G. G. (1979). *In vitro* assembly of microtubules. *In* "Microtubules" (K. Roberts and J. S. Hyams, eds.), pp. 175–254. Academic Press, New York.

Schiff, P. B., Fant, J., and Horowitz, S. B. (1979). Promotion of microtubule assembly *in vitro* by taxol. *Nature (London)* **277**, 665–667.

Schlegel, R. A., and Rechsteiner, M. C. (1978). Red cell mediated microinjection of macromolecules into mammalian cells. *Methods Cell Biol.* **20**, 341–354.

Sloboda, R. D., Dentler, W. L., and Rosenbaum, J. L. (1976). Microtubule associated proteins and the stimulation of tubulin assembly *in vitro*. *Biochem.* **15**, 4497–4505.

Snyder, J. A. (1980). Evidence for a ribonucleoprotein complex as a template for microtubule assembly *in vivo*. *Cell Biol. Int. Rep.* **4**, 859–868.

Snyder, J. A., and McIntosh, J. R. (1975). Initiation and growth of microtubules from mitotic centers in lysed mammalian cells. *J. Cell Biol.* **67**, 744–760.

Summers, K., and Kirschner, M. W. (1979). Characteristics of the polar assembly and disassembly of microtubules observed *in vitro* by dark field light microscopy. *J. Cell Biol.* **83**, 205–217.

Telzer, B. R., and Rosenbaum, J. (1979). Cell cycle-dependent, *in vitro* assembly of microtubules onto the pericentriolar material of HeLa cells. *J. Cell Biol.* **81**, 484–497.

Telzer, B. R., Moses, J., and Rosenbaum, J. (1975). Assembly of microtubules onto kinetochores of isolated mitotic chromosomes in HeLa cells. *Proc. Natl. Acad. Sci. U.S.A.* **72**, 4023–4027.

Travis, J. L. (1980) Preparation and characterization of native, fluorescently labled brain tubulin and microtubule-associated proteins (MAP's). *Exp. Cell Res.* **125**, 421–429.

Timasheff, S. N., and Grisham, L. M. (1980). *In vitro* assembly of cytoplasmic microtubules. *Ann. Rev. Biochem.* **49**, 565–592.

Tippel, D., Pickett-Heaps, J. D., and Leslie, C. (1980). Cell demsion in two large pennate diatoms, *Hantzschin* and *Nitzschin*. *J. Cell Biol.* **86**, 402–416.

Weingarten, M. D., Lockwood, A. H., Hwo, S.-Y., and Kirschner, M. W. (1975). A protein factor essential for microtubule assembly. *Proc. Natl. Acad. Sci. U.S.A.* **72**, 1858–1862.

Weisenberg, R. C. (1972). Microtubule formation *in vitro* in solutions containing low calcium concentrations. *Science* **177**, 1105–1105.

Weisenberg, R. C. (1973). Regulation of tubulin organization during meiosis. *Am. Zool.* **13**, 981–987.

Weisenberg, R. C., and Rosenfeld, A. C. (1975). *In vitro* polymerization of microtubules into asters and spindles in homogenates of surf clam eggs. *J. Cell Biol.* **64**, 146–158.

Wheatley, D. N. (1974). Pericentriolar virus-like particles in chinese hamster ovary cells. *J. Gen. Virol.* **24**, 395–399.

Wickner, W. (1979). The assembly of proteins into biological membranes. *Annu. Rev. Biochem.* **48**, 23–46.

Wieland, Th., and Govindan, V. M. (1974). Phallotoxins bind to actin. *FEBS Lett.* **46**, 351–353.

Zackroff, R., Rosenfeld A., and Weisenberg, R. (1976). Effects of RNase and RNA on *in vitro* aster assembly. *J. Supramol. Struct.* **5**, 577–589.

The Isolated Mitotic Apparatus:
A Model System for Studying
Mitotic Mechanisms

ARTHUR M. ZIMMERMAN and ARTHUR FORER

I. INTRODUCTION

If the mitotic apparatus (the spindle–chromosome–aster complex) could be isolated from a cell in a functional state, one could study directly, *in vitro*, the chemistry of chromosome movement. Almost three decades have passed since Mazia and Dan (1952) first isolated mitotic apparatus from sea urchin zygotes. The chemistry and fine structure of the isolated mitotic apparatus have been studied during the intervening years (for reviews, see Mazia, 1955; Zimmerman, 1963; Forer, 1969; Nicklas, 1971; Hartmann and Zimmerman, 1974; Zimmerman *et al.*, 1977; Sakai, 1978a; McIntosh, 1979; Petzelt, 1979). However, despite reports of improved methods for isolation of labile mitotic apparatus, chromosome movement similar to that observed *in vitro* has not yet been reported in the isolated mitotic apparatus. The purpose of this chapter is to review briefly the methodology used for mitotic isolation and to assess the several approaches that have been designed to study chromosome movement *in vitro*.

MITOSIS/CYTOKINESIS
Copyright © 1981 by Academic Press, Inc.
All rights of reproduction in any form reserved.
ISBN 0-12-781240-7

II. ISOLATION OF THE MITOTIC APPARATUS, 1954–1970

Shortly after Mazia and Dan (1952) reported on the mass isolation of the mitotic apparatus, the composition of the alcohol–digitonin-isolated mitotic apparatus was shown to be composed primarily of protein, with some RNA. Physicochemical analysis of the mitotic apparatus components showed that the material was heterogeneous and consisted of protein and ribonucleic acids with distinct electrophoretic and sedimentation values (Mazia, 1954; Zimmerman, 1960). A significant advance in the study of the mitotic apparatus occurred in the early 1960s when it was reported that mitotic apparatus could be isolated directly from sea urchin zygotes without the intervening stabilization in cold ethanol (Mazia et al., 1961; Kane, 1962). Mazia and co-workers (1961) reported that mitotic apparatus could be isolated directly from sea urchin zygotes if the cells were treated with a dithioglycerol (a penetrating disulfide) solution that was supplemented with dextrose, salts, etc. The method was based on the principle that disulfide protein linkages were essential for mitotic stabilization. It was assumed that a delicate balance between thiol and disulfide was responsible for mitotic apparatus stability. However, sulfur-containing compounds were shown to be non-essential for mitotic apparatus isolation when Kane (1962, 1965) reported that the critical factors for direct isolation of mitotic apparatus were pH and non-electrolyte concentration. Using proper combinations of pH and non-electrolyte, mitotic apparatus were isolated using hexanediol, a penetrating 6-carbon glycol, or hexylene glycol (Kane, 1962, 1965). Hexylene glycol was the preferred isolation/lysis agent because, unlike hexanediol, for example, hexylene glycol does not support bacterial growth.

The original alcohol–digitonin method has also been modified to increase the yield of mitotic isolates. Zygotes were placed in sea water in which the Na^+ was replaced by Li^+. Under these conditions, the in vivo mitotic apparatus was stabilized and did not disperse, and large amounts of metaphase mitotic apparatus were isolated with Triton X following storage in cold ethanol (Mazia et al., 1972).

What have we learned from these studies? Progress was made in analyzing chemically the components present in the isolated mitotic apparatus (see reviews, e.g., in Forer, 1969; Hartmann and Zimmerman, 1974; Petzelt, 1979), and indeed, our present concept that mitotic apparatus are primarily composed of one component derives from analyses of isolated mitotic apparatus in which mainly one component was found. However, little progress was made in obtaining functional mitotic apparatus. We are not aware of any studies in which the authors even reported attempts to induce chromosome movement in these early mitotic isolates. Indeed, the mitotic apparatus as isolated were different from mitotic apparatus in vivo in several important respects. For one, mitotic apparatus responded differently *in vitro* than they did *in vivo*. Cold treatment stabilized isolated mitotic apparatus (e.g., Kane and Forer, 1965) but caused the disappearance of *in vivo* mitotic apparatus (e.g., Inoué, 1964). For another, the isolation procedure allowed recovery of only approximately 10% of the material present in mitotic apparatus *in vivo* (Forer, 1969; Forer and Goodman, 1969, 1972). Mitotic apparatus *in vivo* have more than 20 gm of dry matter per 100 cm^3 of mitotic apparatus, whereas isolated mitotic apparatus have at most 1 gm of dry matter per 100 cm^3 of mitotic apparatus (see the review in Forer *et al.*, 1981). Since isolated mitotic apparatus were not functional and had properties different from those of *in vivo* mitotic apparatus, the question, then, is not only whether the "motor" might be denatured by the isolation procedure but whether or not it might even be lost. More recent experiments have been explicitly directed at modifying the procedures to get functional mitotic apparatus *in vitro*, as we describe in the next section.

III. ISOLATION OF MITOTIC APPARATUS SINCE 1970

Two main approaches were used to try to obtain functional mitotic apparatus *in vitro*: (a) to modify the medium used to stabilize the mitotic apparatus, so that the properties of the mitotic apparatus *in vitro* resembled those of mitotic apparatus *in vivo* (e.g., Forer and Zimmerman, 1974), and (b) to use tubulin monomer to stabilize mitotic apparatus *in vitro*. The latter approach is based on the idea that the mitotic apparatus *in vivo* are in equilibrium between monomer and polymer (e.g., Rebhun *et al.*, 1974; Inoué *et al.*, 1974), so that addition of monomer will help stabilize the polymer in a natural state. We discuss these two approaches in turn.

We have modified the isolation medium to try to get mitotic apparatus *in vitro* that were like mitotic apparatus *in vivo* when challenged with various agents. In particular, we used as our starting point the medium for isolation of cytoplasmic microtubules originally designed by Filner and Behnke (1973); we modified their medium to try to get mitotic apparatus that were cold labile and pressure labile, as are mitotic apparatus *in vivo*. Mitotic apparatus isolated in a glycerol–DMSO–EGTA medium are indeed cold labile and pressure labile, and they maintain these properties (as well as solubility in 0.6 M KCl) during storage (Forer and Zimmerman, 1974, 1975, 1976a,b; Zimmerman and Forer, 1978; Zimmerman *et al.*, 1977). There is *some* decay of properties during storage, but there is quite a difference between these mitotic apparatus and those isolated using hexylene glycol (Forer and Zimmerman, 1978). We have been unable to induce chromosome movement *in vitro*, however.

Other workers have also used glycerol-containing stabilization media; these media used glycerol for stabilization in conjunction with Triton X-100 for lysis (e.g., Sakai and Kuriyama, 1974; Sakai et al., 1975; Sakai et al., 1976). These workers have reported some movement of chromosomes in isolated MA (see reviews in Sakai 1978a,b); the movement was observed in MA at early anaphase and consisted of chromosome movement one-third to one-half of the way to the poles over a period of 1 hour, during which time the poles moved apart approximately 15%. Unfortunately, this movement is quite different from that observed in vivo, in rate, in extent, and in other characteristics (see Chapter 5 by Rickards, this volume; see also Forer, 1978).

The other approach taken was to stabilize isolated mitotic apparatus using tubulin-containing medium (Rebhun et al., 1974; Inoué et al., 1974; Milsted et al., 1977) or medium that supports polymerization of brain tubulin into microtubules (e.g., Silver et al., 1980). In these media the mitotic apparatus birefringence was reduced when the external tubulin concentration was too low, the mitotic apparatus were stabilized when the external tubulin concentration was higher, and the mitotic apparatus birefringence increased when the external tubulin was still higher; one concludes, then, that the external tubulin became incorporated into the mitotic apparatus when the external tubulin concentration was high enough. These mitotic apparatus were like those in vivo in that they were cold labile, at least for short periods (up to 1 hour) after isolation. Here too, though, there was no report that chromosome movement occurred in vitro.

The more recent isolation methods have not resulted in chromosome movement in vitro; this is still an unattained goal, though some progress has been made in defining the conditions under which mitotic apparatus retain in vivo properties. This is true under various conditions, using glycerol and DMSO, glycerol and detergent, or tubulin-containing buffers (see the review in Petzelt, 1979). Conditions in which mitotic apparatus retain Ca^{2+} sensitivity have also been defined (Salmon and Segall, 1980), as have conditions for mitotic apparatus vesicles in sequestering Ca^{2+} (Silver et al., 1980). Some progress has also been made in studying chemical components of the mitotic apparatus. A dynein-like ATPase was found in mitotic apparatus (Pratt et al., 1980). In other studies, various components in isolated mitotic apparatus were identified using SDS–gel electrophoresis, and components found using different isolation methods were compared (Murphy, 1980). In still other studies, quantitative analyses of the tubulin in isolated mitotic apparatus and cell lysate were used to estimate the maximum amount of tubulin in the mitotic apparatus in vivo (Forer et al., 1981). Progress in other aspects of the chemistry of the mitotic apparatus is summarized in Chapter 15 by Nagle and Egrie.

Overall, the progress in using isolated mitotic apparatus to understand mitotic mechanisms has been slow. We have achieved neither functional mitotic apparatus *in vitro* nor a great understanding of the chemistry of the mitotic apparatus. How does one proceed from here to use isolated mitotic apparatus to study mitosis? We discuss this in the next section; before proceeding, it is relevant to point out that though much of the work discussed above uses mitotic apparatus isolated from sea urchin zygotes, mitotic apparatus can also be isolated from cells of other species. These include various marine eggs such as the sea star, *Pisaster* (Bryan and Sato, 1970; Sato *et al.*, 1975) and the surf clam, *Spisula* (e.g., Rebhun and Sharpless, 1964; Murphy, 1980); *Drosophila* cells (Milsted and Cohen, 1973; Milsted *et al.*, 1977); crane fly spermatocytes (Muller, 1970); cultured mammalian cells (Sisken *et al.*, 1967; Sisken, 1970; McIntosh *et al.*, 1979); *Tetrahymena* micronuclei (Davidson and LaFountain, 1975; LaFountain and Davidson, 1979, 1980); and *Lymneae* eggs (Mescheryakov, 1978).

IV. OVERVIEW AND PERSPECTIVES

How does one proceed to obtain chromosome movement *in vitro*? There seem to be two choices. One can either vary the isolation and other conditions, using *in vivo* mitotic apparatus properties as a guide, and hope to find conditions for obtaining chromosome movement, or one can try to develop a bioassay to see if the mitotic apparatus *can* function, and then try to deduce *in vitro* conditions for function. Several groups have tried the first approach (e.g., Sakai and Kuriyama, 1974; see the review in Sakai, 1978a; see also Salmon and Segall, 1980). We have tried the second (Forer *et al.*, 1977; Masui *et al.*, 1978). We have been able to induce cleavage by injecting isolated sea urchin mitotic apparatus into enucleated frog eggs; this occurred only if the mitotic apparatus were not denatured, which suggests that isolated mitotic apparatus *can* function. The ambiguity remains, however, that the frog egg may cause the spindle to break down and may then use the sea urchin chromosomes and centrioles to organize a new spindle (discussed in Forer *et al.*, 1977). For a variety of non-scientific reasons, we have not progressed well in resolving this ambiguity or in circumventing it, but one can think of ways to do so. For example, if one isolates mitotic apparatus from *meiotic* cells (e.g., *Spisula*) and injects anaphase meiotic spindles into frog eggs, then one can distinguish the possibilities by counting the number of chromosomes in the frog embryos. That is, if the *Spisula* meiotic spindle itself were used, one should see n chromosomes per cell, but if the *Spisula* chromosomes were used, after breaking down the spindle, then one should see $2n$ chromosomes per cell. As another example, if the sea urchin mitotic apparatus is indeed functional in the frog egg, then one should be able to add extract of frog egg (or just "ground up" frog egg) directly to an anaphase-isolated mitotic apparatus that one observes under the microscope, and the frog egg extract should cause the chormosomes to move. Tactics of this kind might enable one to determine if the bioassay is indeed a true assay for spindle function, and might enable one to find conditions for obtaining chromosome movement *in vitro*.

Another approach, described in detail in Chapter 13 by Snyder (this volume), is to make cells permeable in such a way that the spindles continue to function (Cande *et al.*, 1974). When this occurs, one can add chemicals that otherwise would not get across the cell membrane. Since the cells are permeable, these components enter the cell and can attack spindle components. Such spindle models do not have a controlled environment, however, so there are still ambiguities of a kind that would not be present in mitotic apparatus *in vitro*, isolated from the rest of the cell. But this semi-lysis approach may lead to understanding that will enable an *in vitro* mitotic apparatus to be obtained.

Besides its function, one wants to study the *chemistry* of the isolated mitotic apparatus. How does one proceed in this study? If material is lost from the mitotic apparatus during isolation, as it seems to be (e.g., Forer and Goldman, 1972), then how can one determine the importance of the residual components? How can one decide whether those components that *are* present are mitotic apparatus components rather than just adhering contaminants? To try to answer the latter question, Murphy (1980) compared amounts of each protein in the mitotic apparatus relative to tubulin for two different methods of isolation. Murphy argued that if the component is a *bona fide* mitotic apparatus component, it should be present in a constant ratio to tubulin. [A similar rationale for actin *not* being a *bona fide* mitotic apparatus component was given by Pratt *et al.* (1980).] This experimental approach is not sufficient to get an unequivocal answer, however, since mitotic apparatus isolated using different methods have different physiological characteristics (e.g., Zimmerman and Forer, 1978; Forer and Zimmerman, 1978), and these differences might result because the various methods of isolation preserve different ratios of proteins. One way to avoid this problem (D. Larson, A. Forer, and A. M. Zimmerman, unpublished) is to study the protein composition of mitotic apparatus under various conditions in which the physiology of the mitotic apparatus changes; in this way, one can compare changes in the proteins (e.g., via SDS–gel electrophoresis) with the changed physiology. For example, one can study (on SDS–gels) the KCl-soluble and KCl-insoluble fractions of freshly isolated mitotic apparatus, and one can compare these fractions with similar fractions of stored mitotic apparatus in which the solubility is visibly changed (Zimmerman *et al.*, 1977; Forer and Zimmerman, 1978). Experiments of this kind could identify those proteins in the mitotic apparatus that are responsible for the various physiological properties.

V. CONCLUDING REMARKS

To summarize, experiments using isolated mitotic apparatus have not as yet contributed much to our understanding of mechanisms of mitosis. This is primarily because isolated mitotic apparatus are not functional, and because it is difficult to distinguish genuine components inherent in the mitotic apparatus from contaminants. Also, much of the mitotic apparatus dry matter is lost during the isolation procedures. However, the potential for use of isolated mitotic apparatus as an *in vitro* model system is great; such a system would greatly aid in understanding the chemistry of mitosis. Nonetheless, until the system is made functional, it is difficult to know how to proceed. We recall that once Niremberg and Matthaei discovered how to get ribosomes to make proteins *in vitro*, then a direct experimental attack on the genetic code began; the same was true for polymerization of microtubules *in vitro*. Until then, one could not use that approach. We expect, similarly, that once one obtains functional mitotic apparatus *in vitro*, a new, important approach to studying mitosis will be opened up. Until then, however, one must be patient in trying to obtain this elusive goal.

REFERENCES

Bryan, J., and Sato, H. (1970). The isolation of the meiosis I spindle from the mature oocyte of *Pisaster ochraceous*. *Exp. Cell Res.* **59**, 371–378.

Cande, W. Z., Snyder, J., Smith, D., Summers, K., and McIntosh, J. R. (1974). A functional mitotic spindle prepared from mammalian cells in culture. *Proc. Natl. Acad. Sci. U.S.A.* **71**, 1559–1563.

Davidson, L., and La Fountain, J. R., Jr. (1975). Mitosis and early meiosis in *Tetrahymena pyriformis* and the evolution of mitosis in the phylum Ciliophora. *Biosystems* **7**, 326–336.

Filner, P., and Behnke, D. (1973). Stabilization and isolation of brain microtubules with glycerol and dimethylsulfoxide (DMSO). *J. Cell Biol.* **59**, 99a.

Forer, A. (1969). Chromosome movements during cell division. *In* "Handbook of Molecular Cytology" (A. Lima-de-Fario, ed.), pp. 553–601. North-Holland Pub., Amsterdam.

Forer, A. (1978). Chromosome movements during cell division: Possible involvement of actin filaments. *In* "Nuclear Division in the Fungi" (I. B. Heath, ed.), pp. 21–88. Academic Press, New York.

Forer, A., and Goldman, R. D. (1969). Comparisons of isolated and *in vivo* mitotic apparatuses. *Nature London* **222**, 689–691.

Forer, A., and Goldman, R. D. (1972). The concentration of dry matter in mitotic apparatuses *in vivo* and after isolation from sea urchin zygotes. *J. Cell Sci.* **10**, 387–415.

Forer, A., and Zimmerman, A. M. (1974). Characteristics of sea-urchin mitotic apparatus isolated using a dimethylsulphoxide/glycerol medium. *J. Cell Sci.* **16**, 481–497.

Forer, A., and Zimmerman, A. M. (1975). Isolation of sea urchin mitotic apparatus using glycerol-dimethyl sulfoxide. *Ann. N. Y. Acad. Sci.* **253**, 378–382.

Forer, A., and Zimmerman, A. M. (1976a). Spindle birefringence of isolated mitotic apparatus analysed by pressure treatment. *J. Cell Sci.* **20**, 309–327.

Forer, A., and Zimmerman, A. M. (1976b). Spindle birefringence of isolated mitotic apparatus analyzed by treatments with cold, pressure and diluted isolation medium. *J. Cell Sci.* **20**, 329–339.

Forer, A., and Zimmerman, A. M. (1978). Stability of isolated mitotic apparatus during storage. *Develop. Growth Differ.* **20**, 93–105.

Forer, A., Larson, D. E., and Zimmerman, A. M. (1981). Experimental determinations of tubulin in the *in vivo* mitotic apparatus of sea-urchin zygotes. *Can. J. Biochem.* **58**, 1277–1285.

Forer, A., Masui, Y., and Zimmerman, A. M. (1977). A possible bio-assay for chromosome movement in isolated mitotic apparatus. *Exp. Cell Res.* **106**, 430–434.

Hartmann, J. F., and Zimmerman, A. M. (1974). The mitotic apparatus. *In* "The Cell Nucleus" (H. Busch, ed.), Vol. 2, pp. 459–486. Academic Press, New York.

Inoué, S. (1964). Organization and function of the mitotic spindle. *In* "Primitive Motile Systems in Cell Biology" (R. D. Allen and N. Kamiya, eds.), pp. 549–598. Academic Press, New York.

Inoué, S., Borisy, G. G., and Kiehart, D. P. (1974). Growth and lability of *Chaetopterus* oocyte mitotic spindles isolated in the presence of porcine brain tubulin. *J. Cell Biol.* **62**, 175–184.

Kane, R. E. (1962). The mitotic apparatus: Isolation by controlled pH. *J. Cell Biol.* **12**, 47–55.

Kane, R. E. (1965). The mitotic apparatus. Physical-chemical factors controlling stability. *J. Cell Biol. Suppl.* **25**, 137.

Kane, R. E., and Forer, A. (1965). The mitotic apparatus. Structural changes after isolation. *J. Cell Biol.* **25**, 31–39.

La Fountain, J. R., Jr., and Davidson, L. A. (1979). An analyses of spindle ultrastructure during prometaphase and metaphase of micronuclear division in *Tetrahymena. Chromosoma* **75**, 293–308.

La Fountain, J. R., Jr., and Davidson, L. A. (1980). An analyses of spindle ultrastructure during anaphase of micronuclear division in *Tetrahymena. Cell Motil.* **1**, 41–61.

Masui, Y., Forer, A., and Zimmerman, A. M. (1978). Induction of cleavage in nucleated and enucleated frog eggs by injection of isolated sea-urchin mitotic apparatus. *J. Cell Sci.* **31**, 117–135.

Mazia, D. (1954). SH and growth. *In* "Glutathione" (S. Colowick, A. Lazarow, E. Racker, D. R. Schwarz, E. R. Stadtman, and H. Waelsch, eds.), pp. 209–228. Academic Press, New York.

Mazia, D. (1955). The organization of the mitotic apparatus. *Symp. Soc. Exp. Biol.* **9**, 335–357.

Mazia, D., and Dan, K. (1952). The isolation and biochemical characterization of the mitotic apparatus of dividing cells. *Proc. Natl. Acad. Sci. U.S.A.* **38**, 826–838.

Mazia, D., Mitchison, J. M., Medina, H., and Harris, P. (1961). The direct isolation of the mitotic apparatus. *J. Biophys. Biochem. Cytol.* **10**, 467–474.

Mazia, D., Petzelt, C., Williams, R. O., and Meza, I. (1972). A Ca-activated ATPase in the mitotic apparatus of the sea urchin egg (isolated by a new method). *Exp. Cell Res.* **70** 325–332.

McIntosh, J. R. (1979). Cell division. *In* "Microtubules" (K. Roberts and J. S. Hyams, eds.), pp. 428–441. Academic Press, New York.

McIntosh, J. R., Sisken, J. E., and Chu, L. K. (1979). Structural studies on mitotic spindles isolated from cultured human cells. *J. Ultrastruct. Res.* **66**, 40–52.

Meshcheryakov, V. N. (1978). Isolation of the mitotic apparatus from the eggs and embryos of pulmonate mollusc *Lymnaea stagnalis* L. *Isitologiya* **20**, 1211–1216.

Milsted, A., and Cohen, W. D. (1973). Mitotic spindles from *Drosophila melanogaster* embryos. *Exp. Cell Res.* **78**, 243–246.

Milsted, A., Cohen, W. D., and Lampen, N. (1977). Mitotic spindles of *Drosophila melanogaster:* A phase-contrast and scanning electron-microscope study. *J. Cell Sci.* **23**, 43–55.

Muller, W. (1970). Interferenzmikroskopische Untersuchungen der Trockenmassenkonzentration in isolierten Mitoseapparaten und lebenden Spermatocyten von *Pales ferruginea* (*Nematocera*). *Chromosoma* **30**, 305–316.

Murphy, D. B. (1980). Identification of microtubule-associated proteins in the meiotic spindle of surf clam oocytes. *J. Cell Biol.* **84**, 235–245.

Nicklas, R. B. (1971). Mitosis. *Adv. Cell Biol.* **2**, 225–294.

Niremberg, M. W., and Matthaei, J. H. (1961). The dependence of cell-free protein synthesis in *E. coli* upon naturally occurring or synthetic polyribonucleotides. *Proc. Natl. Acad. Sci. U.S.A.* **47**, 1588–1602.

Petzelt, C. (1979). Biochemistry of the mitotic spindle. *Int. Rev. Cytol.* **60**, 53–92.

Pratt, M. M., Otter, T., and Salmon, E. D. (1980). Dynein-like Mg^{2+}-ATPase in mitotic spindles isolated from sea urchin embryos. (*Strongylocentrotus droebachiensis*). *J. Cell Biol.* **86**, 738–745.

Rebhun, L. I., and Sharpless, T. K. (1964). Isolation of spindles from the surf clam *Spisula solidissima*. *J. Cell Biol.* **22**, 488–492.

Rebhun, L. I., Rosenbaum, J., Lefebvre, P., and Smith, G. (1974). Reversible restoration of the birefringence of cold-treated, isolated mitotic apparatus of surf clam eggs with chick brain tubulin. *Nature (London)* **249**, 113–115.

Sakai, H. (1978a). The isolated mitotic apparatus and chromosome motion. *Int. Rev. Cytol.* **55**, 23–48.

Sakai, H. (1978b). Induction of chromosome motion in the isolated mitotic apparatus as a function of microtubules. *In* "Cell Reproduction: In Honour of Daniel Mazia" (E. R. Dirkson, D. M. Prescott and C. F. Fox, eds.), pp. 425–432. Academic Press, New York.

Sakai, H., and Kuriyama, R. (1974). The mitotic apparatus isolated in glycerol-containing medium. *Develop. Growth Differ.* **16**, 123–134.

Sakai, H., Hiramoto, Y., and Kuriyama, R. (1975). The glycerol-isolated mitotic apparatus: A response to porcine brain tubulin and induction of chromosome motion. *Develop. Growth Differ.* **17**, 265–274.

Sakai, H., Mabuchi, I., Shimoda, S., Kuriyama, R., Ogawa, K., and Mohri, H. (1976). Induction of chromosome motion in the glycerol-isolated mitotic apparatus: Nucleotide specificity and effects of anti-dynein and myosin sera on the motion. *Develop. Growth Differ.* **18**, 211–219.

Salmon, E. D., and Segall, R. R. (1980). Calcium-labile mitotic spindles isolated from sea urchin eggs (*Lytechinus variegatus*). *J. Cell Biol.* **86**, 355–365.

Sato, H., Ellis, G. W., and Inoue, S. (1975). Microtubular origin of mitotic spindle form birefringence: Demonstration of the applicability of Wiener's equation. *J. Cell Biol.* **67**, 501–517.

Silver, R. B., Cole, R. D., and Cande, W. Z. (1980). Isolation of mitotic apparatus containing vesicles with calcium sequestration activity. *Cell* **19**, 505–516.

Sisken, J. E. (1970). Procedures for the isolation of the mitotic apparatus from cultured mammalian cells. *In* "Methods in Cell Physiol" (D. M. Prescott, ed.), Vol. 4, pp. 71–82. Academic Press, New York.

Sisken, J. E., Wilkes, E., Donnelly, G. M., and Kakefuda, T. (1967). The isolation of the mitotic apparatus from mammalian cells in culture. *J. Cell Biol.* **32**, 212–216.

Zimmerman, A. M. (1960). Physico-chemical analysis of the isolated mitotic apparatus. *Exp. Cell Res.* **20**, 529–547.

Zimmerman, A. M. (1963). Chemical aspects of the isolated mitotic apparatus. *In* "The Cell in Mitosis" (L. Levine, ed.), pp. 159–184. Academic Press, New York.

Zimmerman, A. M., and Forer, A. (1978). Comparative analysis of stability characteristics of hexylene glycol and DMSO/glycerol isolated mitotic apparatus. *In* "Cell Reproduction: In Honour of Daniel Mazia" (E. R. Dirkson, D. M. Prescott, and C. F. Fox, eds.), pp. 477–486. Academic Press, New York.

Zimmerman, A. M., Zimmerman, S., and Forer, A. (1977). The mitotic apparatus: Methods for isolation. *In* "Methods in Cell Biology" (G. Stein, J. Stein, and L. J. Kleinsmith, eds.), Vol. 16, pp. 361–371. Academic Press, New York.

15

Calmodulin and ATPases in the Mitotic Apparatus

BARBARA W. NAGLE and JOAN C. EGRIE

I. INTRODUCTION

Students of the cell cycle have long been interested in the signals which cause the cell to progress from one phase of the cell cycle to the next. As Harris (1978; and Chapter 2, this volume) has described, one model for the regulation of the cell cycle involves fluxes in the free concentration of Ca^{2+}. Calcium ions have been implicated in the regulation of a number of cellular activities, including secretion, contraction, and the response of eggs at fertilization. Research on problems related to mitosis has suggested that Ca^{2+} might play an important role in the assembly and function of the mitotic

MITOSIS/CYTOKINESIS
Copyright © 1981 by Academic Press, Inc.

apparatus and thus might also be an important signal for the mitotic phase of the cell cycle.

While several possible targets for regulation by Ca^{2+} exist in the mitotic apparatus, the exact role(s) of Ca^{2+} in mitosis has been difficult to determine. The demonstration that calmodulin is localized in the mitotic apparatus of dividing tissue culture cells (Welsh *et al.*, 1978; Andersen *et al.*, 1978) suggests that this Ca^{2+}-binding protein may be involved in the mediation of calcium's effects on the mitotic apparatus. Therefore, knowledge of the role of calmodulin in mitosis is central to an understanding of the regulatory actions of Ca^{2+}.

Our knowledge of calmodulin's functions in other systems suggests an almost bewildering number of possibilities for its role(s) in the mitotic apparatus. In this chapter, we will discuss the possible functions of Ca^{2+} and calmodulin in the mitotic apparatus and then focus on the details of our own work, which has specifically addressed the possibility that calmodulin acts to regulate the activity of Ca^{2+}-sensitive ATPases in the mitotic apparatus.

II. CALMODULIN AND THE MITOTIC APPARATUS

A. Characteristics and Functions of Calmodulin

Calcium, like the cyclic nucleotides, is considered to be a second messenger in controlling a number of important physiological reactions. In spite of the extensive evidence for the regulatory role of Ca^{2+}, until recently the mechanism by which it produces its diverse effects was largely unknown. It is now believed that the actions of Ca^{2+} are mediated by Ca^{2+}-binding proteins, the most ubiquitous of which is calmodulin. Although calmodulin does not mediate all Ca^{2+} effects, this multifunctional Ca^{2+} receptor has been shown to be involved in regulating such diverse processes as cyclic nucleotide metabolism, cell motility, microtubule assembly, glycogen metabolism, and Ca^{2+} transport. Several excellent and detailed reviews on the properties and functions of calmodulin have appeared (Wang and Waisman, 1979; Wolff and Brostrom, 1979; Cheung, 1980; Means and Dedman, 1980), and readers are referred to these references for a more comprehensive discussion.

Calmodulin is a 17,000-dalton, heat-stable, acidic protein which is present in micromolar concentrations in all eukaryotic cells examined. It has four high-affinity Ca^{2+}-binding sites per molecule, with dissociation constants in the micromolar range. Calmodulin is a highly conserved protein throughout evolution. Functional assays have demonstrated that there is no tissue or species specificity of calmodulin (Smoake *et al.*, 1974; Waisman *et al.*, 1975;

Waisman *et al.*, 1978), and peptide map analyses have indicated that a variety of vertebrate calmodulins are either identical or very similar (Brooks and Siegel, 1973; Stevens *et al.*, 1976; Vanaman *et al.*, 1977). Calmodulin shares a great deal of sequence homology with troponin C, the Ca^{2+}-binding subunit of the skeletal muscle protein, troponin. Amino acid sequence comparisons of the two proteins have shown that 50% of the amino acids are identical and 25% are functionally conservative replacements (Vanaman *et al.*, 1977; Dedman *et al.*, 1978a; Watterson *et al.*, 1980).

In the majority of systems studied, calmodulin regulates cellular activities by reversibly associating with and activating the rate-limiting enzymes in these systems in a Ca^{2+}-dependent manner (Wang and Waisman, 1979; Wolff and Brostrom, 1979). Increases in the intracellular Ca^{2+} concentration resulting from external stimuli lead to the formation of a Ca^{2+}–calmodulin complex. Upon Ca^{2+} binding, calmodulin changes conformation (Wang and Waisman, 1979; Wolff and Brostrom, 1979), allowing it to associate with the spectrum of calmodulin-regulated proteins in the given cell type. As a consequence of this association, the enzymes are activated and can initiate the biochemical changes that produce the response to the stimulus. Return of the intracellular Ca^{2+} concentration to its resting level is then accompanied by a reverse of the above series of steps, resulting in a decrease in enzyme activity. Although calmodulin is ubiquitous, the profile of calmodulin-regulated proteins differs both qualitatively and quantitatively in different tissues, providing a basis for the specificity of calcium's actions (Kakiuchi *et al.*, 1975; Wang and Desai, 1976; La Porte and Storm, 1978; Grand and Perry, 1979). Specific enzymes shown to be activated by calmodulin in this manner include cyclic nucleotide phosphodiesterase (Kakiuchi and Yamazaki, 1970; Cheung, 1970; Teo and Wang, 1973; Wolff and Brostrom, 1974), brain adenylate cyclase (Brostrom *et al.*, 1975), erythrocyte Ca^{2+}, Mg^{2+}–ATPase (Jarrett and Penniston, 1977; Gopinath and Vincenzi, 1977), dynein ATPase (Blum *et al.*, 1980), phospholipase A_2 (Wong and Cheung, 1979), myosin light chain kinase (Dabrowska *et al.*, 1978; Yagi *et al.*, 1978; Hathaway and Adelstein, 1979), plant NAD kinase (Anderson and Cormier, 1978), and glycogen phosphorylase kinase (Cohen *et al.*, 1978; Shenolikar *et al.*, 1979), as well as a Ca^{2+}-dependent protein kinase having a broad substrate range (Schulman and Greengard, 1978). In the case of glycogen phosphorylase kinase, calmodulin is both an integral subunit (1 mole of calmodulin/mole of enzyme), and a Ca^{2+}-dependent dissociable activator which can produce additional increases in enzyme activity (Cohen *et al.*, 1978; Shenolikar *et al.*, 1979).

Present evidence, therefore, indicates a role for calmodulin as a major intracellular Ca^{2+} receptor which can regulate cell activity in response to changing Ca^{2+} concentrations.

B. Localization of Calmodulin in the Mitotic Apparatus

The first evidence that calmodulin might play a role in mitosis was obtained by calmodulin-localization experiments. The calmodulin distribution in dividing tissue culture cells was examined during interphase and mitosis by indirect immunofluorescence microscopy (Welsh *et al.*, 1978; Anderson *et al.*, 1978; Dedman *et al.*, 1978b) and immunoelectron microscopy (Lin *et al.*, 1979; Means and Dedman, 1980). During interphase, anticalmodulin fluorescence was either diffuse throughout the cytoplasm or associated with actin stress fibers, but was not associated with the nucleus. During all phases of mitosis, calmodulin was localized within the mitotic apparatus. The anticalmodulin fluorescence was brightest near the spindle poles and along the chromosome-to-pole fibers. The fluorescence did not extend to the region of the metaphase plate but terminated on or before the chromosomes. In late anaphase, in addition to being localized to the half-spindles, calmodulin staining also appeared transiently in the interzonal region near the chromosomes. Calmodulin-specific fluorescence was not observed in the cleavage furrow.

The anticalmodulin staining visualized during mitosis was very similar to the staining pattern obtained using antitubulin (Welsh *et al.*, 1979). Treatment of cells with agents known to disrupt microtubules has been used to distinguish calmodulin-specific and tubulin-specific fluorescence patterns (Welsh *et al.*, 1979). Colchicine treatment was seen to disrupt the mitotic apparatus and to abolish both calmodulin and tubulin localization in this organelle. Exposure of cells to a temperature of 4°C, which preferentially depolymerizes pole-to-pole microtubules, resulted in an altered tubulin fluorescence pattern but did not affect the localization of calmodulin. This suggests that calmodulin is associated with chromosome-to-pole microtubules. This parallel localization of calmodulin with chromosome-to-pole microtubules has led some investigators to examine the possibility that calmodulin regulates microtubule depolymerization, as will be discussed below.

Immunoelectron microscopy has been used in an attempt to localize calmodulin more specifically in the mitotic apparatus (Lin *et al.*, 1979; Means and Dedman, 1980). These experiments indicated that calmodulin is associated with the centriole region and kinetochores. The anticalmodulin was not directly associated with the chromosome-to-pole microtubules themselves but, instead, with the smooth endoplasmic reticulum, membrane vesicles, and mitochondrial membranes present in the half-spindles. The association of calmodulin with membranes suggests that calmodulin might also be regulating membrane-bound enzymes.

III. CALCIUM AND THE MITOTIC APPARATUS

A. Microtubule Stability

Calcium has been proposed as an endogenous regulator of mitotic apparatus function and microtubule stability. In support of this proposal, several experiments indicate that spindle microtubules *in situ* are sensitive to micromolar levels of Ca^{2+}. The most direct evidence that the mitotic apparatus is sensitive to Ca^{2+} has been obtained in experiments with live sea urchin eggs. Kiehart (1979) has shown that microinjection of Ca^{2+} solutions buffered at physiological levels ($\sim 5\ \mu M$) locally depolymerizes microtubules in the spindle. In contrast, injection of Mg^{2+} or distilled water has no effect. Sensitivity to Ca^{2+} in the micromolar range has also been shown to be a property of microtubules in the isolated spindle (Salmon and Segall, 1980). Other cytoplasmic microtubules can also be depolymerized by physiological concentrations of Ca^{2+}. Using Ca^{2+} and the ionophore A23187, Schliwa (1976) has shown that $10\ \mu M\ Ca^{2+}$ causes axopodial shortening in the heliozoan *Actinosphaerium eichorni*. In addition, the cytoplasmic microtubule complex of interphase tissue culture cells can be depolymerized when the cells are incubated in a medium containing Ca^{2+} and A23187 (Fuller and Brinkley, 1976).

Although the mechanism for Ca^{2+}-induced microtubule depolymerization is not understood, these experiments suggest that *in vivo*, physiological concentrations of Ca^{2+} may control microtubule stability. In contrast, *in vitro* experiments have not consistently shown regulation by micromolar concentrations of Ca^{2+}. Weisenberg (1972) first showed that microtubule assembly *in vitro* is sensitive to Ca^{2+}, but subsequent work has shown that the levels of Ca^{2+} necessary to depolymerize microtubules vary with the source of the tubulin, the extent and method of tubulin purification, and subsequent assay conditions. For instance, Rosenfeld *et al.* (1976) and Haga *et al.* (1974) have reported that 1–$10\ \mu M\ Ca^{2+}$ is sufficient to depolymerize microtubules in crude brain extracts. However, Rosenfeld *et al.* (1976) found that they observed this Ca^{2+} sensitivity only if approximately physiological levels of Mg^{2+} (5–$10\ mM$) were also present. Olmsted and Borisy (1975) found that brain tubulin purified by assembly–disassembly methods is less sensitive to Ca^{2+}, with 0.1–$1.0\ mM\ Ca^{2+}$ required for depolymerization.

The discrepancy between the Ca^{2+} sensitivity of microtubules *in vivo* and the relative insensitivity of microtubules *in vitro* suggested that some other factor, possibly a Ca^{2+}-binding protein, might be required for Ca^{2+} to produce microtubule disassembly. Evidence for such a factor was obtained by Nishida and Sakai (1977), who showed that a crude brain fraction could restore the Ca^{2+} sensitivity that was lost by brain tubulin during purification. This observation and the presence of calmodulin in the mitotic apparatus led Marcum et al. (1978) to test the possibility that calmodulin could confer Ca^{2+} sensitivity upon the microtubule assembly reaction. These experiments were performed with brain tubulin prepared by three cycles of assembly and disassembly. This preparation was relatively insensitive to Ca^{2+}; addition of 10^{-5} M Ca^{2+} before polymerization resulted in only a 15% decrease in the maximal extent of assembly. While the addition of calmodulin alone did not affect assembly, calmodulin plus 10^{-5} M Ca^{2+} totally inhibited assembly. In addition, calmodulin also enhanced the ability of Ca^{2+} to depolymerize preassembled tubulin. Similar calmodulin sensitivity results were obtained by Nishida et al. (1979). The mechanism of Ca^{2+}–calmodulin inhibition of microtubule assembly in vitro is not known. The calmodulin could be interacting directly with tubulin to block assembly or, additionally, it could be interacting with any of the microtubule-associated proteins (MAPs) to decrease microtubule stability.

Although calmodulin can restore Ca^{2+} sensitivity to the in vitro tubulin polymerization reaction, several features of this system are inconsistent with the usual properties of calmodulin-regulated processes. Marcum and co-workers (1978) originally found that under their conditions of tubulin assembly, complete inhibition by 10^{-5} M Ca^{2+} required a calmodulin:tubulin dimer molar ratio of 8:1. This stoichiometry is considerably greater than that required by other proteins regulated by calmodulin (La Porte et al., 1979) and is also much greater than the calmodulin:tubulin ratio in the cell. Nishida et al. (1979) showed that the calmodulin:tubulin ratio required for Ca^{2+}-induced microtubule disassembly decreases as the salt concentration is raised from 50 to 200 mM, suggesting that the stoichiometry is strongly influenced by the assay conditions. A disturbing aspect of this work is that troponin C, a Ca^{2+}-binding protein closely related to calmodulin, is approximately four times more effective than calmodulin in conferring Ca^{2+} sensitivity upon the microtubule assembly reaction (Marcum et al., 1978). While troponin C has been shown to substitute for some of calmodulin's other functions, such as activation of cyclic nucleotide phosphodiesterase (Dedman et al., 1977) and glycogen phosphorylase kinase (Cohen et al., 1979), it is at least 100- to 1000-fold less effective than calmodulin in these systems. The calmodulin effect on microtubule assembly is also unusual in that it is not inhibited by the drug trifluorperazine (G. Perry and J. Bryan, personal communication), which is known to bind to calmodulin and prevent its Ca^{2+}-dependent interaction with its target proteins. In addition, immunoelectron microscopic localization of calmodulin in the mitotic apparatus has indicated that calmodulin is not directly associated with the microtubule bundles, as would be expected if calmodulin interacted directly with tubulin to effect depolymerization. Instead, calmodulin was localized with the smooth endoplasmic reticulum and vesicles present in the spindle (Lin *et al.*, 1979; Means and Dedman, 1980). These discrepancies suggest that calmodulin's effect on microtubule stability *in vivo* might not be as straightforward as that demonstrated *in vitro*.

Calmodulin may not be required to confer Ca^{2+} sensitivity on the microtubule assembly–disassembly reaction in all systems. The properties of the *in vitro* assembly of purified sea urchin cytoplasmic tubulin are quite different from those of brain tubulin assembly, especially in their Ca^{2+} and calmodulin sensitivity. Assembly of the sea urchin tubulin, which is almost totally free of MAPs and contains no detectable calmodulin, can be inhibited by micromolar free Ca^{2+}; depolymerization of preassembled sea urchin microtubules is also sensitive to physiological levels of free Ca^{2+} (Nishida and Kumagai, 1980; T. Keller, personal communication). The same studies showed that neither brain nor sea urchin calmodulin had a Ca^{2+}-dependent inhibitory effect on sea urchin tubulin polymerization. This difference cannot be due to the lack of MAPs in the sea urchin sample, since the addition of brain MAPs to the sea urchin tubulin did not produce any change in Ca^{2+} sensitivity in the presence of calmodulin (Nishida and Kumagai, 1980). Thus, in this system, calmodulin is not necessary for Ca^{2+}-induced microtubule lability.

Calmodulin may not be the only protein capable of conferring Ca^{2+} sensitivity on purified microtubules. The crude brain factor described by Nishida and Sakai (1977) has physical characteristics very distinct from those of calmodulin. Therefore, the possibility exists that there are several different proteins which increase the Ca^{2+} sensitivity of microtubules, which may differ in different cell types or subcellular localization.

Since calmodulin is concentrated in the mitotic apparatus, it seems worthwhile to consider the possibility that calmodulin could also be regulating other Ca^{2+}-dependent processes in the spindle.

B. Chromosome Movement

One major goal of mitosis research has been the elucidation of the molecular basis of chromosome movement. This movement is accomplished both by a movement of the chromosomes to the poles and a movement of the poles farther apart from each other. Although several models have been suggested to explain these movements, there is still no general agreement as to the mechanism responsible, and, in fact, the real mechanism may actually incorporate aspects of each model. Calcium may control mitotic apparatus function either directly or indirectly, since inherent in all the proposed models is the requirement for controlling the spatial and temporal assembly and disassembly of spindle microtubules.

One theory of chromosome movement, which is based on the assembly–disassembly equilibrium of the mitotic microtubules, suggests that microtubule depolymerization may provide the motive force for chromosome movement (Inoué and Sato, 1967). It has been shown that the depolymerization of microtubules is a rate-limiting factor in the movement of chromosomes during anaphase (Salmon, 1975) and that experimentally induced depolymerization of microtubules can cause movement of chromosomes toward the poles (Inoué and Ritter, 1975; Inoué, 1976). Further support for this theory and for a role for Ca^{2+} has been provided by Salmon and Segall (1980), who have shown that Ca^{2+}-induced depolymerization of microtubules in the isolated mitotic apparatus can produce chromosome movement. Thus, Ca^{2+} (and perhaps calmodulin) could control not only the timing of the appearance and disappearance of the mitotic apparatus but also the force production in chromosome movement via its effect on the microtubule assembly–disassembly equilibrium.

The identification of the dynein–microtubule system as the basis for movement of cilia and flagella (Summers and Gibbons, 1971) has generated speculation that a similar mechanism might be operating in the mitotic apparatus. It has been suggested that chromosome movement could be based upon microtubule sliding (McIntosh *et al.*, 1969; Nicklas, 1971; Margolis *et al.*, 1978) or lateral interactions, termed "zipping," between microtubules (Bajer, 1973). Electron microscopic analysis of serially sectioned diatom spindles has indicated that changes in spindle morphology observed during mitosis are compatible with a sliding mechanism (McDonald *et al.*, 1977, 1979; McIntosh *et al.*, 1979). While no molecular mechanism has been proposed for the zipping theory, microtubule sliding could perhaps be accomplished by cross-bridges with dynein-like ATPase similar to those which are responsible for ciliary and flagellar movement (Summers and Gibbons, 1971). Although it is not yet known whether dynein arms are associated with spindle microtubules, an enzyme with dynein-like properties has been identified in isolated mitotic apparatus from sea urchin eggs (Pratt *et al.*, 1980), and an enzyme with dynein-like properties was found in association with cycle-purified cytoplasmic microtubules of neural origin (Hiebsch *et al.*, 1979). Furthermore, Mohri *et al.* (1976) have localized a dynein-like molecule to the mitotic apparatus by using indirect immunofluorescence. Cande and Wolniak (1978) have shown that vanadate, an inhibitor of several ATPase activities including dynein, inhibits chromosome movement in lysed cell models, suggesting that an ATPase activity may be required for chromosome movement. The low concentration of vanadate required for inhibition of chromosome movement is more compatible with the inhibition of dynein ATPase than with other mitotic apparatus-associated ATPases such as myosin ATPase, which require higher vanadate levels for inhibition (Cande and Wolniak, 1978). In further support of a role for dynein in chromosome movement, Sakai *et al.* (1976) have demonstrated that antidynein antibodies inhibit chromosome movement in isolated sea urchin mitotic apparatus.

Dynein ATPase was originally described as a high-molecular-weight, Mg^{2+}-requiring ATPase responsible for force generation in ciliary and flagellar motility. Although initial characterizations of flagellar dynein did not reveal any activation by Ca^{2+} (Gibbons and Gibbons, 1972), more recent experiments have demonstrated that Ca^{2+} and calmodulin can increase the activity of dynein preparations. Specifically, Blum *et al.* (1980) have shown that 14 and 30 S dynein ATPase isolated from *Tetrahymena* cilia can be activated 10- and 2-fold, respectively, by calmodulin plus 10 μM Ca^{2+}. In addition, they have purified dynein ATPase by affinity chromatography using calmodulin–Sepharose, indicating that dynein interacts directly with calmodulin.

The demonstration of flagellar dynein ATPase regulation by Ca^{2+} and calmodulin raises the possibility that the dynein-like ATPase found in the mitotic apparatus might also be regulated in this manner. If such a control mechanism can be demonstrated, then Ca^{2+} could be having a direct effect on chromosome movement. Studies to determine the Ca^{2+} and calmodulin sensitivities of this ATPase are now in progress.

A third theory of chromosome movement proposes that actin and myosin are the major force-producing elements in the spindle (Forer, 1974). Calcium and calmodulin have been shown to be involved in modulating actomyosin interactions via an effect on myosin light chain kinase (Dabrowska *et al.*, 1978; Yagi *et al.*, 1978; Hathaway and Adelstein, 1979). Although actin and myosin have been localized in the mitotic apparatus by several techniques (reviewed by Forer, 1978), more recent evidence argues against a functional role for actomyosin interactions in chromosome movement. First, the vanadate results described above are more compatible with a role for dynein than for myosin in chromosome movement. In addition, Sakai *et al.* (1976) were not able to inhibit chromosome movement in isolated sea urchin mitotic apparatus by the addition of antimyosin antibodies. Furthermore, microinjection of antimyosin antibodies into live starfish blastomeres was found to inhibit cytokinesis but not nuclear division (Mabuchi and Okuno, 1977). Similar microinjections of antimyosin into live sea urchin eggs did not inhibit chromosome movement (Kiehart *et al.*, 1977). In view of this evidence, we have not pursued the possibility that calmodulin in the mitotic apparatus is interacting with microfilaments.

C. Ca^{2+}- Sequestering System

Implicit in all hypotheses of a role for Ca^{2+} in mitosis is the assumption that the cell has a means of controlling free Ca^{2+} concentrations in a localized and specific manner. Several observations indicate that a specific Ca^{2+}-

sequestering system does exist in the region of the mitotic apparatus. Kiehart (1979) has shown that Ca^{2+} microinjected into sea urchin eggs in the region of the spindle is not freely diffusible. Injection of amounts of unbuffered Ca^{2+} sufficient to depolymerize the entire spindle, if it were freely diffusible, only depolymerizes the microtubules nearest the injection site. In addition, those spindle microtubules depolymerized by the Ca^{2+} microinjection re-form soon afterward, suggesting the further sequestration of the remaining Ca^{2+}.

The presence of a large number of vesicles and an extensive system of endoplasmic reticulum around the mitotic apparatus has led to the suggestion that these vesicles might function to sequester Ca^{2+} in a manner similar to that of the sarcoplasmic reticulum in muscle (Hepler, 1977; Harris, 1978). Wolniak *et al.* (1979) have used chlorotetracycline (CTC), a fluorescent probe sensitive to membrane-associated Ca^{2+}, to show that in living endosperm cells of *Haemanthus*, membrane-bound Ca^{2+} increases at the polar regions during mitosis and then disperses at telophase, with a concomitant increase in fluorescence at the cell plate region. The correspondence of regions of CTC fluorescence with regions of microtubule assembly suggests that sequestration of free Ca^{2+} by membranes does occur at sites of microtubule polymerization.

Although the exact nature of the Ca^{2+}-sequestering system is not well understood, at least some of the Ca^{2+} uptake is thought to be non-mitochondrial. Caffeine, a potent uncoupler of the Ca^{2+} pump–ATPase of sarcoplasmic reticulum (Weber, 1968), also inhibits the Ca^{2+}-sequestering system associated with the mitotic apparatus (Kiehart, 1979) yet does not induce Ca^{2+} release from mitochondria (Weber, 1968). Injection of oxalate into sea urchin eggs resulted in the presence of highly birefringent crystals, presumably Ca^{2+}–oxalate precipitates, contained within a membrane compartment (Kiehart, 1979). Since mitochondria are not permeable to oxalate (Kiehart, 1979), the oxalate injection experiments further support the idea that the Ca^{2+}-sequestering system in sea urchin eggs is, at least in part, non-mitochondrial.

Proposals that the endoplasmic reticulum around the mitotic apparatus could function to sequester Ca^{2+} are supported by *in vitro* experiments with either crude vesicular preparations or mitotic apparatus isolated with intact vesicles. Kinoshita and Yazaki (1967) have shown that vesicles which bind $^{45}Ca^{2+}$ can be isolated from sea urchin eggs by methods routinely used for sarcoplasmic reticulum. More recently, Silver *et al.* (1980) and Egrie and Nagle (1980) have isolated sea urchin mitotic apparatus which contain an extensive system of osmotically intact membrane vesicles capable of sequestering $^{45}Ca^{2+}$ in an ATP-dependent fashion. The accumulated Ca^{2+} is released by either Triton X-100 or the divalent ionophore A23187, indicating that intact membranes are required for the Ca^{2+} sequestration. The Ca^{2+}-sequestering vesicles can be isolated on a sucrose step gradient after ^{45}Ca–oxalate loading of the isolated intact mitotic apparatus preparation (Silver *et al.*, 1980).

By analogy with the sarcoplasmic reticulum, the mitotic Ca^{2+} pump is expected to be associated with an ATPase. Evidence for a Ca^{2+}–ATPase activity in the spindle was first provided by Mazia *et al.* (1972). A substantial proportion of the ATPase fractionates with the isolated sea urchin mitotic apparatus prepared without the use of detergents (Nagle and Egrie, 1979). Treatment of the isolated mitotic apparatus with Triton X-100 releases the enzyme into a supernatant fraction, indicating that it is normally membrane associated (Nagle and Egrie, 1979). The activity of this ATPase fluctuates during the cell cycle. The peak activities occur during interphase and metaphase (Petzelt, 1972), both times at which microtubules are undergoing polymerization and free Ca^{2+} concentrations would be expected to be very low. A similar Ca^{2+}–ATPase showing the same increase in activity in mitosis has been shown to occur in a number of dividing cells including mouse fibroblasts, mouse mastocytoma cells, surf clam eggs, and *Physarum* (reviewed by Petzelt, 1979). Although the Ca^{2+}–ATPase does have some characteristics required of the pumping enzyme, it has not yet been proven that this ATPase is responsible for the ATP-dependent Ca^{2+} uptake system. One problem is that optimal enzyme activity requires millimolar levels of Ca^{2+}, while a role for the ATPase *in vivo* would require that it be sensitive to Ca^{2+} at approximately micromolar levels. It is possible that the ATPase is more sensitive to Ca^{2+} *in situ* and that the Ca^{2+} sensitivity is lost upon isolation.

IV. CALMODULIN AND ATPases IN THE SEA URCHIN MITOTIC APPARATUS

A. Experimental System

The dramatic relocalization of calmodulin from the cytoplasm to the mitotic apparatus during mitosis suggests that calmodulin is of great importance in the process of cell division. One postulated role for calmodulin in conferring Ca^{2+} sensitivity upon spindle microtubules has received some *in vitro* experimental support (Marcum *et al.*, 1978). The interaction of calmodulin with microtubules does not preclude its having additional Ca^{2+}-dependent regulatory functions in this organelle. The very high concentrations of calmodulin in the mitotic apparatus and its localization to endoplasmic reticulum and vesicles, as opposed to microtubule bundles (Lin *et al.*, 1979;

Means and Dedman, 1980), support the suggestion that calmodulin may have multiple functions in the spindle. The localization of calmodulin to membranes and vesicles in the half-spindle suggested to us that calmodulin might operate by regulating the activity of a membrane-associated Ca^{2+}–ATPase responsible for Ca^{2+} sequestration. In this way, calmodulin could be responsible for modulating Ca^{2+} levels compatible with mitotic apparatus functioning and stability. In areas of high Ca^{2+}, the Ca^{2+}–calmodulin complex could both mediate the actions of calcium (for instance, by mediating microtubule depolymerization) and also, by interacting with a Ca^{2+} pump, decrease the Ca^{2+} concentration in the immediate area. This could confine the Ca^{2+} signal both temporally and spatially and allow this region of the cell to recover rapidly from any transient change in Ca^{2+} concentration. Calmodulin has previously been shown to modulate the activity of two other cellular ATPases responsible for controlling intracellular Ca^{2+} concentrations: the erythrocyte plasma membrane Ca^{2+}, Mg^{2+}–ATPase (Jarrett and Penniston, 1977; Gopinath and Vincenzi, 1977) and the sarcoplasmic reticulum Ca^{2+}–ATPase (Le Peuch *et al.*, 1979). We have been studying the Ca^{2+}–ATPases and the Ca^{2+}-sequestering system in the mitotic apparatus to determine if they are regulated by calmodulin. For these experiments we have used isolated mitotic apparatus from the sea urchin, *Strongylocentrotus purpuratus*. These isolates contain osmotically active vesicles and a Ca^{2+}–ATPase whose activity is cell-cycle dependent.

B. Measurement of Calmodulin in Sea Urchin Eggs and Embryos

As a prerequisite to determining a role for calmodulin in the mitotic apparatus, we have measured the levels of calmodulin in sea urchin eggs and embryos. Approximately 0.05–0.1% of the total egg protein is calmodulin, as assayed by its ability to stimulate a calmodulin-dependent cyclic nucleotide phosphodiesterase (J. C. Egrie and B. W. Nagle, unpublished results). The concentration of egg calmodulin is approximately 5 μM, an amount comparable to that found in other cells and organisms (Wolff and Brostrom, 1979). Our results indicate that approximately 80–90% of the calmodulin is in the soluble fraction after centrifugation at 12,000 g and that this accounts for approximately 0.1–0.2% of the soluble egg protein. The amount and subcellular distribution of calmodulin remains constant throughout development of the sea urchin egg from before fertilization until the prism stage (J. C. Egrie and B. W. Nagle, unpublished results). We should note that we cannot accurately quantify calmodulin in the mitotic apparatus by using biochemical and cell fractionation techniques. Isolation of this organelle requires the presence of high concentrations of EGTA, which results in the dissociation of calmodulin from its target proteins: mitotic apparatus localization for calmodulin has been obtained only with indirect immunofluorescence (Welsh *et al.*, 1978; Andersen *et al.*, 1978). Our values for the total amount and subcellular distribution of sea urchin calmodulin are in agreement with those which have been reported by other investigators (Head *et al.*, 1979; Carroll and Longo, 1979). Our finding that calmodulin levels do not change during development contrasts with the results of Carroll and Longo (1979), who found a 40% increase in calmodulin levels in *Lytechinus pictus* eggs after fertilization. We have been unable to observe this increase, and the possibility remains that the difference could be due to species variation.

C. Calmodulin Sensitivity of Mitotic Ca²⁺-ATPase

In order to determine the sensitivity of the mitotic Ca^{2+}-ATPase to calmodulin, we have partially purified the enzyme, using techniques which would separate the enzyme from endogenous calmodulin. At all stages of the purification procedure, we have monitored the effect of the addition of both exogenous calmodulin and trifluoperazine (a calmodulin inhibitor) on enzyme activity.

Most of the experiments to be described were performed starting with mitotic apparatus isolated in media based on microtubule assembly buffers, as outlined in Fig. 1. The mitotic apparatus isolated without the use of Triton X-100 retained osmotically active membranous vesicles and a Ca^{2+} pumping activity (Silver *et al.*, 1980; Egrie and Nagle, 1980). This preparation typically accounts for approximately 10–20% of the total cellular protein and 30% of the cellular Ca^{2+}-ATPase activity. We generally obtain a 2- to 3-fold higher specific activity for Ca^{2+}-ATPase in the mitotic apparatus as compared to the whole egg lysate. The remaining 70% of the enzyme activity could be totally recovered in a 12,000 *g* pellet. In order to determine whether the enzyme activity not isolated with the mitotic apparatus has the same calmodulin sensitivity as that in the mitotic apparatus, we have performed all experiments with both isolated mitotic apparatus and resuspended 12,000 *g* pellets. The properties of the activity from both subcellular fractions are identical.

The crude mitotic Ca^{2+}-ATPase exhibits a pH optimum of 8.5 and a rather sharp Ca^{2+} optimum of 1 mM. The addition of 1–2 mM Mg^{2+} does not significantly affect the Ca^{2+} optimum and either does not affect or slightly decreases total enzyme activity. These characteristics are very similar to those described by Petzelt (1972) for the cell cycle-dependent mitotic Ca^{2+}-ATPase. The Ca^{2+}-ATPase obtained by this procedure can be distinguished from mitochondrial, Na^+/K^+-, and dynein ATPases by its insensitivity to 10^{-4} M oligomycin, 5×10^{-5} M ouabain, and 10^{-4} M vanadate.

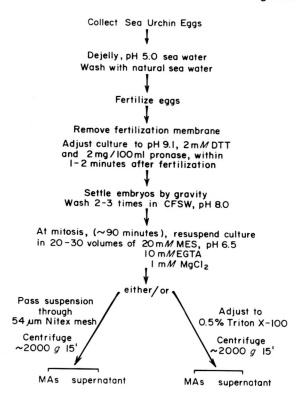

Collect Sea Urchin Eggs

Dejelly, pH 5.0 sea water
Wash with natural sea water

Fertilize eggs

Remove fertilization membrane
Adjust culture to pH 9.1, 2 mM DTT
and 2 mg/100 ml pronase, within
1–2 minutes after fertilization

Settle embryos by gravity
Wash 2–3 times in CFSW, pH 8.0

At mitosis, (∼90 minutes), resuspend culture
in 20–30 volumes of 20 mM MES, pH 6.5
10 mM EGTA
1 mM MgCl$_2$

either/or

Pass suspension
through
54 μm Nitex mesh

Centrifuge
∼2000 g 15'

MAs   supernatant

Adjust to
0.5% Triton X-100

Centrifuge
∼2000 g 15'

MAs   supernatant

Typically, isolate ∼10–20% of the protein and
30% of the Ca^{2+}-ATPase activity in the MA pellet

Fig. 1. Procedure for mitotic apparatus isolation.

Increasing ionic strength inhibits enzyme activity, with 55 and 20% of the activity remaining at 50 and 100 mM NaCl (or KCl), respectively.

The results of adding exogenous calmodulin to whole cell lysates and mitotic apparatus assayed either with or without Triton X-100 are shown in Table I. The addition of 1 mM Ca^{2+} produces a 5- to 7-fold increase in activity. The addition of 100 ng of porcine brain calmodulin plus Ca^{2+}, however, did not produce any further increases in enzyme activity. The approximate 2-fold increase in enzyme activity in the presence of Triton X-100 was consistently observed in all experiments. Triton X-100 has no effect on the ionic requirements or optima of the enzyme and does not interfere with calmodulin activation in other enzyme systems (Wolff and Brostrom, 1979).

TABLE I

Effect of Calmodulin on Mitotic Apparatus Ca^{2+}-ATPase^a

Method of assay	Fraction	ATPase activity (nmoles ATP hydrolyzed/min/mg protein)		
		$-$Ca^{2+}	$+$1mM Ca^{2+}	$+$1 mM Ca^{2+} and 100 ng calmodulin
$-$Triton X-100	Whole cell lysate	4.3	28.8	28.4
	Mitotic apparatus	4.8	20.7	21.8
$+$Triton X-100	Whole cell lysate	6.5	45.3	42.1
	Mitotic apparatus	9.9	69.5	72.4
	Supernatant	6.2	44.1	45.0

^{a} *Strongylocentrotus purpuratus* whole lysates and mitotic apparatus, isolated without Triton X-100, were assayed for ATPase activity. Assays were performed both with and without 1 mM Ca^{2+} and 100 ng of calmodulin.

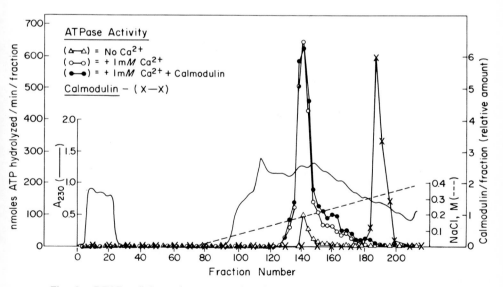

Fig. 2. DEAE-cellulose chromatography of mitotic apparatus Ca^{2+}-ATPase and calmodulin. Ca^{2+}-ATPase activity was assayed both with and without 1 mM Ca^{2+} and 100 ng of calmodulin. Calmodulin was assayed by its ability to activate a calmodulin-deficient brain cyclic nucleotide phosphodiesterase.

As described above, sea urchin eggs contain substantial amounts of calmodulin. While we suspected that most of the calmodulin associated with the mitotic apparatus in the above experiment would have been removed by the high EGTA concentrations and centrifugations used for the isolations, we could not be sure that any remaining calmodulin would not have been sufficient to activate the Ca^{2+}–ATPase to its maximum level. Therefore, we used DEAE–cellulose chromatography to separate the sea urchin Ca^{2+}–ATPase from the remaining endogenous calmodulin, and then tested this calmodulin-depleted Ca^{2+}–ATPase for a response to added calmodulin. Mitotic Ca^{2+}–ATPase was solubilized by the addition of Triton X-100 and chromatographed on DEAE–cellulose, as shown in Fig. 2. The Ca^{2+}–ATPase activity eluted at approximately 0.2 M NaCl and was well separated from calmodulin, which eluted as one sharp peak at 0.3 M NaCl. The Ca^{2+}–ATPase activity was greatly stimulated by the addition of 1 mM Ca^{2+}. The addition of both Ca^{2+} and calmodulin, however, did not produce any further increase in enzyme activity. Separation of Ca^{2+}–ATPase from the remaining endogenous calmodulin, therefore, did not produce an enzyme dependent

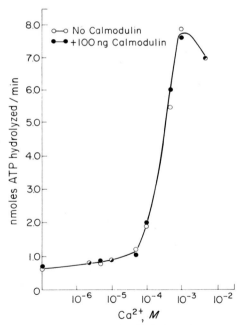

Fig. 3. Calcium optimum of the DEAE–cellulose-purified mitotic Ca^{2+}-ATPase. The DEAE–cellulose-purified Ca^{2+}-ATPase was assayed at Ca^{2+} concentrations ranging from 2 $\times 10^{-6}$ to 5 $\times 10^{-3}$ M in either the presence or absence of 100 ng of calmodulin. Calcium concentrations were maintained by Ca^{2+}-EGTA buffers. Portzehl et al. (1964).

on calmodulin for maximal activity when assayed at 1 mM Ca^{2+}. Furthermore, as shown in Fig. 3, calmodulin did not increase the Ca^{2+} sensitivity of the enzyme; millimolar levels of Ca^{2+} were required for maximal activity even in the presence of added calmodulin.

Lynch and Cheung (1979) have shown that the Ca^{2+}-stimulated activity of the erythrocyte Ca^{2+}, Mg^{2+}–ATPase is highly labile after detergent solubilization. The response to Ca^{2+} of the detergent-solubilized ATPase had a half-life of approximately 2 hours. In order to be sure that the lack of a calmodulin response in our previous experiments was not due to the long time necessary for ion-exchange chromatography, we used gel filtration on Ultrogel AcA 34 to separate the ATPase from calmodulin. This procedure could be completed in approximately 1½ hours from the time the enzyme was solubilized in Triton X-100 until the end of the ATPase assay. Despite the rapid partial purification procedure, however, the enzyme was still unaffected by the addition of calmodulin.

We have also probed for a calmodulin requirement by using an inhibitor of calmodulin action. The phenothiazine drug trifluoperazine (TFP) has been shown to bind calmodulin at two high-affinity Ca^{2+}-dependent sites per molecule ($K_d = 10^{-6}$ M) and thus prevent it from binding to and activating its target enzymes (Levin and Weiss, 1977). Consequently, demonstration of phenothiazine inhibition of an enzyme reaction has sometimes been used as a diagnostic test for a calmodulin-dependent system (Satir *et al.*, 1980; Crimaldi *et al.*, 1980). The effect of TFP on Ca^{2+}–ATPase activity of a 12,000 g pellet fraction of sea urchin eggs is shown in Fig. 4. Similar results were obtained with the DEAE–cellulose-purified enzyme preparation. Although TFP did produce some inhibition of both the crude and DEAE–cellulose-purified enzyme preparations, this inhibition occurred in the presence of either Ca^{2+} or EGTA. Moreover, this inhibition could not be overcome by the addition of exogenous calmodulin. The high concentration of TFP required and the failure of TFP to inhibit enzyme activity to a greater degree in the presence of Ca^{2+} than in the presence of EGTA indicate that the inhibition seen is not due to an effect on calmodulin. This supports our conclusion that calmodulin is not a regulator of Ca^{2+}–ATPase activity. These experiments also illustrate the need for using caution in interpreting the results of inhibitor experiments. Calmodulin, as well as other proteins such as bovine serum albumin, catalase, and cytochrome *c*, also binds TFP with low affinity at Ca^{2+}-independent sites (Levin and Weiss, 1977). The TFP inhibition which we have observed with the Ca^{2+}–ATPase is probably a non-specific effect of TFP binding to a hydrophobic region on the enzyme itself. Thus, TFP inhibition alone is not sufficient to indicate a role for calmodulin; appropriate controls are necessary to determine that any TFP inhibition is, in fact, due to its specific action on calmodulin.

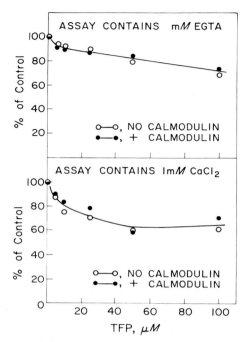

Fig. 4. Effect of TFP on ATPase activity. A resuspended 12,000 g pellet obtained from homogenized sea urchin eggs was assayed in the presence of 1 mM CaCl$_2$ or 1 mM EGTA, with or without 54 μg calmodulin.

D. Calcium Uptake by Isolated Mitotic Apparatus

As another approach to determining calmodulin's possible role in the regulation of Ca^{2+} levels in the region of the mitotic apparatus, we examined the effect of exogenous calmodulin on Ca^{2+} uptake by isolated mitotic apparatus. Although a Ca^{2+}–ATPase has been identified in the mitotic apparatus, it has never been demonstrated that this enzyme is associated with the Ca^{2+} pump. In fact, preliminary experiments on the optimal conditions for Ca^{2+} sequestration suggest that some of the characteristics of the pumping enzyme may be different from those of the previously described Ca^{2+}–ATPase.

We have prepared mitotic apparatus and assayed their ability to sequester ^{45}Ca^{2+}. As shown in Table II, the addition of ATP to the incubation mixture resulted in an approximately 4-fold increase in ^{45}Ca^{2+} accumulation. Similar results were obtained whether or not oxalate was included in the incubation buffer. The accumulated radioactivity could be released from the mitotic

TABLE II

Uptake of $^{45}Ca^{2+}$ by Isolated Mitotic Apparatus[a]

Sample	CPM in pelleted mitotic apparatus
−ATP	88,360
+1 mM ATP	338,830
+1 mM ATP and 350 ng of calmodulin	348,810

[a] Mitotic apparatus was isolated without the use of Triton X-100 and then incubated in 30 mM MES, pH 6.5, 1 mM MgCl$_2$, 1 mM K$_2$-oxalate, and 20 μM CaCl$_2$ (containing 10 μCi ^{45}CaCl$_2$), with or without 1 mM ATP. Incubations were at 14°C for 30 minutes, after which the samples were washed three times and the radioactivity associated with the mitotic apparatus pellet was determined.

The data are expressed as cpm $^{45}Ca^{2+}$/ml of packed sea urchin eggs.

apparatus by the addition of Triton X-100, which destroys vesicle integrity. Addition of calmodulin at this or higher concentrations did not enhance the $^{45}Ca^{2+}$ accumulation by isolated mitotic apparatus. We conclude that calmodulin does not directly stimulate the Ca^{2+} pump.

E. Conclusions and Future Directions

The dramatic reorganization of calmodulin into the mitotic apparatus suggests that calmodulin must be an important mediator in calcium's proposed regulatory functions in cell division. The localization of calmodulin with membranous elements of the mitotic apparatus (Lin *et al.*, 1979; Means and Dedman, 1980) suggested the mitotic Ca^{2+} pump as a prime candidate for regulation by calmodulin, by analogy with calmodulin's stimulation of the erythrocyte plasma membrane Ca^{2+}, Mg^{2+}–ATPase and Ca^{2+} pump.

Our experiments eliminate the possibility that the low-affinity mitotic Ca^{2+}–ATPase and Ca^{2+}-sequestering activity of the isolated mitotic apparatus are regulated by a direct interaction with calmodulin. The possibility still remains that calmodulin might be an indirect regulator of Ca^{2+}–ATPase activity. For instance, although calmodulin has been shown to increase the activity of the erythrocyte Ca^{2+},Mg^{2+}–ATPase by directly binding to it as a Ca^{2+}–calmodulin complex (Niggli *et al.*, 1979), calmodulin regulates the sarcoplasmic reticulum Ca^{2+}–ATPase in an indirect manner. In sarcoplasmic reticulum, regulation is achieved by an effect of calmodulin on a Ca^{2+}-sensitive protein kinase which phosphorylates the protein phospholamban (Le Peuch *et al.*, 1979). Once phosphorylated, phospholamban interacts with the Ca^{2+}–ATPase to stimulate Ca^{2+} uptake. We are presently investigating the possible regulation of the mitotic Ca^{2+}–ATPase by such an indirect protein-phosphorylating mechanism.

Our experiments have shown that calmodulin does not confer increased Ca^{2+} sensitivity upon the low-affinity Ca^{2+}–ATPase. However, it has not been established that this ATPase is the one responsible for Ca^{2+} sequestration in the mitotic apparatus. We now plan to investigate the possibility that an enzyme having a high affinity for Ca^{2+}, but a low turnover rate, is associated with the mitotic apparatus and regulated by calmodulin. An ATPase with these characteristics would be more physiologically relevant and may not have been detected by the procedures used to assay the low-affinity, high-activity ATPase. We are using a calmodulin–Sepharose affinity column to isolate the proposed calmodulin-regulated ATPase, since such an enzyme might have been obscured by other ATPase activity.

Another potential role for calmodulin in the mitotic apparatus has been suggested by the finding that calmodulin binds to and activates *Tetrahymena* flagellar dynein (Blum *et al.*, 1980), another Ca^{2+}-sensitive ATPase. A dynein-like ATPase activity has been identified in sea urchin mitotic apparatus (Pratt *et al.*, 1980), and we are presently investigating whether it is regulated by calmodulin. Preliminary experiments suggest that sea urchin sperm tail dynein is activated by calmodulin (J. C. Egrie and B. W. Nagle, unpublished results).

It is also possible that calmodulin participates in other Ca^{2+}-dependent functions of the mitotic apparatus which are as yet unknown. The determination of the number and cellular location of calmodulin-binding proteins would be of great help in identifying other processes in which Ca^{2+} and calmodulin might be directly involved. Calmodulin–Sepharose affinity columns provide us with a method to isolate specifically those enzymes or proteins which may be the targets of calcium's action in the mitotic apparatus. A knowledge of the role of calmodulin and its target proteins in the mitotic apparatus should advance our understanding of the generation and function of Ca^{2+} fluctuations during the mitotic phase of the cell cycle.

ACKNOWLEDGMENTS

We thank Drs. Daniel Mazia and Fred Wilt, in whose laboratories this work was done, for their support and encouragement. We also thank Drs. K. Sluder, Z. Cande, F. Wilt, and A. S. Goustin for reading the manuscript. We are especially grateful to those investigators who shared their unpublished results with us. This investigation was supported in part by a California Division–Cancer Society Junior Fellowship #J-452 to B. W. N. and by USPHS GM 13882.

REFERENCES

Andersen, B., Osborn, M., and Weber, K. (1978). Specific visualization of the distribution of the calcium-dependent regulatory protein of cyclic nucleotide phosphodiesterase (modulator protein) in tissue culture cells by immunofluorescence microscopy: Mitosis and intracellular bridge. *Cytobiologie* 17, 354–364.

Anderson, J. M., and Cormier, M. J. (1978). Calcium-dependent regulator of NAD kinase in higher plants. *Biochem. Biophys. Res. Commun.* 84, 595–602.

Bajer, A. S. (1973). Interaction of microtubules and the mechanism of chromosome movement (zipper hypothesis). I. General principle. *Cytobios* 8, 139–160.

Blum, J. J., Hayes, A., and Jamieson, G. (1980). Calmodulin confers calcium-sensitivity on dynein ATPase. *Fed. Proc. Fed. Am. Soc. Exp. Biol.* 39, 1626.

Brooks, J. C., and Siegel, F. L. (1973). Purification of a calcium-binding phosphoprotein from beef adrenal medulla: Identity with one of two calcium-binding proteins of brain. *J. Biol. Chem.* 248, 4189–4193.

Brostrom, C. O., Huang, Y. C., Breckenridge, B. McL., and Wolff, D. J. (1975). Identification of a calcium-binding protein as a calcium-dependent regulator of brain adenylate cyclase. *Proc. Natl. Acad. Sci. U.S.A.* 72, 64–68.

Cande, W. Z., and Wolniak, S. M. (1978). Chromosome movement in lysed mitotic cells is inhibited by vanadate. *J. Cell Biol.* 79, 573–580.

Carroll, A. G., and Longo, F. J. (1979). Changes in activity of the calcium dependent regulator protein calmodulin upon fertilization in the sea urchin, *Lytechinus pictus*. *J. Cell Biol.* 83, 212a.

Cheung, W. Y. (1970). Cyclic 3′-5′-nucleotide phosphodiesterase: Demonstration of an activator. *Biochem. Biophys. Res. Commun.* 33, 533–538.

Cheung, W. Y. (1980). Calmodulin plays a pivotal role in cellular regulation. *Science* 207, 19–27.

Cohen, P., Burchell, A., Foulkes, J. G., and Cohen P. T. W. (1978). Identification of the Ca^{2+}-dependent modulator protein as the fourth subunit of rabbit skeletal muscle phosphorylase kinase. *FEBS Lett.* 92, 287–293.

Cohen, P., Picton, C., and Klee, C. B. (1979). Activation of phosphorylase kinase from rabbit skeletal muscle by calmodulin and troponin. *FEBS Lett.* 104, 25–30.

Crimaldi, A. A., Kuo, I. C. Y., and Coffee, C. J. (1980). Effect of trifluoperazine on cyclic nucleotide phosphodiesterase activity in crude homogenates of bovine adrenal medulla. *Ann. N. Y. Acad. Sci.* 356, 367–368.

Dabrowska, R., Sherry, J. M. F., Amatorio, D. K., and Hartshorne, D. (1978). Modulator protein as a component of myosin light chain kinase from chicken gizzard. *Biochemistry* 17, 253–258.

Dedman, J. R., Potter, J. D., and Means, A. R. (1977). Biological cross-reactivity of rat testis phosphodiesterase activator protein and rabbit skeletal muscle troponin-C. *J. Biol. Chem.* 252, 2437–2440.

Dedman, J. R., Jackson, R. L., Schreiber, W. E., and Means, A. R. (1978a). Sequence homology of the Ca^{2+}-dependent regulator of cyclic nucleotide phosphodiesterase from rat testis with other Ca^{2+}-binding proteins. *J. Biol. Chem.* 253, 343–346.

Dedman, J. R., Welsh, M. J., and Means, A. R. (1978b). Ca^{2+}-dependent regulator. Production and characterization of a monospecific antibody. *J. Biol. Chem.* 253, 7515–7521.

Egrie, J. C., and Nagle, B. W. (1980). Are the sea urchin mitotic Ca^{2+}-ATPase and Ca^{2+}-sequestering activity affected by calmodulin? *Ann. N. Y. Acad. Sci.* 356, 376–377.

Forer, A. (1974). Possible roles of microtubules and actin-like filaments during cell division. *In* "Cell Cycle Controls" (G. M. Padilla, I. T. Cameron, and A. M. Zimmerman, eds.), pp. 319–336. Academic Press, New York.

Forer, A. (1978). Chromosome movements during cell division: Possible involvement of actin filament. *In* "Nuclear Division in the Fungi" (I. B. Heath, ed.), pp. 21–88. Academic Press, New York.

Fuller, G. M., and Brinkley, B. R. (1976). Structure and control of assembly of cytoplasmic microtubules in normal and transformed cells. *J. Supramol. Struct.* **5**, 497–514.

Gibbons, B. H., and Gibbons, I. R. (1972). Flagellar movement and adenosine triphosphatase activity in sea urchin sperm extracted with Triton X-100. *J. Cell Biol.* **54**, 75–97.

Gopinath, R. M., and Vincenzi, F. F. (1977). Phosphodiesterase protein activator mimics red blood cell cytoplasmic activator of Ca^{2+}-Mg^{2+} ATPase. *Biochem. Biophys. Res. Commun.* **77**, 1203–1209.

Grand, R. J. A., and Perry, S. V. (1979). Calmodulin-binding proteins from brain and other tissues. *Biochem. J.* **183**, 285–295.

Haga, T., Abe, T., and Kurokawa, M. (1974). Polymerization and depolymerization of microtubules *in vitro* as studied by flow birefringence. *FEBS Lett.* **39**, 291–295.

Harris, P. (1978). Triggers, trigger waves, and mitosis. A new model. *In* "Cell Cycle Regulation" (E. D. Bretow, I. L. Cameron, G. M. Padilla, eds.), pp. 75–104. Academic Press, New York.

Hathaway, D. R., and Adelstein, R. S. (1979). Human platelet myosin light chain kinase requires the calcium-binding protein calmodulin for activity. *Proc. Natl. Acad. Sci. U.S.A.* **76**, 1653–1657.

Head, J. F., Mader, S., and Kaminer, B. (1979). Calcium-binding modulator protein from the unfertilized egg of the sea urchin *Arbacia punctulata*. *J. Cell Biol.* **80**, 211–218.

Hepler, P. K. (1977). Membranes in the spindle apparatus: Their possible role in the control of microtubule assembly. *In* "Mechanisms and Control of Cell Division" (T. L. Rost and E. M. Gifford, eds.), pp. 212–232. Dowden, Hutchinson, and Ross, Stroudsburg, PA.

Hiebsch, R. R., Hales, D. D., and Murphy, D. B. (1979). Identity and purification of the dynein-like ATPase on cytoplasmic microtubules. *J. Cell Biol.* **83**, 345a.

Inoué, S. (1976). Chromosome movement by reversible assembly of microtubules. *In* "Cell Motility, Book C: Microtubules and Related Proteins" (R. Goldman, T. Pollard, and J. Rosenbaum, eds.), pp. 1317–1328. Cold Spring Harbor Lab., Cold Spring Harbor, New York.

Inoué, S., and Ritter, H. (1975). Dynamics of mitotic spindle organization and function. *In* "Molecules and Cell Movement" (S. Inoué and R. E. Stephens, eds.), pp. 3–30. Raven Press, New York.

Inoué, S., and Sato, H. (1967). Cell motility by labile association of molecules. The nature of mitotic spindle fibers and their role in chromosome movement. *J. Gen. Physiol.* **50**, 259–292.

Jarrett, H. W., and Penniston, J. T. (1977). Partial purification of the Ca^{2+}-Mg^{2+} ATPase activator from human erythrocytes: Its similarity to the activator of 3′,5′-cyclic nucleotide phosphodiesterase. *Biochem. Biophys. Res. Commun.* **77**, 1210–1216.

Kakiuchi, S., and Yamazaki, R. (1970). Calcium dependent phosphodiesterase activity and its activating factor from brain. *Biochem. Biophys. Res. Commun.* **41**, 1104–1110.

Kakiuchi, S., Yamazaki, R., Teshima, Y., Uenishi, K., and Miyamoto, E. (1975). Multiple cyclic nucleotide phosphodiesterase activities from rat tissues and occurrence of a calcium-plus-magnesium-ion-dependent phosphodiesterase and its protein activator. *Biochem. J.* **146**, 109–120.

Kiehart, D. (1979). Microinjection of echinoderm eggs. Ph.D. Thesis, Univ. of Pennsylvania, Philadelphia, Pennsylvania. (Available from University Microfilms, Ann Arbor, Michigan)

Kiehart, D., Inoue, S., and Mabuchi, I. (1977). Evidence that force production in chromosome movement does not involve myosin. *J. Cell Biol.* **75**, 258a.

Kinoshita, S., and Yazaki, I. (1967). The behavior and localization of intracellular relaxing system during cleavage in the sea urchin egg. *Exp. Cell Res.* **47**, 449–458.

LaPorte, D. C., and Storm, D. R. (1978). Detection of calcium-dependent regulatory protein binding components using ^{125}I-labeled calcium-dependent regulatory protein. *J. Biol. Chem.* **253**, 3374–3377.

LaPorte, D. C., Toscano, W. A., and Storm, D. R. (1979). Cross-linking of Iodine-125-labeled, calcium-dependent regulatory protein to the Ca^{2+}-sensitive phosphodiesterase purified from bovine heart. *Biochemistry* **18**, 2820–2825.

Le Peuch, C. J., Haiech, J., and Demaille, J. (1979). Concerted regulation of cardiac sarcoplasmic reticulum calcium transport by cyclic adenosine monophosphate-dependent and calcium-calmodulin-dependent phosphorylations. *Biochemistry* **18**, 5150–5157.

Levin, R. M., and Weiss, B. (1977). Binding of trifluoperazine to the calcium-dependent activator of cyclic nucleotide phosphodiesterase. *Mol. Pharmacol.* **13**, 690–697.

Lin, C. T., Dedman, J. R., Welsh, M. J., Brinkley, B. R., and Means, A. R. (1979). Immunoelectron microscopic localization of calmodulin in the mitotic apparatus. *J. Cell Biol.* **83**, 378a.

Lynch, T. J., and Cheung, W. Y. (1979). Human erythrocyte Ca^{2+}-Mg^{2+}-ATPase: Mechanism of stimulation by Ca^{2+}. *Arch Biochem. Biophys.* **194**, 165–170.

Mabuchi, I., and Okuno, M. (1977). The effect of myosin antibody on the division of starfish blastomeres. *J. Cell Biol.* **74**, 251–263.

Marcum, J. M., Dedman, J. R., Brinkley, B. R., and Means, A. R. (1978). Control of microtubule assembly-disassembly by calcium-dependent regulator protein. *Proc. Natl. Acad. Sci. U.S.A.* **75**, 3771–3775.

Margolis, R. L., Wilson, L., and Kiefer, B. I. (1978). Mitotic mechanism based on intrinsic microtubule behavior. *Nature (London)* **272**, 450–452.

Mazia, D., Petzelt, C., Williams, R. O., and Meza, I. (1972). A Ca^{2+}-activated ATPase in the mitotic apparatus of the sea urchin egg (isolated by a new method). *Exp. Cell Res.* **70**, 325–332.

McDonald, K., Pickett-Heaps, J. D., McIntosh, J. R., and Tippit, D. H. (1977). On the mechanism of anaphase spindle elongation in *Diatoma Vulgare*. *J. Cell Biol.* **74**, 377–388.

McDonald, K. L., Edwards, M. K., and McIntosh, J. R. (1979). Cross-sectional structure of the central mitotic spindle of *Diatoma vulgare*. Evidence for specific interactions between anti-parallel microtubules. *J. Cell Biol.* **83**, 443–461.

McIntosh, J. R., Hepler, P. K., and Van Wie, D. G. (1969). Model for mitosis. *Nature (London)* **224**, 659–663.

McIntosh, J. R., McDonald, K. L., Edwards, M. K., and Ross, B. M. (1979). Three-dimensional structure of the central mitotic spindle of *Diatoma vulgare*. *J. Cell Biol.* **83**, 428–442.

Means, A. R., and Dedman, J. R. (1980). Calmodulin - an intracellular calcium receptor. *Nature (London)* **285**, 73–77.

Mohri, H., Mohri, T., Mabuchi, I., Yazaki, I., Sakai, H., and Ogawa, K. (1976). Localization of dynein in sea urchin eggs during cleavage. *Dev. Growth Differ.* **18**, 391–398.

Nagle, B. W., and Egrie, J. C. (1979). Partially purified Ca^{2+}-ATPase from sea urchin mitotic apparatus is not activated by calmodulin. *J. Cell Biol.* **83**, 378a.

Nicklas, R. B. (1971). Mitosis. *Adv. Cell Biol.* **2**, 225–297.

Niggli, V., Penniston, J. T., and Carafoli, E. (1979). Purification of the $(Ca^{2+}Mg^{2+})$-ATPase from human erythrocyte membranes using a calmodulin affinity column. *J. Biol. Chem.* **254**, 9955–9958.

Nishida, E., and Kumagai, H. (1980). Calcium sensitivity of sea urchin tubulin in *in vitro* assembly and the effects of calcium-dependent regulator (CDR) proteins isolated from sea urchin eggs and porcine brain. *J. Biochem.* **87**, 143–151.

Nishida, E., and Sakai, H. (1977). Calcium-sensitivity of the microtubule reassembly system. Difference between crude brain extract and purified microtubular proteins. *J. Biochem.* **82**, 303–306.

Nishida, E., Kumagai, H., Ohtsuki, I., and Sakai, H. (1979). The interactions between calcium-dependent regulator protein of cyclic nucleotide phosphodiesterase and microtubule proteins. I. Effect of calcium-dependent regulator protein on the calcium sensitivity of microtubule assembly. *J. Biochem.* **85**, 1257–1266.

Olmsted, J. B., and Borisy, G. G. (1975). Ionic and nucleotide requirements for microtubule polymerization *in vitro*. *Biochemistry* **14**, 2996–3005.

Petzelt, C. (1972). Ca^{2+}-activated ATPase during the cell cycle of the sea urchin *Strongylocentrotus purpuratus*. *Exp. Cell Res.* **70**, 333–339.

Petzelt, C. (1979). Biochemistry of the mitotic spindle. *Int. Rev. Cytol.* **60**, 53–92.

Portzehl, H., Caldwell, P. C., and Ruëgg, J. C. (1964). The dependence of contraction and relaxation of muscle fibers from the crab *Maia squinado* on the internal concentration of free calcium ions. *Biochem. Biophys. Acta* **79**, 581–591.

Pratt, M. M., Otter, T., and Salmon, E. D. (1980). Dynein-like Mg^{2+}-ATPase in mitotic spindles isolated from sea urchin embryos (*Strongylocentrotus droebachiensis*). *J. Cell Biol.* **86**, 738–745.

Rosenfeld, A. C., Zackroff, R. V., and Weisenberg, R. C. (1976). Magnesium stimulation of calcium binding to tubulin and calcium-induced depolymerization of microtubules. *FEBS Lett.* **65**, 144–147.

Sakai, H., Mabuchi, I., Shimoda, S., Kuriyama, R., Ogawa, K., and Mohri, H. (1976). Induction of chromosome motion in the glycerol isolated mitotic apparatus; nucleotide specificity and effects of anti-dynein and myosin sera on the motion. *Dev. Growth Differ.* **18**, 211–219.

Salmon, E. D. (1975). Spindle microtubules: Thermodynamics of *in vivo* assembly and role in chromosome movement. *Ann. N. Y. Acad. Sci.* **253**, 383–406.

Salmon, E. D., and Segall, R. R. (1980). Calcium-labile mitotic spindles isolated from sea urchin eggs (*Lytechinus variegatus*). *J. Cell Biol.*, **86**, 355–365.

Satir, B. H., Garofalo, R. S., Maihle, N. J., and Gilligan, D. M. (1980). Possible role of calmodulin in secretion. *Ann. N. Y. Acad. Sci.*, in press.

Schliwa, M. (1976). The role of divalent cations in the regulation of microtubule assembly. *In vivo* studies on microtubules of the heliozoan axopodium using the ionophore A23187. *J. Cell Biol.* **70**, 527–540.

Schulman, H., and Greengard, P. (1978). Ca^{2+}-dependent protein phosphorylation system in membranes from various tissues, and its activation by "calcium-dependent regulator". *Proc. Natl. Acad. Sci. U.S.A.* **75**, 5432–5436.

Shenolikar, S., Cohen, P. T. W., Cohen, P., Nairn, A. C., and Perry, S. V. (1979). The role of calmodulin in the structure and regulation of phosphorylase kinase from rabbit skeletal muscle. *Eur. J. Biochem.* **100**, 329–337.

Silver, R. B., Cole, R. D., and Cande, W. Z. (1980). Isolation of mitotic apparatus containing vesicles with calcium sequestration activity. *Cell* **19**, 505–516.

Smoake, J. A., Song, S. Y., and Cheung, W. Y. (1974). Cyclic 3′,5′-nucleotide phosphodiesterase, distribution and developmental changes of the enzyme and its protein activator in mammalian tissues and cells. *Biochim. Biophys. Acta* **341**, 402–411.

Stevens, F. C., Walsh, M., Ho, H., Teo, T. S., and Wang, J. H. (1976). Comparison of calcium binding proteins: Bovine heart and bovine brain protein activators of cyclic nucleotide phosphodiesterase and rabbit skeletal muscle troponin C. *J. Biol. Chem.* **251**, 4495–4500.

Summers, K. E., and Gibbons, I. R. (1971). Adenosine triphosphate-induced sliding of tubules in trypsin-treated flagella of sea urchin sperm. *Proc. Natl. Acad. Sci. U.S.A.* **68**, 3092–3096.

Teo, T. S., and Wang, J. H. (1973). Mechanism of activation of a cyclic adenosine 3',5'-monophosphate phosphodiesterase from bovine heart by calcium ions: Identification of the protein activator as a Ca^{2+}-binding protein. *J. Biol. Chem.* **248**, 5950–5955.

Vanaman, T. C., Sharief, F., and Watterson, D. M. (1977). Structural homology between brain modulator protein and muscle TnCs. *In* "Calcium Binding Proteins and Calcium Function" (R. H. Wasserman *et al.*, eds.), pp. 107–117. Elsevier, North Holland, New York.

Waisman, D., Stevens, F. C., and Wang, J. H. (1975). The distribution of the Ca^{2+}-dependent protein activator of cyclic nucleotide phosphodiesterase in invertebrates. *Biochem. Biophys. Res. Commun.* **65**, 975–982.

Waisman, D. M., Stevens, F. C., and Wang, J. H. (1978). Purification and characterization of a Ca^{2+}-binding protein in *Lumbricus terrestris*. *J. Biol. Chem.* **253**, 1106–1113.

Wang, J. H., and Desai, R. (1976). A brain protein and its effects on the Ca^{2+}- and protein modulator-activated cyclic nucleotide phosphodiesterase. *Biochem. Biophys. Res. Commun.* **72**, 926–932.

Wang, J. H., and Waisman, D. M. (1979). Calmodulin and its role in the second-messenger system. *In* "Current Topics in Cellular Regulation" (B. L. Horecker and E. R. Stadtman, eds.), Vol. 15, pp. 47–107. Academic Press, New York.

Watterson, D. M., Sharief, F., and Vanaman, T. C. (1980). The complete amino acid sequence of the Ca^{2+}-dependent modulator protein (calmodulin) of bovine brain. *J. Biol. Chem.* **255**, 962–975.

Weber, A. (1968). The mechanism of the action of caffeine on the sarcoplasmic reticulum. *J. Gen. Physiol.* **52**, 760–772.

Weisenberg, R. C. (1972). Microtubule formation *in vitro* in solutions containing low calcium concentrations. *Science* **177**, 1104–1105.

Welsh, M. J., Dedman, J. R., Brinkley, B. R., and Means, A. R. (1978). Calcium-dependent regulator protein: Localization in mitotic apparatus of eukaryotic cells. *Proc. Natl. Acad. Sci. U.S.A.* **75**, 1867–1871.

Welsh, M. J., Dedman, J. R., Brinkley, B. R., and Means, A. R. (1979). Tubulin and calmodulin: Effects of microtubule and microfilament inhibitors on localization in the mitotic apparatus. *J. Cell Biol.* **81**, 624–634.

Wolff, D. J., and Brostrom, C. O. (1974). Calcium-binding phosphoprotein from pig brain: Identification as a calcium-dependent regulator of brain cyclic nucleotide phosphodiesterase. *Arch. Biochem. Biophys.* **163**, 349–358.

Wolff, D. J., and Brostrom, C. O. (1979). Properties and functions of the calcium-dependent regulator protein. *Adv. Cyclic Nucleotide Res.* **2**, 27–88.

Wolniak, S. M., Hepler, P. K., Saunders, M. J., and Jackson, W. T. (1979). Changes in the distribution of calcium-associated membranes during mitosis in endosperm cells of *Haemanthus*. *J. Cell Biol.* **83**, 374a.

Wong, P. Y. K., and Cheung, W. Y. (1979). Calmodulin stimulates human platelet phospholipase A2. *Biochem. Biophys. Res. Commun.* **90**, 473–480.

Yagi, K., Yazawa, M., Kakiuchi, S., and Ohshima, M. (1978). Identification of an activator protein for myosin light chain kinase as the Ca^{2+}-dependent modulator protein. *J. Biol. Chem.* **253**, 1338–1340.

Mechanisms of Cytokinesis

Mechanisms of Cytokinesis in Animal Cells

GARY W. CONRAD and RAYMOND RAPPAPORT

I. OVERVIEW

The overall mechanism by which animal cells divide remains poorly understood despite intensive study. A discriminate evaluation of several aspects of the problem seems warranted. Our purpose in what follows has been to distinguish what is clearly known about cytokinesis from what is imperfectly understood.

A. Establishment of the Furrowing Mechanism: Roles Played by Major Cell Regions during Division

Animal cells usually form a cleavage furrow in association with a mitotic apparatus. The apparatus may consist of a spindle with a prominent aster at each end, as in many fertilized eggs and early blastomeres, or it may consist of a spindle with relatively small asters, as in cells from later-stage embryos

MITOSIS/CYTOKINESIS

and adults. The mitotic apparatus expands greatly during the division cycle and is readily apparent to the observer of the living process. However, in the absence of a sharply defined periphery, it is difficult to obtain accurate measurements of any of its dimensions, except those involving the astral centers. When a normal mitotic apparatus is present, the furrow appears on the cell surface in the plane that is normal to the spindle axis and that passes through the midpoint of the spindle length. In the absence of a spindle, the furrow develops in the same relation to a straight line joining the astral centers. The cleavage furrow actively cuts through the subsurface cytoplasm, including the mitotic apparatus, until the daughter cells are completely separated. The geometrical relationship which exists between the mitotic apparatus and the surface region in which the furrow develops suggests that the mitotic apparatus somehow alters or stimulates the cortical cytoplasm to establish the furrow. The cortical cytoplasm, or cortex, is a layer of cytoplasm of varying thickness (2–4 μm) immediately beneath and attached to the plasma membrane, which, according to microdissection experiments (Hiramoto, 1957; Higashi, 1972), appears to be a firmly gelled layer. Ultra-structurally, it consists of an amorphous or microfilamentous network. During cell division, a distinct bundle of linearly aligned microfilaments appears in the cortical cytoplasm associated with the cleavage furrow.

In discussing what the mitotic apparatus does to the cortical cytoplasm and how that superficial part of the cell then forms a furrow, it will be important to remember that in the cells of adults and in the fertilized eggs of echinoderms and mammals, the mitotic apparatus is usually centrally located and the cleavage furrow appears simultaneously around the entire equatorial cell circumference (symmetrical cleavage), whereas in the fertilized eggs of birds, fish, reptiles, amphibians, mollusks, and other yolky cells, the mitotic apparatus is excentric, and the cleavage furrow first appears in the nearest part of the equatorial surface (unilateral or non-symmetrical cleavage). Formation of cleavage furrows in early insect embryos represents a special case of unilateral cleavage and will be discussed below.

Research over the last 70 years has revealed the following information about how a cleavage furrow is established (see also Rappaport, 1971a).

1. In the process of furrow establishment, there is a point after which the mitotic apparatus can be removed physically or chemically without affecting cytokinesis (Yatsu, 1911; Hiramoto, 1956, 1971; Hamaguchi, 1975). This point may be reached while the cell is still spherical, so all the dimensional changes which normally take place during the division process can occur independently of the physical presence of the mitotic apparatus. Thus, the role the mitotic apparatus plays in cytokinesis must be stimulatory, rather than mechanical or physical. Presumably, the mitotic apparatus somehow alters the cortical cytoplasm in such a way that it subsequently accomplishes

division independently of the subsurface cytoplasm. Although the idea that the mitotic apparatus affects the cortex by altering its chemical environment is more than a century old (Bütschli, 1876, in Ziegler, 1898), we do not know as yet whether the alteration is accomplished by transfer of substance to the cortex or removal of substance from it.

2. A mitotic apparatus can cause a cleavage furrow to form on any region of the cell surface. Although predetermination appears to exist in polar lobe formation (see below), all areas of surface seem equally able to respond to a mitotic apparatus by furrowing. This has been determined by moving the mitotic apparatus to different positions within a cell and observing that a furrow forms in a position that corresponds to the new position of the apparatus (Harvey, 1935; Rappaport and Ebstein, 1965; Rappaport, 1976).

3. Although the normal furrowing pattern for a particular cell type may be symmetrical, unilateral furrowing can be elicited simply by moving the mitotic apparatus from the center to an excentric position in the cell (Yatsu, 1908). Conversely, in cells where the furrowing pattern normally is unilateral, symmetrical furrowing can be induced by cutting away the yolky portion of the cell, thereby leaving the mitotic apparatus centrally positioned within the remaining cell fragment (Rappaport and Conrad, 1963). Thus, the furrowing pattern displayed by a particular cell type is determined only by the geometrical position of the mitotic apparatus within the cell.

4. Several furrows can appear in a single cell. This happens normally during cellularization of the blastoderm of early insect embryos (Fullilove and Jacobson, 1971) and cleavage of certain large, yolky coelenterate eggs. It occurs under experimental conditions if the mitotic apparatus is moved to a new location within a cell after stimulation of the cortical cytoplasm in the original position has occurred (Harvey, 1935; Rappaport, 1975a).

5. A single mitotic apparatus can cause the formation of multiple furrows within a single cell, although normally it would induce only one furrow (Harvey, 1935; Rappaport, 1975a).

6. A cleavage furrow can be established not only between two asters connected by a spindle but also between two asters that have never been connected by a spindle. The latter situation arises normally during cellularization of the blastoderm in insects (Fullilove and Jacobson, 1971) and is observed under experimental conditions in fertilized eggs of echinoderms (Wilson, 1901a; Rappaport, 1961).

The observations above do not reveal how the mitotic apparatus affects the cortical cytoplasm, although most current models assume that a stimulus travels from the mitotic apparatus to the cortex. In this instance, the stimulus may take the form of any alteration of the subsurface environment. Whatever form and direction the stimulus actually assumes, several major questions remain unanswered: What is the nature of the stimulus? Where does it come from? When during mitosis is it sent, and when does it arrive? Does the stimulus act on the cortical region which will subsequently become a furrow or on regions which will not be involved in furrowing per se? What is the first response of the cortical cytoplasm to the stimulus (Rappaport and Rappaport, 1976)? We will discuss experimental results relevant to some of these questions in Section II.

B. Furrowing and Cytokinesis: How Is Division Accomplished?

A cleavage furrow is an enormous, moving deformation of cell shape. It forges rapidly through the cell, past and through organelles, pushing things out of the way. It can be stopped physically or slowed only by such relatively nondeformable materials as glass rods or yolk. We consider a cleavage furrow to be an organelle of the same sort as a nucleolus or a Golgi apparatus, i.e., a structure that is visible specifically (and only) because of its own activity and that disappears if that activity ceases. The cleavage furrow, nucleolus, and Golgi apparatus, as organelles, are also similar in that they are dependent for their creation on another organelle (i.e., mitotic apparatus, chromosome, and nucleus, respectively). The extent to which, in each case, they *remain* dependent upon the first organelles during their subsequent period of existence depends on the cell type. Thus, although a cleavage furrow can be represented as a wave of activity whose behavior can be studied, it may be more useful to think of it as a short-lived organelle whose physical and biochemical properties can be related to its motile activity.

The base of a cleavage furrow consists of a plasma membrane together with a portion of cortical cytoplasm containing a band of microfilaments whose long axis coincides with the plane of the cleavage furrow itself (Schroeder, 1968, 1970, 1972; Arnold, 1968, 1969; Szollosi, 1970). The advancing tip of the furrow is of greatest interest, for this specialized area of cortex, isolated by itself in a small fragment of cytoplasm, continues to function until the fragment is divided (Dan and Kojima, 1963; Rappaport, 1969b). It also appears that the base of the furrow is structurally self-sufficient and independent of organized physical structures in the subsurface cytoplasm, for the latter can be replaced or steadily mixed and agitated, but cytokinesis still occurs (Hiramoto, 1965; Rappaport, 1966). Thus, present evidence suggests that furrowing is primarily an activity of the cortical cytoplasm at the base of the cleavage furrow. Analyses of cleavage furrows have revealed the following: (1) The site of force generation, as well as the site where the force is exerted, is in the furrow region (Rappaport, 1966, 1967, 1977; Hiramoto, 1979). (2) The microfilaments in the cortex at the base of the furrow are actin-like on the basis of their ability to bind the heavy meromyosin fragments from skeletal muscle myosin in a manner which suggests periodicity (Schroeder, 1973; Schmidt *et al.*, 1980). (3) Indirect and direct immunofluorescence localizations in extracted cells using reasonably well-characterized antibodies and fluorescent heavy meromyosin staining reveal no greater concentration of actin in the cleavage furrow than in the non-furrow surface (Herman and Pollard, 1978, 1979). This suggests that actin is present everywhere in the cortex but is more easily identified ultrastructurally in the furrow because of the more uniform orientation of the microfilaments in that region. However, immunocytochemical studies with antibodies to myosin (Fujiwara and Pollard, 1976, 1978; Fujiwara *et al.*, 1978), α-actinin (Fujiwara *et al.*, 1978), and tropomyosin (Ishimoda-Takagi, 1979) suggest some degree of enhanced fluorescence in the furrowing region.

Currently, some of the major questions concerning furrowing are the following: What generates the mechanical force that accomplishes furrowing? What relation does the band of microfilaments at the base of the furrow have to the mechanism of force generation? In unilateral furrowing, how does the furrow proceed through cell surface that was too far from the mitotic apparatus to have been directly affected by it, and what determines the apparently straight-line nature of its progress? Why should physical or chemical removal of the mitotic apparatus early in mitosis (e.g., prophase) give rise to a partial furrow? That is, if sufficient stimulation from the mitotic apparatus occurs to initiate furrowing, and if furrows are self-propagating once started, why don't all furrows, no matter how shallow, go to completion once they begin? What roles do Ca^{2+} and other ions play in initiating and regulating furrowing? How is the final cytoplasmic bridge between the two daughter cells broken?

II. INVESTIGATIONAL RESULTS

A. Furrowing Associated with Asters or a Spindle

1. What Region of the Mitotic Apparatus Establishes the Furrow?

Since present data suggest that the mitotic apparatus influences the overlying cell surface to furrow, attempts to identify the essential parts of the mitotic apparatus have been made.

a. Spindle. Cytokinesis in most adult animal tissue cells involves creation of a cleavage furrow around a spindle apparatus; asters are usually small. In grasshopper neuroblasts and adult newt kidney cells, for example, the spindle apparatus normally is the source of the cleavage stimulus (Rappaport and Rappaport, 1974; Kawamura, 1977). Also, in the fertilized eggs of two species of echinoderms, in which the asters are large, cell surface presented to the spindle experimentally will form a furrow (Rappaport and Rappaport, 1974). However, if the two asters are removed with a suction micropipette, leaving only a spindle apparatus in the center of a large spherical cell, no furrowing occurs (Hiramoto, 1971). Thus, the spindle appears capable of inducing furrowing in cell surface which happens to be close enough.

b. Asters. Most evidence from fertilized echinoderm eggs suggests that the position of the early cleavage furrows is determined by the position of the

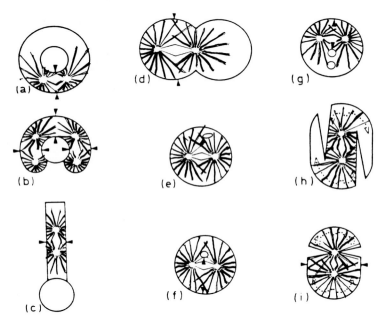

Fig. 1. Furrowing following mechanical distortion. The cell is shown at anaphase, and the subsequent cleavage plane falls between the opposed pairs of black triangles. (a) First cleavage of a torus-shaped egg. (b) Second cleavage of a torus-shaped egg. The central circle represents a glass sphere. (c) Cleavage of an axially loaded, attenuated cell. The circle represents a glass sphere. (d) Furrow orientation after artificial constriction. Most of the mitotic apparatus lies on one side of the constriction. (e) Furrow formation involving one perforation close to the equatorial plane. (f) Furrow formation in a flattened egg involving perforation in the equatorial plane. (g) Furrow formation in a flattened egg involving two perforations in the equatorial plane. (h) Furrowing after deep notches are cut parallel to the spindle axis. (i) Furrowing after deep notches are cut at right angles to the spindle axis. (a,b) From Rappaport (1961); (c) from Rappaport (1965); (d) from Rappaport (1964); (e) from Rappaport (1968); (f) from Dan (1943); (g) from Dan (1943), Rappaport (1968); (h) from Rappaport (1970); (i) from Rappaport (1970, 1975b).

large asters. Aster size remains large, in comparison to spindle size, until the 32- to 64-cell stage (Harris, 1962). Although it was demonstrated above that the spindle is capable of inducing furrowing (Rappaport and Rappaport, 1974), it seems unlikely that this portion of the mitotic apparatus participates significantly in triggering cleavage in very large cells. In fact, several lines of evidence suggest that neither the spindle nor the chromosomes per se are necessary for stimulation of furrowing. Torus-shaped cells tricked into becoming cylindrical and binucleated orient their two mitotic apparatuses in a line, end-to-end, and then proceed to form cleavage furrows, not only between the two sets of asters joined by spindles but also in the space between the backsides of the two asters not joined by a spindle (Fig. 1a,b; Rappaport, 1961). Alternatively, if the spindle is removed by suction micropipette, leaving only the two asters, cells usually cleave completely; if the spindle and one aster are removed, leaving only one aster, cells generally do not cleave (Hiramoto, 1971). Early experimental work also indicated that sperm asters and cytasters were capable of inducing cleavage furrows (Wilson, 1901b).

During cleavage in insect embryos, the first several nuclear divisions occur without accompanying cytokinesis. Subsequently, the nuclei migrate out to the cortical cytoplasm of the multinucleated cell, and as they undergo mitosis in their new location, they become separated by unilateral cleavage furrows ("superficial" cleavage). At least some, and probably as many as one-half, of all the cleavage furrows formed during initial cellularization of these insect blastula-stage embryos arise between the backsides of asters not connected by a spindle (Fullilove and Jacobson, 1971). Otherwise, all cells would be at least binucleate. Thus, furrow establishment between asters not connected by a spindle occurs as a normal part of the development of a very large group of animals.

2. What Region of the Cell is Altered by the Mitotic Apparatus?

In establishing the position of the division plane, the mitotic apparatus determines which parts of the cell periphery will play active roles and which will be passive. Taking a pair of asters as a minimal furrow establishment mechanism, the evidence indicates that there is no polarity in astral function. The backs and sides of the asters are as effective as the regions closest to the equatorial plane, so it seems likely that the delineation of active and passive regions is determined by different levels of the same activity. Presumably that region subject to the highest activity undergoes the changes that precipitate division. The basic question has been whether it is the polar or the equatorial regions which are subjected to the greater activity and undergo the critical changes.

a. Polar Stimulation. According to this hypothetical mechanism, the effect of the mitotic apparatus is greatest outside the cell's equatorial region, presumably in the polar and subpolar areas (Wolpert, 1960; Borisy and White, 1978). The intensity of the mitotic apparatus effect is presumed to be distance related, and the furrow would form where the intensity of the mitotic apparatus effect is lowest.

The polar stimulation mechanism is attractive because at the time the interaction takes place, the astral centers appear to be closer to the cell's polar surface than to the cell's equatorial surface. Under these circumstances, if stimulatory activity followed the inverse square law, the asters would have a greater effect on the cell poles. The ability of the mitotic apparatus to affect the surface operates over a limited range, but there is no evidence that the inverse square law applies within that range.

There is, moreover, no experimental evidence which supports the possibility that a polar stimulation mechanism exists. No manipulation of the geometry of the polar areas or alteration of their dimensions affects furrow establishment (Rappaport and Ratner, 1967). Despite measures in which normally spherical cells are stretched until the poles of the cell are many times more distant from the astral centers than is the equator of the cell, furrow establishment and function are normal (Fig. 1c; Rappaport 1960b, 1978). Blocks in the form of perforations (Rappaport, 1968) or oil drops (R. Rappaport, unpublished) located between the aster and the polar surface are without effect. The same is true of chronic manipulations which greatly disturb the normal structure of the region between the astral center and the polar surface (R. Rappaport, unpublished). According to the polar stimulation hypothesis, the positioning of the furrow precisely between the astral centers is a consequence of the equality of interaction between the asters and the adjacent surfaces. However, in natural circumstances, such as polar body formation during meiosis, in which the mitotic apparatus is normal to the surface and highly excentric, the furrow cuts between the asters, although the distances from the astral centers to their respective poles differ greatly and so, presumably, would the intensities of interaction. The same geometrical relations have been established in otherwise normally cleaving sea urchin eggs with the same results (Fig. 1d; R. Rappaport, 1964 and unpublished). According to the polar stimulation mechanism, conditions necessary for furrow establishment are achieved when an area of the surface characterized by a low level of interaction with the asters lies between regions with higher levels. This condition can be easily produced by placing perforations as blocks between the aster and the surface, but such surface regions never show furrowing activity (Fig. 1e; Rappaport, 1968). If aster–polar surface interaction were essential and distance related, then measures which place the asters closer than normal to the poles of the cell, and at the same time farther apart, would not be expected to prevent division; but, in fact, they do (Rappaport, 1961, 1969a).

b. Equatorial Stimulation. According to this hypothetical mechanism, the maximum effect of the mitotic apparatus is brought to bear directly on the equatorial surface of the cell, and the critical changes which precipitate division occur there rather than at the poles.

The concept of equatorial stimulation is supported by experimental results which reveal that any geometrical alterations between the mitotic apparatus and the surface that alter or prevent furrowing involve the region between the equator and the mitotic apparatus. Perforations placed in the equatorial region prevent furrowing in the equatorial margin (Fig. 1f) and in the walls of more distal perforations (Fig. 1g). In the perforation, the margin closest to the mitotic apparatus forms a furrow, while the opposite margin that faces away from the mitotic apparatus does not (Fig. 1f; Rappaport, 1968). Long cuts through the cell parallel to the axis of the mitotic apparatus block marginal furrowing (Fig. 1h; Rappaport, 1970). On the other hand, furrows form on surfaces of perforations that are in contact with the spindle and lack an appropriate relation to the poles (Rappaport and Rappaport, 1974). When the distance between asters is sufficiently large, no furrowing occurs, but when the surface is brought closer than normal to such asters, the ability to establish furrows is restored (Rappaport, 1969a). In both cases, the manipulations involved relationships in the equatorial region. Although the capacity of the mitotic apparatus to affect the surface is distance limited, within its effective range any surface in the equatorial plane will form a furrow if no block is interposed between it and the mitotic apparatus. On the other hand, the polar surface does not furrow when it is moved into the astral center (R. Rappaport, unpublished). When deep notches are cut in flattened cells so as to isolate the equator partially from one polar area, furrowing takes place, suggesting that no event essential for the process took place in the operated pole region (Fig. 1i; Rappaport, 1970).

Although the experimental evidence appears to support the equatorial stimulus hypothesis, serious questions remain. The most important one concerns the basic mechanism by which the essential pair of asters can accomplish the activity which neither can accomplish alone. According to the general scheme proposed above, the pair of asters, by acting jointly, must establish an area of higher activity under the equatorial surface. Since the visible extent of the mitotic apparatus appears greatest at about the time the interaction takes place, this model has been proposed several times previously, beginning about 100 years ago (Bütschli, 1876, in Ziegler, 1898). However, it now appears that the normal radiate structure of the peripheral part of the mitotic apparatus (as observed with the light microscope) is not necessary (Rappaport, 1978). There is also some question about the normal extent and configuration of the microtubular elements of the mitotic apparatus during the period of furrow establishment, which will be discussed below.

3. When during Mitosis Does the Mitotic Apparatus Establish the Furrow?

Although it would be useful to know when the stimulus actually leaves the mitotic apparatus (if it travels in that direction), it is now only possible to detect when the stimulus has been received in an amount sufficient to elicit a subsequent furrowing response in the surface. The mitotic apparatus behaves as a structurally intact unit loosely suspended inside the cell until it achieves maximum size. Its rigidity allows it to be displaced intact to excentric parts of the cell by micromanipulation (Chambers, 1917) or centrifugation (Harvey, 1935) at different times during mitosis. It can also be removed entirely from the cell, and the degree to which the mitotic apparatus has affected the surface is then judged by the ability of the original equatorial zone to furrow in its absence.

Hiramoto (1956) used a suction micropipette to remove the entire mitotic apparatus from fertilized eggs of the heart urchin, *Clypeaster japonicus*, at various stages of mitosis. Although these eggs are very transparent and the mitotic apparatus is reasonably distinct, chromosome behavior (and, therefore, the stage of mitosis) was not clearly determinable in the living cells. Thus, the mitotic stages designated below should be taken as only good approximations. If the mitotic apparatus was removed at the dumbbell stage of actual furrowing or later, cytokinesis was completed. If the apparatus was removed at an early stage of furrowing or during what appeared to be anaphase, most cells furrowed and some completed cytokinesis. Finally, if the apparatus was removed at metaphase, a stage at which the asters are small, a few eggs initiated furrowing, but presumably none completed cytokinesis. Thus, although complete cytokinesis was seen only if the mitotic apparatus was removed at anaphase or later, the steps by which the apparatus was determining the plane of the future furrow and initiating furrowing had clearly already begun by metaphase. Hiramoto, therefore, concluded that the cleavage plane is determined by metaphase, although continued stimulation from metaphase at least to anaphase was needed to stimulate the surface sufficiently to undergo complete cytokinesis without the mitotic apparatus. In a subsequent study using two-cell stage embryos, Hiramoto (1971) removed the mitotic apparatus, or various parts of it, at different times during the period between nuclear envelope breakdown and half-completed cytokinesis. If the entire mitotic apparatus was removed at stages from prometaphase to anaphase (total time, approximately 8 minutes), half the cells formed a cleavage furrow (one-half of which were completed); the other half of the cells changed shape slightly but did not furrow. Using another approach, Hiramoto (1965) injected solutions of sucrose or seawater into heart urchin eggs, which caused disintegration of the mitotic apparatus. When this was done at anaphase, the cell formed a cleavage furrow and divided completely, even when a droplet of seawater replaced more than 50% of the cellular volume. These manipulations of heart urchin eggs suggest that during late prophase the mitotic apparatus begins to stimulate the cortex and that by metaphase the position of the furrow is determined and furrowing can begin. However, it is not until anaphase that sufficient influence has reached the cortex (or the cortex has had sufficient time to react) to allow complete cytokinesis in the absence of the mitotic apparatus. Deciding when, during mitosis, the influence is sent from the mitotic apparatus is important, for the present data form the basis for at least one evaluation of whether astral microtubules specifically could be the source or conveyor of the influence (Asnes and Schroeder, 1979).

Similar micromanipulative experiments performed on grasshopper neuroblasts provide contrasting data (Kawamura, 1977). These cells differ from echinoderm eggs in that the mitotic apparatus has small asters, and cleavage, which begins during late anaphase (instead of telophase or later in echinoderms), gives rise to cells of unequal size. Kawamura used a needle to move the mitotic apparatus to various excentric positions in the cell at different times during mitosis and then determined the effect on subsequent cytokinesis. The plane of the furrow was not established by early anaphase; it began to be determined only by mid-anaphase. If the mitotic apparatus was removed from beneath an already formed furrow, the furrow regressed. Thus, in these cells the stimulus is sent later in mitosis than in echinoderm zygotes. The cleavage furrow appears to require continued stimulation from the apparatus and may never really become independent of it. Another interesting feature of these cells concerns the adhesion of the mitotic apparatus to the cortex. The mitotic apparatus became attached to one pole of the cell (where the smaller of the two cells is formed), and the cleavage furrow appeared to stick to the spindle at late anaphase.

Probably the most precise determination of the time of mitotic apparatus stimulation of the cortex has involved dissolving the apparatus at specified times before second cleavage by injecting colchicine into one of the two blastomeres of two-cell stage embryos (Hamaguchi, 1975). In blastomeres of *Temnopleurus toreumaticus*, a regular sea urchin, the ability to furrow subsequently across more than one-half of the cell diameter appeared at the beginning of anaphase, and approximately 1 minute later, all cells acquired the ability subsequently to cleave completely. In blastomeres of the heart urchin, *C. japonicus*, however, the ability to form partial furrows was established, at the latest, midway between nuclear envelope breakdown and the beginning of anaphase, that is, perhaps by late prophase or during metaphase. When the operation was performed at the beginning of anaphase, some blastomeres were able to cleave completely. These data from *Clypeaster* therefore agree with those of Hiramoto (1956) and suggest that, in this species at least, stimulation by the mitotic apparatus begins to take effect by metaphase and has been completed by anaphase.

Since none of the studies cited above actually determined the precise stage of mitosis in the cells used, and because there is a disparity in the time when the stimulus becomes effective (echinoderm embryos versus grasshopper neuroblasts), it is not possible to make any assumptions concerning the precise time when the mitotic apparatus of animal cells in general stimulates the cortex to furrow. We also do not know whether in most cells the cleavage furrow becomes independent of the mitotic apparatus, as in sea urchin eggs, or remains continuously dependent upon its presence in order to complete cytokinesis, as in grasshopper neuroblasts. It seems apparent that in large yolky cells, such as fertilized amphibian eggs, the first few cleavage furrows become independent of the mitotic apparatus as they progress, for they steadily move out of range of the apparatus as they advance from the animal hemisphere to the vegetal, and the second cleavage is initiated before the first is finished.

The association between mitosis and cytokinesis is so striking that it is easy to forget that there is no direct causal relationship between the two events, as neither the position of the chromosomes nor their presence is essential (Ziegler, 1898, Rappaport, 1961). Furrow establishment can be accelerated by decreasing the distance between the mitotic apparatus and the equatorial surface and can be delayed by repeatedly moving the mitotic apparatus before the process is completed (Rappaport, 1975a). Since both manipulations move cytokinesis and mitosis out of their usual time relationship, it appears likely that mitosis merely constitutes a convenient clock for estimating the status of yet another event, which is probably the size increase of the achromatic mitotic apparatus.

4. Which Constituents of the Mitotic Apparatus Are Involved?

a. Vesicles. The nuclear envelope and annulate lamellae (Kessel, 1968) disperse into vesicles near the end of prophase, that is, just at the time when, as discussed above, stimulation of the cortex by the mitotic apparatus seems to begin in echinoderms. The vesicle dispersal process has been examined in detail during early cleavage in the sea urchins *Strongylocen-*

trotus purpuratus (Harris, 1975) and *Arbacia punctulata* (Longo, 1972). Vesicles are seen in various parts of the mitotic apparatus in all stages of its formation, from the streak stage to telophase, and are especially prominent in asters (Harris, 1965, 1975).

Evidence from various cell types suggests that the nuclear envelope is reversibly depolymerized during mitosis (Ely *et al.*, 1978; Gerace *et al.*, 1978; Gerace and Blobel, 1980). Antigens from the inner nuclear envelope were used to make antibodies, and fluorescence microscopy indicated that the antigens disperse throughout the cell during prophase and then aggregate again around telophase chromosomes. Phospholipids of the nuclear envelope undergo a similar dispersal during prophase, followed by resequestration during telophase (*Amoebae:* Maruta and Goldstein, 1975; CHO cells: Conner *et al.*, 1980).

The timing of the nuclear envelope and annulate lamellae dispersals during prophase permits speculation that vesicles derived from them may become associated with astral microtubules. Vesicles could be transported distally along the microtubules toward the cortex and participate in stimulating the cortex to furrow. Such vesicles have been shown to sequester Ca^{2+} in the presence of ATP (Silver *et al.*, 1980). On the other hand, many turn-of-the-century cytologists reached the not illogical conclusion that the increase in size of the centrospheres during mitosis resulted from a flow of cytoplasm in the opposite direction, along the astral rays toward the centers (Wilson, 1901a). The role, if any, that vesicles play in establishing the cleavage furrow remains to be demonstrated.

b. Microtubules. Experimental evidence cited above suggests that furrowing can be initiated if the cortex comes close to the spindle or a pair of asters. These structures contain microtubules, although the predominant protein of microtubules, tubulin, represents only one of several proteins of the isolated mitotic apparatus (Silver *et al.*, 1980). Studies of the mechanism of cytokinesis have frequently focused on the roles that microtubules or microtubule-associated molecules and organelles might play in stimulating the cortex to furrow.

i. NORMAL DISTRIBUTION OF MICROTUBULES IN DIVIDING CELLS. The distribution of microtubules in dividing eggs of *S. purpuratus* has been examined in detail by indirect immunofluorescence (Harris *et al.*, 1980) and by electron microscopy (Asnes and Schroeder, 1979). Although these studies were done on eggs of the same organism, raised at almost the same temperatures, there were significant differences in the descriptions of the timing and morphology of certain mitotic stages, especially during the period when the asters may be altering the cortex.

Harris *et al.* (1980) used rabbit antibodies against tubulin to visualize microtubules by indirect immunofluorescence. Descriptions of changes in the microtubule distributions were given according to the number of minutes after fertilization and were correlated with the stage of mitosis by referring to the general times when such stages occur. Chromosomes were not specifically visualized. The photomicrographs show that the microtubules undergo three major phases of polymerization and depolymerization. The monaster forms and breaks down in the first phase. During the second phase, which coincides with the period of existence of the streak (see above), the cell is packed with microtubules, and two very large interphase asters develop which surround the intact nucleus, sending out rays that appear to reach to the periphery of the cell and overlap in their distal regions. The interphase asters, whose function in cell division is not known, then fade as microtubules appear to disassemble from the astral center outward toward the periphery. The few remaining microtubules are attached to the centrospheres and become short and straight. This second phase concludes with the "pause" stage, which is characterized by a pair of asters of distinctly minimum size surrounding a still intact nucleus. During the third phase, which encompasses the remainder of mitosis, new asters form around the preexisting astral centers. The mitotic asters contain microtubules which are distinctly straighter than those of previous stages. The chromosomes probably are in early prophase at the beginning of this third phase of microtubule polymerization, for the nuclear envelope disappears as the mitotic asters grow. The degree to which the distal portions of these two asters overlap at prophase and metaphase cannot be ascertained from the photomicrographs; the microtubules radiate in all directions from the centrosphere of each aster and appear to reach the cell periphery at least in the polar regions, but they may not do so in the equatorial zone at these stages. The astral rays continue to elongate until by telophase they appear to fill the cell. These data suggest that in the period from prophase to metaphase, when cortical stimulation begins and the nuclear envelope disperses into vesicles, the mitotic asters appear and begin a period of continuous enlargement that extends through telophase, apparently not interrupted by intervals of major microtubule depolymerization. In comparing these data with those obtained by electron microscopy, it is necessary to keep in mind that each "microtubule" visualized by indirect immunofluorescence is probably a bundle of several microtubules (Rebhun and Sander, 1971). This technique, therefore, might not detect major changes in the density or distribution of uniformly single microtubules, although it enhances our ability to see the microtubule fasicles or bundles, which by light microscopy appear as an astral ray.

Asnes and Schroeder (1979) used transmission electron microscopy to assess quantitatively the distribution and relative density of microtubules near the cortex of the equatorial zone and polar region during different stages of mitosis. Mitotic stages were determined by direct observation of chromosomes in thick sections. Developmental times (number of minutes postfertilization) were not given. This careful study was designed to determine whether the classical light microscopic observations of fixed and living cells which appeared to reveal numerous astral rays under the equatorial surface, some of which crossed, had any ultrastructural validity when techniques designed to enhance microtubule visibility were applied. The point is important because the light microscopic appearance has, over the years, been used to support the idea that the asters may be able to interact simultaneously with the equatorial surface lying between them and thus accomplish the alteration of the region that becomes the active, contractile furrow region (Ziegler, 1898; Rappaport, 1965). If this relationship exists, one might expect that the extension of astral rays into the subcortical cytoplasm would be maximal during the period that furrow establishment takes place. Asnes and Schroeder assumed that furrow establishment takes place during early to mid-anaphase, and because there was a greater number of microtubules near the polar surface than the equatorial surface during this period, they questioned the validity of the model. However, although stimulation may be sufficiently complete by anaphase to permit the furrow to function independently of the mitotic apparatus, as discussed above, stimulation has in fact begun some time during prophase, when the authors report that the microtubules extend far out into the cytoplasm. There is also a latent period between the termination of a stimulation period long enough to elicit a furrow and the actual appearance of the furrow (Rappaport and Ebstein, 1965). Asnes and Schroeder report that astral microtubules extend far out into the yolky cytoplasm during prophase but are largely withdrawn by metaphase. Early in anaphase, astral microtubules again grow outward but do not radiate far into the yolky cytoplasm until mid-anaphase, reaching the cortex first in the polar regions but not reaching the equatorial cortex until telophase.

The arrangements of microtubules during prophase, metaphase, and anaphase as seen by electron microscopy (Asnes and Schroeder, 1979) are clearly at variance with the arrangements of microtubules as seen by immunofluorescence (Harris *et al.*, 1980). Both studies suggest that astral microtubules grow outward extensively during early prophase. However, shortening of astral microtubules during late prophase and metaphase was seen by electron microscopy but was not seen in the immunofluorescence analysis of the same stages. Either the mitotic stages in the two studies do not correspond, or else the two techniques visualized different populations of microtubules. Because of these uncertainties and because cortical stimulation actually must begin earlier in mitosis than judged by Asnes and

Schroeder, a quantitative assessment of regional differences in microtubule density throughout prophase and metaphase is necessary. Although the studies described above sought to analyze the normal distribution of microtubules or tubulin components, it is now evident from experimental studies that the normal configuration of linear elements between the surface and the mitotic apparatus is not essential for furrow establishment (Rappaport, 1978). Stirring the cytoplasm between the two interacting regions prevents the establishment of the normal radiate appearance beneath the surface but does not prevent establishment of a furrow. It therefore seems likely that neither the normal stable arrangement of linear elements nor their precise quantitative arrangement may be the critical elements which determine the part of the cortex that forms furrow. It is possible that what is delivered by the asters to the cortical region is not represented quantitatively by microtubule density, but rather as vesicle density (Harris, 1965) or concentration of some other substance. The essential material might not be visualized by conventional electron microscopy, and its differential distribution to (or from) the equatorial zone might not require that microtubules per se reach all the way to the cortex (or cross) during prophase and anaphase.

ii. MANIPULATION OF MICROTUBULE CONTENT.   Microtubules appear to be required not only for chromosome separation, in association with the spindle, but also for stimulation of furrowing. Evidence for this comes from microinjection experiments and studies in which microtubules were depolymerized.

In several studies, cellular organelles have been injected into eggs to determine which cellular substituents are necessary for stimulation of cleavage. Generally, injection of isolated mitotic apparatus, partially purified centrioles, or basal bodies elicits furrowing (Masui *et al.*, 1978; Heidemann and Kirschner, 1978; Hirano and Ishikawa, 1979). However, the molecular heterogeneity of the components injected precludes a conclusion that microtubules per se elicit cytokinesis. A somewhat different approach has been followed by Raff *et al.*, (1976), who described a mutant of *Ambystoma mexicanum* in which homozygous females lay eggs that look normal morphologically and display a normal cortical response to sperm, but form neither cytasters nor a mitotic apparatus and do not cleave. Nevertheless, the mutant eggs contain a normal amount of soluble tubulin. If such fertilized eggs are injected with fragments of the outer nine doublet microtubules prepared from sea urchin sperm tail axonemes, they cleave (injection of soluble tubulin does not elicit cleavage). They continue to cleave, at least in the animal hemisphere, until they arrest as partial blastulas. The microtubule fragments injected contained 97% tubulin and no detectable dynein or other high-molecular-weight protein. In this mutant, egg activation is not dependent on the presence of microtubules. On the other hand, initiation of cytokinesis appears to require polymerized tubulin. What role the injected microtubules play in initiating cleavage is not known.

The other major approach used to demonstrate a role for microtubules in cleavage furrow establishment involves colchicine and colcemid. These compounds bind to tubulin subunits and prevent their assembly into microtubules (Borisy and Taylor, 1967; Wilson *et al.*, 1974). Because microtubules are in a state of constant assembly and disassembly (Margolis and Wilson, 1979), inhibition of tubulin polymerization results in their rapid disappearance. Fertilized eggs incubated in solutions of colchicine or colcemid, or injected with these drugs, lose their microtubules and fail to form cleavage furrows (Hamaguchi, 1975), although cycles of nuclear envelope breakdown and reassembly still occur (Sluder, 1979). The fact that lumi-colchicine or lumi-colcemid, the light-inactivated forms of the parent compounds, fail to bind to tubulin and fail to inhibit furrowing suggests that the inhibition of cytokinesis arises from diminished microtubule polymerization, rather than from the common side effects on nucleoside transport. Some results suggest, however, that colchicine may inhibit cell shape changes independently of its effects on tubulin or previously studied transport systems. Nocodazole, for example, causes disorganization of microtubules when applied during lens differentiation, but it does not prevent a cell shape change which was previously thought to be microtubule dependent on the basis of sensitivity to colchicine and insensitivity to lumi-colchicine (Beebe *et al.*, 1979).

It is also important to remember that the ability of microtubule inhibitors to prevent cytokinesis does not prove that microtubules or tubulin subunits are the active agents in furrow establishment. It only suggests that microtubules are somehow involved in the process. Because the stimulus for furrowing may depend upon microtubules for its delivery, any agent which specifically or nonspecifically decreases the size of the asters or disorganizes the spindle will inhibit furrowing. For example, the anesthetics ether and urethane decrease aster size by non-specific mechanisms and concomitantly inhibit furrowing (Wilson, 1901b; Rappaport, 1971b), and their effects can be countered by moving the reduced aster closer to the surface.

The roles of aster and spindle microtubules in the establishment of cleavage furrows remain mysterious. They may transport vesicles, directly stimulate the cortex themselves, or remove something from the cortex. The photomicrographs of Harris *et al.* (1980) suggest that as astral microtubules polymerize, they extend outward from the astral center; they appear to depolymerize by disappearing first near the astral center and then progressively later in distal regions. It will be important to determine at which end of the astral microtubules polymerization is favored and, later, at which end depolymerization is favored (Margolis *et al.*, 1978; Margolis and Wilson, 1979; Summers and Kirschner, 1979; Bergen *et al.*, 1980).

The convenience of existing methods for chemical and structural characterization of microtubules leads almost inevitably to increased investigation of and emphasis on their significance. It may be important, however, to keep in mind that current information does not permit us to conclude with great confidence that microtubules are more important than any other component of the mitotic apparatus in the specific activity of furrow establishment. Present data do not exclude the possibility that although they are essential structural components in astral organization, they play no stimulatory role in furrow establishment. Furrows could be elicited by any of the other complex components of the mitotic apparatus, whether they can be presently visualized or not. In that case, the rough correspondence between the rate of movement of the cleavage stimulus and the elongation of microtubules (Rappaport, 1973a,b) may simply indicate that the effect of the mitotic apparatus which elicits the furrow moves centrifugally with the poorly defined front separating cytoplasm which is incorporated in the mitotic apparatus from that which is not.

B. Furrowing Not Obviously Associated with a Mitotic Apparatus

1. Introduction

In addition to the cases cited above, in which stimulation of cortical cytoplasm to form a cleavage furrow is closely and obviously associated geometrically with asters or a spindle, there are numerous experimental and natural cases in which furrow-like constrictions develop without any close or obvious association with a mitotic apparatus.

Tilney and Marsland (1969) found that if fertilized eggs of the sea urchin *A. punctulata* are centrifuged under conditions of high centrifugal force (41,000 g) and high hydrostatic pressure (10,000 psi) for 5 minutes, the cytoplasm stratifies, and 4–5 minutes after centrifugation ceases, 90% of the eggs form a single furrow parallel to the plane of stratification. In most cases, the furrow completely divides the cell (untreated control eggs do not begin furrows for another 20 minutes). In centrifuged eggs, the furrow cuts through the cell at the level of the hyaline zone, a middle layer of cytoplasm which, under the same centrifugal conditions but at atmospheric pressure, contains the nucleus and the annulate lamellae. When the eggs are centrifuged under high pressure, both the nuclear envelope and the annulate lamellae disappear and no mitotic apparatus is seen. Under these experimental conditions, the timing and position of furrowing match the timing and position of breakdown of the nuclear envelope and annulate lamellae, which may be similar to the events preceding normal cytokinesis except for the lack of a mitotic apparatus. The experimental furrow itself appears similar to a normal cleavage furrow, not only because it generally cleaves the cell completely but also because it is associated with a band of microfilaments at its base. In this system, the need for a mitotic apparatus therefore appears to have been eliminated, but normal stimulation, perhaps from breakdown of the nuclear envelope and annulate lamellae, may still have been delivered to the cortical cytoplasm.

In amphibian eggs, a cleavage furrow-like constriction can be induced to form, apparently without a mitotic apparatus, by an injection of saline (Heidemann and Kirschner, 1978), calcium ion (Hollinger and Schuetz, 1976), or subcortical cytoplasm from beneath a cleavage furrow (Sawai, 1972, 1976, 1979). Subcortical cytoplasm from beneath a nonfurrowing surface does not induce a furrow when injected. Even if subcortical cytoplasm from beneath a furrow is injected, only the cortex in certain areas of a cleaving egg is competent to respond, whereas the cortex of an interphase egg does not respond (Sawai, 1972, 1976). In addition to such cases of experimentally induced furrow formation, there are several examples of furrowing which seem to occur in the absence of a mitotic apparatus. Although they do not cleave the cell completely, they involve extensive constriction of the cell surface. For example, barnacle eggs form several incomplete furrow-like constrictions that move peristaltically toward one end of the egg (Lewis *et al.*, 1973; Schroeder, 1975). Each such constriction is associated with a band of microfilaments at its base, disruptable with cytochalasin B, with concomitant cessation of furrowing (Lewis, 1977).

In the normal development of several animal groups, there is an event which must be considered cytokinesis without karyokinesis. Gammarid crustacea cleavage is initially holoblastic, but shortly before blastoderm formation the nucleus and a surrounding region of yolk-free cytoplasm segregate at the periphery of the blastomeres. The nucleus and associated cytoplasm are then separated from the yolky portion by a constriction which slowly and completely separates the two regions. Except for the time required, the process closely resembles normal cytokinesis, although the nucleus remains at interphase (Rappaport, 1960a).

Tubifex eggs display an elaborate pattern of multiple furrows in concert with polar body formation (Shimizu, 1979), a timing pattern similar to that of polar lobe formation, to be described below.

In many of the examples above, the experimental manipulations elicit a furrow that, like a cleavage furrow, frequently cleaves the cytoplasm completely. Moreover, with the exception of injecting subcortical cytoplasm (Sawai, 1972, 1976), these furrows can form anywhere on a cell surface. The furrow position is determined by the site where the stimulus is applied: The hyaline zone (in centrifuged eggs) or the site of injection, i.e., the furrowing site, is not predetermined.

2. Polar Lobe Formation

The fertilized eggs of some marine mollusks and annelids display a shape change which is an interesting mimic of cytokinesis (Collier, 1965; Clement, 1976). The process, known as "polar lobe formation," involves the formation of a cleavage furrow-like constriction in a plane slightly below the equator of the egg. It may occur during meiotic divisions as well as during the early cleavages. Polar lobe formation occurs spectacularly in the fertilized eggs of the common marine mudsnail, *Ilyanassa obsoleta* Stimpson (*Nassarius obsoletus* Say). The details of the normal shape changes have been described previously (Conrad *et al.*, 1973; Conrad, 1973; Conrad and Williams 1974a). The polar lobe constriction that forms before and during the first cleavage (causing formation of the third polar lobe) is shown in Figs. 2 and 3. There are several remarkable things about a polar lobe constriction: Its position in the cell is fixed; it does not cleave the cell completely; it forms and relaxes at least four times during early development (once with each meiotic division and once with at least each of the first two cleavages), and it bears no obvious geometrical relationship to a mitotic apparatus. What stimulates the formation of the polar lobe constriction and regulates its rate and extent of constriction are not known, but study of its unique features may reveal some elements of control which also operate during cytokinesis.

   a. The Position of the Polar Lobe Constriction Is Fixed. Even when *Ilyanassa* eggs are centrifuged in reverse orientation to alter drastically the distribution of the displaceable cytoplasm, the polar lobe constriction still forms at its normal position in the vegetal hemisphere (as judged by the position of the polar bodies at the opposite, animal pole of the egg) (Morgan,

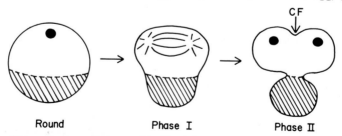

Round   Phase I   Phase II

Fig. 2. Change in the cell shape of fertilized *Ilyanassa* eggs during the first cleavage. Eggs first form phase I polar lobes by steady constriction of a band somewhat below the equator (broken line). Phase II begins as the cleavage furrow (CF) appears, and the polar lobe neck begins to constrict much more rapidly (see Fig. 3). The shaded area designates the polar lobe cytoplasm. From Schmidt *et al.* (1980).

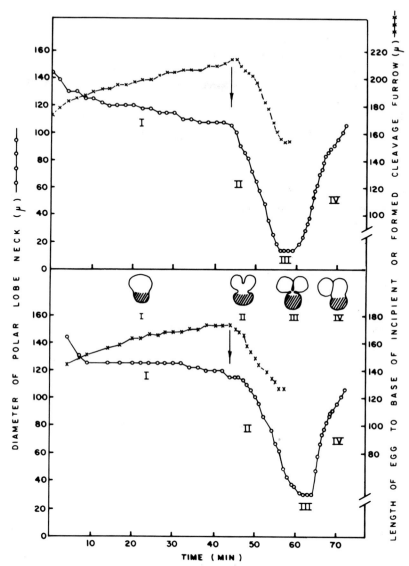

Fig. 3. Rates of change in the diameter of the polar lobe neck, the length of the egg, and the distance to the base of the cleavage furrow during the third and fourth polar lobe formation and resorption. Roman numerals denote phases I–IV of polar lobe activity described in the text. The time of cleavage furrow formation is indicated by arrows. Open circles represent the diameter of the polar lobe neck. *Before* the time of cleavage furrow formation, the crosses represent the length of the egg from animal pole to vegetal pole. After the time of furrow formation, the crosses represent the distance between the vegetal pole and the base of the cleavage furrow and give a measure of the rate at which the cleavage furrow progressed. Top graph, third polar lobe and first cleavage furrow. Bottom graph, fourth polar lobe and second cleavage furrow, CD,blastomere only. From Conrad and Williams (1974a).

1935a; Clement, 1968). Thus, unlike a cleavage furrow, the position of a polar lobe constriction is predetermined, but it is not known whether what is fixed in position (presumably in the cortex) is the stimulatory apparatus/ mechanism, or the materials needed for furrowing, or both. Perhaps only the stimulatory mechanism is fixed in position, because constrictions resembling those of the polar lobe can be induced to form quickly at any point on the surface of these eggs by microiontophoretic injections (see below). Such stimulation experiments suggest that the materials necessary for furrowing are distributed uniformly beneath all regions of the cell surface.

b. Constriction and Relaxation of the Polar Lobe Neck Occurs in the Absence of a Mitotic Apparatus and Is Independent of the Nucleus. The polar lobe constriction forms at a site distant from the mitotic apparatus and in no fixed geometrical relationship to its orientation. The spindle axis of the mitotic apparatus is oriented perpendicular to the plane of the polar lobe constriction during formation of the first and second polar bodies (during which time the first and second polar lobes form), but then shifts by 90° to become parallel with it during the first and second cleavages (during which time the third and fourth polar lobes form).

Ilyanassa eggs are very yolky, which prevents observations of aster behavior in the living state. However, in fixed eggs the mitotic apparatus is near the animal pole. The polar lobe constriction begins as the meiotic or mitotic asters form, apparently during prophase; cytokinesis begins during early telophase (Cather, 1963). The extent to which the astral rays penetrate into the vegetal hemisphere is not clear. There certainly are microtubules in the vegetal hemisphere in the region of the polar lobe constriction, but they lack any discernible order when the polar lobe forms, and it is not known whether they are direct extensions of astral microtubules (Conrad and Williams, 1974a).

If a fertilized egg of *Ilyanassa* or another marine mollusk, *Dentalium*, is physically separated into a nucleated animal half and a non-nucleated vegetal half, the vegetal half forms and relaxes polar lobe constrictions in synchrony with normal embryos at both the first and second cleavages (Wilson, 1904; Morgan, 1933, 1935b; Verdonk *et al.*, 1971). There appears to be a cytoplasmic clock, operating independently of the nucleus, which triggers polar lobe constriction virtually on schedule with nucleated controls. Cytoplasmic clocks also have been demonstrated in early cleavages of echinoderm eggs (Moore, 1933; Yoneda *et al.*, 1978) and amphibian eggs (Sawai, 1979; Hara *et al.*, 1980).

c. The Polar Lobe Neck Constricts at Two Rates. Measurement of the diameter of the lobe neck for second, third, and fourth polar lobes reveals that, unlike a cleavage furrow, they do not constrict at a uniform rate. Instead, they constrict slowly at first (phase I) and then abruptly change to an accelerated rate (phase II) which is maintained until maximal constriction is attained (phase III) (Fig. 3). In the first or second cleavage, the beginning of phase II is closely synchronized with the first appearance of the cleavage furrow at the animal pole. The cleavage furrow completes cytokinesis when the polar lobe neck is maximally constricted (phase III) (Conrad and Williams, 1974a). Thus, during phase II, both the polar lobe constriction and the cleavage furrow are moving at their fastest rates.

   d. **Actin-like Microfilaments Are Associated with Polar Lobe Constriction.** There are circumferentially oriented microfilaments in the cortical cytoplasm at the base of both the polar lobe constriction and the cleavage furrow (Conrad et al., 1973). The microfilament number in the polar lobe constriction is highest during phase II, when the constriction rate is also highest (Conrad and Williams, 1974a). The microfilaments in the polar lobe constriction are actin-like, in that they can be decorated with heavy meromyosin. Decoration is prevented if either ATP or pyrophosphate is also present. These microfilaments are better preserved if tropomyosin from skeletal muscle is present during glycerination and all other incubation steps (Schmidt et al., 1980). Ilyanassa eggs as a whole contain a type of actin whose primary structure, as assessed by two-dimensional tryptic–chymotryptic mapping of ^{125}I-labeled peptides, is virtually indistinguishable from that of chick brain actin, but is distinctly different from that of smooth or skeletal muscle actins or skeletal muscle myosin (Schmidt et al., 1980). These data do not identify the type of actin present specifically in the polar lobe constriction or cleavage furrow.

   If Ilyanassa eggs in spherical stages are treated with cytochalasin B, they neither cleave nor form polar lobe constrictions. If eggs are treated when they are cleaving or constricting a polar lobe, the lobe neck relaxes, the cleavage furrow recedes, and the eggs become spherical. Concomitantly, the band of microfilaments at the base of the polar lobe constriction and that at the base of the cleavage furrow both disappear (Conrad and Williams, 1974a). These results suggest that constriction of the polar lobe neck, like constriction of a cleavage furrow, is in some way dependent upon microfilament integrity.

   e. **Rapid Constriction of the Polar Lobe Neck (Phase II) Is Microtubule Dependent.** When Ilyanassa eggs are incubated with colchicine beginning before formation of the third polar lobe, microtubules disappear, but the eggs form perfectly normal phase I polar lobe constrictions in time with controls. However, when control eggs begin phase II lobe constriction and cleavage furrows, the colchicine-treated eggs neither begin the rapid Phase II constriction nor cleave. Instead, they maintain the polar lobe neck in the Phase I shape for several hours (Conrad and Williams, 1974a). Thus, formation of Phase I constrictions appears to be independent not only of nuclear function (see above) but of microtubule function as well. As judged by sensitivity to colchicine, increase in neck constriction rate (to that of phase II), even neck relaxation, and cytokinesis all appear to depend in some way on microtubules.

f. Exogenous Ca^{2+} Is Not Required, but Endogenous Ca^{2+} May Play a Role.   Normal polar lobe formation and cytokinesis do not appear to require exogenous Ca^{2+} (Fig. 3) (Conrad and Davis, 1980). These events occur in Ca^{2+}, Mg^{2+}-free seawater with or without 10 mM EDTA or in Ca^{2+}-free seawater with or without 10 mM EGTA. However, it seems likely that the normal shape changes involve Ca^{2+}, which must come from internal sources. Three types of evidence suggest this likelihood.

First, *Ilyanassa* eggs are stimulated to form constrictions within 30 seconds in response to microiontophoretic injection of Ca^{2+}, Sr^{2+}, or 3′, 5′-cAMP (Conrad and Davis, 1977, 1980). In contrast, injection of Na^+, K^+, Mg^{2+}, HEPES, or cGMP derivatives do not elicit furrows. Eggs form constrictions in response to Ca^{2+}, Sr^{2+}, or cAMP only if the micropipette tip is inside the cell and situated near the cortical cytoplasm. Eggs form constrictions if injected anywhere on their surface and are responsive to injection at all stages of normal shape change except when they are resorbing a polar lobe (phase IV). Mysteriously, although normal lobe formation and cytokinesis are not dependent upon exogenous Ca^{2+}, injections of Ca^{2+} or of cAMP cause furrowing only if Ca^{2+} is present exogenously (see also Chapter 19 by Sisken, this volume).

A second line of evidence suggesting a role for Ca^{2+} is indirect and comes from treating eggs with membrane-active agents such as ionophores. Ionophores A23187 and X537A, as well as compound 48/80, are all reported to permit entry of exogenous Ca^{2+} into cells (along with some other ions) and, at least in the case of the ionophores, are thought to cause release of Ca^{2+} from intracellular sites. When these three compounds are applied to spherical *Ilyanassa* eggs, shape changes occur which, to varying degrees, resemble at least Phase I of normal polar lobe formation (Conrad and Davis, 1980). Like the normal process, these drug-induced shape changes are not dependent upon exogenous Ca^{2+} and are not inhibited by colchicine. They are prevented by cytochalasin B, however. Because the ionophores used in these investigations transport cations across membranes in exchange for H^+, the above data could alternatively implicate changes in intracellular pH in regulating normal shape changes (Tilney *et al.*, 1978).

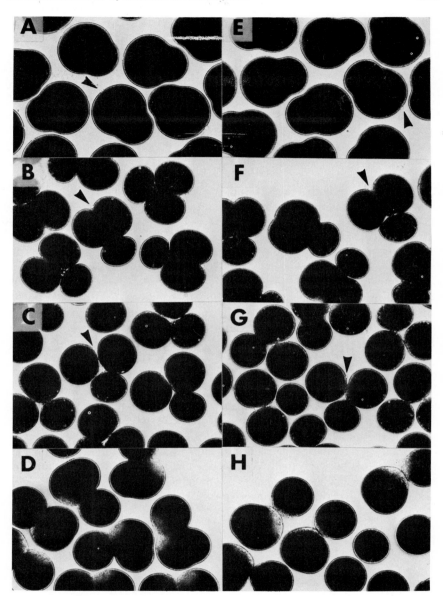

Fig. 4. Polar lobe formation and cytokinesis of *Ilyanassa* eggs in the presence and virtual absence of exogenous Ca²⁺. (A–D) Controls in normal seawater. (E–H) Eggs from the same capsule in Ca²⁺,Mg²⁺− free seawater containing 10 mM EDTA (pH 7.9). Eggs in the latter solution undergo normal polar lobe formation and resorption as well as cytokinesis on

The third piece of evidence implicating Ca^{2+} arises from experiments in which eggs that would normally remain spherical for 40 minutes are placed in solutions containing only one or two salts, isotonic to seawater, to determine which salts rapidly (within 10 minutes) elicit constrictions. Application of high levels of exogenous $CaCl_2$ rapidly caused constrictions, whereas isotonic solutions of NaCl, KCl, $MgCl_2$, $MnCl_2$, sucrose, or mannitol did not cause constrictions. As with the ionophore-induced shape changes, constrictions induced by high exogenous $[Ca^{2+}]$ are inhibited by cytochalasin B, but not by colchicine, and are associated with a band of microfilaments at their base (Conrad and Williams, 1974b).

g. Do Ionic Activities and Plasma Membrane Properties Remain Constant during Cytokinesis? Intracellular ionic activities for Na^+, K^+, Cl^-, and most recently, Ca^{2+} have been reported to remain constant during the first cleavage of amphibian eggs (DeLaat *et al.*, 1974, 1975; Rink *et al.*, 1980). Nevertheless, because it still seems so likely that changes in intracellular ionic activities, however localized and transitory, could regulate contractile mechanisms, it is important to repeat these determinations on a variety of other cell types, including *Ilyanassa* eggs. As a first step in the analysis, the membrane potential of *Ilyanassa* eggs during cytokinesis and polar lobe formation was measured and found to remain constant (Conrad *et al.*, 1977). This potential and its constancy are required data when measuring intracellular ionic activities with ion-selective microelectrodes. As in most animal cells, the membrane potential of *Ilyanassa* eggs is predominantly K^+ dependent. Eggs were placed in seawater containing various concentrations of K^+, up to and including isotonic KCl (0.53 M). The measured membrane potentials ranged from -3 mV in isotonic KCl to -80 mV in seawater containing 9 mM K^+. Despite this wide range of experimentally induced membrane potential, the eggs formed polar lobes and underwent normal cytokinesis in synchrony with control eggs in seawater. Measurements of membrane resistance and changes in specific ionic activities during lobe formation and cytokinesis are in progress.

h. Why Doesn't a Polar Lobe Constriction Cleave? The remarkable thing about a cleavage furrow is that it cuts completely through the cytoplasmic bridge, despite the presence of the spindle apparatus. The remark-

---

the same time schedule as controls. Phase I, third polar lobe: A and E. Phase II, third polar lobe and half-completed first cytokinesis: B and F. Phase III, third polar lobe and completed first cytokinesis (trefoil stage): C and G. Two-cell stage, third polar lobe resorption is complete: D and H. Arrowheads indicate polar bodies (animal pole) X120. From Conrad and Davis (1980).

able thing about a polar lobe constriction in *Ilyanassa* in phase III is that the constriction does *not* cut through the cytoplasmic bridge, despite the absence of a spindle apparatus in the neck and a neck diameter of only approximately 10 μm. Microfilaments seem to be absent from the neck of a lobe constriction in phase III (Conrad *et al.*, 1973). What controls their disappearance and what stops the constriction are not known.

i. What Information Is Needed? In addition to trying to identify and localize proteins involved in furrowing by immunofluorescence and immunoelectron microscopy, it would seem useful to obtain answers to the following specific questions during shape changes in *Ilyanassa* eggs: (1) What changes occur in intracellular ionic activities? (2) Do any proteins undergo changes in degree of phosphorylation? (3) Where is the intracellular pool of Ca^{2+} which may be used during shape changes? (4) What controls the biphasic constriction rate of polar lobe necks? (5) What accounts for the synchrony between Phase II and cytokinesis and the apparent dependence of both upon microtubules?

ACKNOWLEDGMENTS

Original investigations reviewed in this chapter were supported by grants from the National Institutes of Health (GWC) and the National Science Foundation (RR). We thank Barbara N. Rappaport for her careful and dedicated assistance during the preparation of the manuscript.

REFERENCES

Arnold, J. M. (1968). Formation of the first cleavage furrow in a telolecithal egg. *Biol. Bull.* (Woods Hole, Mass.) **135**, 408–409.

Arnold, J. M. (1969). Cleavage furrow formation in a telolecithal egg (*Loligo pealii*). I. Filaments in early furrow formation. *J. Cell Biol.* **41**, 894–904.

Asnes, C. F., and Schroeder, T. E. (1979). Cell cleavage. Ultrastructural evidence against equatorial stimulation by aster microtubules. *Exp. Cell Res.* **122**, 327–338.

Beebe, D. C., Feagans, D. E., Blanchette-Mackie, E. J., and Nau, M. E. (1979). Lens epithelial cell elongation in the absence of microtubules: Evidence for a new effect of colchicine. *Science* **206**, 836–838.

Bergen, L. G., Kuriyama, R., and Borisy, G. G. (1980). Polarity of microtubules nucleated by centrosomes and chromosomes of Chinese hamster ovary cells *in vitro*. *J. Cell Biol.* **84**, 151–159.

Borisy, G. G., and Taylor, E. (1967). The mechanism of action of colchicine; binding of colchicine ^{3}H to cellular protein. *J. Cell Biol.* **34**, 525–533.

Borisy, G. G., and White, J. (1978). A model for cytokinesis in animal cells. *J. Cell Biol.* **79**, 14a.

Cather, J. N. (1963). A time schedule of the meiotic and early mitotic stages of *Ilyanassa*. *Caryologia* **16**, 663–670.

Chambers, R. (1917). Microdissection studies. II. The cell aster: A reversible gelation phenomenon. *J. Exp. Zool.* **23**, 483–505.

Clement, A. C. (1968). Development of the vegetal half of the *Ilyanassa* egg after removal of most of the yolk by centrifugal force, compared with the development of animal halves of similar visible composition. *Dev. Biol.* **17**, 165–186.

Clement, A. C. (1976). Cell determination and organogenesis in molluscan development: A reappraisal based on deletion experiments in *Ilyanassa*. *Am. Zool.* **16**, 447–453.

Collier, J. R. (1965). Morphogenetic significance of biochemical patterns in mosaic embryos. *In* "The Biochemistry of Animal Development" (R. Weber, ed.), Vol. 1, pp. 203–249. Academic Press, New York.

Conner, G. E., Noonan, N. E., Noonan, K. D. (1980). Nuclear envelope of Chinese hamster ovary cells. Re-formation of the nuclear envelope following mitosis. *Biochemistry* **19**, 277–289.

Conrad, G. W. (1973). Control of polar lobe formation in fertilized eggs of *Ilyanassa obsoleta* Stimpson. *Am. Zool.* **13**, 961–980.

Conrad, G. W., and Davis, S. E. (1977). Microiontophoretic injection of calcium ions or of cyclic AMP causes rapid shape changes in fertilized eggs of *Ilyanassa obsoleta*. *Dev. Biol.* **61**, 184–201.

Conrad, G. W., and Davis, S. E. (1980). Polar lobe formation and cytokinesis in fertilized eggs of *Ilyanassa obsoleta*. III. Large bleb formation caused by Sr^{2+}, Ionophores X537A and A23187, and compound 48/80. *Dev. Biol.* **74**, 152–172.

Conrad, G. W., and Williams, D. C. (1974a). Polar lobe formation and cytokinesis in fertilized eggs of *Ilyanassa obsoleta*. I. Ultrastructure and effects of cytochalasin B and colchicine. *Dev. Biol.* **36**, 363–378.

Conrad, G. W., and Williams, D. C. (1974b). Polar lobe formation and cytokinesis in fertilized eggs of *Ilyanassa obsoleta*. II. Large bleb formation caused by high concentrations of exogenous calcium ions. *Dev. Biol.* **37**, 280–294.

Conrad, G. W., Kammer, A. E., and Athey, G. F. (1977). Membrane potential of fertilized eggs of *Ilyanassa obsoleta* during polar lobe formation and cytokinesis. *Dev. Biol.* **57**, 215–220.

Conrad, G. W., Williams, D. C., Turner, F. R., Newrock, K. M., and Raff, R. A. (1973). Microfilaments in the polar lobe constriction of fertilized eggs of *Ilyanassa obsoleta*. *J. Cell Biol.* **59**, 228–233.

Dan, K. (1943). Behavior of the cell surface during cleavage. V. Perforation experiment. *J. Fac. Sci., Tokyo Imp. Univ.* **6**, 297–321.

Dan, K., and Kojima, M. K. (1963). A study on the mechanism of cleavage in the amphibian egg. *J. Exp. Biol.* **40**, 7–14.

DeLaat, S. W., Buwalda, R. J. A., and Habets, A. M. M. C. (1974). Intracellular ionic distribution, cell membrane permeability and membrane potential of the *Xenopus* egg during first cleavage. *Exp. Cell Res.* **89**, 1–14.

DeLaat, S. W., Wouters, W., Guarda, M. M. M. D. S. P., and Guarda, M. A. D. S. (1975). Intracellular ionic compartmentation, electrical membrane properties, and cell membrane permeability before and during first cleavage in the *Ambystoma* egg. *Exp. Cell Res.* **91**, 15–30.

Ely, S., D'Arcy, A., and Jost, E. (1978). Interaction of antibodies against nuclear envelope-associated proteins from rat liver nuclei with rodent and human cells. *Exp. Cell Res.* **116**, 325–331.

Fujiwara, K., and Pollard, T. D. (1976). Fluorescent antibody localization of myosin in the cytoplasm, cleavage furrow, and mitotic spindle of human cells. *J. Cell Biol.* **71**, 848–875.

Fujiwara, K., and Pollard, T. D. (1978). Simultaneous localization of myosin and tubulin in human tissue culture cells by double antibody staining. *J. Cell Biol.* **77**, 182–195.

Fujiwara, K., Porter, M. E., and Pollard, T. D. (1978). Alpha-actinin localization in the cleavage furrow during cytokinesis. *J. Cell Biol.* **79,** 268–275.

Fullilove, S. L., and Jacobson, A. G. (1971). Nuclear elongation and cytokinesis in *Drosophila montana. Dev. Biol.* **26,** 560–577.

Gerace, L., and Blobel, G. (1980). The nuclear envelope lamina is reversibly depolymerized during mitosis. *Cell* **19,** 277–287.

Gerace, L., Blum, A., and Blobel, G. (1978). Immunocytochemical localization of the major polypeptides of the nuclear pore complex-lamina fraction. *J. Cell Biol.* **79,** 546–566.

Hamaguchi, Y. (1975). Microinjection of colchicine into sea urchin eggs. *Dev. Growth Differ.* **17,** 111–117.

Hara, K., Tydeman, P., and Kirschner, M. (1980). A cytoplasmic clock with the same period as the division cycle in *Xenopus* eggs. *Proc. Natl. Acad. Sci. U.S.A.* **77,** 462–466.

Harris, P. (1962). Some structural and functional aspects of the mitotic apparatus in sea urchin embryos. *J. Cell Biol.* **14,** 475–487.

Harris, P. (1965). Some observations concerning metakinesis in sea urchin eggs. *J. Cell Biol.* **25,** 73–77.

Harris, P. (1975). The role of membranes in the organization of the mitotic apparatus. *Exp. Cell Res.* **94,** 409–425.

Harris, P., Osborn, M., and Weber, K. (1980). Distribution of tubulin-containing structures in the egg of the sea urchin *Strongylocentrotus purpuratus* from fertilization through first cleavage. *J. Cell Biol.* **84,** 668–679.

Harvey, E. G. (1935). The mitotic figure and the cleavage plane in the egg of *Parechin microtuberbulatus,* as influenced by centrifugal force. *Biol. Bull.* (Woods Hole, Mass.) **69,** 287–297.

Heidemann, S. R., and Kirschner, M. W. (1978). Induced formation of asters and cleavage furrows in oocytes of *Xenopus laevis* during in vitro maturation. *J. Exp. Zool.* **204,** 431–443.

Herman, I. M., and Pollard, T. D. (1978). Actin localization in fixed dividing cells stained with fluorescent heavy meromyosin. *Exp. Cell Res.* **114,** 15–25.

Herman, I. M., and Pollard, T. D. (1979). Comparison of purified anti-actin and fluorescent-heavy meromyosin staining patterns in dividing cells. *J. Cell Biol.* **80,** 509–520.

Higashi, A. (1972). The thickness of the cortex and some analytical experiments of thixotropy in sea urchin eggs. *Annot. Zool. Jpn.* **45,** 119–144.

Hiramoto, Y. (1956). Cell division without mitotic apparatus in sea urchin eggs. *Exp. Cell Res.* **11,** 630–636.

Hiramoto, Y. (1957). The thickness of the cortex and the refractive index of the protoplasm in sea urchin eggs. *Embryologia* **3,** 361–374.

Hiramoto, Y. (1965). Further studies on cell division without mitotic apparatus in sea urchin eggs. *J. Cell Biol.* **25,** 161–167.

Hiramoto, Y. (1971). Analysis of cleavage stimulus by means of micromanipulation of sea urchin eggs. *Exp. Cell Res.* **68,** 291–298.

Hiramoto, Y. (1979). Mechanical properties of the dividing sea urchin egg. *In* "Cell Motility: Molecules and Organization" (S. Hatano, H. Ishikawa, and H. Sato, eds.), pp. 653–663. Univ. of Tokyo Press, Tokyo.

Hirano, K.-I., and Ishikawa, M. (1979). Induction of aster formation and cleavage in eggs of the sea urchin *Hemicentrotus pulcherrimus* by injection of sperm components. *Dev. Growth Differ.* **21,** 473–481.

Hollinger, T. G., and Schuetz, A. W. (1976). "Cleavage" and cortical granule breakdown in *Rana pipiens* oocytes induced by direct microinjection of calcium. *J. Cell Biol.* **71,** 395–401.

Ishimoda-Takagi, T. (1979). Localization of tropomyosin in sea urchin eggs. *Exp. Cell Res.* **119**, 423–428.

Kawamura, K. (1977). Microdissection studies on the dividing neuroblast of the grasshopper, with special reference to the mechanism of unequal cytokinesis. *Exp. Cell Res.* **106**, 127–137.

Kessel, R. (1968). Annulate lamellae. *J. Ultrastruct. Res. Suppl.* **10**, 1–82.

Lewis, C. A. (1977). Ultrastructure of a fertilized barnacle egg (*Pollicipes polymerus*) with peristaltic contrictions. *Wilhem Roux' Arch. Entwicklungsmech. Org.* **181**, 333–355.

Lewis, C. A., Chia, F. S., and Schroeder, T. E. (1973). Peristaltic constrictions in fertilized barnacle eggs. *Experimentia* **29**, 1533.

Longo, F. J. (1972). An ultrastructural analysis of mitosis and cytokinesis in the zygote of the sea urchin, *Arbacia punctulata. J. Morphol.* **183**, 207–237.

Margolis, R. L., and Wilson, L. (1979). Regulation of the microtubule steady state *in vitro* by ATP. *Cell* **18**, 673–679.

Margolis, R. L., Wilson, L., and Kiefer, B. I. (1978). Mitotic mechanism based on intrinsic microtubule behavior. *Nature (London)* **272**, 450–452.

Maruta, H., and Goldstein, L. (1975). The fate and origin of the nuclear envelope during and after mitosis in *Amoeba proteus*. I. Synthesis and behavior of phospholipids of the nuclear envelope during the cell life cycle. *J. Cell Biol.* **65**, 631–645.

Masui, Y., Forer, A., and Zimmerman, A. M. (1978). Induction of cleavage in nucleated and enucleated frog eggs by injection of isolated sea-urchin mitotic apparatus. *J. Cell Sci.* **31**, 117–135.

Moore, A. R. (1933). Is cleavage rate a function of the cytoplasm or of the nucleus? *J. Exp. Biol.* **10**, 230–236.

Morgan, T. H. (1933). The formation of the antipolar lobe in *Ilyanassa. J. Exp. Zool.* **64**, 433–467.

Morgan, T. H. (1935a). Centrifuging the eggs of *Ilyanassa* in reverse. *Biol. Bull. (Woods Hole, Mass.)* **68**, 268–279.

Morgan, T. H. (1935b). The rhythmic changes in form of the isolated antipolar lobe of *Ilyanassa. Biol. Bull. (Woods Hole, Mass.)* **68**, 296–299.

Raff, E. C., Brothers, A. J., and Raff, R. A. (1976). Microtubule assembly mutant. *Nature (London)* **260**, 615–617.

Rappaport, R. (1960a). The origin and formation of blastoderm cells of gammarid crustacea. *J. Exp. Zool.* **144**, 43–59.

Rappaport, R., (1960b).. Cleavage of sand dollar eggs under constant tensile stress. *J. Exp. Zool.* **144**, 225–231.

Rappaport, R. (1961). Experiments concerning the cleavage stimulus in sand dollar eggs. *J. Exp. Zool.* **148**, 81–89.

Rappaport, R. (1964). Geometrical relations of the cleavage stimulus in constricted sand dollar eggs. *J. Exp. Zool.* **155**, 225–230.

Rappaport, R. (1965). Geometrical relations of the cleavage stimulus in invertebrate eggs. *J. Theoret. Biol.* **9**, 51–66.

Rappaport, R. (1966). Experiments concerning the cleavage furrow in invertebrate eggs. *J. Exp. Zool.* **161**, 1–8.

Rappaport, R. (1967). Cell division: Direct measurement of maximum tension exerted by furrow of echinoderm eggs. *Science* **156**, 1241–1243.

Rappaport, R. (1968). Geometrical relations of the cleavage stimulus in flattened, perforated sea urchin eggs. *Embryologia* **10**, 115–130.

Rappaport, R. (1969a). Aster-equatorial surface relations and furrow establishment. *J. Exp. Zool.* **171**, 59–68.

Rappaport, R. (1969b). Division of isolated furrows and furrow fragments in invertebrate eggs. *Exp. Cell Res.* **56**, 87–91.

Rappaport, R. (1970). An experimental analysis of the role of cytoplasmic fountain streaming in furrow establishment. *Dev. Growth Differ.* **12**, 31–40.

Rappaport, R. (1971a). Cytokinesis in animal cells. *Int. Rev. Cytol.* **31**, 169–213.

Rappaport, R. (1971b). Reversal of chemical cleavage inhibition in echinoderm eggs. *J. Exp. Zool.* **176**, 249–255.

Rappaport, R. (1973a). On the rate of movement of the cleavage stimulus in sand dollar eggs. *J. Exp. Zool.* **183**, 115–119.

Rappaport, R. (1973b). Cleavage furrow establishment—A preliminary to cylindrical shape change. *Am. Zool.* **13**, 941–948.

Rappaport, R. (1975a). Establishment and organization of the cleavage mechanism. *In* "Molecules and Cell Movement" (S. Inoué and R. E. Stephens, eds.), pp. 287–304. Raven, New York.

Rappaport, R. (1975b). The biophysics of cleavage and cleavage of geometrically altered cells. *In* "The Sea Urchin Embryo" (G. Czihak, ed.), pp. 308–332. Springer-Verlag, Berlin and New York.

Rappaport, R. (1976). Furrowing in altered cell surfaces. *J. Exp. Zool.* **195**, 271–277.

Rappaport, R. (1977). Tensiometric studies of cytokinesis in cleaving sand dollar eggs. *J. Exp. Zool.* **201**, 375–378.

Rappaport, R. (1978). Effects of continual mechanical agitation prior to cleavage in echinoderm eggs. *J. Exp. Zool.* **206**, 1–11.

Rappaport, R., and Conrad, G. W. (1963). An experimental analysis of unilateral cleavage in invertebrate eggs. *J. Exp. Zool.* **153**, 99–112.

Rappaport, R., and Ebstein, R. P. (1965). Duration of stimulus and latent periods preceding furrow formation in sand dollar eggs. *J. Exp. Zool.* **158**, 373–382.

Rappaport, R., and Rappaport, B. N. (1968). Cytokinesis in cultured newt cells. *J. Exp. Zool.* **168**, 187–196.

Rappaport, R., and Rappaport, B. N. (1974). Establishment of cleavage furrows by the mitotic spindle. *J. Exp. Zool.* **189**, 189–196.

Rappaport, R., and Rappaport, B. N. (1976). Prefurrow behavior of the equatorial surface in *Arbacia lixula* eggs. *Dev. Growth Differ.* **18**, 189–193.

Rappaport, R., and Ratner, J. H. (1967). Cleavage of sand dollar eggs with altered patterns of new surface formation. *J. Exp. Zool.* **165**, 89–100.

Rebhun, L. I., and Sander, G. (1971). Electron microscope studies of frozen-substituted marine eggs. III. Structure of the mitotic apparatus of the first meiotic division. *Am. J. Anat.* **130**, 35–54.

Rink, T. J., Tsien, R. Y., and Warner, A. E. (1980). Free clacium in *Xenopus* embryos measured with ion-selective microelectrodes. *Nature (London)* **283**, 658–660.

Sawai, T. (1972). Roles of cortical and subcortical components in cleavage furrow formation in amphibia. *J. Cell Sci.* **11**, 543–556.

Sawai, T. (1976). Induction of a furrow-like dent on the surface of the newt egg before the onset of cleavage. *Dev. Growth Differ.* **18**, 357–361.

Sawai, T. (1979). Cyclic changes in the cortical layer of non-nucleated fragments of the newt's egg. *J. Embryol. Exp. Morphol.* **51**, 183–193.

Schmidt, B. A., Kelly, P. T., May, M. C., Davis, S. E., and Conrad, G. W. (1980). Characterization of actin from fertilized eggs of *Ilyanassa obsoleta* during polar lobe formation and cytokinesis. *Dev. Biol.* **76**, 126–140.

Schroeder, T. E. (1968). Cytokinesis: Filaments in the cleavage furrow. *Exp. Cell Res.* **53**, 272–276.

Schroeder, T. E. (1970). The contractile ring. I. Fine structure of dividing mammalian (Hela) cells and the effects of cytochalasin B. Z. *Zellforsch. Mikrosk. Anat.* **109**, 431–449.

Schroeder, T. E. (1972). The contractile ring. II. Determining its brief existence, volumetric changes, and vital role in cleaving *Arbacia* eggs. *J. Cell Biol.* **53**, 419–434.

Schroeder, T. E. (1973). Actin in dividing cells: Contractile ring filaments bind heavy meromyosin. *Proc. Natl. Acad. Sci. U.S.A.* **70**, 1688–1692.

Schroeder, T. E. (1975). Dynamics of the contractile ring. In "Molecules and Cell Movement" (S. Inoué and R. E. Stephens, ed.), pp. 305–332. Raven, New York.

Shimizu, T. (1979). Surface contractile activity of the *Tubifex* egg: Its relationship to the meiotic apparatus functions. *J. Exp. Zool.* **208**, 361–377.

Silver, R. B., Cole, R. D., and Cande, W. Z. (1980). Isolation of mitotic apparatus containing vesicles with calcium sequestration activity. *Cell* **19**, 505–516.

Sluder, G. (1979). Role of spindle microtubules in the control of cell cycle timing. *J. Cell Biol.* **80**, 674–691.

Summers, K., and Kirschner, M. W. (1979). Characteristics of the polar assembly and disassembly of microtubules observed *in vitro* by darkfield light microscopy. *J. Cell Biol.* **83**, 205–217.

Szollosi, D. (1970). Cortical cytoplasmic filaments of cleaving eggs: A structural element corresponding to the contractile ring. *J. Cell Biol.* **44**, 192–209.

Tilney, L. G., and Marsland, D. (1969). A fine structural analysis of cleavage induction and furrowing in the eggs of *Arbacia punctulata*. *J. Cell Biol.* **42**, 170–184.

Tilney, L. G., Kiehart, D. P., Sardet, C., and Tilney, M. (1978). Polymerization of actin. IV. Role of Ca^{++} and H^+ in the assembly of actin and in membrane fusion in the acrosomal reaction of echinoderm sperm. *J. Cell Biol.* **77**, 536–550.

Verdonk, N. H., Geilenkirchen, W. L. M., and Timmermans, L. P. M. (1971). The localization of morphogenetic factors in uncleaved eggs of *Dentalium*. *J. Embryol. Exp. Morphol.* **25**, 57–63.

Wilson, E. B. (1901a). Experimental studies in cytology. I. A cytological study of artificial parthenogenesis in sea-urchin eggs. *Arch. Entwicklungsmech. Org.* **12**, 529–596.

Wilson, E. B. (1901b). Experimental studies in cytology. II. Some phenomena of fertilization and cell division in etherized eggs. *Arch. Entwicklungsmech. Org.* **13**, 353–373.

Wilson, E. B. (1904). Experimental studies on germinal localization. I. The germ-regions in the egg of *Dentalium*. *J. Exp. Zool.* **1**, 1–72.

Wilson, L., Bamburg, J. R., Mizel, S. B., Grisham, L. M., and Creswell, K. M. (1974). Interaction of drugs with microtubule proteins. *Fed. Proc.,Fed. Am. Soc. Exp.* **33**, 158–166.

Wolpert, L. (1960). The mechanics and mechanism of cleavage. *Int. Rev. Cytol.* **10**, 163–216.

Yatsu, N. (1908) Some experiments on cell-division in the egg of *Cerebratulus lacteus*. *Annot. Zool. Jpn. Pt. 4*, **6**, 267–276.

Yatsu, N. (1911). Observations and experiments on the sea urchin egg. II. Notes on early cleavage stages and experiments on cleavage. *Annot. Zool. Jpn. Pt. 5*, **7**, 333–346.

Yoneda, M., Ikeda, M., and Washitani, S. (1978). Periodic change in the tension at the surface of activated non-nucleate fragments of sea urchin eggs. *Dev. Growth Differ.* **20**, 329–336.

Ziegler, H. E. (1898). Experimentelle Studien über Zelltheilung. I. Die Zerschnurung der Seeigelier. II. Furchung ohne Chromosomen. *Arch. Entwicklungsmech. Org.* **6**, 249–293.

Mechanical Properties of Dividing Cells

Yukio Hiramoto

MITOSIS/CYTOKINESIS
Copyright © 1981 by Academic Press, Inc.
ISBN 0-12-781240-7

I. FURROWING AND GROWTH OF THE CELL SURFACE DURING CYTOKINESIS

There are two different modes of cytokinesis: one by furrowing or constriction and the other by forming a new cell surface between potential daughter cells. The first is characteristic of the cytokinesis in higher animals and the second is characteristic of the cytokinesis in higher plants. From the viewpoint of mechanics, the cytokinesis of the first type can be regarded as a deformation of the cell in which some parts of the cell contract or expand actively while other parts are stretched or compressed passively. Therefore, investigations of the mechanical properties of dividing cells are of primary importance in understanding the mechanism of cytokinesis. In this chapter, I describe the mechanics of three types of division: the equal symmetrical division of the sea urchin egg, the equal unilateral division of the amphibian zygote, and the unequal symmetrical division of the starfish oocyte at the time of polar body formation. I also discuss the structural changes that underlie these division processes.

Movements of the cell surface during cytokinesis have been extensively investigated in sea urchin eggs (for reviews, see Dan, 1948; Hiramoto, 1958; Wolpert, 1960). The result is schematically shown in Fig. 1. During cytokinesis, the equatorial (furrow) surface, a bandlike area occupying the equatorial (furrow) region of the cell, contracts in the direction of the equator, while the equatorial surface expands in the meridian direction during most stages of cytokinesis except for an early period during which the surface contracts. Rappaport and Rappaport (1976) reported that the equatorial surface contracts in the meridian direction before the appearance of the cleavage furrow. The polar surface facing the spindle pole expands uniformly in all directions during most stages of cytokinesis except for a late period, during which it contracts.

In amphibian eggs, the cleavage furrow first appears at the animal pole of the cell and progresses toward the vegetal pole. The behavior of the cell surface during cytokinesis was investigated by Selman and Waddington (1955) and Sawai (1976). During early stages of cytokinesis, the cell surface near the cleavage furrow is stretched in the direction perpendicular to the furrow and shrinks in the direction parallel to the furrow. During late stages of cytokinesis, an unpigmented surface appears on both walls of the furrow. It is believed that this unpigmented surface is formed *de novo*, rather than by stretching the preexisting surface.

In polar body formation by the starfish oocyte, the cell surface at the furrow region contracts during furrow formation and expands at later stages and after formation of the polar body (cf. Hamaguchi and Hiramoto, 1978).

Remarkable expansion of the cell surface near the cleavage furrow is ob-

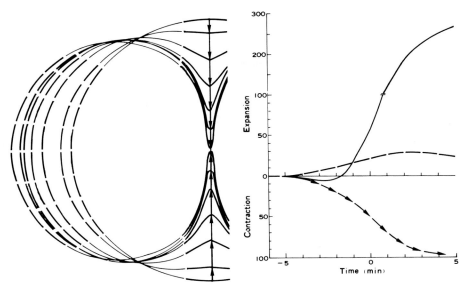

Fig. 1. Expansion and contraction of the cell surface during cytokinesis in the sea urchin zygote. Left: contours of a dividing zygote at successive stages. Right: linear change of the polar surface (dash line), the furrow surface in the meridian direction (solid line), and the furrow surface in the equatorial direction (arrow line) during cytokinesis. Drawn after Hiramoto's (1958) data and unpublished ones in zygotes of *Clypeaster japonicus*.

served in sea urchin eggs during late stages of cytokinesis and after cytokinesis has ended. The expansion of the furrow surface observed in sea urchin eggs and starfish oocytes, mentioned above, is believed to correspond to the formation of the unpigmented surface in amphibian zygotes. It is considered that both the furrowing and formation of the new cell surface are processes commonly occurring in all types of cells, and that furrowing is more conspicuous in animal cells than in plant cells (cf. Roberts, 1961; Hiramoto, 1968).

II. STRUCTURE OF THE DIVIDING CELL

A. The Endoplasm

When the cell is put in a centrifugal field of an appropriate magnitude, particles in the endoplasm (such as vacuoles, granules, and oil droplets) are moved centrifugally or centripetally in accordance with the difference in their densities from the density of the surrounding cytoplasmic matrix. Heilbrunn and collaborators used Stokes' formula to determine the viscosity of the endoplasm from the size, density, and speed of the particles and the magnitude of the centrifugal force (for a review, see Heilbrunn, 1958). They obtained several centipoises, i.e., several times the viscosity of water, for values of the protoplasmic viscosity in various cells including sea urchin eggs. Hiramoto (1967) obtained a much larger value (30 poises) for the protoplasmic viscosity from the speed of the nucleus in the endoplasm in sea urchin eggs measured with a centrifuge microscope. Visco-elasticity of the endoplasm was shown by the movement of a nickel or iron particle in the endoplasm by a magnetic field in chick fibroblasts (Crick and Hughes, 1950) and sea urchin zygotes (Hiramoto, 1969a,b) and by the movement of the endoplasm in starfish oocytes under a pressure gradient (Shôji et al., 1978). The above discrepancy of the results among previous investigations may be due to the existence of a network structure in the endoplasm, as suggested by Hiramoto (1967, 1970).

The mitotic apparatus is one of the most conspicuous structures in dividing cells. Fibrous structures such as spindle fibers and astral rays have been observed in living cells as well as fixed cells by light microscopy. Electron microscopy has revealed that a number of microtubules are arranged in the direction of the spindle axis and in radial directions from the center of the aster. Polarization microscopy has shown that the spindle and aster are optically anisotropic, indicating that they contain oriented structures at a submicroscopic level. Chambers (cf. Chambers and Chambers, 1961) showed by micromanipulation that the aster and the metaphase spindle are gelled bodies, while the interzonal region of the spindle has reverted to a sol state at telophase. This result was confirmed by quantitative determinations of visco-elasticity of the endoplasm by the magnetic particle method (Hiramoto, 1969b). The result mentioned above indicates that the mitotic apparatus is a relatively rigid structure. Therefore, this rigidity should be taken into consideration in measuring the mechanical properties of the cell, though in some cases the stresses developed by the deformation of endoplasmic structures are small compared with the stresses developed by the structure at the cell surface, as mentioned in Section V,A.

B. The Cortex

The existence of a gelated layer (cortex) underneath the cell membrane has been recognized in many cells (cf. Chambers and Chambers, 1961). It is presumably responsible for most of the mechanical properties of the cell surface (e.g., the tension and elasticity of the cell surface). The fact that particles near the cell surface are hardly displaced by centrifugal force which is sufficient to move particles in the endoplasm suggests the existence of a gelated layer (cortex) at the cell surface. The cortex can be detected by probing with a microneedle inserted into the cell, from distortion of the shape of microinjected oil drops, and from the movement of an iron particle in the endoplasm by a magnetic field (cf. Chambers and Chambers, 1961; Hiramoto, 1974).

In some cases, the cortex is morphologically distinguished from the endoplasm by the presence of granules, vacuoles, or filaments, or by its birefringence, while in other cases no morphological difference is found between the cortex and the endoplasm at the light microscopic as well as the electron microscopic level (e.g., Mercer and Wolpert, 1962). It is considered that in some cases, the high consistency of the cortex is due to special submicroscopic structure such as actin filaments, while in other cases the difference between the cortex and the endoplasm is merely due to the colloidal state of the cytoplasm.

C. The Cell Membrane

It is possible that the cell membrane contributes in part to the mechanical properties of the cell surface. The degree of this contribution cannot be estimated at present, because no reliable information is available on the mechanical properties of the cell membrane in its physiological condition. Since the cell surface is often folded when observed by electron microscopy, it is expected that the stretching of the cell surface causes the unfolding of the cell membrane without actual stretching. Therefore, it is questionable whether the cell membrane contributes much to the mechanical properties of the cell surface.

D. Extraneous Coats

Living cells are often surrounded by various membranes and layers, e.g., the fertilization membrane and the hyaline layer in fertilized sea urchin zygotes, the capsule and the vitelline membrane in amphibian eggs, and the vitelline membrane in starfish oocytes. Some of these coats are quite rigid compared with the cell membrane and the cortex, and therefore must be removed beforehand in order to determine the properties of the cell surface. Various methods have been reported for removing these coats without affecting the physiological functions of the cell, especially division activity. For example, the fertilization membrane and the hyaline layer in sea urchin zygotes can be removed by treating the zygotes with a nonelectrolyte solution, e.g., 1 M urea solution, and the capsule and vitelline membrane in amphibian zygotes can be removed mechanically with forceps.

III. THE SITE OF GENERATION OF THE MOTIVE FORCE OF CYTOKINESIS

Formation of the cleavage furrow begins during the early telophase of mitosis at the cell surface close to the equator of the mitotic spindle. In sea urchin zygotes, furrowing starts at the equatorial surface of the cell as determined by the position and direction of the spindle, and the cell is eventually divided into two equal daughter cells. In amphibian zygotes, the furrowing starts at the animal pole, progresses toward the vegetal surface, and finally divides the cell into two equal daughter cells. In starfish oocytes, the cleavage furrow for polar body formation appears as a ring-shaped region near the animal pole and eventually divides the cell into a small polar body and a large main egg cell body. It is believed that the site of the cleavage furrow is determined by the stimulus arising from the mitotic apparatus (see the review by Rappaport, 1971; see also Chapter 16 in this volume by Conrad and Rappaport).

It is believed that the motive force of cytokinesis is located in the cortex of the cell, because cytokinesis occurs in cells in which mitotic apparatus have been removed or destroyed before the onset of cytokinesis (cf. Rappaport, 1971, for a review). It is supposed that the differentiation of the cortex is completed because of a cleavage stimulus that arose from the mitotic apparatus before the onset of cytokinesis, after which the processes of cytokinesis can proceed without collaboration of the mitotic apparatus.

IV. MORPHOLOGICAL AND PHYSICAL CHANGES IN THE CELL CORTEX DURING CYTOKINESIS

No conspicuous morphological difference was found for many years at the light microscopic level between the cortex at the cleavage furrow and the cortex at other regions of the cell. By electron microscopy, Mercer and Wolpert (1958) discovered a layer consisting of fine filaments in the furrow cortex in dividing sea urchin eggs. Development of electron microscopic techniques during the following decade enabled us to observe a layer consisting of many filaments in the furrow cortex in various dividing cells (for a review, see Schroeder, 1975, 1978; and Chapter 18 in this volume by Karasiewicz). This layer, which is called "the contractile ring," consists of many microfilaments 5–9 nm in diameter arranged almost parallel to the cleavage furrow. The cross section of the contracticle ring is 0.1–0.2 μm thick and several micrometers (more than 10 μm in some materials) wide. In symmetrically dividing cells such as sea urchin zygotes, the contractile ring stretches over the entire length of the cleavage furrow, while it does not do so in unilaterally dividing cells such as amphibian eggs. Evidence that the microfilaments of the contractile ring are made of actin is demonstrated in some materials by the formation of arrowhead structures following the application of heavy meromyosin (cf. Schroeder, 1975) and by the fluorescent heavy meromyosin technique (e.g., Sanger, 1975). The localization of myosin at the cleavage furrow is shown by the fluorescent antibody technique (e.g., Fujiwara and Pollard, 1976). These facts strongly suggest that cytokinesis is caused by the activity of an actin–myosin system localized at the furrow cortex.

Differences in physical propoerties between the furrow cortex and the cortex at other regions was shown by Marsland (1939). He determined the "cortical gel strength," which was measured from the magnitude of centrifugal force required to dislodge granules from the cortex in sea urchin zygotes under a high hydrostatic pressure, and he found that the granules in the equatorial (furrow) cortex are more firmly held than those in the cortex at other regions in sea urchin zygotes prior to or during cleavage.

According to Zimmerman et al. (1957), three peaks of cortical gel strength are observed in sea urchin zygotes at the one-cell stage, and the last peak persists until the furrow of the first cleavage is completed. It is difficult to compare this result with the results of the rigidity of the cell and the surface force mentioned in Sec. V,A because the physical significance of the cortical gel strength measurement is by no means clear, though it may indicate some physical changes in the cortex before as well as during cytokinesis.

The birefringence of the cortex in sea urchin zygotes, which is positive with respect to the normal line to the cell surface and the birefringence of the cortex, increases almost uniformly over the cell surface before the onset of cytokinesis (e.g., Mitchison and Swann, 1953; Shôji et al., 1981). According to Mitchison and Swann (1953), the cortical birefringence decreases first at the polar region, and the wave of decrease propagates toward the equatorial region during cytokinesis. According to Shôji et al. (1981), the cortical birefringence decreases at the equatorial region during early stages of cytokinesis and increases during later stages, while the cortical birefringence at the polar region increases during early stages of cytokinesis and decreases during later stages. The birefringence change of the furrow cortex during cytokinesis resembles the length change of the furrow surface in the meridian direction mentioned above, whereas the birefringence change of the polar cortex is not parallel to the length change of the polar surface. Because both the cortical gel strength and the birefringence of the cortex are manifestations of the structure of the cortex, these results may indicate that the structure of the cortex changes uniformly over the cell surface before the onset of cytokinesis and that a difference arises in the cortical structure between the equatorial region and the other regions of the cell surface. This difference may be responsible for the active and/or passive contraction and expansion of the cortex and *de novo* formation of the cell surface during cytokinesis mentioned in the preceding section.

V. METHODS FOR DETERMINING MECHANICAL PROPERTIES OF THE CELL

A. Introduction

Because the cell consists of the endoplasm containing various cell organelles, the cortex, the cell membrane, and extraneous coats which have mechanical properties different from one another, the mechanical properties of the cell as a whole depend on the properties of all of these components. If it is assumed that the total thickness of the structure at the cell surface including the cortex, the cell membrane, and the extraneous coats is sufficiently small compared with the cell size, and that the stress developed by the deformation of the endoplasm is negligibly small compared with the total stress developed in the cell, the "membrane theory" (Timoshenko and Woinowsky-Krieger, 1959) for thin-walled shells is applicable in calculating stresses in the cell. In this case, the internal pressure of the cell balances the tangential force of the surface structure, and when no external force is applied to the cell

$$P_i = T_1/R_1 + T_2/R_2 \tag{1}$$

where P_i is the internal pressure, T_1 and T_2 are tangential forces per unit length of the cell surface perpendicular to each other, and R_1 and R_2 are radii of curvature of the cell surface in corresponding directions (cf. Fig. 2b). The tangential force per unit length of the surface, which is the sum of the forces per unit length of the cortex, the cell membrane, and the extraneous coats, is called "surface force" (Cole, 1932) or "tension at the cell surface" (Harvey, 1931).

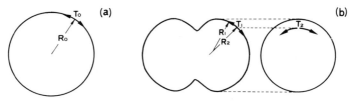

Fig. 2. Surface forces and the curvature of the cell surface. (a) Spherical cell; (b) symmetrically dividing cell. R_0, R_1, and R_2, radii of curvature of the cell surface. T_0, T_1, and T_2, surface forces.

The mechanical properties of the cell are determined, in principle, from the relation between the deformation of the cell and the force applied to the cell. Various methods have been reported to determine the mechanical properties of the cell, and some mechanical characteristics of the cell, such as the surface force and the intracellular pressure, have been determined from the membrane theory and the results of the measurements.

Like most problems in biological measurement, the measurement of mechanical properties of the cell presents complex problems that cannot be completely solved by simple physical concepts. Whether or not the membrane theory is applicable is not always strictly checked in individual measurements, and the reliability of the mechanical characteristics obtained depends on the applicability of the theory. In some measurements of sea urchin zygotes and starfish oocytes, stresses developed by the deformation of the endoplasm are generally small compared with the total stress developed by the deformation of the cell from the results of visco-elastic properties of the endoplasm (Hiramoto, 1969a,b; Shôji et al., 1978) unless the rate and the degree of deformation are very large (Hiramoto, 1970; Nakamura and Hiramoto, 1978). However, such a check is not made in other cells because no quantitative data are available for mechanical properties of the endoplasm.

B. Compression Method

In this method, the cell is compressed between two parallel plates, and the relation between the applied force and the deformation of the cell is determined (cf. Fig. 3a). For a fixed value of the applied force, the greater the rigidity of the cell, the thicker (D in Fig. 3a) the compressed cell. Therefore, the relative value of the rigidity of the cell can be represented by D, provided the shape and size of the cell and the applied force are fixed. Because the cell is generally visco-elastic, the rate and duration of the applied force should be constant in comparing rigidity values in this way.

The surface force and the intracellular pressure are determined assuming that the membrane theory is applicable for the compressed cell (cf. Fig. 3a). If the area of the cell surface in contact with the compressing plate is A, the internal pressure (P_i) is represented by

$$P_i = F/A \qquad (2)$$

where F is the applied force. Equation (1) is applicable for the part of the cell not in contact with the plate. When the cell shape is symmetrical with respect to a line that passes through the center of the cell and is perpendicular to the compressing plates, as shown in Fig. 3a, A, R_1 and R_2 are determined from the shape of the compressed cell. P_i and T_1 and T_2 at any point

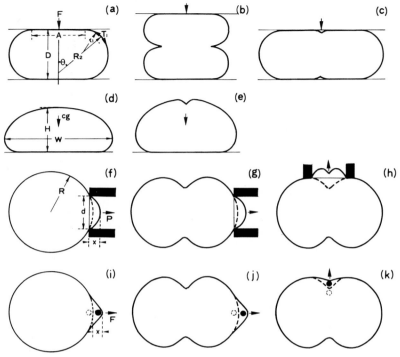

Fig. 3. Methods for determining mechanical properties of the cell surface. For symbols, see text. (a,b,c) Compression method; (d,e) sessile drop method; (f,g,h) suction method; (i,j,k) magnetic particle method.

on the cell surface are calculated using equations (1) and (2) and another equation derived from the balance of forces acting on the membrane (cf. Timoshenko and Woinowsky-Krieger, 1959, p. 435)

$$2\pi R_2 T_1 \sin^2\theta = \pi R_2^2 P_i \sin^2\theta - F \qquad (3)$$

where θ is the angle formed by the axis of symmetry and the normal to the cell surface at the point in question on the cell surface (cf. Fig. 3a). Cole (1932) and Hiramoto (1963a) calculated the surface force by methods similar in principle to the one above. Hiramoto (1968) used a modified compression method, in which the cell was compressed between a plate and a rod, in order to determine A with exactitude.

Yoneda (1964, 1972) presented a method to determine the surface force from the force–deformation relation and the change in surface area (S) of the cell by compression using the following equation, assuming that the surface force (T) is uniform over the entire surface of the cell.

$$-T\frac{dS}{dD} = F \tag{4}$$

where D is the thickness of the cell (Fig. 3a). It has been pointed out that a serious error is introduced in the calculation of the surface force by this method if the surface force is different in different directions (cf. Hiramoto, 1967, 1976b).

C. Sessile Drop Method

Large cells such as amphibian zygotes are flattened under the influence of gravity when extraneous coats are removed. Small cells, e.g., sea urchin zygotes, which are not appreciably deformed by gravity, can be flattened when a centrifugal force of an appropriate magnitude is applied (Hiramoto, 1967). The relative stiffness of the cell as a whole can be represented by the height (H) or the ratio (H/W) of the height to the width (W) of the cell under gravity or a centrifugal field of a definite magnitude (cf. Fig. 3d).

If it is assumed that the membrane theory is applicable and that the surface force is uniform over the entire surface, the surface force can be calculated using equations for a sessile drop (e.g., Dorsey, 1928; Hiramoto, 1967) from some parameters characterizing cell shape, density, and size. If the surface forces are not uniform as expected from the elasticity of the cell surface, the calculation becomes complicated.

D. Suction Method

The cell surface can be deformed when a negative pressure is partially applied with a micropipette in contact with the surface (Mitchison and Swann, 1954; cf. Fig. 3f). Because the height (x in Fig. 3f) of the bulge generally increases linearly with the increases in applied negative pressure, the slope of the curve is a parameter suitable for characterizing the rigidity (stiffness) of the cell surface (Mitchison and Swann, 1954). The negative pressure required to form a bulge of a definite height is also used as a measure of the rigidity of the cell surface (Hiramoto, 1979). This method is advantageous in comparing regional differences in the rigidity of various points on the cell surface, because the negative pressure can be applied locally on different parts of the surface of the same cell.

The surface force is calculated from the rigidity of the cell surface and the size of the micropipette. If it is assumed that the surface force (T) is uniform over the cell surface and that the membrane theory is applicable, T is approximately represented by

$$T = (x_o^2 + d^2/4)^2 s/(d^2 - 4x_o^2) \tag{5}$$

where x_0 is the height of the bulge when the negative pressure is not applied, d is the diameter of the pipette, and s is the rigidity of the cell surface, i.e., the slope of the negative pressure–deformation curve mentioned above (Hiramoto, 1970; Nakamura and Hiramoto, 1978).

Although Mitchison and Swann (1954) and Selman and Waddington (1955) assumed that the rigidity of the cell surface determined by the suction method is due to the flexural rigidity of the cortex, it should really be considered to be due to the surface force (the tangential force at the cell surface), as pointed out by Wolpert (1960) and Hiramoto (1963a).

E. Magnetic Particle Method

The cell surface can be deformed locally by applying a force with an electromagnet to an iron particle embedded in the cortex of the cell (Hiramoto, 1974). The rigidity of the cell surface is qualitatively represented by the reciprocal of the height (x in Fig. 3i) of the bulge formed by the application of a force of a definite magnitude. The surface force is calculated from the applied force and the deformation of the cell surface if it is assumed that the surface force is unchanged by the deformation of the cell surface and that the membrane theory is applied in this case (Hiramoto, 1974).

VI. CHANGES IN MECHANICAL PROPERTIES OF THE CELL BEFORE AND DURING CYTOKINESIS

A. Introduction

The cell is generally visco-elastic, and it takes a long time to obtain data on its visco-elastic properties covering a wide range of relaxation times. On the other hand, in dividing cells the measurements can be carried out within a period sufficiently short that the change in the properties of the cell accompanying the division process can be neglected. Therefore, in some experiments, measurements made by applying a force to the cell for a short period are repeated, and in other experiments, measurements are carried out by applying a force to the cell continuously throughout the division cycle. From the results of such experiments, it is possible to deduce the general tendency of the mechanical change of the cell during cytokinesis.

In measurements on the dividing cell, the direction of the applied force and the part of the cell where the force is applied are important. In the compression method, the force is usually applied either parallel (Fig. 3b) or perpendicular (Fig. 3c) to the spindle axis. Amphibian zygotes settle on a substratum, keeping their animal poles upward as a result of a density difference in the cell. Spindles in sea urchin zygotes in a centrifugal field rest at the centripetal region of the cell, keeping their axes perpendicular to the direction of the centrifugal force (Hiramoto, 1967). Therefore, the directions of gravity or centrifugal force in the above cases are always perpendicular to the spindle axis (cf. Fig. 3e). In the suction method and the magnetic particle method, local deformations can be made at either the polar surface (Figs. 3g,j) or the furrow surface (Figs. 3h,k).

B. Change in the Rigidity of the Cell before and during Cytokinesis

The rigidity of the cell as a whole increases before the onset of cytokinesis in all types of cytokinesis mentioned in this chapter irrespective of the method of measurement (cf. Fig. 4 and Table I). In some cases, the rigidity reaches a maximum at about the onset of cytokinesis, the moment when the cell begins to deform in preparation for formation of the cleavage furrow (cf. Fig. 4), and gradually decreases during cytokinesis (curves A and A' in Fig. 4). This type of change was observed in sea urchin zygotes by the compression method (Cole and Michaelis, 1932; Danielli, 1952; Hiramoto, 1963a,b; Wolpert, 1963, 1966), in amphibian eggs by the sessile drop method (Harvey and Fankhauser, 1933; Selman and Waddington, 1955; Sawai, 1979) and the suction method (Selman and Waddington, 1955; Sawai and Yoneda, 1974; Sawai, 1979), and in starfish oocytes by the compression method (Hiramoto, 1976a; Ikeda *et al.*, 1976) and the suction method (Nakamura and Hiramoto, 1978) (cf. Table I). Whether or not the moment of maximum rigidity coincides exactly with the onset of cytokinesis is uncertain in most cases, because the deformation of the cell during and after application of force for measurement is not clearly distinguished from the inherent change in cell shape at

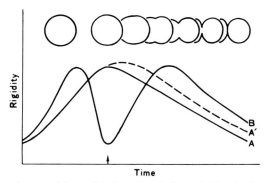

Fig. 4. Rigidity change of the cell before and during cytokinesis. Successive stages of cytokinesis are shown at the top of the figure. The arrow indicates the onset of cytokinesis.

TABLE I

Changes in the Rigidity of the Cell before and during Cytokinesis

Type of division	Method of measurement[a]	Type of rigidity change[b]	References
Cleavage of the	C	A[c]	Cole and Michaelis (1932)
Sea urchin zygote	C	A	Danielli (1952)
	C	A	Hiramoto (1963a,b)
	C	B	Hiramoto (1968)
	C	B	Yoneda and Dan (1972)
	SD	B	Hiramoto (1967)
	S	A	Mitchison and Swann (1955)
	S	A	Wolpert (1963, 1966)
	S	B	Hiramoto (1979)
	MP	B	Hiramoto (1974)
Cleavage of the	SD	A[c]	Harvey and Fankhauser (1933)
amphibian zygote	SD	A	Selman and Waddington (1955)
	SD	A	Sawai (1979)
	S	A	Selman and Waddington (1955)
	S	A	Sawai and Yoneda (1974)
Polar body	C	A	Hiramoto (1967a)
formation of the	C	A	Ikeda et al. (1976)
starfish oocyte	S	A	Nakamura and Hiramoto (1978)

[a] C, compression method; SD, sessile drop method; S, suction method; MP, magnetic particle method.
[b] A, type A or A' in Fig. 4; B, type B in Fig. 4. Type A and type A' are not clearly distinguished in most experiments.
[c] Only the rigidity increase before the onset of cleavage is described.

the onset of cytokinesis. In amphibian zygotes, the height of the zygotes deprived of the capsule and the vitelline membrane reaches its maximum shortly after the onset of furrow formation at the animal pole (curve A' in Fig. 4), while the rigidity of the cell surface measured by the suction method reaches its maximum almost at the onset of furrow formation at the animal pole (curve A) (Harvey and Fankhauser, 1933; Selman and Waddington, 1955; Sawai and Yoneda, 1974; Sawai, 1979).

Some experiments in sea urchin zygotes show a different type of rigidity change upon cytokinesis (Hiramoto, 1967, 1968, 1974, 1979; Yoneda and Dan, 1972; cf. curve B in Fig. 4 and Table I). The rigidity reaches a maximum shortly before the onset of cytokinesis, then decreases, reaches a minimum at the onset of cytokinesis, increases again, reaches the second maximum during cytokinesis, and then decreases during the late stage of cytokinesis. Yoneda and Dan (1972) reported that there are two maxima in the thickness of the sea urchin zygote compressed between two parallel plates with a constant force, and that the time of the second maximum coincides with the *onset of cleavage*, the time when a definitive furorw is recognized at the cell surface that is in contact with the plates compressing the egg. Because the thickness of the compressed zygote in their experiment was about 60% of the original zygote diameter, it is obvious that the rigidity reached the second maximum *during* cytokinesis. Judging from the time schedule of cytokinesis shown in their paper, it seems likely that the first maximum thickness is reached shortly before the onset of cytokinesis as defined in this chapter. The above conclusion has been confirmed by Dr. M. Yoneda (personal communication). In some zygotes, a small rigidity peak is observed between the two peaks in curve B in Fig. 4 (Hiramoto, 1979).

C. Regional Difference in Rigidity in the Dividing Cell

It might be expected that variations in rigidity would occur between different regions of the cell surface. However, Mitchison and Swann (1955) could not find such variations at any stage of cleavage in sea urchin zygotes. Hiramoto (1974) found that a characteristic change in the rigidity of the cell surface shown by curve B in Fig. 4 occurs almost simultaneously at both the polar surface and the furrow surface.

Selman and Waddington (1955) found that rigidity of the cell surface is almost uniform in dividing amphibian zygotes. Measurements during the same stage of cytokinesis give similar values regardless of whether they are made on pigmented cortex near the furrow, on unpigmented new cortex, on the surface opposite the spindle, or on animal or vegetal surfaces, although a smaller value of the rigidity is measured on the surface of the forming furrow between the daughter cells.

According to Sawai and Yoneda's (1974) experiment using two micropipettes for measurement, the rigidity of the cell surface in amphibian eggs changes as shown by curve A in Fig. 4, and the rigidity change travels meridionally along the cell surface, with the progress of the furrow toward the vegetal region. Their figures 9 and 10 indicate that the rigidity changes occur almost simultaneously at surfaces on the same latitude, irrespective of whether the surface is close to the furrow or away from it. The maximum rigidity at the onset of furrowing is larger at the vegetal surface than at the animal surface. The rigidity of the unpigmented surface appearing on both sides of the deepening furrow is smaller than that of the other surface.

The similarity of the rigidity of the furrow surface to that of the polar surface does not imply that there is no structural difference between them. It may be interpreted that the structural difference is difficult to be found as a difference of the rigidity in this case, because the rigidity is proportional to the surface force [cf. equation (5)] and the surface force balances the intracellular pressure [cf. equation (1)], which is almost uniform within the cell as a result of stretching of weak region and contraction of strong region. As described in Sections VII,A,B the surface force and the rigidity of the cell surface at the trough of the furrow containing the contractile ring are different from those at other regions.

D. The Surface Force and Intracellular Pressure of the Dividing Cell

Before the onset of cytokinesis, both the intracellular pressure and the surface force, which are calculated from measurements of mechanical properties of the cell (assuming that the membrane theory is applicable), change in parallel with the rigidity change of the cell; this is because both shape and size are practically unchanged at this stage. It is expected, from the membrane theory, that the surface force is uniform over the entire surface in spherical cells such as sea urchin zygotes and starfish oocytes before the onset of cytokinesis and that the intracellular pressure (P_i) is $2T_0/R_0$ from equation (1), where T_0 is the surface force and R_0 is the radius of the cell (cf. Fig. 2a).

The surface forces are different at different regions in cells during cytokinesis. In cells cleaving symmetrically with respect to the spindle axis, such as sea urchin zygotes and starfish oocytes (cf. Fig. 2b), the surface forces at any points on the cell are calculated from the intracellular pressure (P_i) and the principal radii of curvature $(R_1$ and R_2 in Fig. 2b), using the following equations (cf. Hiramoto, 1968):

$$T_1 = P_i R_2 /2 \tag{6}$$

and

$$T_2 = P_i R_2 (2 - R_2/R_1)/2. \tag{7}$$

The intracellular pressure of the dividing sea urchin zygote was determined from measurements at the polar surface by the compression method (Hiramoto, 1968) or the suction method (Hiramoto, 1979), assuming that the membrane theory was applicable. As a result, it was found that on cytokinesis the intracellular pressure changes almost in parallel with the rigidity change (in this case, type B in Fig. 4). The surface forces calculated from data on the intracellular pressure and curvatures of the cell surface using Eqs. (6) and (7) change on cytokinesis as shown by curve B in Fig. 4 (Hiramoto, 1968, 1974, 1979) because the intracellular pressure is one of the principal factors determining the surface force [cf. Eqs. (6) and (7)]. Similar calculations of the surface forces and the intracellular pressure may be possible in amphibian cells and starfish oocytes from the data on the rigidity of the cell surface.

In most calculations of surface force and intracellular pressure mentioned above, it is assumed that the surface force does not appreciably change with the deformation of the cell. If the surface force increases by stretching due to the elasticity of the structures at the cell surface (the cortex, cell membrane, and extraneous coats), the actual surface force is smaller than the surface force calculated assuming no elasticity, as discussed by Mitchison and Swann (1954), because the rigidity of the cell surface depends not only on the value of the surface force but also on the value of the elasticity of the surface structure. Although the elasticity of the cell surface has been demonstrated in unfertilized sea urchin eggs (Cole, 1932; Hiramoto, 1963a, 1967, 1976b) and in starfish oocytes (Nakamura and Hiramoto, 1978), detailed investigations have never been made in dividing cells.

VII. MECHANICAL PROPERTIES OF THE CLEAVAGE FURROW

A. The Force Exerted by the Cleavage Furrow

The cell surface at the cleavage furrow responds to external stress differently from the cell surface at other regions. Wolpert (1966) observed that when the furrow surface of the dividing sea urchin zygote was sucked with a micropipette, it was drawn into the pipette keeping the characteristic shape of the cleavage furrow (cf. Fig. 3h). Yoneda and Dan (1972) reported that the diameter of the cell at the cleavage furrow was scarcely changed when the force was suddenly removed from a dividing sea urchin zygote compressed between two parallel plates, suggesting that the rigidity of the furrow surface is greater than the rigidity of the surface at other regions.

The force exerted by the cleavage furrow in sea urchin zygotes can be measured directly from the bending of a glass needle inserted into the cleavage plane, which (together with another needle inserted into the cleavage plane) arrests the advance of the cleavage furrow (cf. Fig. 5a). The stalk connecting potential daughter cells finally attains the shape of a flattened tube whose dimensions are determined by the diameters of the needles and the distance between them (cf. Fig. 5a). A more or less constant force is developed when the distance between the needles is kept unchanged. This "isometric" force is 1.4–4.2×10^{-3} dynes, corresponding to 0.78–1.3×10^5 dynes cm^2 of cross section of the contractile ring (Rappaport, 1977; Hiramoto, 1979).

The force exerted by the cleavage furrow is also determined from the

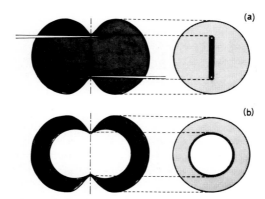

Fig. 5. Methods for determining the force exerted by the cleavage furrow in dividing cells. (a) The advance of the cleavage furrow is arrested with a pair of microneedles inserted through the polar surfaces into the cell. The cross section of the cell at the equatorial plane is shown on the right. The force exerted by the furrow is determined from the bending of one of the microneedles, which is calibrated before experimentation. (b) The advance of the furrow is arrested with an oil drop microinjected at the center of the cell. The force exerted by the furrow is determined from the degree of deformation of the oil drop and the interfacial tension between the oil and the protoplasm.

deformation of an oil drop microinjected into the cell by the advancing furrow in sea urchin zygotes (cf. Fig. 5b; Hiramoto, 1975). Values of the force similar to those determined with glass needles (mentioned above) were obtained. In this and the above cases, the measured force is the force acting on the microneedle or the oil drop arresting the advance of the furrow, which is the force developed by the contractile ring minus the force required to deform the cell to a dumbbell shape against the intracellular pressure and the surface force of other regions of the cell. The force generated by the furrow surface (including the contractile ring) in the normally dividing cell is calculated from the surface force in the direction of the equator and the width of the furrow surface. Assuming the width of the furrow surface to be 10 μm and using data on the surface force in sea urchin zygotes (Hiramoto, 1968), the force is calculated to be 3×10^{-3} dyne, which coincides fairly well with values obtained by the direct measurements mentioned above. It is considered that this force is generated by the contractile ring formed in the cortex at the trough of the furrow.

B. Visco-elastic Properties of the Cleavage Furrow

Hiramoto (1979) measured the force–length relation of the furrow surface by changing the distance between two microneedles thrust into the cleavage plane of a dividing sea urchin zygotes after the isometric force had developed. The force–length relation displays a marked hysteresis curve. The slope value of the curve, which largely depends on the mode of stretch and release, is approximately 10^{-2} dyne/100% stretch, corresponding to 10^6 dyne/cm^2 of Young's modulus of the contractile ring.

VIII. RIGIDITY CHANGE OF NON-NUCLEATED FRAGMENTS

Rigidity changes were investigated by Yoneda *et al.* (1978) in activated non-nucleated fragments of sea urchin zygotes and by Sawai (1979) and Hara *et al.* (1980) in non-nucleated fragments isolated from fertilized amphibian eggs. In both cases, the rigidity changes characteristic of the division cycle were observed (type B in Fig. 4 in sea urchin zygote fragments, and type A in amphibian egg fragments). This fact indicates that these rigidity changes (and probably the changes in surface force and intracellular pressure) are caused by periodic changes in the cell surface characteristic of a cytoplasmic rhythm which is triggered by the activation of the egg. It is considered that the structural change occurs over the entire surface of the cell in this case, judging from the fact that the shape of the fragment does not markedly change during the periodic change.

IX. STRUCTURAL CHANGES UNDERLYING CYTOKINESIS

The results from experiments with non-nucleated fragments and the properties of the furrow surface mentioned in the above sections suggest that two kinds of structural changes occur on cytokinesis. One is the change due to the cytoplasmic rhythm and occurs before and during cytokinesis over the entire surface of the cell. This is characterized by the rigidity change shown by the A, A', or B curve in Fig. 4. The other is the change due to the activity of the contractile ring which is formed at the time of cytokinesis. The contractile ring is formed at a particular region of the cortex under the stimulus arising from the mitotic apparatus, and is, of course, not formed in non-nucleated fragements. Because this change is localized at a particular region of the cortex, the balance of the surface forces at different parts of the cell induces the formation of the cleavage furrow.

Although it has been widely accepted that the activity of the contractile ring is caused by an actin–myosin system, the biochemical basis of the former structural change is not yet established. It seems likely that the contractile protein found by Sakai (for a review, see Sakai, 1968) is responsible for this structural change, because (1) the rigidity change on cytokinesis is similar to the changes in the contractility and the SH content of this protein on cytokinesis and (2) the change in SH content occurs in cells in which cytokinesis is inhibited by suppressing the formation of the miotic apparatus by ultraviolet irradiation and some chemical reagents (Ikeda, 1965; Dan and Ikeda, 1971). In order to confirm this idea, the causal relationship between the mechanical change of the cell and the change of this protein should be investigated in detail.

ACKNOWLEDGMENT

I thank Professor M. Yoneda, Kyoto University, for his invaluable criticisms and reading the manuscript.

REFERENCES

Chambers, R., and Chambers, E. L. (1961). "Explorations into the Nature of the Living Cell. Harvard University Press, Cambridge, Massachusetts.

Cole, K. S. (1932). Surface forces of the *Arbacia* egg. *J. Cell. Comp. Physiol.* 1, 1–9.

Cole, K. S., and Michaelis, E. M. (1932). Surface forces of fertilized *Arbacia* eggs. *J. Cell. Comp. Physiol.* 2, 121–126.

Crick, F. H. C., and Hughes, A. F. W. (1950). The physical properties of cytoplasm. A study by means of the magnetic particle method. *Exp. Cell Res.* 1, 37–80.

Dan, J. C. (1948). On the mechanism of astral cleavage. *Physiol. Zool.* 21, 191–218.

Dan, K., and Ikeda, M. (1971). On the system controlling the time of micromere formation in sea urchin embryos. *Dev. Growth Differ.* 13, 285–301.

Danielli, J. F. (1952). Division of the flattened egg. *Nature (London)* 191, 496.

Dorsey, N. E. (1928). A new equation for the determination of surface tension from the form of a sessile drop or bubble. *J. Wash. Acad. Sci.* 18, 505–509.

Fujiwara, K., and Pollard, T. D. (1976). Fluorescent antibody localization of myosin in the cytoplasm, cleavage furrow, and mitotic spindle of human cells. *J. Cell Biol.* 71, 848–875.

Hamaguchi, M. S., and Hiramoto, Y. (1978). Protoplasmic movement during polar-body formation in starfish oocytes. *Exp. Cell Res.* 112, 55–62.

Hara, K., Tydeman, P., and Kirschner, M. (1980). A cytoplasmic clock with the same period as the division cycle in *Xenopus* eggs. *Proc. Natl. Acad. Sci. U.S.A.* 77, 462–466.

Harvey, E. N. (1931). The tension at the surface of marine eggs, especially those of the sea urchin *Arbacia. Biol. Bull (Woods Hole, Mass.)* 61, 273–279.

Harvey, E. N., and Fankhauser, G. (1933). The tension at the surface of the eggs of the salamander, *Tritrus (Diemyctylus) viridescens. J. Cell. Comp. Physiol.* 3, 463–475.

Heilbrunn, L. V. (1958). The viscosity of protoplasm. *Protplasmatologia* 2, C1.

Hiramoto, Y. (1958). A quantitative description of protoplasmic movement during cleavage in the sea-urchin egg. *J. Exp. Biol.* 35, 407–424.

Hiramoto, Y. (1963a). Mechanical properties of sea urchin eggs. I. Surface force and elastic modulus of the cell membrane. *Exp. Cell Res.* 32, 59–75.

Hiramoto, Y. (1963b). Mechanical properties of sea urchin eggs. II. Changes in mechanical properties from fertilization to cleavage. *Exp. Cell Res.* 32, 76–89.

Hiramoto, Y. (1967). Observations and measurements of sea urchin eggs with a centrifuge microscope. *J. Cell. Physiol.* **69**, 219–230.

Hiramoto, Y. (1968). The mechanics and mechanism of cleavage in sea-urchin eggs. "Aspects of Cell Motility" (22nd Symp. *Soc. Exp. Biol.*), pp. 311–327. Cambridge Univ. Press, London and New York.

Hiramoto, Y. (1969a). Mechanical properties of the protoplasm of the sea urchin egg. I. Unfertilized egg. *Exp. Cell Res.* **56**, 201–208.

Hiramoto, Y. (1969b). Mechanical properties of the protoplasm of the sea urchin egg. II. Fertilized egg. *Exp. Cell Res.* **56**, 209–218.

Hiramoto, Y. (1970). Rheological properties of sea urchin eggs. *Biorheology* **6**, 201–234.

Hiramoto, Y. (1974). Mechanical properties of the surface of the sea urchin egg at fertilization and during cleavage. *Exp. Cell Res.* **89**, 320–326.

Hiramoto, Y. (1975). Force exerted by the cleavage furrow of sea urchin eggs. *Dev. Growth Differ.* **17**, 27–38.

Hiramoto, Y. (1976a). Mechanical properties of starfish oocytes. *Dev. Growth Differ.* **18**, 205–209.

Hiramoto, Y. (1976b). Mechanical properties of sea urchin eggs. III. Visco-elasticity of the cell surface. *Dev. Growth Differ.* **18**, 379–388.

Hiramoto, Y. (1979). Mechanical properties of the dividing sea urchin egg. "Cell Motility: Molecules and Organization" (S. Hatano, H. Ishikawa, and H. Sato, eds.), pp. 653–663. Univ. of Tokyo Press, Tokyo.

Ikeda, M. (1965). Behaviour of sulfhydryl groups of sea urchin eggs under the blockage of cell division by UV and heat shock. *Exp. Cell Res.* **40**, 282–291.

Ikeda, M., Nemoto, S., and Yoneda. M. (1976). Periodic changes in the content of protein-bound sulfhydryl groups and tension at the surface of starfish oocytes in correlation with the meiotic division cycle. *Dev. Growth Differ.* **18**, 221–225.

Marsland, D. A. (1939). The mechanism of cell division. Hydrostatic pressure effects upon dividing egg cells. *J. Cell. Comp. Physiol.* **13**, 15–22.

Mercer, E. H., and Wolpert, L. (1958). Electron microscopy of cleaving sea urchin eggs. *Exp. Cell Res.* **14**, 629–632.

Mercer, E. H., and Wolpert, L. (1962). An electron microscope study of the cortex of the sea urchin (*Psammechinus miliaris*) egg. Exp. Cell Res. **27**, 1–13.

Mitchison, J. M., and Swann, M. M. (1953). Optical changes in the membranes of the sea-urchin egg at fertilization, mitosis and cleavage. *J. Exp. Biol.* **29**, 357–362.

Mitchison, J. M., and Swann, M. M. (1954). The mechanical properties of the cell surface. I. The cell elastimeter. *J. Exp. Biol.* **31**, 443–460.

Mitchison, J. M., and Swann, M. M. (1955). The mechanical properties of the cell surface. III. The sea-urchin egg from fertilization to cleavage. *J. Exp. Biol.* **32**, 734–750.

Nakamura, S., and Hiramoto, Y. (1978). Mechanical properties of the cell surface in starfish eggs. *Dev. Growth Differ.* **20**, 317–327.

Rappaport, R. (1967). Cell division: Direct measurement of maximum tension exerted by furrow of echinoderm eggs. *Science* **156**, 1241–1243.

Rappaport, R. (1971). Ctyokinesis in animal cells. *Int. Rev. Cytol.* **31**, 169–213.

Rappaport, R. (1977). Tensiometric studies of cytokinesis in cleaving sand-dollar eggs. *J. Exp. Zool.* **201**, 375–378.

Rappaport, R., and Rappaport, B. N. (1976). Prefurrow behavior of the equatorial surface in *Arbacia lixula* eggs. *Dev. Growth Differ.* **18**, 189–193.

Roberts, H. S. (1961). Mechanisms of cytokinesis: A critical review. *Q. Rev. Biol.* **36**, 155–177.

Sakai, H. (1968). Contractile properties of protein threads from sea urchin eggs in relation to cell division. *Int. Rev. Cytol.* **23**, 89–112.

Sanger, J. W. (1975). Changing patterns of actin localization during cell division. *Proc. Natl. Acad. Sci. U.S.A.* **72**, 1913–1916.

Sawai, T. (1976). Movement of the cell surface and change in surface area during cleavage in the newt's egg. *J. Cell Sci.* **21**, 537–551.

Sawai, T. (1979). Cyclic changes in the cortical layer of non-nucleated fragments of newt's egg. *J. Embryol. Exp. Morphol.* **51**, 183–193.

Sawai, T., and Yoneda, M. (1974). Wave of stiffness propagating along the surface of the newt egg during cleavage. *J. Cell Biol.* **60**, 1–7.

Schroeder, T. E. (1975). Dynamics of contractile ring. *In* "Molecules and Cell Movement" (S. Inoué and R. E. Stephens, eds.), pp. 305–334. Raven, New York.

Schroeder, T. E. (1978). Cytochalasin B, cytokinesis, and the contractile ring. *In* "Cytochalasins Biochemical and Cell Biological Aspects" (S. W. Tanenbaum, ed.), pp. 91–112. North-Holland Publ., Amsterdam.

Selman, G. G., and Waddington, C. H. (1955). The mechanism of cell division in the cleavage of the newt's egg. *J. Exp. Biol.* **32**, 700–733.

Shôji, Y., Hamaguchi, M. S., and Hiramoto, Y. (1978). Mechanical properties of the endoplasm in starfish oocytes. *Exp. Cell Res.* **117**, 79–87.

Shôji, Y., Hamaguchi, Y., and Hiramoto. Y. (1981). Polarization optical properties of living cells. III. Cortical birefringence of the dividing sea urchin egg. *Cell Motility* **1**, 387–397.

Timoshenko, S. P., and Woinowsky-Krieger, S. (1959). "Theory of Plates and Shells," 2nd Ed. McGraw-Hill, New York.

Wolpert, L. (1960). The mechanics and mechanism of cleavage. *Int. Rev. Cytol.* **10**, 163–216.

Wolpert, L. (1963). Some problems of cleavage in relation to the cell membrane. *Int. Soc. Cell Biol.* **2**, 277–298.

Wolpert, L. (1966). The mechanical properties of the membrane of the sea urchin egg during cleavage. *Exp. Cell Res.* **41**, 385–396.

Yoneda, M. (1964). Tension at the surface of sea-urchin egg: A critical examination of Cole's experiment. *J. Exp. Biol.* **41**, 893–906.

Yoneda, M. (1972). Tension at the surface of sea-urchin eggs on the basis of 'liquid-drop' concept. *Adv. Biophys.* **4**, 153–190.

Yoneda, M., and Dan, K. (1972). Tension at the surface of dividing sea-urchin egg. *J. Exp. Biol.* **57**, 575–587.

Yoneda, M., Ikeda, M., and Washitani, S. (1978). Periodic change in the tension at the surface of activated non-nucleate fragments of sea-urchin eggs. *Dev. Growth Differ.* **20**, 329–336.

Zimmerman, A. M., Landau, J. V., and Marsland, D. (1957). Cell division: A pressure-temperature analysis of the effects of sulfhydryl reagents on the cortical plasmagel structure and furrowing strength of dividing eggs (*Arbacia* and *Chaetopterus*). *J. Cell. Comp. Physiol.* **49**, 395–435.

18

Electron Microscopic Studies of Cytokinesis in Metazoan Cells

Jolanta Karasiewicz*

I. OBSERVATION, OR: WHICH ORGANELLES MIGHT BE ESSENTIAL FOR CYTOKINESIS?

Light microscopic observations, measurements, and experiments (reviewed by Wolpert, 1960; Rappaport, 1971, 1975), completed mainly before the epoch of the electron microscope, culminated in the following conclusion: Of the various hypotheses, the equatorial contraction hypothesis (Marsland and Landau, 1954) appears best to explain cytokinesis. This hypothesis requires that a contractile organelle be present at the equatorial surface of a dividing cell.

*Previous papers of the author have been published under the name "Opas."

MITOSIS/CYTOKINESIS

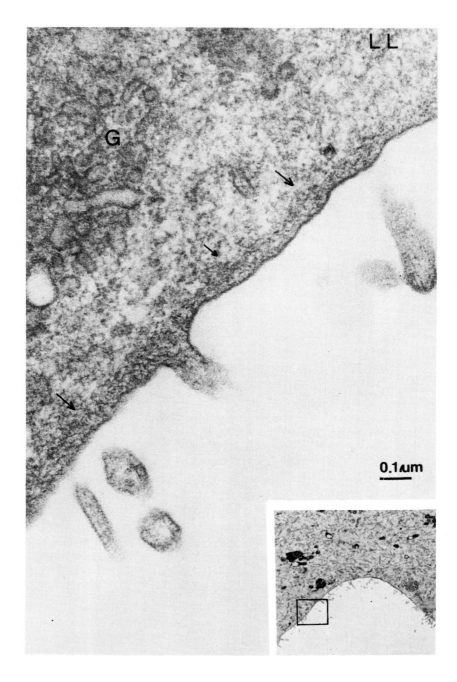

0.1μm

The electron microscope (EM) provides the opportunity of searching for organelles and establishing their positions within cells at a resolution below that of the light microscope. The EM is thus an ideal tool to test the equatorial contraction hypothesis, or, in other words, to look for the presumed contractile organelle ("looking for" approach).

A. "Looking for" Approach

Indeed, except for sporadic earlier papers (cited by Arnold, 1976), it appears that the interest of electron microscopists in cytokinesis centers on finding the contractile organelle. This proved to be a fruitful interest. At present, we have ultrastructural evidence that the filamentous ring or belt (contractile ring) does exist in the cleavage furrow of the cells of such diverse phyla as *Cnidaria* (Schroeder, 1968; Szollosi, 1970), *Annelida* (Szollosi, 1970), *Echinodermata* (Schroeder, 1969, 1972), *Mollusca* (Arnold, 1969; Longo, 1972; Conrad and Williams, 1974; Burgess, 1977), *Arthropoda* (Forer and Behnke, 1972), and, among vertebrates in amphibians (Bluemink, 1970), birds (Gipson, 1974; Bellairs *et al.*, 1978; Hurle and Lafarga, 1978), and mammals (Schroeder, 1970; Szollosi, 1970; Gulyas, 1973; Ducibella *et al.*, 1977; Opas and Soltyńska, 1978; Zeligs and Wollman, 1979). It is worth pointing out that, in addition to differences in their systematic position, cells in which the contractile ring has been seen also differ in many cytophysiological respects.

For example, the contractile ring is present in the furrow of cells that are in contact with glass (e.g., HeLa cells) or with other cells (e.g., thyroid follicular epithelial cells), as well as in single, isolated cells (e.g., fertilized eggs). The contractile ring is present in cells living at changing temperatures (marine invertebrate eggs, amphibian eggs) as well as in cells living at constant, rather high temperatures around 37°C (e.g., mammalian eggs or HeLa cells). Finally, the contractile ring is present with various types of furrowing, such as in equilateral furrows during symmetrical division [e.g., in sea urchin eggs or mammalian blastomeres (Fig. 1) and in tissue cells]; in equilateral furrows during unequal division (e.g., during extrusion of polar bodies); in unilateral furrows (e.g., those in coelenterates or squid eggs); and in the furrows at the edges of open cells (e.g., in cleavage in bird embryos).

Recognition of the contractile ring as the predicted (and presumed) organelle of equatorial contraction has been possible not only because of the resolving power of the EM but also because of accumulation of knowledge

Fig. 1. Contractile ring in a mouse blastomere from the two-cell stage. The inset shows the location of the enlarged area within the furrow. Arrows, contractile ring; G, Golgi vesicles; LL, lattice-like structures. Pictures obtained in collaboration with Dr. M. S. Soltyńska.

about the contractile properties of actin filaments. This knowledge stemmed from muscle research, which has also supplied a valuable method of probing ultrastructurally to determine whether or not filaments contain actin. The method is reaction with heavy meromyosin (in application to non-muscle cells: Ishikawa *et al.*, 1969). This method has been applied to cells during furrowing, and the actin character of the contractile ring filaments has been demonstrated at the EM level in various cells (amphibian eggs: Perry *et al.*, 1971; cranefly spermatocytes: Forer and Behnke, 1972; HeLa cells: Schroeder, 1973; PtK$_2$ cells: Sanger and Sanger, 1980). This has confirmed that contractile rings may be contractile organelles and as such may account for equatorial contraction.

All these data taken together strongly suggest that the contractile organelle—the contractile ring—is ubiquitous in the furrow of metazoan cells. This, in turn, appears to have led authors to refer to the contractile ring as "an organelle responsible for cleavage" (Schroeder, 1975, p. 306), which indicates that it might explain *all* stages of cytokinesis. It should be noted, however, that although the contractile ring is ubiquitous in the furrow, i.e., during the constriction stage of cytokinesis, the contractile

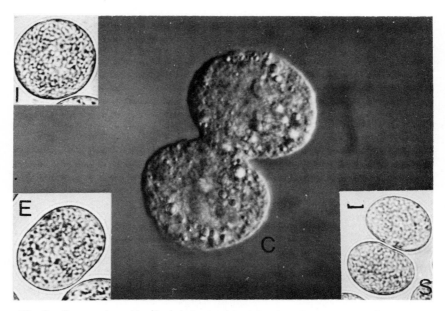

Fig. 2. Stages of cytokinesis. I, induction; E, elongation; C, constriction; S, separation. This picture also shows "the proportion of interest" the constriction stage has obtained in EM studies. Mouse blastomeres; I, E, S, living cells; C, cell extracted with glycerol (see Section III, A, 2); differential interference contrast. Scale = 10 μm.

ring is absent from the equatorial region of cells during elongation or separation (Schroeder, 1970, 1972; Forer and Behnke, 1972; Opas and Soltyńska, 1978). In their study of the induction stage of cytokinesis, Asnes and Schroeder (1979) do not mention any equatorial specializations of the submembrane region either. This means that all the aforementioned studies lend morphological support to the equatorial contraction hypothesis as applied to the constriction stage of cytokinesis, but they do not support a role for the contractile ring in the remaining stages, which are induction, elongation, and separation (Fig. 2). This, in turn, creates two possibilities: (1) either the equatorial contraction hypothesis applies only to the constriction stage and other stages need other mechanisms, or (2) the general equatorial contraction mechanism has a different morphological basis (organelles?) at each stage of cytokinesis.

B. "Just Watching" Approach

I have referred to observations that are restricted to the contractile ring as the "looking for" approach; other studies take another approach, the "just watching" approach, in which the authors describe organelles found during cytokinesis whether or not they are predicted by the theory; that is to say, organelles are studied even though they cannot be considered as candidates for accomplishing equatorial contraction, being either not present in the equator or containing no microfilaments.

Judging from the number of papers published, this approach—what is found when there is no contractile ring? and what else is found when the contractile ring is present?—has been much less common. These data appear somewhat scattered throughout the literature on cytokinesis, which is apparently due to their unknown status in relation to the theory. Two reports analyze in detail the cytoplasmic bridge with the midbody in tissue cells at the separation stage of cytokinesis (Byers and Abramson, 1968; Mullins and Biesele, 1977). A few reports show what appear to be special structures related to a sophisticated way of cleaving, i.e., the furrow base body and the "cytoplasmic root" in open cells during chick embryo cleavage (Gipson, 1974; Bellairs et al., 1978). Finally, there is a very interesting paper describing a regularly ordered cortical layer of actin filaments throughout the cell membrane in both the elongation and constriction stages of cytokinesis in cranefly spermatocytes (Forer and Behnke, 1972). It should be stressed that in this latter paper, visualization of the cortical layer has been possible due to pretreatment of cells with glycerol solution before fixation. Similar observations on the cortical layer have been reported in mouse blastomeres (Opas and Soltyńska, 1978).

Scarce as they are, the aforementioned reports pose the following questions:

1. Is the cytoplasmic bridge with the midbody an organelle of separation? How could it accomplish separation? Earlier data have not reported actin microfilaments within the bridge. Schroeder (1975) shows actin filaments in contact with the midbody, which supports the idea of contraction.

2. Is the cortical layer an organelle of elongation? How could it accomplish elongation?

3. Are organelles that are found during contraction (i.e., in the presence of the contractile ring) also essential for furrowing? This question applies mainly to the cortical layer, which appears to be associated with the contractile ring during furrowing, which contains actin filaments, and which, as suggested by Opas and Sołtyńska (1978), appears to give rise to the contractile ring by local reorganization.

These questions—developed from the "just watching" approach—might be generalized as follows: (1) Do the different stages of cytokinesis require many different organelles? (This is the same general problem as the one originated in the "looking for" approach, but it is brought closer to experimental testing by suggesting candidate organelles). (2) Is the contractile ring the only organelle necessary for furrowing, i.e., is the equatorial contraction phase of cytokinesis accomplished by this single organelle?

II. EXPERIMENT, OR: ARE THE CANDIDATE ORGANELLES REALLY ESSENTIAL?

Having established the presence of the contractile ring in the cleavage furrow, electron microscopists continued to verify the equatorial contraction hypothesis. The theoretical assumption was as follows: If the contractile ring is the crucial organelle for furrow formation, then receding of the furrow should be accompanied by the receding (disappearance) of the contractile ring. In other words: If furrow, then contractile ring, but if no furrow, no contractile ring. The observation of Carter (1967) that the drug cytochalasin B interferes with cytokinesis but does not interfere with mitosis provided a seemingly ideal probe for experimentally testing the role of the contractile ring in cytokinesis.

Two controversies have developed from this research, which are extensively discussed by Schroeder (1978). First, does the contractile ring disappear or remain in place after cytochalasin B treatment, i.e., is the contractile ring affected by cytochalasin B? Second, does cytochalasin B affect the furrow or rather the cytoplasmic bridge with the midbody?

A. Is the Contractile Ring Affected by Cytochalasin B?

It now appears proved, as a result of studies on amphibian eggs (Bluemink, 1971; Luchtel *et al.*, 1976), that the contractile rings, which remain in place after cytochalasin B treatment, become disorganized, i.e., they *are* affected by cytochalasin B. Since all the contractile rings studied up to now fall into one of the two categories, they either disappear or undergo disorganization; the controversy no longer exists (reviewed by Schroeder, 1978). Whether the contractile ring is absent from the cell or just disorganized, it is no longer a ring or belt of filaments, i.e., it is no longer an intact organelle. Thus, either disappearance or disorganization of the contractile ring might confirm the causal role of the organelle contractile ring in furrow formation.

B. Is the Furrow Affected by Cytochalasin B?

The second controversy is of crucial importance for understanding the roles of the contractile ring in cytokinesis. There is a group of light microscopic results showing that mammalian tissue cells *in vitro* do undergo furrowing in the presence of cytochalasin B (discussed in detail by Schroeder, 1978). At the heart of the controversy are data from HeLa cells. Carter (1967, 1972) and Sanger and Holtzer (1972) have documented at the light microscopic level that furrowing occurs in these cells in the presence of cytochalasin B. On the other hand, Schroeder has found no evidence of furrowing either at the light microscopic (1978) or the EM level (1972, 1978). All the results on HeLa cells indicate that eventually cytochalasin B-treated cells become spherical, with no indication of the furrow. In the opinion of Sanger and Holtzer (1972), this is due to impairment of the separation and regression of the furrow and not to impairment of the constriction stage of cytokinesis.

The described differences in the response of the furrowing HeLa cells to cytochalasin B have been attributed to presumably different strains of HeLa cells used and to some differences in experimental procedure (Schroeder, 1978). However, from light microscopic studies on mouse blastomeres during furrowing, it also appears that the furrow quickly deepens in the presence of cytochalasin B (cf. Fig. 3) before it finally recedes (Opas, 1977). Because of these light microscopic results, one cannot draw the general conclusion that cytochalasin B prevents furrowing. The final result cannot be unequivocally attributed to a direct effect of cytochalasin B on furrowing, as illustrated by the aforementioned examples.

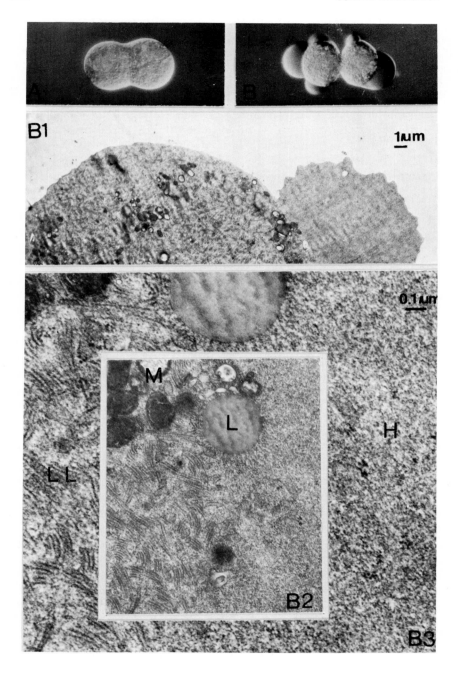

C. Are the Effects of Cytochalasin B Confined to the Contractile Ring?

An unexpected problem arising from EM studies of cytochalasin B-affected cells in cytokinesis concerns the question, what *else* changes besides the contractile ring? i.e., what else is different from the control after cytochalasin B treatment? In fact, if something else does change, this would seem immediately to make the cytochalasin B results ambiguous in terms of the causal role of the contractile ring in cytokinesis. Indeed, other components *are* altered by cytochalasin B, for Schroeder states: "It is also necessary to point out, *of course*, that the loss of contractile ring microfilaments and absence of a properly formed cleavage furrow are not the only alterations induced by cytochalasin B [in Fig. 2]" (my italics and interpolation) (Schroeder, 1978, p. 102). In HeLa cells, which are the subject of the quoted sentence, the texture of the cytoplasm and mitochondria is altered by treatment with cytochalasin B. Comparison of Figs. 2 and 3 in the discussed paper also suggests that the distribution of mitochondria differs from the control, i.e., they no longer occupy a peripheral position, about 1 μm under the cell membrane. Blebs and surface irregularities are the next ultrastructural changes found in both HeLa cells (Schroeder, 1970, 1978) and *Arbacia* eggs (Schroeder, 1972) treated with cytochalasin B during cytokinesis. In our EM studies of mouse blastomeres treated with cytochalasin B during furrowing, we have found blebs containing hyaloplasm, which suggests that cytochalasin B induces stratification of the cytoplasm (Fig. 3) and disorganization of the cortical layer (Fig. 4), as well as disorganization of the contractile ring (Karasiewicz and Soltyńska, in preparation).

The data summarized above suggest that cytochalasin B affects not only the contractile rings but also two overall aspects of ultrastructure: (1) organization of the non-furrow surface (irregularities, evaginations in the form of blebs, disorganization of the cortical layer) and (2) organization of the cytoplasm (stratification into hyaloplasm and organelle-containing cytoplasm, possibly displacement of mitochondria). At least one of these effects is independent of the impairment of the contractile ring, since blebs like those found in mouse blastomeres during furrowing have also been ultrastructurally shown in the same blastomeres at interphase (Perry and Snow, 1975).

Fig. 3. Effects of cytochalasin B on mouse blastomers during furrowing. Blebs and stratification of cytoplasm observed within the first minutes of treatment with cytochalasin B (10 μg/ml). A–B: living cells, differential interference contrast; A, blastomere just before adding cytochalasin B; B, during cytochalasin B treatment. B1–B3: subsequent EM enlargements to show the boundary between the bleb and the remaining cytoplasm. H, hyaloplasm within the bleb; L, lipid droplet; LL, lattice-like structures; M, mitochondrium. Pictures obtained in collaboration with Dr. M. S. Soltyńska.

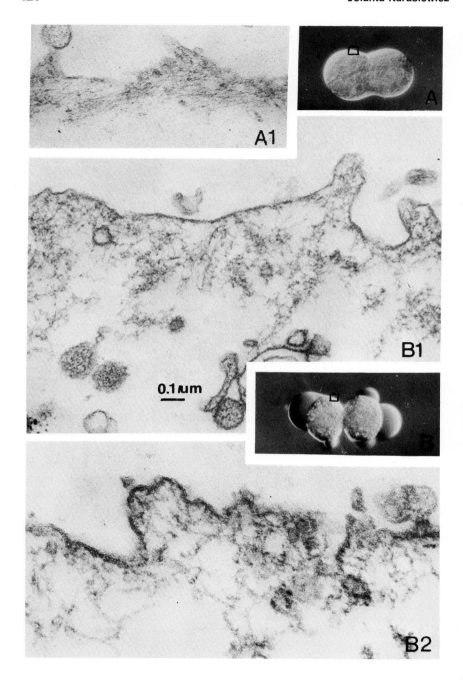

0.1 μm

As well as making it difficult to reach conclusions about the causal role of the contractile ring, these observations suggest that EM studies might allow one to distinguish between the contributions of the contractile ring and the cortical layer and cytoplasm to cytokinesis. In such experiments, cells treated with cytochalasin B during elongation (no contractile ring) might serve as a control for cells treated during the constriction stage (contractile ring present). The need for testing the roles of the cortical layer and the cytoplasm is an obvious prerequisite for drawing conclusions about the causal role of the contractile ring. On the other hand, the cortical layer and the organization of the cytoplasm might be interesting in other respects, which will be discussed briefly in Section III,A,1 and 2.

D. Conclusions

From the discussion of EM experiments with cytochalasin B (Section II,B and C), it appears that, as an organizing approach leading to verification of a preexisting hypothesis, these experiments have produced more questions than answers. These questions are relevant to (and reinforce those of) the other approaches, as follows.

First, the cytochalasin B data reinforce doubts created by the "looking for" approach, i.e., regarding the general mechanisms of cytokinesis and the role of the contractile ring.

Second, the cytochalasin B data strengthen doubts created by the "just watching" approach, i.e., regarding the roles of the cortical layer in cytokinesis.

Third, the cytochalasin B data emphasize the possibilities and limitations of EM studies in research on cytokinesis. This will be discussed in the last section.

III. FUTURE, OR: HOW ARE THE ESSENTIAL ORGANELLES ESSENTIAL?

A. Contractile Ring–Cortical Layer–Cytoplasm System in Cytokinesis

The light microscopic data suggest the following. (1) Cytochalasin B prevents cytokinesis (i.e., cells remain spherical) when applied before visible

---

Fig. 4. Effects of cytochalasin B on mouse blatomers during furrowing. Disorganization of the cortical layer in blastomeres treated with cytochalasin B, like those shown in Fig. 3. Cells extracted with glycerol solution before fixation for EM. A–B as in Fig. 3; location of the enlarged areas of the cortical layer indicated. A1, arrangement of the cortical layer filaments in the control (parallel to the long axis of the cell and to each other). B1, disorganization during cytochalasin B treatment. B2, as B1, after the reaction with heavy meromyosin. Pictures obtained in collaboration with Dr. M. S. Sołtyńska.

signs show that it has started (HeLa cells: Schroeder, 1970; *Arbacia* eggs: Schroeder, 1972; mouse blastomeres: Snow, 1973; Opas, 1977). From this, it may be inferred that cytochalasin B prevents elongation (affects induction?). (2) Cytochalasin B causes final rounding up when applied during elongation (*Arbacia* eggs: Schroeder, 1972; mouse blastomeres: Opas, 1977), which again suggests that cytochalasin B acts upon elongation. (3) Cytochalasin B causes final founding up or relaxation of the furrow when applied during constriction (*Arbacia* eggs, mouse blastomeres—*op. cit.*), which suggests that cytochalasin B affects constriction–separation (with the precautions concerning discrimination between the two, as discussed in Section II,B. (4) Cytochalasin B causes rounding up of mouse blastomeres (*op. cit.*) when applied during separation, which suggests that it affects separation itself.

This list shows that cytochalasin B may affect *all* stages of cytokinesis, which indicates that there is a common mechanism for the whole process of cytokinesis. This is discouraging when we look for a single organelle of cytokinesis, and the contractile ring is present during only one stage of the process. On the other hand, the contractile ring, cortical layer, and cytoplasm *are* affected by cytochalasin B, which shows that they are in some way related to this mechanism. It seems, then, that the contractile ring–cortical layer–cytoplasm system deserves further attention. The reasons are presented in the next two subsections.

1. Cortical Layer

The cortical layer of actin filaments is in contact with the whole cell membrane, except for the area occupied by the contractile ring. Because of its localization in the cell, the cortical layer is very likely to be an organelle which organizes local configurations of the cell membrane (microvilli, folds). This appears to be closely connected to the surface-to-volume ratio, which has to increase when making two spheres from one (reviewed by Wolpert, 1960). One of the ways to achieve this is to unfold the preexisting surface (reviewed by Arnold, 1976). The cortical layer would be a good candidate to accomplish or control such a task.

The cortical layer is closely related to the contractile ring both structurally and biochemically (see Section I,B), which poses the question of the *functional* relationship between the cortical layer and the contractile ring and whether the latter might arise from the former. Finding filamentous masses in contact with the membrane of the cytoplasmic bridge during separation (Mullins and Biesele, 1977) opens the possibility that the cortical layer is present in the area previously occupied by the contractile ring. There are claims that the contractile ring exists before furrowing [e.g., during elongation (Zeligs and Wollman, 1979) and during separation (Hurle and Lafarga, 1978)], but these need closer examination. In summary, it appears that studying the dynamics of membrane-associated microfilaments during cytokinesis might shed some light on the roles and relationships of the contractile ring and the cortical layer, both in time and in space. This suggestion is in agreement with what has already been generally recommended as essential in EM studies of cell division (Schroeder, 1976) and of cytokinesis itself (Schroeder, 1978).

So far, the best technique for studying this relationship seems to be that of reacting the cortical layer filaments with heavy meromyosin. In this technique, cells are treated with solutions of glycerol or detergents (see Fig. 2), in order to render them permeable to heavy meromyosin. This extraction with glycerol of cells in cytokinesis (Hoffmann-Berling, 1954; Forer and Behnke, 1972; Schroeder, 1973; Opas and Soltyńska, 1978) allows one to visualize in the EM thin filaments (see Fig. 4), which are otherwise masked by the cytoplasm.

2. Cytoplasm

It should be recalled here that the contracting gel theory (Marsland and Landau, 1954) that underlies the interest in the contractile ring assumes that the cortical gel, a layer about 6 μm *thick* and present throughout the *whole cell periphery* during *all stages* of cytokinesis, is the motive force for all the changes in cell shape during cytokinesis (cf. Fig. 6 in Marsland and Landau, 1954). Neither the contractile ring, which is about 0.1 μm thick and a transitory structure, nor the cortical layer, which is also about 0.1 μm thick (though present throughout the whole membrane), can satisfy the conditions of this theory. They more closely fit the rejected expanding membrane theory (Swann and Mitchison, 1958), which defines "membrane" as being no thicker than about 2 μm.

Evidence from light microscopic measurements repeatedly show that the gelled cortex is a few micrometers thick in various marine eggs (reviewed by Rappaport, 1971, 1975). Some EM data (Section II,C) show that cytochalasin B, which impairs cytokinesis, affects the spatial organization of the cytoplasm. It thus appears that we might return to the idea of the gelled cortex (see Fig. 5) and try to test it experimentally using cytochalasin B and the EM. Simple observation has repeatedly failed to reveal any differences in the texture of peripheral versus inner cytoplasm of cells in cytokinesis, i.e., it has not substantiated any indication of a gelled cortex. Judging from the displacements of organelles induced by cytochalasin B at various stages of cytokinesis, one might use cytochalasin B to obtain some information on the presumed existence and extent of the gelled cortex. Persisting interest of molecular biologists in the mechanism of cytochalasin B action (reviewed by Tannenbaum, 1978) may quickly yield data which will help interpretation of EM results.

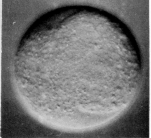

Fig. 5. Mouse blastomere treated with distilled water during furrowing. Swelling results in rounding up of the surface, but the cytoplasm retains the cytokinetic shape. This may suggest the existence of the rigid, compact (gelled?) peripheral cytoplasm.

B. Models

As may be inferred from Section III,A, cytokinesis is postulated to be due to a series of changes within the contractile ring–cortical layer–cytoplasm system, at least at the present level of our knowledge. How are these changes related to each other and to the initial stimulus? How are they causally related to subsequent stages of cytokinesis, as well as to the changes of cell shape and reorganization of the surface? Answers to these questions remain to be elucidated.

The present problems arising from EM research and from cytochalasin B experiments might appear to concern mainly (or exclusively) only a few types of cells. In fact, only sea urchin eggs, HeLa cells, and mouse blastomeres have been discussed in detail in this chapter. Two other groups of cytochalasin B-treated cells that have been studied with the EM are molluscan and amphibian eggs (reviewed by Schroeder, 1978). Data from these groups in no case appear to contradict what has been discussed above. On the contrary, they illustrate additional theoretical and practical problems, such as changes in permeability to cytochalasin B, new membrane formation during cytokinesis (amphibian eggs: Bluemink and de Laat, 1973), and organization of the contractile apparatus when polar lobes accompany cytokinesis (snail eggs: Conrad and Williams, 1974).

If we assume that cytokinesis is a series of changes within the contractile ring–cortical layer–cytoplasm system, we must also assume that the particular relations within this system will differ depending on various properties of the cells studied. For example, in cases in which the gelled cortex seems to be essential, one might expect differences between cells living at various temperatures, since temperature affects the gel–sol state, at least in the cytoplasmic extracts (reviewed by Clarke and Spudich, 1977). In cases in which the surface contractile apparatus (contractile ring–cortical layer) is essential, then one might expect differences in different-size cells, because of different surface-to-volume ratios, and one might expect different relationships of cortical layer to contractile ring in cells that have different types of furrowing.

From such considerations, it appears obvious that cells should be chosen carefully for further study of cytokinesis with the EM. Two systems appear favorable. One is the sea urchin egg, which has been extensively studied up to now at both the EM and light microscopic experimental levels. The sea urchin egg provides a model of an isolated, equilaterally cleaving cell adapted to change temperatures. The second system might be the mouse egg (blastomere), which is also an isolated, equilaterally cleaving cell, comparable in size to sea urchin eggs (blastomers) but adapted to 37°C. Both technical advantages (obtaining cells, observing them individually, and performing micromanipulations on them) and the similarity to the first model make mouse eggs superior to HeLa cells. Both proposed models allow one to study equilateral cytokinesis. It appears convenient in that this type of cytokinesis might be the simplest choice, being characteristic of cells not overloaded with yolk and not complicated by any other special developmental adaptations (such as polar lobes in molluscan eggs).

C. Potential of EM Studies

It has been suggested in this chapter that EM studies of cytokinesis, which involve looking for the contractile organelle and watching cells in cytokinesis, together with the cytochalasin B experiments testing the causal role of the contractile organelle, have produced creative confusion by obtaining data not compatible with the initially tested general theory of cytokinesis; indeed, the data are not compatible with any *single* theory, though discussion of this point is beyond the scope of this chapter. It has been further suggested that it is the *experimental* EM approach (e.g., extraction with glycerol and the use of cytochalasin B) that may now further clarify our understanding of cytokinesis.

Besides its ability to visualize organelles and their relative positions within cells at high resolution, the EM offers the advantage of making states from what is otherwise a process, i.e., a continuous series of changes. This, in turn, allows one to compare chosen aspects of ultrastructure in subsequent states, and to draw conclusions about presumed relationships. To benefit the researcher, such an approach demands the fulfillment of some conditions, which might be called the limitations of the EM approach. These limitations are that both the organelles to be followed and the stages of the process have to be properly chosen. It seems that up to now, the EM results on cytokinesis have not fulfilled both of these conditions.

Finally, some actual mechanisms of cytokinesis surely lie below the organelle level. This appears to hold when the contractile ring is considered, since its ultrastructure (i.e., the arrangement and density of filaments) does not change during contraction (Schroeder, 1972, 1975). This same condition may exist with the presumed gelled cortex, since it cannot be even visualized by standard EM methods. Nevertheless, it might be expected that our strong need to *see* things will stimulate development of specific markers which will eventually allow one to visualize in the EM the sequence of changes during cytokinesis. The level of EM remains the lowest cellular level of research, and cytokinesis remains a cellular process.

ACKNOWLEDGMENTS

I am grateful to Drs. M. S. Soltyńska and J. A. Modliński, and to Prof. A. K. Tarkowski for critical reading of the manuscript. Special thanks are due to Dr. Modliński for stimulating discussions and for sharing with me his unpublished observations derived from applying the micromanipulation technique to mouse eggs and blastomeres in cytokinesis.

REFERENCES

Arnold, J. M. (1969). Cleavage furrow formation in a telolecithal egg (*Loligo pealii*). I. Filaments in early furrow formation. *J. Cell Biol.* **41**, 894–904.

Arnold, J. M. (1976). Cytokinesis in animal cells: New answers to old questions. *In* "The Cell Surface in Animal Embryogenesis and Development" (G. Poste and G. L. Nicolson, eds.), pp. 55–80. Elsevier Amsterdam.

Asnes, C. F., and Schroeder, T. E. (1979). Cell cleavage. Ultrastructural evidence against equatorial stimulation by aster microtubules. *Exp. Cell Res.* **122**, 327–338.

Bellairs, R., Lorenz, F. W., and Dunlap, T. (1978). Cleavage in the chick embryo. *J. Embryol. Exp. Morphol.* **43**, 55–69.

Bluemink, J. G. (1970). The first cleavage of the amphibian egg: An electron microscope study of the onset of cytokinesis in the egg of *Ambystoma mexicanum. J. Ultrastruct. Res.* **32**, 142–166.

Bluemink, J. G. (1971). Effects of cytochalasin B on surface contractility and cell junction during egg cleavage in *Xenopus laevis. Cytobiologie* **3**, 176–187.

Bluemink, J. G., and de Laat, S. W. (1973). New membrane formation during cytokinesis in normal and cytochalasin B-treated eggs of *Xenopus laevis.* I. Electron microscope observations. *J. Cell Biol.* **59**, 89–108.

Burgess, D. R. (1977). Ultrastructure of meiosis and polar body formation in the egg of the mud snail (*Ilyanassa obsoleta*). *In* "Cell Shape and Surface Architecture" (J. P. Revel, U. Henning, and C. F. Fox, eds.), pp. 569–579. Alan R. Liss, Inc., New York.

Byers, B., and Abramson, D. H. (1968). Cytokinesis in HeLa: Post-telophase delay and microtubule associated motility. *Protoplasma* **66**, 413–435.

Carter, S. B. (1967). Effects of cytochalasins on mammalian cells. *Nature (London)* **213**, 261–264.

Carter, S. B. (1972). The cytochalasins as research tools in cytology. *Endeavour* **31**, 77–82.

Clarke, M., and Spudich, J. A. (1977). Nonmuscle contractile proteins: The role of actin and myosin in cell motility and shape determination. *Annu. Rev. Biochem.* **46**, 797–822.

Conrad, G. W., and Williams, D. C. (1974). Polar lobe formation and cytokinesis in fertilized eggs of *Ilyanassa obsoleta*. *Dev. Biol.* **36**, 363–378.

Ducibella, T., Ukena, T., Karnovsky, M., and Anderson, E. (1977). Changes in cell surface and cortical cytoplasmic organization during early embryogenesis in the preimplantation mouse embryo. *J. Cell Biol.* **74**, 153–167.

Forer, A., and Behnke, O. (1972). An actin-like component in spermatocytes of a cranefly (*Nephrotoma suturalis* Loew). II. The cell cortex. *Chromosoma* **39**, 175–190.

Gipson, I. (1974). Electron microscopy of early cleavage furrows in the chick blastodisc. *J. Ultrastruct. Res.* **49**, 331–347.

Gulyas, B. (1973). Cytokinesis in rabbit zygote: Fine structural study of the contractile ring and the mid-body. *Anat. Rec.* **177**, 195–207.

Hoffmann-Berling, H. (1954). Die glyzerin-wasserextrahierte Telophasezelle als Model der Zytokinese.*Biochim. Biophys. Acta* **15**, 332–339.

Hurle, J. M., and Lafarga, M. (1978). Cytokinesis in developing cardiac muscle cells. An ultrastrucutral study in the chick embryo. *Biol. Cellulaire* **33**, 195–198.

Ishikawa, H., Bischoff, R., and Holtzer, M. (1969). Formation of arrowhead complexes with heavy meromyosin in a variety of cell types. *J. Cell Biol.* **43**, 312–328.

Longo, F. J. (1972). The effects of cytochalasin B on the events of fertilization in the surf clam (*Spisula solidissima*). I. Polar body formation. *J. Exp. Zool.* **182**, 321–344.

Luchtel, D., Bluemink, J. G., and de Laat, S. W. (1976). The effects of injected cytochalasin B on filament organization in the cleaving egg of *Xenopus laevis*. *J. Ultrastruct. Res.* **54**, 406–419.

Marsland, D., and Landau, J. (1954). The mechanisms of cytokinesis: Temperature pressure studies on the cortical gel system in various marine eggs. *J. Exp. Zool.* **125**, 507–539.

Mullins, J. M., and Biesele, J. J. (1977). Terminal phase of cytokinesis in D-98S cells. *J. Cell Biol.* **73**, 672–695.

Opas, J. (1977). Effects of cytochalasin B on cytokinesis in mouse blastomeres. I. Light microscopic study. *Dev. Biol.* **61**, 373–377.

Opas, J., and Soltyńska, M. S. (1978). Reorganization of the cortical layer during cytokinesis in mouse blastomeres. *Exp. Cell Res.* **113**, 208–211.

Perry, M. M., and Snow, M. H. L. (1975). The blebbing response of 2-4 cell stage mouse embryos to cytochalasin B. *Dev. Biol.* **45**, 372–377.

Perry, M. M., John, H. A., and Thomas, N. S. T. (1971). Actin-like filaments in the cleavage furrow of newt eggs. *Exp. Cell Res.* **65**, 249–253.

Rappaport, R. (1971). Cytokinesis in animal cells. *Int. Rev. Cytol.* **31**, 169–213.

Rappaport, R. (1975). Establishment and organization of the cleavage mechanism. *In* "Molecules and Cell Movement" (S. Inoué and R. E. Stephans, eds.), pp. 287–303. Raven, New York.

Sanger, J. W., and Holtzer, H. (1972). Cytochalasin B: Effects on cytokinesis, glycogen and ³H-D glucose incorporation. *Am. J. Anat.*, **135**, 293–298.

Sanger, J. M., and Sanger, J. W. (1980). Banding and polarity of actin filaments in interphase and cleaving cells. *J. Cell Biol.* **86**, 568–575.

Schroeder, T. E. (1968). Cytokinesis: Filaments in the cleavage furrow. *Exp. Cell Res.* **53**, 272–276.

Schroeder, T. E. (1969). The role of contractile ring filaments in dividing *Arbacia* eggs. *Biol. Bull. (Woods Hole, Mass.)* **137**, 413–414.

Schroeder, T. E. (1970). The contractile ring. I. Fine structure of dividing mammalian (HeLa) cells and the effect of cytochalasin B. *Z. Zellforsch. Mikrosk. Anat.* **109**, 431–449.

Schroeder, T. E. (1972). The contractile ring. II. Determining its brief existence, volumetric changes and vital role in cleaving *Arbacia* eggs. *J. Cell Biol.* **53**, 419–434.

Schroeder, T. E. (1973). Actin in dividing cells: Contractile ring filaments bind heavy meromyosin. *Proc. Natl. Acad. Sci. U.S.A.* **70**, 1688–1692.

Schroeder, T. E. (1975). Dynamics of the contractile ring. *In* "Molecules and Cell Movement" (S. Inoué and R. E. Stephens, eds.), pp. 305–332. Raven, New York.

Schroeder, T. E. (1976). Actin in dividing cells: Evidence for its role in cleavage but not mitosis. *In* "Cell Motility" (R. Goldman, T. Pollard, and J. Rosenbaum, eds.), pp. 265–277. Cold Spring Harbor Lab. Cold Spring Harbor, New York.

Schroeder, T. E. (1978). Cytochalasin B, cytokinesis and the contractile ring. *In* "Cytochalasins—Biochemical and Cell Biological Aspects" (S. W. Tanenbaum, ed.), pp. 91–112. North-Holland Publ., Amsterdam.

Snow, M. H. L. (1973). Tetraploid mouse embryos produced by cytochalasin B during cleavage *Nature (London)* **244**, 513–514.

Swann, M. M., and Mitchison, J. M. (1958). The mechanism of cleavage in animal cells. *Biol. Rev. Cambridge Philos. Soc.* **33**, 103–135.

Szollosi, D. (1970). Cortical cytoplasmic filaments of cleaving eggs: A structural element corresponding to the contractile ring. *J. Cell Biol.* **44**, 192–209.

Tannenbaum, J. (1978). Approaches to the molecular biology of cytochalasin action. *In* "Cytochalasins—Biochemical and Cell Biological Aspects" (S. W. Tanenbaum, ed.), pp. 547–559. North-Holland Publ., Amsterdam.

Wolpert, L. (1960). The mechanics and mechanisms of cleavage. *Int. Rev. Cytol.* **10**, 163–216.

Zeligs, J. D., and Wollman, S. H. (1979). Mitosis in thyroid follicular epithelial cells *in vivo*. III. Cytokinesis. *J. Ultrastruct. Res.* **66**, 288–303.

19

Inhibitors and Stimulators in the Study of Cytokinesis

Jesse E. Sisken

I. INTRODUCTION

The aim of research on cytokinesis, as in many other areas of biology, is to understand the nature of the process at the molecular level and the mechanisms which regulate its onset and progress. As may be seen from Chapter 16 by Conrad and Rappaport in this volume, cytokinesis has been studied in a number of different cells using a variety of methodologies. One approach commonly used to try to understand biological processes, particularly in systems where one cannot work with isolated and purified material, is to treat the living cell with selected agents and observe how the process of interest responds to the treatment. What is learned from such studies is a function of a number of factors, especially the specificity of the perturbing agent. Ideally, an agent should react only with a single, well-characterized molecule which is directly and uniquely involved in the process of interest, and its effects should be easily measured.

In general, and specifically in the case of cytokinesis, the agents available frequently either lack the desired specificity or their specificities are as yet

MITOSIS/CYTOKINESIS
Copyright © 1981 by Academic Press, Inc.
All rights of reproduction in any form reserved.
ISBN 0-12-781240-7

unknown. In addition, it is often difficult to distinguish between primary and secondary effects. Nevertheless, the judicious use of various types of agents can be instructive, and in this chapter I will discuss what I believe to be representative examples from the literature and from my own laboratory to illustrate some of the work that has been done in studying cytokinesis with this approach. The objective will be to evaluate the approach and not to review our knowledge of cytokinesis; this has been done elsewhere in this volume. In general, the questions I will try to answer are: (1) What has the approach contributed? (2) Why hasn't it contributed more? (3) What might be done to make the approach more productive? (4) What opportunities lie in the future?

II. RESULTS OF STUDIES WITH SELECTED AGENTS

A. Scope

Since the initiation of cytokinesis is dependent upon the successful completion of the earlier events of mitosis, any agent which interfers with the structure or function of the mitotic apparatus will also have secondary effects on cytokinesis. While such agents and their effects are interesting and important from other points of view, they are generally irrelevant to cytokinesis except as they can be utilized to examine the effects of inhibition of chromosome movement on the onset and velocity of cytokinesis (see Section II,B,5). However, the emphasis in this chapter will be on agents whose effects on cytokinesis are separable from effects on earlier stages of division.

Much of the work done in recent years in which agents were used to try to understand the mechanisms of cytokinesis involved two groups of substances: (a) the cytochalasins, mainly cytochalasin B, and (b) agents believed to either increase or decrease the intracellular availability of calcium ions. Discussion of these two groups takes up the greater portion of this chapter. Several polyamines have also been reported to affect cytokinesis and cortical contractions, and these constitute the third group to be discussed. The fourth group of agents, chosen because of their great specificity, are antibodies against contractile proteins. To my knowledge, effects of only a single representative, antimyosin, have been published, but further use of this type of agent would seem to have great potential. The final agents discussed, p-fluorophenylalanine and abnormal temperatures, were chosen because they slow the rate of anaphase chromosome movement and thus provide information concerning the dependency of cytokinesis on the prior functioning of the mitotic apparatus.

B. The Effects of the Agents

1. *The Cytochalasins*

The cytochalasins are a group of fungal metabolites that were first observed by Carter (1967) to inhibit cytokinesis and other movements in cultured mammalian cells without affecting karyokinesis. Since Carter's initial report, these substances (mainly cytochalasin B) have been used in an enormous number of studies in many organisms dealing with various aspects of motility and membrane activities (for a review, see Tannenbaum, 1978). Their effects on cytokinesis have been a subject of some disagreement. I will summarize here only some of the main observations. A more detailed discussion of these effects can be found in Bluemink (1978) and Schroeder (1978).

In his original study on mouse L cells, Carter (1967) observed that deep cleavage furrows formed in the prescence of cytochalasin B until the daughters were connected only by a narrow intercellular bridge. This bridge failed to break as it should have, and the daughters reunited to form a single unit. Similar observations were reported by Krishan (1972) in mouse L cells and by Hammer *et al.* (1971) and Bluemink (1971a,b) in amphibian eggs. That is, during exposure to cytochalasin B, cleavage furrows formed, progressed almost to completion, and then regressed, causing the formation of binucleate cells.

On the other hand, Schroeder (1969, 1970) reported that in sea urchin eggs and HeLa cells, the process of furrowing was actually prevented by application of cytochalasin B and that those furrows in progress when treatment began halted and regressed. Ultrastructural studies of the treated cells indicated that the microfilament-containing contractile ring was either prevented from forming or was disorganized by the drug. Subsequent observations made by other workers in cleaving marine eggs were in agreement with this interpretation (see Schroeder, 1978).

However, in later experiments with HeLa cells, Schroeder (1978) obtained results like those of Carter (1967) and Krishan (1972) in mouse L cells and not like his earlier findings. That is, furrowing did occur and did progress in the presence of cytochalasin B, but final separation of the daughters failed to occur, causing the formation of binucleate cells. Thus, different cell types and even two lines of HeLa cells studied by the same investigator yielded different results.

A number of proposals have been made to account for these results. In his original paper, Carter (1967) suggested that the cytochalasins act at the cell surface to increase adhesion to glass substrates and to increase surface viscosity, and that both effects could retard progression of the cleavage furrow. He further suggested that if the completion of furrowing is delayed, the mechanisms which complete the separation of daughter cells will no longer operate and the furrow will regress. Others have also suggested that the lesion induced by cytochalasin B occurs in the final separation of the daughter cells, involving either their movement apart (Krishan, 1972) or the approximation and fusion of the cell membranes of the isthmus connecting the daughter cells (e.g., Hammer *et al.*, 1971; Estensen, 1971).

Bluemink (1971a) found that lysolecithin and phospholipase C could also produce furrow regression and pointed out that Kuno (1954) and Kuno-Kojima (1957) also observed that membrane-destabilizing agents could cause dividing sea urchin eggs to form syncitia. Inhibition of cytokinesis by a membrane mobility agent has also been more recently reported in Friend leukemia cells (Lustig *et al.*, 1977). Bluemink (1971a) suggested that the primary site of action of cytochalasin B was at the cell membrane, altering the process of membrane ingrowth and interfering with the formation of interblastomeric junctions. Bluemink also suggested that the effects on the microfilament system were only indirect, possibly resulting from changes in cell permeability. However, in later studies, Bluemink and co-workers (de Laat *et al.*, 1973) observed that *Xenopus* eggs become sensitive to externally applied cytochalasin B only about 7 minutes following the onset of furrowing but that furrowing would regress within 1 minute if cytochalasin B was injected directly beneath the membrane in the furrow. This was true even if injection occurred immediately after the beginning of cleavage. In addition, they could find no cytochalasin B-induced alterations in membrane permeability. They concluded that the reason cleavage continues in the presence of external cytochalasin B almost to completion before regressing is that cytochalasin B does not enter the egg until a natural change in permeability occurs about 7 minutes after the onset of furrowing but that the cytokinetic machinery is sensitive to cytochalasin B any time during the process, as indicated by the injection experiments. That microinjection of cytochalasin B could have immediate effects on furrowing in *Xenopus* was confirmed in a later study, which also showed that the agent could cause disorganization of the contractile ring microfilaments (Luchtel *et al.*, 1976) but not necessarily their decomposition. Luchtel *et al.* (1976) thus proposed that the delayed furrow regression seen in their earlier experiments was due to the delayed entry of the agent into the cell. This raises the question of whether some experimentally induced alteration in cell permeability might account for the differences between Schroeder's earlier (1970) studies and both his later studies (see Schroeder, 1978) and those of Krishan (1972).

The possibility was also raised that, since cytochalasin B could inhibit hexose transport (e.g., Estensen and Plagemann, 1972), its effects on cytokinesis and other cell movements might be indirect and caused by an energy deficiency. However, this does not seem to be the case, since growth in glucose-free solutions does not inhibit motility in other systems (Yamada and Wessells, 1973; Taylor and Wessells, 1973) and since cytochalasin D and dihydrocytochalasin B, which do not inhibit sugar transport, are also strong inhibitors of motility (Miranda *et al.*, 1974a,b; Atlas and Lin, 1978; see also the review by Godman and Miranda, 1978).

More recent studies on the effects of cytochalasins at the molecular level provide further clues to the mechanisms by which cytokinesis is inhibited. Plasma membrane fractions have been observed to have high-affinity binding sites for the cytochalasins, and evidence suggests that these sites involve proteins, such as actin, bound to the inner side of the plasma membrane (Lin and Spudich, 1974; Tannenbaum *et al.*, 1975). Since scattered filaments were observed in regressed furrows (Bluemink, 1971a,b; Luchtel *et al.*, 1976), Bluemink (1978) suggested that it might be the membrane anchorages for the actin-containing microfilaments which are affected by cytochalasin B, allowing furrow regression to occur because the microfilaments are thus released from the membrane.

Further studies with purified systems indicate that actin does have a high-affinity binding site for cytochalasins, and that the agents can inhibit actin polymerization (Lin and Lin, 1979; Brown and Spudich, 1979; Lin *et al.*, 1980; Brenner and Korn, 1980; MacLean-Fletcher and Pollard, 1980), break actin filaments (Hartwig and Stossel, 1979), and/or reduce actin–actin interactions (MacLean-Fletcher and Pollard, 1980). Since actin is believed to exert the force which brings about furrowing, an inhibition of this process by cytochalasin B could result from any of these phenomena. To reconcile the difference between Schroeder's earlier observations on HeLa cells with the observations in mouse L cells (Carter 1967; Krishan, 1972) and with later work on HeLa cells in more difficult, but some possibilities do exist. It is known that not all cells are equally sensitive to cytochalasin B, possibly resulting from differences in permeability, and that even the same cell at different times in the cell cycle may respond differently to the agent (see Godman and Miranda, 1978). In addition, it has been reported that the integrity of microfilament bundles under normal conditions is dependent upon ATP levels (Bershadsky *et al.*, 1980) and that inhibitors of energy metabolism can prevent cytochalasin D action (Miranda *et al.*, 1974a,b). Thus the degree to which cleavage furrow microfilaments might be affected by cytochalasin B could well vary depending upon differences in cell type and metabolic state of the cell. In some cells or under some conditions, the effects of a given concentration of cytochalasin B could be drastic enough to stop the furrow from forming. In other cases, however, a small inhibition of polymerization or small amounts of breakage and/or reduced actin–actin interaction might not have significant effects on furrowing until the later stages of cytokinesis, when endogenous mechanisms might be more active in breaking down the contractile ring. There thus might be a synergistic effect between the endogenous mechanisms and the drug effects, leading to premature breakdown of the contractile ring. This would allow the early stages of furrowing to proceed, but premature breakdown of the contractile ring would cause regression later on.

To return to the theme of this chapter, we have to ask what we have learned about cytokinesis from studies on the effects of cytochalasin B and what we might learn in the future. It seems fair to say that, apart from generating a lot of interest in the problem and reaffirming a role for intact actin filaments in the process, the use of cytochalasin B has so far taught us very little about cytokinesis. This is at least partly due to the kinds of problems discussed earlier, i.e., lack of knowledge of the mechanisms of action, lack of specificity, and variable responses by different cells. However, these problems are not insurmountable. The mechanisms of action of the cytochalasins are becoming clearer, and this knowledge is helping us understand actin assembly (see earlier references). This is a fundamental step toward understanding the formation of the contractile ring. Also, one can work with cytochalasins which are more specific [such as cytochalasin D and dihydrocytochalasin B, which also inhibit cytokinesis (Atlas and Lin, 1978; D. C. Lin and S. Lin, personal communication)], and one can concentrate, as a first step at least, on a single, convenient cell type grown under carefully controlled conditions. Nevertheless, at this time, it seems difficult to design experiments using cytochalasins which would indicate how actin filaments become organized into the contractile ring and how they interact with other substances to initiate the process of cytokinesis and bring it to conclusion. While such problems may be approachable by *in vitro* studies, it is not clear at the present time how further studies on the effect of cytochalasins in live cells could contribute to the problem.

2. Agents Affecting the Availability of Calcium Ions

Since the contractile ring contains actin and myosin (see Fujiwara and Pollard, 1976), and since actin structure and function depend upon calcium ions (see, e.g., Barany *et al.*, 1962; Weber *et al.*, 1964; Hasselbach, 1965; Ebashi and Endo, 1968; Condeelis and Taylor, 1977; Hellewell and Taylor, 1979), it is reasonable to expect that calcium might play a role in cytokinesis. This resemblance of contractile systems in non-muscle cells to those of muscle, particularly during cleavage, and the possible role of calcium in these systems was considered before the association of actin with the cleavage furrow had been established (see Hofman-Berling, 1960). The earlier literature has been reviewed, e.g., by Wolpert (1960), Mazia (1961), Marsland (1970), Bluemink (1970), Rappaport (1971), Schroeder (1975), Rebhun (1977), Conrad and Davis (1977), and Harris (1978). More recent evidence that calcium plays a role in the process of cleavage has been obtained from a number of studies on the effects of agents which either increase or decrease the intracellular availability of calcium ions. Most of this work has been done in non-mammalian systems, and I will review these studies first. I will then consider our own work with mammalian cells in culture.

a. Studies on Nonmammalian Systems

i. AGENTS WHICH DECREASE CALCIUM AVAILABILITY. Ethylene-diaminetetraacetic acid (EDTA), which can chelate calcium and magnesium ions, and ethylene glycol-bis(β-aminoethyl ether)-N,N'-tetraacetic acid (EGTA), which specifically chelates calcium, have been used in a number of studies to examine the calcium requirements for cytokinesis. A study by Hanson (cited by Timourian *et al.*, 1972), who showed that cleavage was blocked when sea urchin eggs were placed in seawater containing EDTA, was followed by experiments in which EDTA was locally applied to the surface of the egg via a capillary pipette (Timourian *et al.*, 1972). This treatment had no effect on cells once cleavage had begun but did arrest cells in metaphase if applied prior to mid-metaphase. Arrested cells were able to recover from the block, but the type of division that occurred depended upon where on the cell surface the EDTA was applied. Cleavage was normal when EDTA was applied to the polar zones, but abnormal cleavages were observed when EDTA was administered at the site of the presumptive cleavage furrow or between the presumptive furrow and the pole. Based upon these findings, the authors suggested that calcium may be required for the determination of the cleavage furrow, i.e., for the organization and siting of the contractile ring, rather than for the process of cleavage itself. Unfortunately, it is not clear from this study how much of the effect of the EDTA was due to metaphase blockage or other effects such as the reported "fading" of the chromosomes, and how much was due to effects on the contractile ring itself.

In a later X-ray microanalytical study of calcium localization in cleaving sea urchin zygotes, Timourian *et al.* (1974) reported that, in some of the cells, calcium was more highly concentrated in the presumptive furrow region than in the polar regions during metaphase, but not anaphase, and was more highly concentrated at the mitotic centers than in surrounding cytoplasm during both anaphase and metaphase. They suggested that calcium released by the aster during metaphase was responsible for determining the site of the cleavage furrow.

In a more direct study, Baker and Warner (1972) found that *Xenopus* eggs immersed in EGTA-containing solutions retained the capacity to cleave but that injection of EGTA into the cortical region of the cell could block cytokinesis without affecting nuclear division. This indicates that calcium is required for cytokinesis but could be supplied from internal stores.

ii. AGENTS WHICH INCREASE CALCIUM AVAILABILITY: INJECTION OF CALCIUM.   Several workers have found that they could induce cortical contractions or cleavage-like constrictions by injecting calcium into eggs. Gingell (1970) was able to do this by iontophoretically injecting calcium ions directly beneath the cell membrane of *Xenopus* eggs, but contractions did not occur when calcium was injected deeper into the cell or just outside the membrane or when other ions, including magnesium, were substituted for calcium. These observations were taken as evidence that the calcium-sensitive region responsible for contraction lies just beneath the plasma membrane. Hollinger and Schuetz (1976) also observed cleavage-like constrictions when they injected calcium ions into oocytes of *Rana pipiens*. In addition, they found that the orientation of the furrow was determined by the location of the injection.

Similar findings were obtained by Conrad and Davis (1977), who found that microiontophoretic injection of calcium ions into fertilized *Ilyanassa* eggs caused polar lobe-like protuberances at the site of injection shortly after injection, but this occurred only when exogenous calcium ions were available. They also found that simply placing fertilized *Ilyanassa* eggs into 0.34 M CaCl could cause a fraction of them to form protuberances resembling normal polar lobes (Conrad and Davis, 1980). They suggested that the requirement for exogenous calcium could be due to either an alteration in membrane permeability caused by the injection or a deficiency in intracellular calcium levels due to prior exposure of the eggs to calcium-free seawater. In either case, it seems to be a peculiarity of the experimental treatment since normal cleavage can occur in the absence of exogenous calcium (see Conrad and Davis, 1980). They also suggested that the injection of calcium could trigger a localized polymerization and contraction of filament bundles beneath any area of the cell surface (Conrad and Davis, 1977)

iii. IONOPHORES AND OTHER AGENTS.   Ionophores are substances which insert into membranes and allow the diffusion of ions along concentration gradients (see Pressman, 1976). A23187 is an ionophore for divalent cations whose biological effects are generally believed to be related to its capacity to transport calcium ions (see, e.g., Pressman, 1976; Rasmussen and Goodman, 1977; Pfeiffer *et al.*, 1978). There are, however, some complexities to its mode of action, which may vary in different cell types, and these are discussed below.

Schroeder and Strickland (1974) found that local administration of A23187 to the surface of *R. pipiens* eggs caused local cortical contractions which were independent of external calcium. They also found that these contractions did not occur in the vicinity of wounds in eggs exposed to EDTA or EGTA. Their interpretation was that the ionophore caused release of calcium from intracellular pools but that the chelating agents which diffused into the cell in the region of the wounds could bind the released calcium and inhibit the calcium-induced contractions.

In a later study, Osborn *et al.* (1979) induced cortical contractions in *Xenopus* embryos with A23187 and examined the effects of this treatment on the intracellular distribution of calcium ions in cells exposed to calcium- and magnesium-free saline. They observed increases in mitochondrial and cytosolic levels of calcium and decreases in calcium levels in fractions containing yolk platelets and pigment granules. Their interpretation of these and other findings was that the ionophore caused release of calcium from the platelets and/or granules, producing an increase in cytosolic levels, which, in turn, led to the contractions. The mitochondria attempted to correct for this increase by taking up calcium.

Similarly, Conrad and Davis (1980) found that eggs of *Ilyanassa* placed in seawater containing A23187, X537A, or compound 48/80 would form polar lobelike protuberances resembling those produced by injected calcium (Conrad and Davis, 1977) and that the formation of these protuberances would occur more rapidly if calcium but not magnesium ions were added to the seawater. X537A is also a calcium ionophore, but it is much less specific in that it can also transport other divalent cations, monovalent cations, and some organic substances (Pressman, 1976; Pfeiffer *et al.*, 1978). The effects of compound 48/80, a synthetic polyamine, are not as well understood but, as reviewed by Conrad and Davis (1980), it also appears to increase the level of calcium ions in a variety of cells. Cell shape changes induced by all three agents occurred even in seawater free of both calcium and magnesium in the presence of EDTA, suggesting again that intracellular calcium could be made available by treatment with these agents. Further, normal cytokinesis and development were found to occur in many eggs in calcium-free and magnesium-free seawater. Taken together, their data support the idea that calcium ions are involved in polar lobe formation and cytokinesis and indicate that the calcium can be derived from internal sources.

Although it might be argued that cortical contraction and polar lobe formation are not exactly the same as cytokinesis, cytokinesis appears to react in the same way. Arnold (1975) has reported that in squid embryos, A23187 could cause the premature appearance of incipient furrows, increased visibility of furrows and, interestingly, increased speed of the undercutting furrows. In the following section, I will discuss a similar acceleration of cytokinesis in mammalian cells which is related to increased calcium levels.

b. Studies on Mammalian Cells.   Our own work in this area began with a time-lapse cinemicrographic study of the effects of nicotine on HeLa cells which showed that the alkaloid in the concentration range of 10–100 μg/ml speeded up the process of cytokinesis by about 20–25% (VedBrat et al., 1979). At 100 μg/ml, the agent also produced a small but statistically significant increase in the duration of metaphase. Since nicotine is known to act by increasing the availability of calcium in muscle and other cells (Ahmad and Lewis, 1962; Naylor, 1963; Weiss, 1966), we suggested that it might be doing the same thing in HeLa (VedBrat et al., 1979). To determine whether this might be the case, we also examined the effects of A23187, X537A, and caffeine [which can also release calcium ions from internal pools (Weber and Herz, 1968)] on metaphase and cytokinesis. The effects were similar to those of nicotine (Sisken et al., in preparation). At appropriate concentrations, all of these agents were observed to shorten the duration of cytokinesis (e.g., Table I), and A23187 and caffeine also significantly prolonged metaphase (Table II). These findings supported our earlier interpretation of the effects of nicotine and suggested that increased levels of intracellular Ca^{2+} can speed up cytokinesis and increase the duration of metaphase (i.e., delay the onset of anaphase chromosome movement) in mammalian cells.

We then tried to determine whether the above agents would produce the same effects in cells growing in calcium-free media but found in control experiments that calcium-free media themselves tended to speed up

TABLE I

Effects of Agents Which Increase Cytosolic Levels of Calcium Ions on the Duration of Cytokinesis in HeLa Cells[a]

Agent	Concentration	Cytokinesis duration (min)	± SE[b]	n[c]	p[d]	Ref[e]
Nicotine	0	3.06	0.074	290		a
	10 μg/ml	2.33	0.127	96	0.0008	
	100 μg/ml	2.32	0.146	45	0.0015	
Caffeine	0	2.80	0.09	180		2,3
	1×10^{-3} M	2.18	0.09	140	0.0009	
A23187	0	3.00	0.08	514		2,3
	1×10^{-6} M	2.01	0.15	200	0.0001	

[a] Obtained from M. A. Bioproducts, Walkersville, Maryland.

[b] Each value is a mean derived from a number of separate experiments. The calculation of standard error included variance components for both intraexperiment and interexperiment variation (VedBrat et al., 1979).

[c] Number of cells analyzed, n.

[d] Significance level, p.

[e] Key to references: (1) VedBrat et al. (1979); (2) Sisken and VedBrat (1976); (3) Sisken et al. (in preparation).

TABLE II

Effects of Caffeine and A23187 on the Duration of Metaphase in HeLa Cells[a]

Agent	Concentration	Metaphase duration (min)	± SE	n	p	Ref
Caffeine	0	19.7	2.12	160		2,3
	1×10^{-3} M	27.5	2.26	104	0.032	
A23187	0	21.67	0.85	485		2,3
	1×10^{-6} M	26.93	1.38	210	0.01	

[a] Calculations and references as in Table I.

cytokinesis (though without noticeable effects on metaphase duration). We therefore did a series of experiments in which the only treatment was to limit the availability of calcium (Sisken *et al.*, in preparation). This was done by (a) adding 2 m*M* EGTA to the medium, (b) removing calcium and magnesium from the medium with the ion-exchange resin Chelex-100 and adding back Mg^{2+}, and (c) treating the cell with lanthanum chloride, which is thought to bind to calcium-binding sites on cell membranes and to inhibit calcium transport (see Weiss, 1974). All of these treatments produced essentially the same shortening of the duration of cytokinesis (e.g., Table III), leaving us with the conclusion that treatments which either increase intracellular calcium levels or limit the availability of extracellular calcium can speed up the rate of furrowing in HeLa cells. In an attempt to reconcile these findings, we pointed out (Sisken *et al.*, in preparation) that the effects of low-calcium media and lanthanum on cytokinesis seem analogous to observations that some muscle cells which are dependent upon intracellular calcium for contraction can be stimulated to contract by exposure to calcium-free solutions (Hurwitz *et al.*, 1967; Bianchi, 1969). An explanation proposed to account for

TABLE III

Effects of Ethylene Glycol-*bis*(β-aminoethyl ether)-*N,N*'-tetraacetic Acid (EGTA) on the Duration of Cytokinesis in HeLa Cells[a]

Concentration	Cytokinesis duration (min)	± SE	n	p	Ref
0	2.91	0.06	432		2,3
2 m*M*	2.49	0.08	114	0.001	

[a] Calculations and references as in Table I.

this was that a pool of calcium ions is bound to the inner surface of the plasma membrane and stabilized in place by a layer of calcium ions bound to the outer surface of the membrane. Removal of this outer layer causes release of the internally bound pool, which in turn stimulates contraction (Bianchi, 1969). Evidence concerning the existence of a similar pool in other systems has been reviewed by Bolton (1979), and the same situation could hold for HeLa cells as well.

In sum, the data from HeLa cells indicated that calcium is involved in cytokinesis, and since increased levels can increase the velocity of furrowing, it is possible that under normal conditions the level of available calcium might even be rate limiting (other possibilities are considered in the next section). The findings are also consistent with the existence of a plasma membrane-bound calcium pool which might or might not normally have a role in cytokinesis.

   c. **Summary of Studies with Calcium Effectors.**   Since agents which are known or believed to increase intracellular calcium levels can either stimulate cleavage in dividing cells or induce cortical contractions in cells not yet ready to divide, and since treatments which reduce intracellular calcium levels can inhibit cleavage and other cortical activities, one may suggest that calcium has a role in controlling the activity of cortical microfilaments and thus in controlling the process of cytokinesis. Further, since quantitative studies indicate that increased calcium levels can increase the speed at which furrowing occurs, it might be that calcium levels in the untreated cell are ratelimiting with respect to this process.

A third conclusion from these studies is that since cytokinesis in both mammalian and nonmammalian cells can occur in calcium-deficient media, it seems clear that many cells can use intracellular calcium stores for this purpose. Whether or not mammalian or other cells actually do so when they also have the choice of utilizing external calcium remains to be determined. Fourth, treatments which limit the availability of extracellular calcium suggest that in mammalian cells a pool of calcium ions may be bound to the inner surface of the plasma membrane, and that at least under experimental conditions this pool may be available for use by the contractile ring.

Thus the use of calcium effectors has provided evidence for a role for calcium in cytokinesis. However, some major questions remain unanswered. One of these concerns the regulation of calcium availability. As noted earlier, it has been proposed that a release of calcium from the astral region or from vesicles in the region of the presumptive furrow following anaphase movement is a key event leading to the organization and/or stimulation of the contractile ring (e.g., Hepler, 1977; Rebhun, 1977; Harris, 1978). While these are very reasonable hypotheses, direct evidence that such release in fact occurs is lacking, and the source of the presumptive released calcium is still unknown. Further, the regulatory events which lead to the release and the specific role of the released calcium remain to be determined. So, although the use of various agents and treatments has contributed some information and has helped to define questions to be asked about the role of calcium in cytokinesis, this approach has not provided, and may not be able to provide, detailed answers. These will probably have to come from other kinds of studies.

A final point is that even though calcium may be involved in the structure and function of the actin-containing microfilaments, it does not necessarily mean that the increased calcium levels induced by the treatments accelerate cytokinesis by acting directly on the microfilaments. Among the difficulties in this area is the fact that calcium ions are involved in a number of important systems in the cell and might play more than one role in cytokinesis. It is in fact possible that the functional role of calcium could be met by baseline levels of calcium and that higher levels induced by the various treatments may involve some other mechanism. For example, calcium ions are involved in the regulation of activity of the enzymes which control cyclic nucleotide levels and activities, and increased levels of calcium could be affecting this important regulatory system (e.g., Rasmussen and Goodman, 1977; Whitfield et al., 1980). Increased levels of free calcium ions could also cause a loss of hydrogen ions from the cell and thus an increase in intracellular pH (see Epel, 1978; Steinhardt et al., 1978). That alterations in pH can stimulate actin polymerization has been indicated by observations of Tilney et al. (1978) on the acrosome reaction of invertebrate sperm cells and by Begg and Rebhun (1979) in isolated egg cortices, and pH could affect other systems as well. That calcium ions and pH might be involved in the regulation of actin polymerization has been discussed by Spudich et al. (1979).

3. Polyamines

In a few studies, cells were treated with polyamines and observed for their effects on either cytokinesis (in mammalian cells) or cortical contractions (in oocytes and cleaving embryos). These studies fall into two groups. In the first group, the agents studied were large polycations and included ribonuclease, polylysine, and compound 48/80, which is a synthetic polyamine of heterogeneous molecular weight (see Conrad and Davis, 1980). In this group of studies, the agents stimulated cortical contractions in Xenopus oocytes (Gingell and Palmer, 1968; Gingell, 1970) or caused cell shape changes resembling polar lobe formation in Ilyanassa eggs (Conrad and Davis, 1980). In both cases, calcium ions again seem to be involved.

In the case of Xenopus, Gingell and Palmer (1968) and Gingell (1970) found that the stimulation produced by ribonuclease and polylysine (in the molecular weight range of 2600 to 150,000) could occur only in the presence of extracellular calcium ions. Since these agents only bind to the cell surface and do not enter the cell, one interpretation offered was that adsorption of the polycation to the cell surface caused an increase in membrane permeability which allowed calcium ions to enter the cell and trigger the cortical contractions. As a possible alternative explanation, they suggested that such treatment in the presence of external divalent cations might cause a release of such cations from the inner surface of the cell membrane. It was also pointed out that although the membrane might thus have the capacity to act in this way as a transducer between the interior of the cell and its exterior, internally bound calcium is more likely involved in normal cleavage (Gingell, 1970).

As noted earlier, Conrad and Davis (1980) observed that compound 48/80, which also seems to act on the cell surface, caused changes in cell shape of *Ilyanassa* eggs, apparently by increasing the availability of intracellular calcium ions, even in calcium-free solutions. It appears that binding of the polycation to the cell surface caused a release of calcium from an internal pool. One possibility is that this is a pool localized on the inner surface of the plasma membrane, as hypothesized to exist in some muscles (Bianchi, 1969), HeLa cells (Sisken *et al.*, in preparation), and early *Xenopus* embryos (Gingell and Palmer, 1968).

The second group of studies involves the low-molecular-weight polyamines spermine, spermidine, and putrescine, which appear to be required for cytokinesis. Experiments which fit into this group come from the work of Sunkara *et al.* (1979), who observed that 48–72-hour treatments with methylglyoxal-bis(guanylhydrazone) (MGBG), an inhibitor of spermidine and spermine biosynthesis, and α-methylornithine (α-MO), an inhibitor or ornithine decarboxylase, which leads to reduced levels of putrescine and spermidine, increased the percentage of binucleate cells in a number of mammalian cell lines.

Sunkara *et al.* (1979) also noted that the microfilamentous cytoskeleton, as observed by indirect immunofluorescence, became disorganized, had a diffuse appearance, and did not display the distinct actin cables seen in controls. Addition of spermine or spermidine to MGBC-treated cells reduced the frequency of binucleate cells to control levels. Since their data indicated that the degree of binucleation most closely paralleled the reduction in spermidine levels and that addition of spermidine was most effective in counteracting the inhibitors, they suggested that spermidine might play a role in the structure and function of microfilaments. The nature of the role that spermidine or other intracellular polyamines might have with respect to cytokinesis is suggested by other studies which showed that both high- and low-molecular-weight polyamines, including those mentioned above, can stimulate the polymerization of actin *in vitro*. Oriol-Audit (1978) found that putrescine, spermidine, and spermine, as well as several other diamines and guanidine derivatives, could do this and that the polymerization "yield" was related to chain length. For example, both spermine (molecular weight 144) and spermidine (molecular weight 202) were nearly twice as effective in stimulating polymerization as was putrescine (molecular weight 88). With respect to high-molecular-weight polyamines, observations by Magri *et al.* (1978) indicated that histone H1 and protamine can stimulate polymerization of rabbit muscle actin in a biphasic manner, which was interpreted to mean that these basic proteins can both induce nucleation of actin and stimulate its rate of polymerization. In addition, Brown and Spudich (1979) have shown that polylysine-coated beads, as well as free polylysine, can stimulate the polymerization of actin purified from the slime mold *Dictyostelium discoideum* and suggested that positively charged regions might serve as nucleation sites for actin polymerization *in vivo*.

Lin *et al.* (1980) also showed that polylysine can stimulate the *in vitro* polymerization of monomeric actin molecules. They proposed that the polylysine increased the rate at which nuclei for polymerization of G actin molecules were formed and that it did so by promoting the association of negatively charged actin monomers. They also found that this stimulation was dependent upon the size of the polylysine polymer and that low-molecular-weight polylysines (approximately 3000) were much less effective than those of higher molecular weight.

In sum, then, the work with high-molecular-weight polyamines points in two directions. When used in studies of live cells, they appear to confirm a role for calcium in cortical contractions and cytokinesis and have again raised the questions of both the source of the calcium used for these processes and the role of the plasma membrane in regulating the release of this calcium. Since both high- and low-molecular-weight polyamines can stimulate actin polymerization *in vitro*, and since reduced levels of natural low-molecular-weight polyamines lead to an apparent failure of cytokinesis, these studies are consistent with the idea that spermidine or one of the other polyamines might be involved in the formation and function of the contractile ring. More detailed studies of the effects of the inhibitors used by Sunkara *et al.* (1979) would seem to be in order.

4. Antibodies against Contractile Proteins

Specific antibodies against selected proteins are another class of agents which has considerable potentail for identifying the macromolecules involved in the process of cytokinesis or in its regulation. They have been used in many immunocytochemical studies to localize specific proteins in fixed cells, but the relevance of antibodies to this chapter is that they can be injected into live cells and used as inhibitors of cytokinesis (Mabuchi and Okuno, 1977). So far, only antimyosin antibody has been used in this manner. Mabuchi and Okuno (1977) purified starfish egg myosin, injected it into a rabbit, and obtained an antimyosin-containing gamma globulin fraction. This gamma globulin fraction, which inhibited actin-activated myosin AT-Pase activity *in vitro*, did not affect anaphase chromosome movement when injected into cleaving starfish blastomeres, but it did inhibit cytokinesis depending upon the quantity injected and the position of the cell in the cell cycle when it was injected. Injection of appropriate amounts during interphase inhibited the next and successive cleavages without affecting nuclear divisions, while injection during or just prior to cytokinesis did not affect that particular cleavage but did affect the subsequent one. As suggested by the authors (see Mabuchi, 1978), a reasonable interpretation of the data is that myosin which is not yet built into the contractile ring can be inactivated by the antibody, but once the myosin molecule becomes part of the contractile ring complex, the antibody can no longer inhibit its function.

Thus, as a means of identifying macromolecules involved in a process such as cytokinesis, the injection of specific antibodies against selected isolatable macromolecules would seem to represent an excellent experimental approach. Antiactin has also been injected into cells, but the effects on cytokinesis were not studied (Rungger, 1979).

5. Effects of an Amino Acid Analog and Temperature

We have done some other studies which relate to the interlock between mitosis and cytokinesis. It is known that the polar regions of the spindle determine the location of the cleavage furrow (see Rappaport, 1971; Conrad and Rappaport, Chapter 16, this volume). It is also known that the initiation of furrowing is generally dependent upon the prior activity of the spindle but that once cytokinesis gets beyond a certain point, the mitotic apparatus is no longer required. It can, in fact, be removed from the cell at this point, and cytokinesis will continue (Hiramoto, 1956).

In a study of the effects of an amino acid analog, p-fluorophenylalanine, on cell division of a line of human amnion cells, it was noted that the analog caused an increase in the duration of metaphase to the point of total blockage depending upon the duration of treatment (Sisken and Wilkes, 1967). An increase in the duration of anaphase also occurred, but there was no change in the duration of cytokinesis. Since the analog is incorporated into all proteins containing phenylalanine, it seems clear that replacement of phenylalanine with p-fluorophenylalanine led to functional alterations in proteins associated with the mitotic spindle but did not alter mechanisms associated with cytokinesis.

When we analyzed the time of onset of cytokinesis relative to the distance the chromosomes had moved, it was found that the onset of cytokinesis occurred when the chromosomes reached an almost constant distance apart.

That is, irrespective of how long cells were delayed in metaphase or the rate at which the chromosomes moved, cytokinesis began when leading edges of the chromosome groups had moved, on the average, 85% of their final distance apart (Sisken, 1973) (Fig. 1). Having made this observation, the question was whether this was a general phenomenon or was true only of p-fluorophenylalanine-treated cells. Since we knew from other work (Ris, 1949; Sisken *et al.*, 1965) that the rate of anaphase movement was affected by temperature, we reanalyzed films made in our earlier study and again found that cytokinesis began when the chromosomes had moved, on the average, 85% of their final distance apart, irrespective of the time of onset or the velocity of chromosome movement (Table IV). In other words, in at least this line of cells, the time of onset of cytokinesis is somehow related to or dependent upon the spindle's having performed a specific proportion of its function.

The association between spindle function and cytokinesis has previously been discussed in some detail (see e.g., Ris, 1949; Mazia, 1961; Rappaport, 1971), and it is clear from a number of studies that the location and time of onset of cytokinesis are determined by the spindle. Observations of Ris

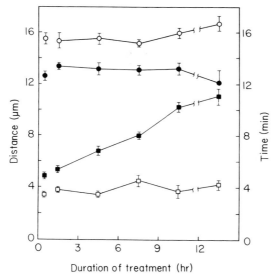

Fig. 1. Some mitotic parameters of human amnion cells as a function of exposure to 1.mM p-fluorophenylalanine. Open circles: first distance attained between leading edges of chromosome groups; closed circles: distance between leading edges of chromosome groups at the beginning of cytokinesis; closed square: time from the beginning of anaphase to the beginning of cytokinesis; open squares: time required to complete cytokinesis (with the exception of the persisting intracellular bridge). From Sisken (1973).

TABLE IV

Dependence of Onset of Cytokinesis on Degree of Chromosome Movement[a]

Temperature (°C)	Time to begin cytokinesis (min ± SE)	Time to complete cytokinesis (min ± SE)	Distance between chromosomes		
			(a) At beginning of cytokinesis (μm ± SE)	(b) At end of cytokinesis (μm ± SE)	
34	6.8 ± 0.3	5.1 ± 0.2	12.6 ± 0.2	14.6 ± 0.5	0.86
37	4.8 ± 0.2	3.2 ± 0.2	13.4 ± 0.5	15.3 ± 0.5	0.88
40.5	4.7 ± 0.2	3.1 ± 0.1	13.5 ± 0.4	15.9 ± 0.4	0.84

[a] From Sisken (1973).

(1949) that the onset of cleavage in grasshopper spermatocytes is delayed or completely blocked if spindle elongation is inhibited led a step further and indicated that a particular aspect of spindle function, elongation, is somehow involved. The data derived from both the p-fluorophenylalanine and temperature studies discussed here are consistent with Ris' findings and indicate that the signal initiating the onset of cytokinesis is given when elongation has progressed to a specific point. Models presented by a number of investigators would suggest that this critical point might be the time at which calcium is released into the cell. The released calcium would simultaneously initiate breakdown of the mitotic spindle, which has completed most of its work, and trigger the onset of furrowing.

For the purposes of this chapter, two points are illustrated by these studies. One is that quantitative analysis of the effects of interference in chromosome movement on the onset of cytokinesis has demonstrated that the interlock between spindle function and the initiaton of cytokinesis may be much tighter than has generally been suspected. Another is that the use of agents which affect another process, i.e., chromosome movement, is useful in gaining information on the regulation of cytokinesis.

III. CONCLUSIONS

As stated earlier, the intent of this chapter was to discuss the value of treating live cells with inhibitors and other agents as a means of helping us understand the mechanisms of cytokinesis. I chose some examples to illustrate the various degrees of success which have been and/or might be achieved by this methodology. What has been seen, I believe, is that what can be learned by treating cells with an agent depends upon what we know about the agent and its degree of specificity, as well as upon the experimental design with which it is used. Since, with the possible exception of antibodies, agents tend to be less than ideal, their main use may thus be in helping to dissect and define problems rather than to solve them.

It must be remembered that this kind of treatment represents only one approach, and any one approach can be most useful when combined with other techniques and approaches. The use of this methodology does have an advantage in that it allows one to work with live cells and to try to understand what happens in them, rather than working with *in vitro* systems to try to deduce what can happen. The main disadvantages include lack of specificity and the difficulty of distinguishing between primary and secondary effects. It is hoped that the future will offer new types of agents and better knowledge about the molecular components and regulatory systems involved in cytokinesis upon which further experiments can be built.

ACKNOWLEDGMENTS

The author is indebted to Drs. Arthur Forer and Janet Morgan for their helpful suggestions and criticisms of the chapter and to Miss Pamela Campbell for her typing and assistance with the manuscript. Thanks are also due to the publishers of *Chromosoma* for allowing the author to adopt the data presented in Fig. 1 and Table IV. Some of the author's studies cited in this chapter were supported by Grant KTRB 24111 from the University of Kentucky Tobacco and Health Research Institute.

REFERENCES

Ahmad, K., and Lewis, J. J. (1962). The influence of drugs which stimulate skeletal muscle and of their antagonists on flux of calcium, potassium, and sodium ions. *J. Pharmacol. Exp. Ther.* **136**, 298.

Arnold, J. M. (1975). An effect of calcium in cytokinesis as demonstrated with ionophore A23187. *Cytobiologie* **11**, 1–9.

Atlas, S. J., and Lin, S. (1978). Dihydrocytochalasin B. Biological effects and binding to 3T3 cells. *J. Cell Biol.* **76**, 360–370.

Baker, P. F., and Warner, A. E. (1972). Intracellular calcium and cell cleavage in early embryos of *Xenopus laevis*. *J. Cell Biol.* **53**, 579–581.

Barany, M., Finkleman, F., and Therattil-Antony, T. (1962). Studies on the bound calcium of actin. *Arch. Biochem. Biophys.* **98**, 28–45.

Begg, D. A., and Rebhun, L. I. (1979). pH regulates the polymerization of actin in the sea urchin egg cortex. *J. Cell Biol.* **83**, 241–249.

Bershdasky, A. D., Gelfand, V. I., Svitkina, T. M., and Tint, I. S. (1980). Destruction of microfilament bundles in mouse embryo fibroblasts treated with inhibitors of energy metabolism. *Exp. Cell Res.* **127**, 421–429.

Bianchi, C. P. (1969). Pharmacology of excitation-contraction coupling in muscle. *Fed. Proc. Fed. Am. Soc. Exp.* **28**, 1624–1628.

Bluemink, J. G. (1970). The first cleavage of the amphibian egg: An electron microscope study of the onset of cytokinesis in the egg of *Ambystoma mexicanum. J. Ultrastruct. Res.* **32**, 142–166.

Bluemink, J. G. (1971a). Effects of cytochalasin B on surface contractility and cell junction formation during egg cleavage in *Xenopus laevis. Cytobiologie* **3**, 176–187.

Bluemink, J. G. (1971b). Cytokinesis and cytochalasin-induced furrow regression in the first-cleavage zygote of *Zenopus laevis. Z. Zellforsch.* Mikrosk Anat. **121**, 102–126.

Bluemink, J. G. (1978). Use of cytochalasins in the study of amphibian development. *In* "Cytochalasins-Biochemical and Cell Biological Aspects" (S. W. Tannenbaum, ed.), pp. 113–142. Elsevier, Amsterdam.

Bolton, T. B. (1979). Mechanisms of action of transmitters and other substances on smooth muscle. *Physiol. Rev.* **59**, 606–718.

Brenner, S. L., and Korn, E. D. (1979). Substoichiometric concentrations of cytochalasin D inhibit actin polymerization. *J. Biol. Chem.* **254**, 9982–9985.

Brown, S. S., and Spudich, J. A. (1979). Cytochalasin inhibits the rate of elongation of actin filament fragments. *J. Cell Biol.* **83**, 657–663.

Carter, S. B. (1967). Effects of Cytochalasins on mammalian cells. *Nature (London).* **213**, 261–264.

Condeelis, J. S., and Taylor, D. L. (1977). The contractile basis of amoeboid movement. V. The control of gelation, solution and contraction in extracts from *D. discoideum. J. Cell Biol.* **74**, 901–927.

Conrad, G. W., and Davis, S. E. (1977). Microiontophoretic injection of calcium ions or of cyclic AMP causes rapid shape changes in fertilized eggs of *Ilyanassa obsoleta. Dev. Biol.* **61**, 184–201.

Conrad, G. W., and Davis, S. E. (1980). Polar lobe formation and cytokinesis in fertilized eggs of *Ilyanassa obsoleta.* III. Large bleb formation caused by Sr^{2+}, ionophores X537A, A23187, and compound 48/80. *Dev. Biol.* **74**, 152–172.

de Laat, S. W., Luchtel, D., and Bluemink, J. G. (1973). The action of cytochalasin B during egg cleavage in *Xenopus laevis:* Dependence on cell membrane permeability. *Dev. Biol.* **31**, 163–177.

Ebashi, S. and Endo, M. (1968) Calcium ion and muscle contraction. *Prog. Biophys. Mol. Biol.* **18**, 123–183.

Epel, D. (1978). Intracellular pH and activation of the sea urchin egg at fertilization. *In* "Cell Reproduction: In Honor of Daniel Mazia" (E. Dirksen, D. M. Prescott, and C. F. Fox, eds.), pp. 367–378. Academic Press, New York.

Estensen, R. D. (1971). Cytochalasin B. I. Effects on cytokinesis of Novikoff hepatoma cells. *Proc. Soc. Exp. Biol. Med.* **136**, 1256–1260.

Estensen, R. D., and Plagemann, P. G. W. (1972). Cytochalasin B: Inhibition of glucose and glucosamine transport. *Proc. Natl. Acad. Sci. U.S.A.* **69**, 1430–1434.

Fujiwara, K., and Pollard, T. D. (1976). Fluorescent antibody localization of myosin in the cytoplasm, cleavage furrow, and mitotic spindle of human cells. *J. Cell Biol.* **71**, 848–875.

Gingell, D. (1970). Contractile responses at the surface of an amphibian egg. *J. Embryol. Exp. Morphol.* **23**, 583–609.

Gingell, D., and Palmer, J. F. (1968). Changes in membrane impedence associated with a cortical contraction in the egg of *Xenopus laevis. Nature (London)* **217**, 98–102.

Godman, G. C., and Miranda, A. F. (1978). Cellular contractility and the visable effects of cytochalasin. *In* "Cytochalasins—Biochemical and Cell Biological Aspects" (S. W. Tannenbaum, ed.), pp. 277–429. Elsevier, Amsterdam.

Hammer, M. G., Sheridan, J. D., and Estensen, R. D. (1971). Cytochalasin B. II. Selective inhibition of cytokinesis in *Xenopus laevis*. *Proc. Soc. Exp. Biol. Med.* **136**, 1158–1162.

Harris, P. (1978). Triggers, trigger waves, and mitosis: A new model. *In* "Cell Cycle Regulation" (J. R. Jeter, Jr., I. L. Cameron, G. M. Padilla, and A. M. Zimmerman, eds.), pp. 75–104. Academic Press, New York.

Hartwig, J. H., and Stossel, T. P. (1979). Cytochalasin B and the structure of actin gels. *J. Mol. Biol.* **134**, 539–553.

Hasselbach, W. (1965). Relaxing factor and the relaxation of muscle. *Prog. Biophys. Mol. Biol.* **14**, 167–222.

Hellewell, S. B., and Taylor, D. L. (1979). The contractile basis of amoeboid movement. VI. The solution-contraction coupling hypothesis. *J. Cell Biol.* **83**, 633.

Hepler, P. K. (1977). Membranes in the spindle apparatus: Their possible role in the control of microtubule assembly. *In* "Mechanisms and Control of Cell Division" (T. L. Rost and E. M. Gifford, Jr., eds.), pp. 212–232. Dowden, Hutchinson and Ross, Inc., Stroudsburg, Pennsylvania.

Hiramoto, Y. (1956). Cell division without mitotic apparatus in sea urchin eggs. *Exp. Cell Res.* **11**, 630–636.

Hoffman-Berling, H. (1960). Other mechanisms producing movements. *Comp. Biochem.* **2**, 341–370.

Hollinger, T. G., and Schuetz, A. W. (1976). "Cleavage" and cortical granule breakdown in *Rana pipiens* oocytes induced by direct microinjection of calcium. *J. Cell Biol.* **71**, 395–402.

Hurwitz, L., Von Hagen, S., and Joiner, P. D. (1967). Acetylcholine and calcium on membrane permeability and contraction of intestinal smooth muscle. *J. Gen. Physiol.* **50**, 1157–1172.

Krishan, A. (1972). Cytochalasin B: Time-lapse cinematographic studies on its effects on cytokinesis. *J. Cell Biol.* **54**, 657–664.

Kuno, M. (1954). Comparative studies on experimental formation and multinucleated eggs of sea urchins by means of various agents. *Embryologia* **2**, 43–49.

Kuno-Kojima, M. (1957). On the regional difference in the nature of the cortex of the sea urchin egg during cleavage. *Embryologia* **2**, 279–293.

Lin, D. C., and Lin, S. (1979). Actin polymerization induced by a motility-related high affinity cytochalasin binding complex from human erythrocyte membrane. *Proc. Natl., Acad. Sci. U.S.A.* **76**, 2345–2349.

Lin, S., and Spudich, J. A. (1974). Biochemical studies on the mode of actin of cytochalasin B. Cytochalasin B binding to red cell membrane in relation to glucose transport. *J. Biol. Chem.* **249**, 5778–5783.

Lin, D. C., Tobin, K. D., Grumet, M., and Lin, S. (1980). Cytochalasins inhibit nuclei-induced actin polymerization by blocking filament elongation. *J. Cell Biol.* **84**, 455–461.

Luchtel, D., Bluemink, J. G., and de Laat, S. W. (1976). The effect of injected cytochalasin B on filament organization in the cleaving egg of *Xenopus laevis*. *J. Ultrastruct. Res.* **54**, 406–419.

Lustig, S., Kosower, N. S., Pluznik, D. H., and Kosower, E. M. (1977). Inhibition of cytokinesis in Friend leukemia cells by membrane mobility agents. *Proc. Natl. Acad. Sci. U.S.A.* **74**, 2884–2888.

Mabuchi, I. (1978). Role of myosin and actin in cell division of echinoderm eggs. *In* "Cell Motility: Molecules and Organization" (S. Hatano, H. Ishikawa, and H. Sato, eds.), pp. 147–163. Univ. Park Press, Baltimore, Maryland.

Mabuchi, I., and Okuno, M. (1977). The effect of myosin antibody on the division of starfish blastomeres. *J. Cell Biol.* **74**, 251–264.

MacLean-Fletcher, S., and Pollard, T. D. (1980). Mechanism of action of cytochalasin B on actin. *Cell* **20**, 329–341.

Magri, E., Zaccarini, M., and Grazi, E. (1978). The interaction of histone and protamine with actin. Possible involvement in the formation of the mitotic spindle. *Biochem. Biophys. Res. Commun.* **82**, 1207–1210.

Marsland, D. (1970). Pressure-temperature studies on the mechanisms of cell division. *In* "High Pressure Effects on Cellular Processes" (A. M. Zimmerman, ed.), pp. 259–312. Academic Press, New York.

Mazia, D. (1961). Mitosis and the physiology of cell division. *In* "The Cell" (J. Brachet and A. E. Mirsky, eds.). pp. 77–412. Academic Press, New York.

Miranda, A., Godman, G., Deitch, A., and Tannenbaum, S. W. (1974a). Action of cytochalasin D on cells of established lines. I. Early events. *J. Cell Biol.* **61**, 481–500.

Miranda, A., Godman, G., and Tannenbaum, S. W. (1974b). Action of Cytochalasin D on cells of established lines. II. Cortex and microfilaments. *J. Cell Biol.* **62**, 406–423.

Naylor, W. G. (1963). Effect of nicotine on cardiac muscle contractions and radiocalcium movements. *Am. J. Physiol.* **205**, 890–896.

Oriol-Audit, C. (1978). Polyamine-induced actin polymerization. *Eur. J. Biochem.* **87**, 371–376.

Osborn, J. C., Duncan, C. J., and Smith, J. L. (1979). Role of calcium in the control of embryogenesis of *Xenopus*. Changes in the subcellular distribution of calcium in early cleavage embryos after treatment with the ionophore A23187. *J. Cell Biol.* **80**, 589–605.

Pfeiffer, D. R., Taylor, R. W., and Lardy, H. A. (1978). Ionophore A23187: Cation binding and transport properties. *Ann. N. Y. Acad. Sci.* **307**, 402–423.

Pressman, B. C. (1976). Biological applications of ionophores. *Annu. Rev. Biochem.* **45**, 501–530.

Rappaport, R. (1971). Cytokinesis in animal cells. *Int. Rev. Cytol.* **31**, 169–208.

Rasmussen, H., and Goodman, B. P. (1977). Relationships between calcium and cyclic nucleotides in cell activation. *Physiol. Rev.* **57**, 421–509.

Rebhun, L. I. (1977). Cyclic Nucleotides, calcium, and cell division. *Int. Rev. Cytol.* **49**, 1–54.

Ris, H. (1949). The anaphase movement of chromosomes in the spermatocytes of the grasshopper. *Biol. Bull. (Woods Hole, Mass.)* **96**, 90–106.

Rungger, D., Rungger-Brändle, E., Chaponnier, C., and Gabbiani, G. (1979). Intranuclear injection of anti-actin antibodies into *Xenopus* oocytes blocks chromosome condensation. *Nature (London)* **282**, 320–321.

Schroeder, T. E. (1969). The role of "contractile ring" filaments in dividing *Arbacia* eggs. *Biol. Bull. (Woods Hole, Mass.).* **137**, 413–414.

Schroeder, T. E. (1970). The contractile ring I. Fine structure of dividing mammalian (HeLa) cells and the effects of cytochalasin B. *Z. Zellforsch. Mikrosk. Anat.* **109**, 431–449.

Schroeder, T. E. (1975). Dynamics of the contractile ring. *In* "Molecules and Cell Movement" (S. Inoue and R. E. Stephens, eds.), pp. 305–334. Raven, New York.

Schroeder, T. E. (1978). Cytochalasin B, cytokinesis, and the contractile ring. *In* "Cytochalasins—Biochemical and Cell Biological Aspects" (S. W. Tannenbaum, ed.), pp. 91–112. Elsevier, Amsterdam.

Schroeder, T. E., and Strickland, D. L. (1974). Ionophore A23187, calcium and contractility in frog eggs. *Exp. Cell Res.* **83**, 139–142.

Sisken, J. E. (1973). The effects of p-DL-fluorophenylalanine on chromosome movement and cytokinesis of human amnion cells in culture. *Chromosoma* **44**, 91–98.

Sisken, J. E., and VedBrat, S. S. (1977). On the effects of variations in intracellular and extracellular calcium ions on mitosis and cytokinesis of HeLa cells. *J. Cell Biol.* **75**, 263a.

Sisken, J. E., and Wilkes, E. (1967). The time of synthesis and the conservation of mitosis-related proteins in cultured human amnion cells. *J. Cell Biol.* **34**, 97–110.

Sisken, J. E., Morasca, L., and Kibby, S. (1965). Effects of temperature on the kinetics of the mitotic cycle of mammalian cells in culture. *Exp. Cell Res.* **39,** 103–116.

Sisken, J. E., VedBrat, S. S., and Nasser, M., in preparation. Effects of caffeine, ionophores and calcium deficient media on mitosis and cytokinesis of HeLa cells.

Spudich, J. A., Spudich, A., and Amos, L. (1979). Actin from the cortical layer of sea urchin eggs before and after fertilization. *In* "Cell Motility: Molecules and Organization" (S. Hatano, H. Ishikowa, and H. Sato, eds.), pp. 165–187. Univ. Park Press, Baltimore, Maryland.

Steinhardt, R. A., Shen, S. S., and Zucker, R. S. (1978). Direct evidence for ionic messengers in the two phases of metabolic derepression at fertilization of the sea urchin egg. *In* "Cell Reproduction: In Honor of Daniel Mazia" (E. Dirkson, D. M. Prescott, and C. F. Fox, eds.), pp. 415–424. Academic Press, New York.

Sunkara, P. S., Rao, P. N., Nishioka, K., and Brinkley, B. R. (1979). Role of polyamines in cytokinesis of mammalian cells. *Exp. Cell Res.* **119,** 63–68.

Tannenbaum, J. (1978). "Cytochalasins—Biochemical and Cell Biological Aspects." Elsevier, Amsterdam.

Tannenbaum, J., Tannenbaum, S. W., Lo, L. O., Godman, G. C., and Miranda, A. F. (1975). Binding and subcellular localization of tritated cytochalasin D. *Exp. Cell Res.* **91,** 47–56.

Taylor, E. L., and Wessells, N. K. (1973). Cytochalasin B: Alteration in salivary gland morphogenesis not due to glucose depletion. *Dev. Biol.* **31,** 421–425.

Tilney, L. G., Kiehart, D. P., Sardet, C., and Tilney, M. (1978). Polymerization of actin. IV. Role of Ca^{++} and H^+ in the assembly of actin and in membrane fusion in the acrosomal reaction of echinoderm sperm. *J. Cell Biol.* **77,** 536–550.

Timourian, H., Clothier, G., and Watchmaker, G. (1972). Cleavage furrow: Calcium as determinant of site. *Exp. Cell Res.* **75,** 296–298.

Timourian, H., Jotz, M. M., and Clothier, G. E. (1974). Intracellular distribution of calcium and phosphorous during the first cell division of the sea urchin. *Exp. Cell Res.* **83,** 380–386.

VedBrat, S. S., Sisken, J. E., and Anderson, R. L. (1979). The effects of nicotine on cell division of HeLa cells. *Eur. J. Cell Biol.* **19,** 250–254.

Weber, A., and Herz, R. (1968). The relationship between caffeine contraction of intact muscle and the effect of caffeine on reticulum. *J. Gen. Physiol.* **52,** 750–759.

Weber, A., Herz, R., and Reiss, I. (1964). The regulation of myofibrillar activity by calcium. *Proc. R. Soc. London, B* **160,** 489–501.

Weiss, G. B. (1966). The effect of pH on nicotine-induced contracture and Ca^{45} movements in frog sartorius muscle. *J. Pharmacol. Exp. Ther.* **154,** 605–612.

Weiss, G. B. (1966). The effect of potassium on nicotine-induced contracture and Ca^{45} movements in frog sartorius muscle. *J. Pharmacol. Exp. Ther.* **154,** 595–604.

Weiss, G. B. (1974). Cellular pharmacology of lanthanum. *Annu. Rev. Pharmacol.* **14,** 343–354.

Whitfield, J. F., Boynton, A. L., MacManua, J. P., Rixon, R. H., Sikorska, M., Tsang, B., Walker, P. R., and Swierenga, S. H. H. (1980). The roles of calcium and cyclic AMP in cell proliferation. *Ann. N. Y. Acad. Sci.* **339,** 216–240.

Wolpert. L. (1960). The mechanics and mechanism of cleavage. *Int. Rev. Cytol.* **10,** 163–216.

Yamada, K. M., and Wessells, N. K. (1973). Cytochalasin B: Effects on membrane ruffling, growth cone and microspike activity, and microfilament structure not due to altered glucose transport. *Dev. Biol.* **31,** 413–420.

Cell Division: A Commentary

E. W. Taylor

When faced with another book on cell division, the reader is entitled to ask whether it contains the answers to any of the well-known questions. Are chromosomes moved by microtubules or actomyosin? What initiates the beginning of cleavage, and is the plane determined by the poles or the equatorial spindle? Does calcium release trigger any of the events in mitosis and cytokinesis? None of these problems has been settled, but once one has gotten over this initial disappointment, it is evident that considerable progress has been made in the past half dozen years. Chromosome movement has been obtained using lysed cells, which is perhaps a beginning in reconstructing the steps in mitosis *in vitro*. Fluorescence techniques using either antibodies to locate components in fixed cells or the injection of labeled proteins into living cells have yielded important findings on the organization of the components, and these methods have only begun to be exploited. the use of mutants to dissect the steps in mitosis has been advocated before, but some useful mutations have finally been obtained. The slow improvement in electron microscopic techniques and the use of computers to store and manipulate information from serial sections have allowed spindle microtubule structure to be determined with a precision which seemed unattainable a few years ago.

Improvements in technique will generally lead to an increase in our understanding. Progress can also be made if one stops asking the wrong questions. Anaphase chromosome movement has fascinated biologists for 100 years, yet the physical mechanism by which a chromosome moves 5 or 10 μm at the very conservative speed of a few micrometers per minute is probably one of the least interesting events in mitosis.

Cell movement may once have been a mysterious process, but it is no longer so. While we do not yet have a complete molecular explanation for

MITOSIS/CYTOKINESIS
Copyright © 1981 by Academic Press, Inc.
ISBN 0-12-781240-7

the sliding filament mechanism of actomyosin, one of the proposed mechanisms will probably turn out to be correct in principle. In practice, the answer may require the elucidation of the three-dimensional structure of myosin, a myosin–nucleotide intermediate, and actin. The microtubule–dynein mechanism in flagella is much more difficult because of the complexity of the structure, but a rotating "arm" mechanism is appealing. If the results on the decoration of microtubules with dynein are correct, an "in principle" solution might be obtained first in this system since there is no convincing evidence for two orientations of a myosin cross-bridge.

The mechanism of chromosome movement continues to hold the center of attention and is discussed by several authors in this volume. The possibilities are active sliding of microtubules, polymerization–depolymerization of microtubules, and actomyosin contraction. Each model has been in and out of favor over the past few years. All of them may be correct.

The participation of actomyosin seems to require an interaction between actin filaments and microtubules through a linking protein such as one of the microtubule-associated proteins. The tenuous interactions described by Griffith and Pollard (1978) are the first evidence in favor of such a mechanism. However, the observation that actin is present in spindles is not convincing evidence for an actomyosin mechanism. I suggest that a stringent test is the presence of myosin-thick filaments at the site of force generation. In view of the difficulties in finding thick filaments in motile systems, this test may be considered too stringent. There are two reasons for the suggestion. Myosin–nucleotide intermediate states are very weakly bound to actin at physiological ionic strength. While most of the evidence comes from striated muscle myosin, there are no clear differences for a smooth muscle protein which is a closer analog of cellular myosin (Marston and Taylor, 1980). The dissociation constant is probably greater than 10^{-3} M for heavy meromyosin. Consequently, individual myosin molecules will scarcely be bound to actin whose concentration in the cell is 10 times less than the dissociation constant. Studies with smooth muscle and cellular myosins have shown that thick filaments do not form unless myosin is phosphorylated (Scholey et al., 1980), which suggests that this is a mechanism of regulation. Consequently, interaction with actin filaments to generate force probably requires the formation of a myosin-thick filament.

The evidence for actomyosin as the mover of chromosomes in anaphase is not convincing on this criterion. However, very few strategically placed thick filaments would generate the small forces which are necessary. As stressed by Rickards (Chapter 5, this volume), considerable chromosome and particle movement occurs before anaphase, and the velocities are often much higher. The actomyosin system is present in the cell; it is known to produce saltations or to interact with vesicles; and it would be surprising if some of the activity which is seen in a mitotic cell was not caused by the possible interaction of myosin with astral and spindle microtubules.

The hypothesis that motion is produced by polymerization–depolymerization of microtubules has been discussed by MacIntosh (1979). The difficulty is in explaining how such activity causes motion rather than accompanies it. Evidence for active sliding of microtubules does not appear to be much better, although blocking movement by antidynein antibodies is an intriguing result. The only example of sliding of microtubules in flagella appears to require parallel orientation. The decoration of cytoplasmic microtubules is polar, and microtubule pairs appear to be in parallel orientation. A direct determination of microtubule polarity for kinetochore and central spindle microtubules is an important experiment. The evidence based on polymerization by end addition of subunits suggests that kinetochore and central spindle microtubules are antiparallel, and so are the microtubules in the overlap region at the equator. At first, the mitotic apparatus seems to be poorly designed for sliding tubule mechanisms, but the difficulty could be circumvented by suitable hypotheses.

The force generator for anaphase chromosome movement is probably not unique to mitosis. Important as this step may be, the more interesting question is the integration of the sequence of events necessary for successful cell division. Several processes are running on parallel tracks, including chromosome condensation and maturation of the kinetochores, nuclear envelope, microtubules and activity of MTOC, actomyosin, cell membrane, the cleavage organelle, and calcium concentration. How is the timing of these processes kept in synchrony? The one signal that has been studied is the release of calcium. Harris, in Chapter 2, this volume, has reviewed the elegant studies on calcium in the fertilization and division of eggs. The processes may be more easily studied in a large cell, but similar events probably occur in ordinary mitosis. The finding of calmodulin in centriole and kinetochore regions and the activation of dynein by calmodulin emphasize the importance of studies on calcium regulation. Calmodulin usually acts as a subunit of the enzymes involved in phosphate group transfer and is known to be a subunit of myosin protein kinase, which activates cellular actomyosin. It also appears to sensitize microtubules to calcium depolymerization. While calcium probably does act as a trigger for some of the processes, we have hardly begun to ask questions about control of the parallel events. Calcium-binding proteins as regulators of actin gelation and actomyosin interactions are being intensely investigated, and these studies are probably relevant to mitosis.

The old question of how the cleavage furrow is initiated in the plane transverse to the spindle equator is discussed by Conrad and Rappaport in Chapter 16 of this volume. It is probably the best example of integration of parallel events in both space and time. The beautiful studies of Rappaport and collaborators have defined the problem, but we still do not have a molecular explanation.

The problem of cell division is one of the most complex in cell biology. Even a process such as the assembly of a T2 phage requires the participation of at least 50 genes, and cell division is probably even more complex. Most of the information is needed for the control of the sequences of events. Future studies will probably make greater use of lysed cell models and temperature-sensitive mutants blocked at definable steps in the process. Complementation of mutants by cell fusion would allow the connections between events to be investigated. To a large extent, we are still trying to answer the same questions that were asked 20 years ago. The next decade should bring a radical change in emphasis in this subject.

REFERENCES

Griffith, L. M., and Pollard, T. D. (1978). Evidence for actin filament–microtubule interaction mediated by microtubule-associated proteins. *J. Cell Biol.* **78,** 958.

Marston, S. B., and Taylor, E. W. (1980). Comparative studies of actomyosin ATPase mechanisms of four types of muscles. *J. Mol. Biol.* **139,** 573.

Scholey, J. M., Taylor, K. A., and Kendrick-Jones, J. (1980). Regulation of non-muscle myosin assembly by calmodulin dependent light chain kinase. *Nature (London)* **287,** 233.

MacIntosh, J. R. (1979). *In* "Microtubules" (K. Roberts and J. S. Hyams, eds.), Academic Press, New York.

Index

A

Acanthamoeba, 234
Acheta domesticus, 110, 111, 116
Acrosomal process, 61, 75, 79
Actin, 30, 34
  antibody, 234–235, 317
  chromosome movement and, 121–122, 136,
    141, 316–318, 345, 403
  microfilament localization of, 214, 215, 218
  nuclear content, 106
  pH dependency, 78
  polymerization, 35, 450, 452
  spindle content, 47, 183, 223, 249, 316–318
Actin filament, 75, 121, 122
  contractile properties, 422
  cortical layer, 429–431
  cytochalasin effect on, 442
  cytokinesis, role in, 423
  polarity, 127
  ultrastructure, 263–264
Actin-like molecule, 106
Actin-like protein, 118, 121
Actin–myosin interaction, 121–122, 136, 141,
  403
Actinomycin, 51
Activation
  of quiescent cells, 31–35
Actomyosin, 317, 319, 345
  sliding filament mechanism of, 462
Adenosine triphosphatase, 35, 319
  chromosome movement and, 344–345
  mitotic apparatus, function in, 339, 344–345
Adenosine triphosphate, 303
  calcium ion sequestration by, 377
  chromosome movement and, 303, 313, 314,
    315, 319–320
  cytoplasmic streaming and, 34
Aggregation
  of chromosomes, 201
Akinetochoric body, 125

Alcohol–digitonin method, of mitotic apparatus isolation, 328
Allele
  conditional-for-synthesis, 16
  execution point, 16–17
  first cycle arrest, 16
  gene product and, 18
  of temperature-sensitive mutants, 10, 12
Allium, 86
Allium cepa, 107
Allium fistulosum, 86, 87
Ambystoma mexicanum, 380
Amino acid analog, 453–455
Amphibian cell
  in micromanipulation studies, 169
Amphibian egg
  cleavage furrow, 383
  cytokinesis in, 398–399, 402, 432, 433
  non-nucleated fragment, 415
Anaphase
  aster during, 40, 45
  bivalent attachment during, 166–167
  chromosome movements during, 103, 115,
    117, 118, 123, 137, 141, 160–162,
    291–294, 302, 312–313, 461
  chromosome strength during, 168
  kinetochore during, 162
  microtubule assembly–disassembly during,
    280
  microtubule polymerization during, 45–49
  spindle arrangement during, 379
Animal cell, *see also* individual species and
  genera
  cytokinesis in, 365–396
Anisolabis maritima, 109, 110
Annelid egg, polar lobe formation in, 384
Annelida, 421
Annulate lamellae, 44, 376–377, 383
Antibody
  actin, 234–235, 317
  contractile protein, 452–453

CELL BIOLOGY: A Series of Monographs

EDITORS

D. E. BUETOW

Department of Physiology
and Biophysics
University of Illinois
Urbana, Illinois

I. L. CAMERON

Department of Anatomy
University of Texas
Health Science Center at San Antonio
San Antonio, Texas

G. M. PADILLA

Department of Physiology
Duke University Medical Center
Durham, North Carolina

A. M. ZIMMERMAN

Department of Zoology
University of Toronto
Toronto, Ontario, Canada

G. M. Padilla, G. L. Whitson, and I. L. Cameron (editors). THE CELL CYCLE: Gene-Enzyme Interactions, 1969

A. M. Zimmerman (editor). HIGH PRESSURE EFFECTS ON CELLULAR PROCESSES, 1970

I. L. Cameron and J. D. Thrasher (editors). CELLULAR AND MOLECULAR RENEWAL IN THE MAMMALIAN BODY, 1971

I. L. Cameron, G. M. Padilla, and A. M. Zimmerman (editors). DEVELOPMENTAL ASPECTS OF THE CELL CYCLE, 1971

P. F. Smith. The BIOLOGY OF MYCOPLASMAS, 1971

Gary L. Whitson (editor). CONCEPTS IN RADIATION CELL BIOLOGY, 1972

Donald L. Hill. THE BIOCHEMISTRY AND PHYSIOLOGY OF *TETRAHYMENA*, 1972

Kwang W. Jeon (editor). THE BIOLOGY OF AMOEBA, 1973

Dean F. Martin and George M. Padilla (editors). MARINE PHARMACOGNOSY: Action of Marine Biotoxins at the Cellular Level, 1973

Joseph A. Erwin (editor). LIPIDS AND BIOMEMBRANES OF EUKARYOTIC MICROORGANISMS, 1973

A. M. Zimmerman, G. M. Padilla, and I. L. Cameron (editors). DRUGS AND THE CELL CYCLE, 1973

Stuart Coward (editor). DEVELOPMENTAL REGULATION: Aspects of Cell Differentiation, 1973

I. L. Cameron and J. R. Jeter, Jr. (editors). ACIDIC PROTEINS OF THE NUCLEUS, 1974

Govindjee (editor). BIOENERGETICS OF PHOTOSYNTHESIS, 1975

James R. Jeter, Jr., Ivan L. Cameron, George M. Padilla, and Arthur M. Zimmerman (editors). CELL CYCLE REGULATION, 1978

Gary L. Whitson (editor). NUCLEAR–CYTOPLASMIC INTERACTIONS IN THE CELL CYCLE, 1980

Danton H. O'Day and Paul A. Horgen (editors). SEXUAL INTERACTIONS IN EUKARYOTIC MICROBES, 1981

Ivan L. Cameron and Thomas B. Pool (editors). THE TRANSFORMED CELL, 1981

Arthur M. Zimmerman and Arthur Forer (editors). MITOSIS/CYTOKINESIS, 1981

In preparation

Ian R. Brown (editor). MOLECULAR APPROACHES TO NEUROBIOLOGY, 1982

Henry C. Aldrich and John W. Daniel (editors), CELL BIOLOGY OF *PHYSARUM* AND *DIDYMIUM*, Volume 1: Organisms, Nucleus, and Cell Cycle, 1982

John A. Heddle (editor). MUTAGENICITY: New Horizons in Genetic Toxicology, 1982

Puto N, Rao, Robert T. Johnson, and Karl Sperling (editors). PREMATURE CHROMOSOME CONDENSATION: Application in Basic, Clinical, and Mutation Research, 1982

George M. Padilla and Kenneth S. McCarty, Sr. (editors). GENETIC EXPRESSION IN THE CELL CYCLE, 1982